add

ENVIRONMENTAL POLLUTION BY PESTICIDES

ENVIRONMENTAL SCIENCE RESEARCH

ENVIRONMENTAL POLLUTION BY PESTICIDES

Edited by

C. A. Edwards

Rothamsted Experimental Station
Harpenden, Hertfordshire
England

PLENUM PRESS • LONDON AND NEW YORK

Library of Congress Catalog Card Number: 72-95067

ISBN 0-306-36303-8

Copyright © 1973 by Plenum Publishing Company Ltd
Reprinted 1974
Plenum Publishing Company Ltd
4a Lower John Street
London W1R 3PD
Telephone 01–437 1408

U.S. Edition published by
Plenum Publishing Corporation
227 West 17th Street
New York, New York 10011

Printed in Great Britain by R. & R. Clark Ltd, Edinburgh

Contributors

J. E. Davies — Department of Epidemiology and Public Health, University of Miami, Miami, Florida 33125, U.S.A.

M. B. Duggan — Department of Health Education and Welfare, F.D.A. Rockville, Maryland, U.S.A.

R. E. Duggan — 6319 Anneliese Drive, Falls Church, Virginia 22044, U.S.A.

C. A. Edwards — Rothamsted Experimental Station, Harpenden, Hertfordshire, England

D. G. Finlayson — Agriculture Canada Research Station, 6660 N.W. Marine Drive, Vancouver, B.C., Canada

A. V. Holden — Freshwater Fisheries Laboratory, Faskally, Pitlochry, Perthshire, Scotland

D. W. Johnson — Department of Biology, Idaho State University, Pocatello, Idaho 83201, U.S.A.

S. R. Kerr — Ministry of Natural Resources, Fish and Wildlife Research Branch, Maple, Ontario, Canada

H. R. MacCarthy — Agriculture Canada Research Station, 6660 N.W. Marine Drive, Vancouver, B.C., Canada

F. Matsumura — Department of Entomology, University of Wisconsin, Madison, Wisconsin, U.S.A.

J. Robinson — Shell Research, Woodstock Agricultural Research Centre, Sittingbourne, Kent, England

J. H. Ruzicka — Laboratory of the Government Chemist, Cornwall House, London SE1, England

Lucille F. Stickel — Patuxent Wildlife Research Center, Laurel, Maryland 20810, U.S.A.

A. R. Thompson — National Vegetable Research Station, Wellesbourne, Warwickshire, England

W. P. Vass — Fisheries Research Board of Canada, Marine Ecology Laboratory, Bedford Institute, Dartmouth, Nova Scotia, Canada

G. A. Wheatley — National Vegetable Research Station, Wellesbourne, Warwickshire, England

Preface

The persistent organic pesticides have saved millions of lives by controlling human disease vectors and by greatly increasing the yields of agricultural crops. However, in recent years man has become ever more conscious of the way in which his environment is becoming increasingly polluted by chemicals that may harm plants, animals or even himself. Amongst these chemicals the organochlorine insecticides have been well to the fore as a major cause of anxiety to ecologists, not only because they persist so long, but also because of the readiness with which they are taken up into the bodies of living organisms, especially the fatty tissues of both animals and humans.

The extent and seriousness of the potential hazards due to these chemicals still remains to be fully defined. Our information on the occurrence of residues in the various parts of the environment is very uneven and localized. For instance, whereas we have a great deal of data on residues in North America, we know virtually nothing about the extent of pesticide contamination in Africa, South America and much of Asia, although large amounts of organochlorine insecticides have been used in these areas.

The emotional impact of the possibility of serious contamination of the human environment by extremely persistent chemicals has resulted in vociferous clashes of opinion between those ecologists who take the view that all pesticides are bad and should be banned, and agriculturalists and others who consider that continued use of large quantities is essential to the survival of humanity. It seems most likely that between these two extremes lies the saner approach, of controlling pests by alternative means, minimizing the uses of persistent pesticides and investigating more fully the possible hazards caused by their continued use. Such a balanced and considerate approach to the problem necessitates the availability of as much information as possible in a concise form, because the available literature on the subject is enormous. This book, *Environmental Pollution by Pesticides,* attempts to do this, by bringing together the available data on pesticide residues, not only in plants and animals, but also in air, water and soil, each chapter being written by a recognized authority in his or her field. Wherever possible, information is provided in each chapter on the following aspects: the amounts of residues commonly occurring, the principal sources of the residues, factors influencing the persistence of the residues, hazards caused by the residues, methods of minimizing or removing residues from the environment and possible further investigations.

I should like to thank Drs D. C. Abbot, J. S. Alabaster, J. M. Anderson, A. W. Breidenbach, P. A. Butler, O. S. Cope, L. M. Dickie, R. W. Edwards, R. Goulding, W. A. Hayes Jr., C. G. Hunter, H. P. Nicholson, G. G. Roebeck and J. C. Street for their advice and help in selecting the chapter authors, and Mr D. J. Holdaway for valuable assistance in reading and editing the manuscripts and index. This book owes its origin to Mr R. H. Leech, who suggested that it should be written as a fuller, more up-to-date and authoritative version of my earlier book *Persistent Pesticides in the Environment* (1970).

If we have succeeded in presenting a balanced picture of possible environmental hazards due to the persistent pesticides, to a wide audience of scientists and interested laymen, we shall have achieved our aim.

CLIVE A. EDWARDS
Rothamsted Experimental Station

Contents

Introduction

C. A. Edwards

With the slow development of civilisation, so has man gradually realised the extent to which pests harm his crops, annoy him and transmit diseases, both human and those of domestic animals. The use of chemicals to kill pests is not a new concept; about A.D. 70 Pliny the Elder recommended that arsenic could be used to kill insects, and the Chinese used arsenic sulphide as an insecticide as early as the late sixteenth century. The use of arsenical compounds has continued, and, during the early part of the twentieth century, large quantities of such compounds as lead arsenate were used to control insect pests. Another arsenical compound, Paris green (copper aceto-arsenite), was extensively applied to pools and standing water in the tropics, in attempts to control malaria-transmitting mosquitoes.

It was not realised at the time how persistent arsenical pesticides were, although it is now known that they can persist in soil for up to 40 years, and many orchard soils still contain large amounts of these chemicals. For instance, in a recent survey of arsenic residues in arable soils in Canada, residues of arsenic ranging from 1.1 to 121.0 ppm were reported (Miles, 1968). Other inorganic compounds used as insecticides and fungicides containeu antimony, boron, copper, fluorine, manganese, mercury, selenium, sulphur, thallium and zinc as their active ingredients. Although these compounds were not very effective as insecticides, many were so persistent in soil that there were instances of crops being damaged by their residues in soil.

The era of synthetic organic pesticides began about 1940. These chemicals were so successful in controlling pests that there was extremely rapid and general adoption of them and development of new ones. This has progressed so rapidly, that today about 1,000 pesticide chemicals are in common use around the world, of which about 250 are commonly used in agriculture, including about 100 insecticides and acaricides, 50 herbicides, 50 fungicides, 20 nematicides and 30 other chemicals. Only a few of these chemicals persist for more than a few weeks or at most months, in soils or water, and of those that do, most are the organochlorine insecticides, which include aldrin, dieldrin, chlordane, dicofol, endo-

1

sulfan, endrin, lindane, DDT, heptachlor and toxaphene. All these insecticides are toxic to insects and other arthropods at very low doses, and are so persistent that few annual treatments are necessary to maintain pests at low levels; thus costs are low. As more of these chemicals were developed and the use of others extended, it seemed likely that many of the more serious agricultural and medical pest control problems would be solved, particularly because the mammalian toxicity of most of these insecticides was low, and there were few hazards to the pesticide operators.

It is now well established that the benefits the organochlorine insecticides conferred on mankind during the early years of their use were great. Knipling (1953) estimated that during the first decade of use, DDT saved five million lives and prevented 100 million serious illnesses due to malaria, typhus, dysentery and more than 20 other insect-borne diseases. In agriculture, it has been calculated that, even after the effective use of these pesticides, pests still cause annual losses of about $4,000,000,000 in the United States, and as much as $21,000,000,000 on a world scale, so these losses, if persistent chemicals were not being used, would be astronomical.

Although it was known that organochlorines were very persistent, up till the early 1950's there was little anxiety as to possible long-term ecological hazards caused by their use. There was some evidence that large residues in soil could be phytotoxic, small quantities of some were reported from plant and animal tissues and in cows' milk, and there were some instances of fish being killed when water was sprayed in anti-malarial and other pest campaigns, but all of these side-effects were accepted as slight, but unavoidable hazards, and of little concern.

However, during the 1950's and the early 1960's reports of large residues of these insecticides in soils, and small amounts in water and at the bottom of streams began to appear in the literature. Dead birds were sometimes found close to sprayed fields and woodlands, or fields planted with insecticide-treated seed, and numbers of dead fish were sometimes seen on the surface of water after spraying operations. There were ever-increasing reports that organochlorine insecticides were stored, not only in invertebrate and vertebrate tissues, but were also concentrated into the upper trophic levels of food chains. These discoveries began to cause concern about possible long-term ecological effects of the large quantities of insecticides being used, sometimes quite indiscriminately.

Eventually, came the realisation that there were appreciable quantities of pesticide residues in the biota and physical environment of the polar regions; that there were large residues in certain birds, mammals and human beings; and that measurable quantities were present in the atmosphere; and these findings produced a wave of reaction against the persistent pesticides, resulting in quite stringent restrictions or bans on their use in many countries in Europe and North America. In spite of this, the overall amounts of pesticide chemicals manufactured have continued to rise steadily.

In the United States, the annual production of pesticides has risen from about 300 million pounds in 1954 to over 1,200 million pounds in 1973. Of these amounts, about 30 million pounds were organochlorine insecticides, and since other countries also manufacture these chemicals, there is a continual addition to the environment of such large amounts of these extremely persistent compounds that it is giving rise to considerable anxiety among many ecologists.

Certainly, there is now little doubt that the organochlorines, especially DDT and to a lesser extent dieldrin, are major long-term contaminants of the total environment. Thus, global pesticide contamination is so very extensive that Gunther (1966) claimed that 'a qualified pesticide residue analyser with proper equipment could find measurable DDT in any nonfossil sample presented to him, and with enough time and patience could find several other pesticides as well'.

This demonstrates that, currently, scientists are much more competent to confirm the presence of pesticides in the environment than to assess the significance of such residues. Nevertheless, we still have not completed the essential task of establishing the seriousness of the problem. In view of this, there is little doubt about the need for extensive monitoring programmes for pesticide residues in the environment; these already exist to monitor radioisotopes, and there is no reason that they should not be extended to pesticides and their possible transport between different areas of the world.

The Fish and Wildlife Service, the U.S. Department of Health, Education and Welfare, and the U.S. Department of Agriculture in the United States, and the Nature Conservancy and the Ministry of Agriculture, Fisheries and Food in Great Britain, and equivalent organisations in many other countries, are now regularly monitoring residues of persistent pesticides in soil, water air and in the flora and fauna. Many other scientists are also working on these problems, either independently or in association with government organisations.

Nevertheless, the monitoring of residues in various parts of the physical environment is still inadequate, particularly in Great Britain. The major rivers in the United States are now being monitored annually for pesticide residues at a large number of sampling stations, but in Great Britain, only one or two small-scale surveys have been made. The monitoring of soils for pesticides is much more sporadic, both in the U.S. and in Great Britain, and so far has been mainly confined to agricultural soils, so that there is little information on the amounts of pesticides that occur in the large areas of untreated land. When we have more information on the amounts of pesticide residues in untreated areas, we can assess the possibilities of global transport much better, and such data is essential for adequately assessing potential hazards due to pesticides. Moreover, we have little data on pesticide residues in the under-developed countries, although large quantities of persistent pesticides are regularly used in these areas.

We now have some information on amounts of pesticide residues in the biota, but we have much more data on residues in fish, birds and birds' eggs than those in other vertebrates and invertebrates. Pesticide residues have been found in the air and rainwater, but there is still no regular monitoring programme for these media. Enough data on residues in the human diet and in the bodies of human beings in many parts of the world have been collected to assess their possible hazards to man.

Even when we have established adequate monitoring programmes, there still remains the problem of assessing the significance of the residues. Is it most important to preserve the flora and fauna as far as possible in its present state, or are hazards to human health more important?

Much nonsense has been talked about avoiding changes in balance of nature; even if such an essentially static and idealistic concept were ever a reality, most of man's activities combine to disrupt this relationship. Pesticides have often been blamed for catastrophic declines in numbers of wildlife, but probably changes in land and water development and usage exert a much greater influence than do pesticides. It is clearly undesirable to place the existence of certain species of animals and plants at hazard, but it is important to remember that only species that were close to local or general extinction are likely to be eliminated by pesticides.

It is essential to avoid the trap of establishing a simple causal relationship in a particular situation, and then extending it to be an inviolate principle. It is easy to show well-authenticated instances where organochlorine pesticides have caused serious local ecological disturbances, but we have to be cautious in interpreting such instances as a general principle that all organochlorine insecticides are bad, and that all their uses should be outlawed. With a fuller understanding of their environmental impact and potential hazards, they may yet continue to be used for particular purposes and greatly benefit mankind. An unreasoning ban on their use may be just as unwise as a continued thoughtless and widespread over-use.

Since the ecological hazards of persistent pesticides have been fully realised, there has been considerable expansion of research into control of pests by such means as biological control, insect hormones, chemosterilants, attractants and repellents, but, although appreciable success has been attained, it still seems likely that large quantities of pesticides will be needed for some years to come.

Possibly the importance of DDT in the environment has been exaggerated; for instance, it has been suggested that DDT could aversely affect photosynthesis in the oceans, and that this might reduce their productivity or even ultimately the world's oxygen supply. However, it has been shown that it is extremely unlikely and that even if pesticides did appreciably affect photosynthesis, which itself is extremely improbable, there need be little anxiety about oxygen supply.

It seems probable that the greatest hazards of pesticides are to aquatic organisms, which seem to be much more susceptible to them and also to concentrate them into their tissues more readily than do terrestrial organisms. Since the seas are a sink, which ultimately receives a large proportion of industrial chemicals used, perhaps our anxiety should be more directed to possible effects of pesticides on marine organisms. However, it seems likely that life in the sea is endangered much more by other forms of pollutants than by pesticides, which represent only a minute fraction of the total pollutants that reach the oceans of the world.

We certainly need to know much more about global transport of pesticides and we hope that the comprehensive discussion of this phenomenon in Chapters 10 and 12 of this book adequately summarises the present state of knowledge of this phenomenon.

It seems inevitable that the ultimate solution to our environmental pesticide problems must be a compromise which will use the smallest possible quantities of pesticides, combined with other control measures so that environmental pollution by pesticides is kept at a minimum. There seems little likelihood of being able to dispense with the use of pesticides in the foreseeable future but intelligent use of them will greatly reduce the hazards implicit in their continued use.

One of the main difficulties in interpreting environmental residue data is the variability in analytical techniques and sampling procedures. Frequently, data are given in a form which is very difficult to summarise or compare with those produced by other workers. Furthermore, as analytical methods have improved and become more sensitive, so the possible misinterpretations that can be placed upon the data they provide have multiplied. This is particularly true of gas-liquid chromatographic techniques, that rely on relating the position of a peak traced on a chart to one given by a particular compound which is to be estimated. Many different compounds can produce peaks very close together; when peaks are large, well-defined and clearly separated, they usually provide a valid estimate of the amount and kind of residue present if they can be confirmed by paper chromatography or other analytical techniques. However, this is still not invariably so: for instance, when 34 soil samples dating back to 1909-1911 (long before the advent of organochlorines) were analysed, 32 showed apparent insecticide residues. These were eventually attributed to certain interfering soil constituents, but this incident clearly demonstrates the care needed in such analysis (Frazier *et al.*, 1970).

In analyses of extremely small concentrations of residues such as those in air or rainwater, there is much more possibility of error, not only by misinterpreting chromatographic peak traces, but also from contamination of the analytical equipment with small traces of pesticides or related compounds.

For the purpose of this book, it has not been possible to differentiate between analytical data for reliability, and some data may well be suspect; this should be remembered when the data is considered, quoted or discussed. What is more, explanations of contaminations of soil, water or the biota with pesticides may not always be complicated or ecological. For instance, it has been suggested that the minute amounts of chlorinated hydrocarbon insecticides reported from fish and vertebrates in the Antarctic, may be accounted for by insecticides imported into the area with various expeditions; this may well not be so, but such possibilities should be borne in mind when interpreting residue data.

It is important to remember that residues are usually analysed at some particular point in time, and such data can easily be misleading. For example, it is common practice to take only a single set of samples from some compartment of the physical or biological environment, and determine the residues they contain. Such a spot check may have very little value, however, unless reinforced by other data, and may bear little relation to the average amounts that occur in that medium. In the physical environment, such residues can be the remains of a much larger dose, whereas in the biota, they can be the end result of a long process of absorption and concentration. The whole system of movement of pesticides through the environment is essentially a dynamic one, and for a meaningful interpretation, residue data must be seen and studied as such. The most difficult thing to assess from the residue data that is currently available is not the *status quo,* but the changes that will occur if certain actions, for instance banning a particular pesticide, are taken. The information we need is how rapidly residues will disappear from the different compartments of the environment when they are no longer added to, and other methods of pest control are adopted. We also need to known if the residues that exist are of any real importance, or if the organisms living in a contaminated environment can successfully adapt to such chemicals. Our environment is already full of alien chemicals, but we still do not know whether the further addition of these pesticide residues is important.

One of the more important ways in which persistent pesticide residues can be prevented from becoming a major world problem, is by legislation whenever excessive amounts of these chemicals are reported in food, animal tissues, soil or water as a result of monitoring surveys. So far, the more stringent legislative regulations have involved the permitted amounts of pesticides in human foodstuffs. U.S.A., Canada, U.S.S.R., Japan and most European countries have enacted extensive legislation, and many countries have some tolerance limits on the permissible amounts of pesticides in foods. Such tolerance limits often have a built-in safety factor of at least 100 times known toxic levels. Naturally, the enforcement of such legislation is expensive, involving spot analyses of large numbers of samples, for a wide range of pesticides. Fortunately, although there are very large numbers of pesticides in use, only a few chemicals are extensively used and likely to occur in foods, so the analyses can be restricted to those likely to have been used on the crop or animal from which the food was produced. With more international cooperation, hazards due to pollution by persistent pesticides can be kept to a minimum. It is clear that if, as seems likely, the use of persistent pesticides continues, then it is essential that adequate monitoring of residues is maintained. Although there is little evidence that the small amounts of persistent pesticides in the atmosphere or other parts of the environment are harmful, we cannot be sure that there are no harmful side-effects and, wherever possible, less persistent materials should be used.

How then can we assess the future hazards due to persistent pesticides? Even if no more were to be used, it seems probable that there would be some residues present in the environment for several decades to come. However, although their use is decreasing in Europe and North America due to legislation and various restrictions, their total production is still increasing, and it is difficult to foresee how long this will continue. As long as the demand for more food continues no doubt they will still be used. If, as seems probable, at least some degree of global transport of pesticides is a reality, it seems likely that they will remain a potential environmental hazard.

Fortunately there are extensive programmes of research into amounts of residues present and their long-term effects, in many countries, with a resultant increase in our knowledge of the hazards involved. For instance, we still have no evidence of them causing any serious harm to human beings, and there are now indications that human body burdens of persistent pesticides do not continue to increase indefinitely, but, instead, tend to level off even after excessive exposure to these chemicals as by pesticide workers.

There is evidence that the persistent pesticides are sometimes harmful to some species of wildlife and may even threaten the existence of others. However, these effects do not seem to be drastic and it seems unlikely that really serious damage will occur to any species of wild animal, other than those that are already under severe environmental stress for other reasons.

There is an extensive scientific literature on pesticide levels in, and effects of them on, various components of the environment. Many of the more important papers are summarised in the various chapters of our book, but since many of the subjects discussed are still controversial, some discussion of the earlier books and reviews of the subject may be of value.

The first major publication was the book *Silent Spring* (1962), which, by greatly exaggerating the potential hazards of persistent pesticides, focussed much more attention on the problem, and in this way helped to stimulate an awareness of the need for research, and thus indirectly contributed to our knowledge of the status of pesticide residues in the various sections of the environment, although we still know little more of the significance of these residues. This book was followed by a variety of very relevant reviews and symposia concerning the persistence of pesticides in the environment and their ecological effects. These include *Pesticides and the Living Landscape,* a much more balanced review of the influences of pesticides than *Silent Spring*. In 1964 a comprehensive annotated bibliography entitled *Pesticides in Soils and Water* was published by the U.S. Department of Health, Education and Welfare; in the following year another excellent review, *Residues of Chlorinated Hydrocarbon Insecticides in Biological Material* (Marth, 1965), concerned residues in plant material, and in 1965, Moore reviewed the effects of pesticides on birds in Great Britain. Also in 1965, there was a symposium entitled *Research in Pesticides* (Chichester, 1965) which contained several interesting papers on the persistence of pesticides in the environment. In 1966 there were three symposia, *Pesticides and their Effects on Soils and Water* (Anon, 1966), *Agriculture and the Quality of Our Environment* (Brady, 1966), and a third, *Organic Pesticides in the Environment* (Gould, 1966), published by the Chemical Society of America. Papers read at an international symposium organised by the British Ecological Society were published with the title *Pesticides in the Environment and their Effects on Wildlife* (Moore, 1966), and a book, *That We May Live* (Whitten, 1966), was as much biassed in favour of pesticides as *Silent Spring* was against them.

A review of persistent soil insecticides also appeared in 1966, particularly emphasising factors influencing their persistence in soil (Edwards, 1966). The following year, Moore (1967) published *A Synopsis of the Pesticide Problem*, which dealt with ecological aspects of pesticide usage, and a review by Newsom reviewed the consequences of insecticide usage on non-target organisms, but this review, although good, was limited in scope and coverage. A popular book by Mellanby (1967), *Pesticides and Pollution*, discussed the more general aspects of pesticide pollution. In 1968, Stickel produced a review, *Organochlorine Pesticides in the Environment*, which briefly discussed the ecological hazards to wild animals of pesticides. A good review on *Pesticides and Fishes* (Johnson, 1968) had an excellent bibliography. The proceedings of a symposium with the title *Chemical Fallout—Current Research on Persistent Pesticides* (Miller and Berg, 1969) contained some excellent papers, especially one, 'Organochlorine insecticides and bird populations in Britain'; and in another, Robinson (1969) summarised the uptake of organochlorine insecticides into human beings under the title 'The burden of chlorinated hydrocarbon pesticides in man', which particularly emphasised the deficiencies in sampling and analytical techniques and produced a mathematical model to account for the metabolism of pesticides in the human body.

Frost (1969) reviewed global aspects of pesticide contamination, and in 1970 the proceedings of a symposium *Pesticides in the Soil* was also published. Edwards (1970) produced a book *Persistent Pesticides in the Environment*, which attempted to assess and evaluate the amounts of pesticides in all compartments of the environment. The same author (1970, 1973) reviewed the effects of pesticides residues on the whole soil ecosystem. A book *Pesticides and Freshwater Fauna* (Muirhead-Thompson, 1971) and the first of a five-part series edited by Robert White-Stevens, entitled *Pesticides in the Environment*, appeared in 1971.

There have been several recent governmental reports, including the 'Report of the U.S.D.A. Committee on Persistent Pesticides' (1969), and the 'Report of the Secretary's Commission on Pesticides and their Relationship to Environmental Health' (Mrak Report, 1969) from the U.S.D.H.E.W., a 'Further Review of Certain Persistent Organochlorine Pesticides used in Great Britain' (1969), and the 'Third Report of the Agricultural Research Committee on Toxic Chemicals' (1970), also from Great Britain.

One of the main aims of our book is to present factual information on the seriousness and extent of ecological hazards due to pesticides, carefully assessed by leading authorities in the various fields of pesticide pollution. With such information, the scientist and layman may be able to make a much more balanced judgement than he could from some of the articles and books that appear in the popular press and are often emotive and written with an appeal to the sensational. Already, too many far-reaching decisions have had to be made based on inadequate information.

The compilation of data and treatment of the subject in this book is made much more comprehensive by having the chapters written by different authors, each of whom is a recognised authority in his or her field. Moreover, one of the benefits of a book written by a range of different chapter authors, is the variety of viewpoints and emphasis that is given, and it is hoped that in this way the reader will obtain a much broader and more balanced picture than he could from a book written by a single author, however unbiassed he might be. To retain an international flavour to the data, chapter authors have been recruited almost equally from Europe and North America, so there should be no undue emphasis, other than that inevitable where more data are available from North America than from other areas of the world.

We hope that the data presented will not only illustrate the current state of our knowledge of persistent pesticides in the environment, but will pinpoint those areas where more information and research is needed. If it serves to provide those bodies concerned with legislation and environment pollution, with a factual basis for their deliberations and to present a balanced picture for the interested scientist and layman, its main purpose will have been fulfilled.

For references: see Appendix (General references) page 515.

Chapter 1

Methods and Problems in Analysing for Pesticide Residues in the Environment

J. H. Ruzicka

Laboratory of the Government Chemist,
Cornwall House, London S.E.1.

1.1   INTRODUCTION

Pesticide is the general term for insecticides, acaricides, rodenticides, molluscicides, herbicides, fungicides and similarly active compounds. A wide range of compounds are used as pesticides and a correspondingly wide range of methods are used in the analysis of their residues. They can be broadly classified according to their general chemical nature into several principal types as shown in Table 1.1.

TABLE 1.1.   Chemical classification of pesticides.

Chemical type	Example	Structure	Typical action
Organophosphorus	Malathion	CH_3O\\$CH_3O$/$P\text{-}S\text{-}CH\text{-}CO_2C_2H_5$ with $CH_2\text{-}CO_2C_2H_5$, $\parallel S$	Insecticide
Organochlorine	p,p'-DDT	Cl-C$_6$H$_4$-CH(CCl_3)-C$_6$H$_4$-Cl	Insecticide
Chlorophenoxyacid	2,4-D	Cl,Cl-C$_6$H$_3$-$O\text{-}CH_2CO_2H$	Herbicide
Heterocyclic	Simazine	C_2H_5NH / triazine with Cl / $NH\text{-}C_2H_5$	Herbicide
Carbamate	Carbaryl	naphthalene-$O\text{-}CO\text{-}NH\text{-}CH_3$	Insecticide
Dithiocarbamate	Thiram	$(CH_3)_2N\text{-}CS\text{-}S\text{-}SCSN(CH_3)_2$	Fungicide
Organometallic	Phenylmercury acetate	C$_6$H$_5$-$Hg\text{-}OCO\text{-}CH_3$	Fungicide

The analyses for inorganic pesticides do not differ significantly from standard trace element analyses and will not be discussed here. The use of organic pesticides in agricultural practice is now firmly established and is becoming more extensive. Despite the undoubted advantages in their usage, concern has been expressed at the potentially harmful effects of using pesticides which are stable and can accumulate in man and his environment. New and improved compounds are continually being evaluated to overcome this difficulty. The work of the analyst is to detect and determine pesticide residue in directly treated crops and animals, and also in crops subsequently grown in the same area, animal produce (domestic and wild) and in man himself. It will be appreciated that these analyses need to be carried out on a variety of commodities, each with its own extraction and clean-up problems, and the most efficient must be used because the quantity as well as the identity of any residue present is usually required. The analysis is often complicated by chemical changes undergone by pesticides when absorbed into living tissue, adsorbed onto the soil, or exposed to ultraviolet light or sunlight. These changes often produce compounds which are more toxic than the original pesticides and the analyst has to determine the rate of the breakdown, the nature and quantity of these metabolites and end-products, as well as the pesticide residue.

The stages of analysis are as follows:

(1) The extraction of the residue from the simple matrix using an efficient and selective solvent.

(2) The removal of interfering substances from the extract—usually referred to as the clean-up procedure. This often involves either chromatography or solvent partition.

(3) The estimation of the quantity of pesticide residues, together with metabolites and breakdown products, in the cleaned-up extract. These are determined at very low levels (as low as 10^{-12} g), and to obtain this sensitivity, stringent requirements of selectivity are imposed on the selected method.

(4) The confirmation of the presence of the residue by, for example, using a different method or the formation and identification of a derivative.

There are five basic methods of analysis:

(a) Functional group analysis. This method involves, for example, the colorimetric assay of a particular group or element in a compound. It does not give precise identification, requires thorough clean-up and consequently is not now used very frequently.

(b) Biological test methods. These show the presence of toxically significant residues, by, for example, inhibition of the enzyme cholinesterase in animals by certain classes of pesticide. As it is possible to use these methods without clean-up they are useful as screening tests but are non-specific.

(c) Chromatographic methods. These include thin-layer and gas-liquid chromatography which give separation and accurate identification and estimation of a wide range of residues.

(d) Spectroscopic methods. These provide evidence of identity and can also in ideal conditions be used in quantitative analysis.

(e) Radiochemical methods. Included under this heading are neutron activation analysis, direct isotope dilution methods and the more sophisticated double isotope derivative analysis technique.

For all methods of analysis there is a detection limit below which it is impracticable to differentiate between the noise level of the background and the analytical signal. These levels range from 10^{-6} g for chemical methods to 10^{-14} g for radiochemical procedures. The importance of modern analytical instruments in pesticide analysis cannot be exaggerated. Their sensitivity and versatility have made it possible to solve complex problems which were insoluble only a few years ago. This progress is especially noticeable in gas chromatographic detection where the electron-capture and phosphorus-sensitive flame ionisation detectors have lowered the detection limits to the nanogram (10^{-9} g) and picogram (10^{-12} g) ranges. Techniques such as nuclear magnetic resonance, radioisotope labelling and mass spectrometry are proving increasingly useful for the identification of unknown metabolites and breakdown products and for confirming identification. In recent years there has been a need for continuous monitoring of pesticide residues, and the use of automated analysis is being investigated.

Each stage of the analytical procedure is dictated both by the chemical nature of the pesticides under investigation and by the nature of the substrate. Separate discussions of the extraction, clean-up and end method stage will now follow.

1.2 EXTRACTION PROCEDURES

The extraction procedure to be employed is governed by the type of pesticide and the nature of the sample under examination. The extraction procedure should be at least 80% efficient and sufficiently selective to require only the minimum amount of clean-up prior to the estimation. It is important that reagents used in the extraction and clean-up do not in any way interfere with the end detection method. A highly selective extraction procedure is desirable for pesticide residues since the concentration of residue in the substrate is normally so low and it may be useful if selectivity can be achieved, even at some loss of extraction efficiency. Because of the high sensitivity of the end methods used, great care must be taken to avoid contamination by contact with plastics or rubber especially from bungs or caps, and all solvents used must be sufficiently pure to present no interference problems to the end method. Particularly when gas-liquid chromatography is to be used as an end determination method, it may be necessary to redistil all solvents used. This is especially important with the electron-capture detector, which, being somewhat unselective and highly sensitive, requires an almost pure solution of the pesticide in a suitable solvent.

Solvent extraction is the normal extraction method and gives great scope for choice of suitable polarity of solvent or solvent mixture, time of contact of solvent with sample and the manner of contact, so that optimal extraction conditions can be found. When pesticides are found only on the surface of vegetables and fruit it is often possible to effectively remove all the pesticide residues from the sample by simply washing the whole sample with a suitable solvent. This method gives very low co-extractive levels and enables the analysis to be carried out with little clean-up. However, it would be unwise to use this procedure without evidence that good recoveries are obtained in the particular circumstances pertaining. Maceration of samples with a solvent followed by centrifuging or filtering is an efficient extraction system particularly for vegetables. The water in a sample may produce emulsions with non-polar solvents and this may be avoided by the use of a drying agent such as anhydrous sodium sulphate or isopropyl alcohol together with, or prior to, adding solvents.

Meat samples may contain too much connective tissue for a macerator to deal with effectively and it is usually preferable for such samples to be comminuted prior to maceration. Soxhlet extraction of a finely comminuted sample is often used as an extraction method for meat samples. An even simpler method can often be used, namely, heating the minced sample in a beaker on a steam bath with the extraction solvent, possibly after grinding the sample with sodium sulphate and sharp sand which helps to break down some of the connective tissue. This technique can be used for organochlorine pesticides for which n-hexane has been found to be a suitable solvent. Care must be exercised, however, to ensure that the more volatile pesticides, e.g. lindane (γ-BHC), are not lost in the process.

The sampling of air for pesticides is made somewhat complex by the fact that the pesticides may exist in aerosol, vapour or particulate forms. A variety of techniques have been used for the extraction of pesticides from air, including adsorption on packed columns (Hornstein and Sullivan, 1953; Mattson et al., 1960; Simpson and Beck, 1965); absorption by a liquid phase after passage through scrubbers fitted with fritted discs or gas dispersion tubes to break up the air stream (Abbott et al., 1966; Caplan et al., 1956; Hirt and Gisclard, 1951; Jegier, 1964a, b); freezing out in traps filled with glass helices (Shell, 1959); extraction by means of glass fibre filters (Tabor, 1966) or cellulose filter packs (Simpson and Beck, 1965; Wolfe et al., 1966, 1967). A procedure for continuous or intermittent air sampling together with methods for recovery and determination of the pesticide residue has been described (Miles et al., 1970). A system for monitoring atmospheric concentrations of field-applied pesticides has been proposed (Willis et al., 1969) in which ethylene glycol is used as the trapping solvent and the air sucked through by vacuum pump at a controlled rate. By means of a rotameter-type flow meter and regulating valve used in conjuction with the vacuum pump, accurate measurement and control of the air flow at a wide range of flow rates can be maintained. After the desired volume of air has passed through the system, the trapping solvent is partitioned with hexane. The hexane fraction is collected, washed with water, dried, concentrated and then an end determination method applied.

It is advisable when using sintered glass gas dispersion tubes to incorporate a filter, for example, a glass filter type, to remove dust which could clog the sinters of the traps. For completeness of analysis, these filter membranes can be examined for pesticides. Stanley *et al.* (1971) measured levels of pesticides in the atmosphere. The air was drawn through a collection train consisting of a glass-cloth filter, an impinger with 2-methyl 2,4-pentanediol as trapping liquid and an adsorbent column of alumina. After extraction and clean-up, the pesticides in all parts of the collection train were determined by gas chromatography. The chlorinated pesticides were determined with an electron capture detector using 5% SE-30 and mixed 2.5% SE-30, 3.5% OV-17 columns. Organophosphorus pesticides were determined with a flame photometric detector using columns of 5% QF-1 and 5% OV-1.

Water samples are normally shaken with the solvent to extract any pesticide residues. Hexane is the usual solvent for organochlorine compounds and for many of the others chloroform is useful. It is strongly advisable to wash out the sample container with solvent since a large part of the residue may be in the film of grease etc. on the vessel walls. Extraction of pesticides from air or water samples may be effected by adsorption on to a solid such as carbon, whence it can be removed by elution. This method has been used in surveys (U.S. Dept. Health, Education and Welfare, 1964) and gives a means by which large samples may be taken (1,000 litres or more), and is effective at the very low levels $\approx 10^{-12}\,g\,l^{-1}$ at which pesticides may be found in water. The desorption is carried out by Soxhlet extraction using a suitable solvent or combination of solvents. The adsorption of pesticides has sometimes been found to be very high (Rosen and Middleton, 1959) but the recovery on desorption may be low in some cases (Burtchell and Boyle, 1964).

The organochlorine pesticides and their initial breakdown products are fat-soluble and therefore tend to accumulate or be concentrated in the lipoid portions of food, such as the cream of milk and the fat of meats. Extraction of these pesticides from foods rarely presents serious difficulties. Nevertheless, careful choice of solvents and method will often minimise the amount of co-extracted material and thereby reduce the need for a more stringent clean-up procedure with its concomitant lower recovery factors. Hexane or acetone-hexane mixtures, in combination with either Soxhlet extractors, blenders of the Waring type, or simple shaking devices, are usually employed, but other solvents such as acetonitrile, benzene, dimethyl, formamide, dimethyl sulphoxide and propan-2-ol have also been used.

The extraction of residues of organophosphorus pesticides and their metabolites from different sample substrates poses several difficulties. Not only does the nature of the sample under examination govern the process to be applied, but the wide range of polarities encountered in these compounds makes it essential to use a general solvent for preliminary extraction if residues of unknown compounds are being sought. Such solvents also remove considerable amounts of co-extractives that are apt to complicate clean-up stages. Extraction solvents are usually chosen from benzene, chloroform, dichloromethane, acetonitrile, often in the presence of anhydrous sodium sulphate to aid in the liberation of the more water-soluble compounds. The chopped or shredded sample is usually treated with the chosen solvent under vigorous agitation, in a top-drive macerator or similar apparatus; frequently, centrifugal action is required to break the emulsion formed, and filtration of the solvent extract may also be essential. Deep-freezing of the produce, followed by rapid thawing in the presence of solvent, may be used to shatter the cell structure, thus giving ready release of the contents, which may hold systemic pesticide residues. A more efficient clean-up technique may subsequently be required when such methods are used.

Some attention must be paid to the purification of solvents to be used for the extraction of pesticide residues from sample substrates. Thornburg (1966) has reviewed the various methods used for purification, including distillation, adsorption and chemical treatment. The need to avoid contact with rubber, plastic or similar materials which can introduce contamination must be stressed. Beckman and Gauer (1967) have also discussed non-distillation methods of solvent purification.

Two interesting methods of extraction are those of Grussendorf *et al.* (1970) and Johnsen and Starr (1967, 1970). The former method used a rapid sample preparation by ball grinding, extraction freeze-out and semimicro-column clean-up. Johnsen and Starr extracted insecticides from soil samples by the use of ultrasonics and they concluded that it was superior to the roller and blender methods in extraction efficiency and as efficient as an 8-hour Soxhlet extraction.

1.3 CLEAN-UP PROCEDURES

Before the quantitative determination of a pesticide residue can be carried out, the residue must be extracted and, generally, this extract must be purified or cleaned up. Even when a highly selective extraction procedure has been used, the extract will normally contain co-extracted matter sufficient to interfere with the end determination. Different end methods will differ in the amounts and types of co-extractives which can be tolerated. For example, the electron-capture detector is extremely intolerant to halogenated impurities, and thin-layer chromatography cannot be done in the presence of fatty material, which tends to affect the rate of migration of a compound grossly. Other types of co-extractives can also interfere with thin-layer chromatography because of reaction with, or the masking of, the visualising agents. Similarly, these co-extractives can affect spectrophotometric methods. Only in a few instances, such as the waters from reasonably clean rivers, can good gas chromatograms be obtained from simple hexane extracts without further treatment. The presence of co-extracted material in injections made on to the gas chromatograph columns not only cause non-pesticidal peaks to appear on the resultant chromatograms but constant injection of such material damages the separatory power of the columns, peaks begin to broaden, tail, and ultimately to overlap. This is particularly true for extracts containing much fatty material. Constant injection of small amounts of fat also leads to saturation of the first few millimetres of the support material with fat, which tends to act as another stationary phase and can radically alter the characteristics of the column.

The extent and nature of the co-extractives is largely governed by the nature of the sample, and apart from judicious choice of extraction procedure, the analyst must choose an appropriate clean-up procedure. Broadly, the methods available are: adsorption methods (column and thin-layer chromatography), solvent partition, distillation methods (including sweep co-distillation) and gel chromatography. The type of clean-up used will obviously depend on the nature of the co-extracted material. If low in fat, passage of the extracts through a short column of prepared alumina will often suffice; silica gel, charcoal and Florisil have also been recommended for this purpose. The co-extractive materials are removed from the extract solution mostly by virtue of their greater adsorption on the column. The extract solution, after concentration to a small volume, is applied to the top of the prepared column so that it is eluted off by the solvent or solvents and can be collected in fractions. Relatively large amounts of co-extractives can be tolerated by columns and very effective clean-up can be achieved, particularly when the pesticides under investigation can be collected in discrete fractions. It is important that the activity of the adsorbent used is constant because variation in different batches is often found.

The activity of adsorbents can be checked by recovery studies, using a prepared column or by using standard dye materials. Adsorbents have been classified into grades of activity as weak, medium or strong, depending on their power of adsorption. It is also possible to activate or deactivate an adsorbent. Most materials may be activated by strong heating and some may be activated for a particular purpose by pretreatment with acids or bases (such as alumina) or organic solvents (such as charcoal). Deactivation of alumina has been effected by the addition of water, and the different grades of alumina thus produced are standardised according to the Brockmann activity standard, which is based on the absorptive capacity for various azo dyes (Brockmann and Schodder, 1941). Table 1.2 shows the relationship between activity and water content.

The Florisil column clean-up procedure for pesticides is almost universally used. Beasley and Ziegler (1970) compared Florisil and Chrom AR sheet clean-up for organochlorine pesticides in spinach. The activity of Florisil varies from batch to batch and Mills (1968) described a simple method to standardise Florisil compounds. Florisil columns were used by Sans (1967) to fractionate insecticidal compounds that are difficult to resolve by gas chromatography. Fatty materials may be removed by a solvent partition method such as the dimethylformamide-hexane method (de Faubert Maunder *et al.,* 1964). The hexane extract of the sample is extracted three times with dimethylformamide saturated with hexane. The combined dimethylformamide phases are washed with hexane saturated with dimethylformamide and then shaken with a large volume of 2% aqueous sodium sulphate. On standing, the hexane previously held in solution rises to form an upper layer containing the pesticides and can be separated off.

Solvents used for partition methods include acetonitrile (Jones and Ruddick, 1952; Mills, 1959), dimethylformamide (DMF) (de Faubert Maunder *et al.*, 1964; Burchfield and Storrs, 1953) and dimethyl sulphoxide (D.M.S.O.) (Haenni *et al.*, 1962; Eidelman, 1962, 1963). Wood (1969) favoured the use of D.M.S.O. because it was a good solvent for chlorinated pesticide residues and dissolved less oil or fat than either DMF or acetonitrile. The column chromatographic method of Wood was applicable to a wide range of fatty samples. The sample was mixed with celite, packed into a small column and the chlorinated pesticides eluted

TABLE 1.2. Relation between water content and activity of alumina (Hesse *et al.*, 1952).

Activity (Brockmann number)	Water content (%)
I	0
II	3
III	6
IV	10
V	15

with D.M.S.O. The eluate was then adsorbed on Florisil and the residues recovered from the D.M.S.O. by hexane elution. This method gives a good clean-up, the author claiming that the amount of material passing through the clean-up is usually less than 0.01% of the sample. Carbon columns have been used for the clean-up of samples for the determination of both organochlorine and organophosphorus pesticides as they have a high capacity for retaining plant pigments and some waxes and oils.

McLeod *et al.*, (1967) compared various carbon adsorbents and also proposed a mixed column of carbon (Darco G60) and cellulose (Solka Floc). Thin-layer chromatography can be used for clean-up providing that there is not much fatty material present in the extract. The procedure differs from the thin-layer end detection method in that thicker adsorbent layers are used (1-5 mm), and normally the extract is applied as a streak. Standard compounds developed on the same plate are used to indicate the region of the plate on which the pesticide will be found. The band is removed by scraping and extracted to give a suitable solution for an end method of determination.

To increase the amount of co-extractive material that can be applied to thin-layer plate, modified systems such as multiband chromatoplates and wedge-layer plates (Abbott and Thomson, 1964a, b) have been devised. In wedge-layer plates the sample is applied to the thicker end of the plate which can take a heavier loading. After passing along a decreasing thickness of material it reaches the thinner layer which has a lower clean-up potential but an improved resolution. This type of plate can be used as a combined clean-up and end detection method.

The sweep co-distillation technique (Storherr and Watts, 1965) is a useful method, because it eliminates the need for specialised adsorbents and equipment, large volumes of costly purified solvents and laborious clean-up methods. The samples are extracted with suitable solvent, for example, ethyl acetate for organophosphorus pesticides. An aliquot is then concentrated, and clean-up by injection into a heated long glass tube packed with glass wool followed at 3-min intervals by injections of ethyl acetate. Nitrogen carrier gas sweeps the vaporised volatile components through the tube, the organic interferences remaining on the glass wool whilst the pesticides are collected for analysis. An improved version of this technique was claimed by Kim and Wilson (1966), who used hexane as carrier gas in the place of nitrogen, when determining organochlorine residues. A forced volatilisation technique was used by Mestres and Barthes (1966) to obtain vegetable tissue extracts in a suitable state for electron-capture detection.

Gel filtration or gel chromatography has been widely used in the biochemical field for a number of years, but its use in the analysis of pesticides has been limited. Recently, however, fresh interest in the technique has arisen following the introduction of Sephadex LH-20, a lipohilic modified dextran gel, which can achieve separations in non-aqueous systems.

This mode of separation, based on molecular size, offered a potential clean-up system for distinguishing organophosphorus pesticides (molecular weight 200-350) from the common co-extractives chlorophyll (molecular weight 906) and carotene (molecular weight 536). Gel filtration using Sephadex LH-20 in ethanol was shown by Horler (1968) to be a practical method for the single stage clean-up of insecticide residues from grain. Such clean-up was considered adequate for thin-layer chromatography. The possible use of the technique for the determination of organophosphorus pesticides was investigated by Ruzicka *et al.* (1968) with the gel swollen and eluted with either acetone or ethanol. Efficient separation of these pesticides from co-extractives could not be achieved, much of the pigment material eluting in the range in which pesticides were found to emerge. However, they found the observed elution volumes of the compounds examined were constant and distinctive, and the sequential order differed for the two solvents. The technique is therefore useful for the identification of organophosphorous pesticides at the low residue level and it has been incorporated in an analytical scheme by Askew *et al.* (1969) in the qualitative confirmation of the presence of these compounds in river waters. A representative selection of relative elution volumes of a number of organophosphorus compounds is given in Table 1.3. A useful

TABLE 1.3. Separation of some organophosphorus compounds on Sephadex LH-20 (Ruzicka *et al.*, 1968).

Compound	Mol. wt.	Relative elution volume (Parathion = 100)	
		Acetone	Ethanol
Malathion	330	95	86
Mecarbam	329	96	85
Diazinon	304	98	77
Disulfoton	274	100	87
Parathion	291	100*	100†
Phorate	260	103	89
Demeton-O-methyl	230	103	93
Phorate sulphone	309	104	98
Thionazin	248	108	89
Fenchlorphos	321	109	101
Chlorfenvinphos	359	114	77
Phosphamidon	264	117	72
Dichlorvos	221	122	83
Demeton-S-methyl	230	130	86
Dimethoate	229	148	95
Demeton-S-methyl sulphoxide	246	200	70

* Parathion elution volume = 43.5 ml.
† Parathion elution volume = 75.5 ml.

feature of gel chromatography is that the column can be used repeatedly over long periods without any detectable change in the elution volumes and the recoveries. Application of this technique to organochlorine pesticides has not been successful as the separation is poor.

Calderbank (1966) has reviewed the use of ion exchange resins as a clean-up procedure. Cationic resins are particularly useful in the determination of diquat in potato tubers (Calderbank *et al.*, 1961) and paraquat in fruit (Calderbank and Yuen, 1965).

1.4 END METHODS OF DETERMINATION

The main end determination methods used in residue analysis fall into three groups, namely, biological, spectrophotometric and chromatographic. Other methods used, such as electrochemical (chiefly polarographic) and radiochemical, are extremely limited in use.

1.4.1 Bioassay

The determination of pesticide residues by bioassay is based on the measurement of growth, death or some other physiological change in animals, plants or micro-organisms. Any organism which is susceptible to a pesticide may be used for the bioassay of its residues, but some organisms are more sensitive than others. The response of test organisms to toxicants vary with species, stage, age and sex. The susceptibility of the test organism and the consistency of its response affect the sensitivity and accuracy of bioassay results.

The main disadvantages of bioassay are its lack of specificity and the need for the isolation of extremely small quantities of toxicants in the presence of large amounts of plant or animal materials. However, it is possible to separate pesticide residues before bioassay by biological, physical or chemical means so that in certain cases one can determine one pesticide in the presence of another. For example, in the determination of organophosphorus pesticides in the presence of organochlorine pesticides, one could either use insects resistant to the latter or measure cholinesterase activity, which is only inhibited by organophosphorus compounds. The non-specificity does, however, permit the method to be adapted for residues of many toxicants and residue determination in a variety of animal and plant tissue is possible with only minor modifications to the basic procedure. This, together with the simplicity of the method, makes the technique readily acceptable to biologists in particular.

Bioassay usually gives good agreement with colorimetric analysis but has the disadvantage that relatively large numbers of animals are required for the preparation of adequate standard curves. Generally, the sample must be tested with the same culture as the standards at about the same time, and often several replicates are necessary. The sensitivity of bioassay is highly dependent on the toxicity of the pesticide to the organism under test but may be improved by (a) increasing the sample weight, (b) isolating the pesticide from the extract, (c) increasing the quantity of extract used for each test, (d) increasing the contact time between sample and organisms, (e) use of smaller and younger animals. Tissue co-extractives can mask the toxicity of the residues by absorption as well as dilution, and may even increase the toxicity if they contain natural toxic materials. The masking effect is less serious than the toxicity of co-extractives. The results of the assay alter according to the physical conditions, for example, the amount of water present, the form in which the toxicant is present, temperature, and lighting conditions all affect results. Also, the response of the organism varies from time to time, but this can be reduced by inclusion of a standard toxicant.

The methods used for bioassaying pesticide residues can be grouped together according to the method of exposure.

1.4.1.1   Dry methods

A portion of the test solution is evaporated to dryness and used as a dry film to which the test organisms (such as house flies (*Musca*) and fruit flies (*Drosophila*)) are exposed. This method has two main disadvantages. Firstly, volatile pesticides may be lost when the test solution is evaporated to dryness and, secondly, co-extractives may inhibit the pesticidal action.

1.4.1.2   *Wet methods*
Aqueous organisms such as fish or mosquito larvae are exposed to a solution or suspension of the toxicant. The organisms are thus completely externally and partially internally exposed to the toxicant.

1.4.1.3   *Diet methods*
The test animals are fed the sample either directly or mixed in the diet. This method has been used for the determination of pesticide in milk and macerated plant tissues.

1.4.1.4   *Enzymatic methods*
The prime example of this method is the inhibition of cholinesterase by organophosphate and carbamate insecticides. The test animals are fed with the sample and the degree of inhibition of the enzyme is proportional to the sample toxicity. Direct inhibition measurements are discussed in the section on automated analysis.

1.4.2   Spectrophotometric determination
Spectrophotometric determination of pesticide residues does not achieve the sensitivity of thin-layer and gas chromatographic techniques. It may not be able to distinguish between the parent compound, metabolites and hydrolysis products but can be utilised with chromatography as a confirmatory technique. There are three types of spectrophotometric determination based on (a) ultraviolet and visible (200-700 nm), (b) infra-red (2-15μ) and (c) fluorescence and phosphorescence methods.

1.4.2.1   *Ultraviolet and visible*
Ultraviolet methods require a rigorous clean-up of the extract to ensure that the final solution is free from any material that will absorb light in the region of the spectrum where the pesticide content is to be measured. The solvent used for the initial extraction of a crop material should be chosen with care to ensure extractability of the pesticide and with consideration of its ultimate effect on the analysis in the ultraviolet region.

Since most pesticides are not coloured, a chromophoric structure must be created or introduced into the pesticide molecule before it can be determined spectrophotometrically in the visible region. This colorimetric method of detection is specific to the functional group or other fraction involved in the colour-producing reaction.

Ultraviolet and visible spectrophotometry are commonly used in the determination of compounds which are difficult to chromatograph, such as the acids 2,4-D and 2,4,5-T, and the ionic bipyridillium herbicides diquat and paraquat. A distinct advantage of colorimetric detection methods is that they are readily adaptable to automated analysis.

Many of the chlorinated pesticides have been determined by colorimetric methods, but, with the advent of gas-liquid chromatography with its superior detection limits, such methods have lost favour. Fresh interest has been shown recently in colorimetry of organophospherous pesticides using an automated process. This method determines total phosphorus, so a suitable preliminary extraction and clean-up procedure is necessary to remove phosphorus-containing compounds. Direct spectrophotometric techniques can be used for water quality monitoring or for confirmatory evidence after chromatographic separation. In many cases the sensitivity and selectivity of spectrophotometry after clean-up is sufficient for identification and quantification of organic pesticides, especially aquatic herbicides, in water.

1.4.2.2 *Infrared*
This absorption spectrophotometric technique is useful in pesticide residue analysis because it is possible to obtain qualitative identification as well as quantitative determination with one physical measurement. Generally, relatively large pure quantities are required to obtain a spectrum. Good spectra can be obtained with microgram quantities of pesticide providing background adsorption due to extraneous extractives is low. The problem of obtaining a clean extract is the major disadvantage. Quantitative infrared analysis also requires scrupulously water-free solutions so the drying of reagents and final aliquots is necessary. Unfortunately, some of the common solid drying agents may absorb or react with pesticides and most do not remove water completely.

Pesticides can be separated and collected by gas-liquid chromatography with a fraction collector or by preparative thin-layer chromatography, and the infrared spectrum obtained by micro-infrared techniques. This procedure is particularly useful when multiple residues are present.

1.4.2.3 *Fluorescence and phosphoresence*
Generally, fluorometric methods are preferable to absorption measurements because usually both their selectivity of measurement and sensitivity of determination are greater. Their principal disadvantages are the limited applicability and the difficulty in obtaining suitable clean-up procedures that will remove those naturally-occurring biological materials which fluoresce. The relationship between fluorescence and concentration is usually linear only over a limited concentration and great care must be taken to ensure measurements are made in this range.

A survey of the phosphorescence characteristics of certain pesticides and potential utility of the technique was made by Moye and Winefordner (1965), who obtained excitation and emission spectra, decay times, analytical curves and limits of detection for 32 pesticides in 10^{-2} M ethanol solution. Based on the fluorescence and phosphorescence spectra obtained by Bowman and Beroza (1966a) for methylendioxyphenyl synergists, fluorescent methods were developed to determine piperonyl butoxide residues in commercial fly-sprays, stored grain and milled products. Fluorescence analysis provides a rapid and sensitive method for the determination of carbamate residues which are not readily determined by gas chromatography, and has been used for screening milk samples for carbamates. The fluorescence in methanol-tetramethylammonium hydroxide solution was measured (Bowman and Beroza, 1966b). Carbaryl and 1-naphthol have been determined in bees, pollen and honey by measurement of the fluorescence in 0.25N NaOH solution (Argauer *et al.*, 1969). Freed and Hughes (1959) suggested the technique for determining low concentrations of the herbicide diquat. They found a linear relationship between the fluorescent intensity and concentration over the range 0.1-6.0 μg/ml. Unfortunately, the procedure is inapplicable when small concentrations of the compound are to be determined in plant material, owing to interference from naturally-occurring substances remaining after clean-up.

Fluorescent intensity is dependent upon a variety of environmental conditions, the most important being the solvent effect and pH, and their factors are discussed in the review articles by MacDougall (1962, 1964).

1.4.3 Chromatographic methods
The chromatographic methods used in pesticide analysis may be divided into four groups:

(1) paper chromatography
(2) thin-layer chromatography
(3) gas-liquid chromatography
(4) liquid-liquid chromatography.

1.4.3.1 *Paper chromatography*
This is the simplest method of identification and estimation. It has a reasonable degree of sensitivity and selectivity for chlorinated and organophosphorus insecticides and chlorophenoxy-acid type herbicides. Before the introduction of gas and thin-layer chromatography, paper chromatography was the only process generally applicable to the separation and identification of many pesticide residues. The greater sensitivity and concurrent quantitative estimation possible with gas chromatography has led to paper chromatography being used now mostly for confirmation of the relatively non-specific gas chromatography compounds. Thin-layer chromatography has replaced paper chromatography in pesticide residue analysis because of its increased resolution and shorter development time.

Generally, when paper chromatography has been used for pesticide residue analysis no chemical modification of the paper has been made. But fibreglass papers have been used for the reverse-phase chromatography of organophosphorus compounds (McKinley and Read, 1963) and dithiocarbamates (McKinley and Magarvey, 1960). In addition, acetylated papers have been used for the reverse-phase separation of organophosphorus pesticides (McKinley and Read, 1962). Paper chromatography has been used for the determination of organochlorine herbicides in soil and water (Abbott *et al.*, 1964), substituted urea herbicides and their trichloroacetates (Mitchell, 1966) and for the determination of organophosphorus residues in foodstuffs (Bates, 1965). The use of paper chromatography in pesticide residue analysis has been reviewed by several authors (Getz, 1963; Coffin, 1966).

1.4.3.2   *Thin-layer chromatography*

Thin-layer chromatography has grown rapidly in importance in recent years and is now widely accepted as a quick and efficient technique for the detection and determination of very small quantities of the majority of pesticides. While lacking the precise specificity of gas-liquid chromatography, thin-layer is more precise and more sensitive than paper chromatography. Thin-layer chromatography can be used in pesticide analysis for diagnostic work, for accurate quantitative evaluation and as a clean-up procedure for sample extracts. The adsorbent layer on the glass plate should present an even, firm and continuous surface. For most pesticide residue work, a layer 250 μm thick of alumina or silica gel gives the best results, but other adsorbents such as Kieselguhr and magnesium oxide are also useful sometimes. Recent developments have made use of polyamide layers and micro-crystalline cellulose Avicel (Ragab, 1968).

The applied layer after air-drying is completely dried and activated by heating the plate in an oven at $120°C$ for two hours and then cooling before use. The solvent in the solutions to be spotted on the plate should be volatile, and care must be taken that the micro-pipettes used for spotting do not penetrate the surface layer. Standard solutions of pesticides must be developed on the same plate as the sample, preferably on both sides of the sample spot. For the thin-layer chromatography of organochlorine compounds, there are a number of mobile solvents that may be used, but it is important that the developing tank contains a saturated and stabilised atmosphere. The chromatogram is developed by the ascending solvent technique, with the plate vertical, and the origin line approximately 1 cm above the solvent surface. After development, the plate is air dried and then the pesticides may be seen by various means. One method is to spray the plate with ethanolic silver nitrate and irradiate with ultraviolet light, the pesticide spots appearing as dark areas against a white background. This process is not particularly sensitive, the detection limit being approximately 0.5 μg and discolouration of large areas of the plates occurring unless precautions and care are taken. If the silver nitrate is incorporated into the adsorbent layer, the process is improved and organochlorine pesticides in the 5-200 ng range can be detected (Fehringer and Ogger, 1966). Other visualising agents that have been used are aromatic amines, bromine vapour, fluorescent indicators such as fluorescein, rhodamine B and dichlorofluorescein and also the optical brighteners, the calcofluor whites.

Two-dimensional development of chromatograms using hexane and cylcohexane on silica gel chromatoplates has been used for the identification of organochlorine pesticides in blood and tissues (Eliakis *et al.,* 1968).

Another development has been the use of single-dimensional chromato-graphy using multiple development. This technique has been used for the separation of thirteen commonly occurring organochlorine pesticides (Szokolay and Madaric, 1969). The thin-layer chromatographic behaviour of 90 pesticides on Florisil using five solvent systems has been studied (Hamilton and Simpson, 1969). A rapid screening test for detecting organochlorine pesticides present in fats and vegetables, using thin layer has been published and may be useful for surveys (Abbott *et al.,* 1969).

Carbamates that are difficult to analyse by gas chromatographic techniques have been identified and determined by thin-layer. Ramasamy (1969) published residue data for seven carbamates using a combination of two different adsorbents, three solvent systems and three chromogenic sprays. Mendoza and Shields (1970) developed a thin-layer chromato-graphic-enzyme inhibition technique, which they used to detect carbamate and organophosphorus pesticides. This technique enabled carbaryl to be detected at the low level of 0.1 ng. *N*-phenyl, *N*-methylcarbamates and related ureas have been detected, separated and identified by thin-layer chromatography (El-Dib, 1970). Using polyamide layers, Nagasawa and his co-workers (1970) separated and determined carbamates and related compounds. They compared the detection limits on polyamide and silica gel layers, and confirmed the superiority of polyamide layers. Abbott *et al.* (1967) have described a thin-layer chromatographic procedure for the separation, identification and estimation of some carbamates in soil and water.

The organophosphorus pesticides are also susceptible to analysis by thin-layer chromatography using adsorbents and mobile solvents similar to those used for organochlorine pesticides and carbamates. The merits of thin-layer chromatography in the separation of organophosphorus pesti-cide residues have been reviewed (Abbott and Thomson, 1965; Abbott and Egan, 1967). Getz and Wheeler (1968) reported the separation and select-ive visualisation of 42 organophosphorus compounds. They used five ternary solvent systems and three selective chromatogenic sprays for the identification of the migrated spots. The use of polyamide layers has been recommended by many workers (Wang and Chou, 1969; Huang *et al.,* 1968). Watts (1967) reviewed the use of chromatogenic spray reagents for the detection of organophosphorus pesticides. The separation and identifi-cation of organophosphorus pesticides from tissues for medico-legal purposes has been achieved by two-dimensional thin-layer chromatography (Koutselinis *et al.,* 1970). Fluorescent silica gels can be used to separate and detect organophosphorus pesticides. Villeneuve *et al* (1970) made an assessment of the applicability of these gels. Because the organo-phosphorus pesticides are inhibitors of esterases, the combination of thin-layer chromatography with cholinesterase detection methods permits very rapid and sensitive procedures for separation and detection. The work done by Mendoza *et al.* (1968a, b, 1969) shows examples of this technique. Thin-layer chromatography is particularly applicable to herbicides, most of which are of a polar nature and not susceptible to gas chromatography unless first converted to a suitable ester or derivative. Thin-layer chromatography has been used for the detection and determination of phenoxyaliphatic acids (Erne, 1966) and phenylurea herbicides (Geissbühler and Gross, 1967). Cellulose layers have been used for the separation of dinitrophenols and their methyl ethers (Yip and Howard, 1966). The various aspects of the detection and determination of residues of triazine herbicides have been reviewed (Frei and Nomura, 1968; Frei and Freeman, 1968; Frei and Duffy, 1969).

For a simple, rapid and inexpensive procedure determining minute amounts of some herbicides, Coha and Kljajic (1969) have suggested a combination of thin-layer and ring-oven. Homans and Fuchs (1970) have used direct biautography on thin-layer chromatograms for the detection of fungitoxic substances. Thin-layer chromatography has been used for organomercurials; Takeshita *et al.* (1970) detected mercury and alkyl mercury compounds by reverse phase thin-layer chromatography. It has also been used for estimating methyl mercury in fish, animal liver and eggs (Westoo, 1966, 1967). The thin-layer characteristics of the dithizonates of a number of organomercurial fungicides in common use have also been studied (Tatton and Wagstaffe, 1969).

1.4.3.3   *Gas-liquid chromatography*

The most versatile and sensitive end method for pesticide residue analysis is undoubtedly gas-liquid chromatography. Following considerable research to perfect this technique, some features of the apparatus have become standard. The column tubing is usually made of glass 1 or 2 metres in length, although Crossley (1970) has suggested the replacement of glass or metal by columns of Teflon. Columns of copper and stainless steel should be avoided as both can cause decomposition of compounds on the column unless precautions are taken.

The most commonly used stationary phases for the gas-liquid chromatography analysis of pesticides are the organosilicones SE30, QF1 or DC200; mixtures of QF1 and DC200, Apiezon, butane-1, 4-diol succinate, Versamid 900, Carbowax 20M and GE-X60. A recent introduction is a phenyl methyl silicone, OV-17 (Leoni and Puccetti, 1969), which has been used by itself or mixed with the fluorosilicone, QF1. After examining five different stationary phases, Taylor (1970) found that the most suitable one for chlorinated pesticides was a mixture of QF1 and neo-pentyl glycol succinate. The retention times and responses for 60 organophosphorus pesticides and metabolites have been reported using three different stationary phases (Watts and Storherr, 1969).

It cannot be over-emphasised that all samples should be examined on at least two different types of column and no conclusions should be drawn unless the results agree. If necessary, injections should be made on a third type of column with a different composition to the other two. The two, or three, columns used should incorporate stationary phases of differing polarities so as to produce different sets of relative retention times for the various pesticide residues.

In Great Britain, the electron-capture (electron-affinity) detector has become the standard detector for organochlorine compounds, with either tritiated titanium on copper foil, or a nickel 63 foil as the source of beta radiation. The latter detector is more expensive but can be used at temperatures above 250°C, which would damage the tritiated detector. The nickel 63 detector can be used up to a maximum of 400°C, and its use at high temperatures reduces the possibility of contamination of the source from sample impurities and from the bleeding of the liquid phases from columns. In the United States of America, the most commonly used detector for organochlorine pesticides was originally the microcoulometric detector, but except for certain special purposes it has now been replaced by the electron-capture detector. The electron-capture detector responds to all electron-capturing materials whilst the microcoulometric detector can be operated to be specific for halogenated compounds. The microcoulometric detector originally lacked sensitivity, but modifications have increased the detection limit of sulphur or halogen to about 10^{-9} g. This detector is still relatively expensive for residue work despite its high degree of specificity. The theory and modes of operation of electron-capture gas chromatography have been reported by many authors. Amongst these are Lovelock and Lipsky (1960), Lovelock (1961a, b, 1963), Clark (1964), Dimick and Hartman (1963), Gaston (1964) and Peters and Schmidt (1964). These papers should be studied for a detailed understanding of electron-capture detection and the effect of various gas-chromatographic parameters.

Another type of detector that has been used for pesticides, including the halogenated ones, is the emission spectrometric detector. This detector was first described by McCormack and his co-workers (1965) for the gas chromatographic detection of organic compounds containing organic bromine, chlorine, iodine, phosphorus and sulphur. The effluent from the gas chromatographic column passes into an intense microwave-powered argon discharge in which fragmentation and excitation occur. The atomic emission produces sensitive spectral lines which can be used for the quantitative determination of the organic compounds. The detector was used by Bache and Lisk (1965, 1966a) for the determination of organophosphorus and organic iodine pesticides. It was found that if the argon plasma was used at a reduced pressure, a tenfold increase in selectivity and sensitivity was observed for organophosphorus pesticides (Bache and Lisk, 1966b). A low-pressure helium system was used for the detection of sulphur, bromine, chlorine, iodine and phosphorus (Bache and Lisk, 1967a) and for the determination of these major heteroelements in crops containing organic pesticides (Bache and Lisk, 1967b). The technique allowed the pesticides to be determined at sub-nanogram levels. The sensitivity of the emission spectrometric detector was improved by Moye

(1967) who used argon-helium mixtures. He compared the electron-capture detector and the microwave-emission detector responses for celery extracts containing 10 μg/Kg of parathion. Ten nanograms of parathion gave the same response with the microwave instrument as one nanogram with the electron-capture instrument. However, there was a complete absence of extraneous peaks with the microwave detector. The microwave detector has also been used for the determination of carbamate and triazine pesticides (Bache and Lisk, 1968).

It is only in very recent years, that the gas-liquid chromatography of organophosphorus pesticides has reached a state comparable with that achieved in the organochlorine field. The electron-capture detector can be used for a few organophosphorus compounds such as parathion and fenchlorphos, which contain certain electrophoric groups such as Cl and NO_2, but the sensitivity is generally rather low, requiring about 5 or 10 ng to produce a reasonable response. The flame-ionization detector has a similar response to organophosphorus compounds, and because of the lack of a highly sensitive detector, only limited progress was made in this field. In 1964, the sodium-thermionic detector was introduced by Giuffrida (1964) and was an important development because the detector showed a special sensitivity to phosphorus and nitrogen compounds and was comparatively insensitive to chlorine compounds. This consisted of a flame-ionisation detector with the cathode coated with a sodium or other alkali salt. Unfortunately, the sodium-thermionic detector proved unstable in practice because its sensitivity declined rapidly and it required constant recharging with fresh sodium salt. Hartmann (1966) fitted a tip to the burner made of caesium bromide and celite prepared under high pressure. This tip was positioned on the burner so that the gas flow passed through a hole in it and the flame was produced at the exit hole; thus the flame produced contained caesium ions.

The caesium tip enhanced the stability of readings and also had a reasonably long working life. The sensitivity of this detector is of the order of 2 ng for full-scale deflection, with no base-line noise; but for compounds with shorter retention times, such as dichlorvos, the sensitivity is increased and a sensitivity of 0.4 ng for full-scale deflection can be readily obtained. Conversely, the detector is less sensitive to compounds with long retention times, such as carbophenothion, which requires 5 ng to give a full-scale deflection. The sensitivity of the instrument can be improved by re-injecting the sample extracts on shorter columns at higher temperatures. This considerably shortens the retention times, with corresponding narrowing and sharpening of the chromatogram peaks, coupled with loss of resolution. A very comprehensive review of thermionic detectors used in gas chromatography has been recently published (Brazhnikov et al., 1970). Under specific instrumental conditions, the alkali flame detector will give negative peaks for chlorine-containing compounds and positive peaks for bromine, iodine, nitrogen and phosphorus-containing compounds. The detector has been used in the nitrogen form for determining carbaryl directly (Riva and Carisano, 1969) without converting it into a derivative. The negative response was used to detect chlorinated pesticides in soil, at levels between 0.01-10 mg/kg, without purification of an exhaustive hexane extract (Lakota and Aue, 1969).

Recent developments in detectors specific for phosphorus compounds have resulted in the flame photometric detector. Draegerwerk was granted a German patent in 1962 for a method of detection of sulphur and/or phosphorus compounds in air, by flame photometry in a hydrogen-rich flame. Using this principle, Brody and Chaney (1966) produced a detector sensitive to sub-nanogram amounts of phosphorus compounds, and/or sub-microgram amounts of sulphur compounds, which is relatively insensitive to all other organic compounds. The principle upon which the detector is based is the photometric detection of flame emission of phosphorus and sulphur compounds in the hydrogen-air flame. The flame is monitored by a photomultiplier tube which has an interference filter for spectral isolation of the particular emission. The filter at 526 nm is used to sense phosphorus compounds, and a secondary photomultiplier tube, with a 394-nm interference filter for sulphur compounds, provides an additional check on pesticides based on phosphorothioic acids. The usefulness of the detector for measuring organophosphorous pesticide residues in extracts of plants and animal products, with and without clean-up, has been reported (Getz, 1967). The detector has been used for the determination of organophosphate pesticides in cold-pressed citrus oils (Stevens, 1967), oranges, apples, sugar beet and potatoes (Maitlen *et al.*, 1968). Beroza and Bowman (1968) concluded that gas chromatography with flame photometric detection was a versatile, sensitive, reliable and efficient means of analysing pesticides and their metabolites containing phosphorus or sulphur. The detector also proved useful for analysing insecticides such as carbamates, in the form of derivatives containing phosphorus or sulphur. By the use of two columns of differing polarity, together with the flame photometric detector, Stanley and Morrison (1969) identified organophosphate pesticides by comparison of the peak retention times of the unknown sample with those of a known sample. The flame detector has been improved by combining the phosphorus and sulphur detectors into a single unit (Bowman and Beroza, 1968). Since the two detectors monitor the same burning effluent from the gas chromatographic column, precise determination of their relative response to various compounds may be made. Using 21 pesticides, at three concentrations, Bowman and Beroza (1968) found that the phosphorus response was linear with concentration but the sulphur response was not. They defined the response ratio as the phosphorus response divided by the square root of the sulphur response. This ratio provided a means of estimating the atomic ratio of phosphorus to sulphur in a molecule. The operating temperature of the flame photometric detector was approximately 160°C, but by replacing the original air-cooled heat sink with a water-cooled heat sink, the operating temperature could be increased to 250°C (Dale and Hughes, 1968).

For the detection of halogenated compounds, Bowman and Beroza (1969) used a gas chromatographic photometric detector which responded to the green flame obtained when compounds containing chlorine, bromine and iodine burn in the presence of copper. The sensitivity of the detector to organochlorines was usually 100 ng, but for the analysis of chlorinated pesticides in crops, a thorough clean-up was required because the copper screen used in the detector could easily be deactivated. Unlike the normal flame photometric detector, it was not highly selective. A recent development has been the combination of the two different flame photometric detectors to give the dual-flame photometric detector (Versino and Rossi, 1971). In the lower flame, the compounds emerging from the column are burnt and monitored for sulphur and phosphorus. Between the upper and lower flames is a stainless-steel net holding indium pellets. The chlorine-containing compounds react with indium, and the resulting emission is monitored at a wavelength of 360 nm. Therefore the simultaneous and selective determination of phosphorus, sulphur and chlorine-containing compounds eluted from a gas-liquid chromatography column can be obtained.

Another specific detector is the Coulson electrolytic conductivity detector (Coulson, 1965), the principle of which is as follows. The column effluent passes into an aluminium block, where it mixes with a reactant gas of either oxygen or hydrogen. The mixture passes into a pyrolysis tube containing a combustion catalyst. The pyrolysis conditions are then fixed to produce a water-soluble material containing only the desired element. Thus for halogen detection, the pyrolysis unit contains a platinum gauze which gives oxidising conditions, the products being sulphur dioxide and hydrochloric acid. The sulphur dioxide can be removed by a scrubber of calcium oxide, and the hydrochloric acid can then pass into conductivity water. The change in conductivity of this water is measured and can be related to organochlorine pesticide content. For nitrogen detection, the column effluent is reduced, using a nickel catalyst, and the interfering materials are removed with a strontium hydroxide scrubber. The Coulson detector can therefore be useful as a specific detector for nitrogen-containing pesticides such as the triazines. It is also useful for the determination of chlorinated pesticide residues when the solution has not been satisfactorily cleaned up for electron capture. The evaluation of the electrolytic conductivity detector for residue analysis of nitrogen-containing pesticides has been made. The wide distribution of naturally occurring compounds did not present any serious obstacles in the use of this detector, when used for the analyses of crops in the 0.02 mg/kg range (Patchett, 1970).

A new flameless ionization (Chemi-ionization) detector has been recently proposed (Scolnick, 1970). When phosphorus compounds from the column pass into a heated atmosphere of caesium bromide vapour and inert gas, ionisation takes place and can be detected. Results from this detector, using several organophosphorus pesticides, were compared with results obtained using an alkali flame ionization detector. Although the flameless ionization detector was less sensitive to phosphorus compounds than the alkali flame ionization detector it was also much less sensitive to the hydrocarbon solvent. This gives the flameless ionization detector a certain degree of specificity to non-hydrocarbons, the resulting chromatograms being cleaner, with fewer interfering peaks.

Derivatives of pesticides are prepared for two main reasons. They are used to confirm the presence of a particular pesticide, a technique discussed in the section covering confirmatory techniques. Secondly, they are formed to aid the determination, for example, to make chromatography of the pesticide possible. It may be necessary to decrease the pesticide's volatility for paper or thin-layer chromatography, increase the volatility and/or stability for gas-liquid chromatography, increase the sensitivity of the pesticide to a particular detector, or to avoid non-gaussian peaks.

Derivative formation is clearly needed for the carbamate series of compounds, which are used as herbicides and insecticides. These are based on carbamic acid, and carbaryl is probably the best known example. Few carbamate compounds may be analysed by gas chromatography directly because of breakdown and poor detector response. This has led to an interest in the formation of derivatives from hydrolysis products of the pesticide. The method of Cohen and Wheals (1969) involved hydrolysis and formation of the 2,4-dinitrophenyl derivative of the amine on a silica gel thin-layer chromatograph. The derivative was extracted and injected into a gas-liquid chromatographic column. This method of forming the 2,4-dinitrophenyl ethers has been extended to phenols in general, which usually show tailing characteristics and poor detection sensitivities (Cohen *et al.,* 1969). Herbicide acids also have poor chromatography characteristics but the problem has been overcome by the preparation of the methyl derivatives (Stanley, 1966; Scoggins and Fitzgerald, 1969).

1.4.3.4   *Liquid-liquid chromatography*
Considerable advances have been made in the separating technique of liquid-liquid chromatography. Separation depends on the compounds having sufficiently different partition coefficients in the selected solvent system and by this technique, an extremely wide range of compounds can be separated, since there are no limitations set by volatility requirements. The problem of the leaching off of the stationary phase has been overcome with the introduction of new packing materials, in which the stationary phase is chemically bonded to the support, and the use of controlled surface-porosity supports (Kirkland, 1969a, b; Felton, 1969). Available detectors include refractometers, and those based on electrolytic conductivity, heat of adsorption, spectrophotometry and flame ionisation. Lambert and Porter (1964) reported the application of the technique to the analysis of insecticides, using water-ethylene glycol as the stationary phase, adsorbed on siliconised fire-brick and hexane-carbon tetrachloride

as the mobile phase. A pressurised flow scheme, incorporating an automatic recording differential refractometer as the detector, was employed.

Bombaugh *et al.* (1970), using a high performance liquid chromatograph, determined organochlorine pesticides which were separated on a Corasil II column using *n*-hexane as a solvent. The separated insecticides were detected with a differential refractometer and the sensitivity was of the order of 1 μg. Liquid chromatography offers a means of carrying out a rapid and direct analysis of such compounds as the insecticidal carbamates (carbaryl, butacarb etc.) and their hydrolysis products, with minimal interference from the biological matrix.

1.4.3.5 *Polarography*

Polarography has been applied to the detection and determination of several pesticides. Although classical polarography has poor selectivity and low sensitivity, it can be used for the analysis of nitro-compounds such as commercial parathion products. The fast sweep used in cathode ray oscillographic polarography enables the instrument to be particularly useful in rapid determinations. To obtain a polarogram, the compound must contain an oxidisable or reducible group such as nitro, halogen, carbonyl, etc. Pesticides not containing such a group can usually be determined by formation of a suitable derivative. From the published literature on polarographic behaviour, solvents and electrolytes may be selected and their suitability can be tested in the laboratory.

Brand and Fleet (1970a) used polarography and related techniques in the determination of ethylene bisdithiocarbamates, and found that the most suitable method was cathodic stripping analysis. The same authors (1970b) have also evaluated methods for the determination of the fungicide thiram. They used d.c. and a.c. polarography, and obtained linear calibration graphs over the concentration range 10^{-4} to 10^{-3} M for d.c., and 10^{-6} to 10^{-4} for a.c. polarography. By the use of a.c. polarography and cyclic voltammetry, Booth and Fleet (1970a) determined the following methyl carbamate insecticides: carbaryl (detection limit 10^{-5} M), aldicarb (10^{-4} M), butacarb (10^{-5} M) and methiocarb (10^{-5} M). They have also determined trace quantities of triphenyl tin fungicides by anodic stripping voltammetry, at a detection limit of 10^{-8} M and have developed a procedure for the determination of these fungicides in potato crops, at a sensitivity a hundred times better than spectrophotometric methods (Booth and Fleet, 1970b). A fully automatic apparatus for stripping-voltammetry has been described (Booth *et al.,* 1970).

Several workers have investigated the use of polarography and associated techniques to detect parathion and its metabolites, e.g. Hearth *et al.* (1968) determined them in canned peaches. The pesticides were extracted from the fruit by stripping with methylene chloride, and isolated using thin-layer chromatography. They were removed from the plate, dissolved in ethanol, and mixed with a base electrolyte of potassium chloride in acetic acid. Using a cathode ray polarograph, they were able to determine 0.5 mg per kg, starting from 10g of peaches. Results obtained by multiple detection procedures may be verified by oscillographic polarography. Gajan (1969) published the results of a collaborative study for the polarographic detection and determination of parathion, methyl parathion, diazinon and/or malathion at 0.5 and 2 mg per kg levels in apples and lettuces. A recent and sensitive method is that used by Koen *et al.* (1970), who constructed a polarographic micro-detector and coupled this to a high-speed, efficient liquid chromatograph. Detection limits of approximately 10^{-8} M were found when the method was used for the analysis of mixtures of parathion, methyl parathion and 4-nitrophenol. Koen and Huber (1970) determined parathion and methyl parathion on crops. Residues were extracted from appropriate parts of the crop with ethanol, diluted with an aqueous acetate buffer solution and analysed by means of liquid chromatography with polarographic detection. Clean-up and concentration were unnecessary because of the high selectivity and low detection levels.

Gajan (1964) and Allen (1967) have reviewed the application of the polarograph to the detection and determination of pesticide residues.

1.4.3.6 *Radiochemical techniques*

In neutron activation analysis, radioactivity is induced into trace elements. The concentration of these trace elements (as low as 1 μg per kg) is determined by measuring the radiation so induced and comparing with standard samples. The technique has found only limited use because expensive equipment is involved, but its potential in the analysis of pesticides has been discussed by Bogner (1966). Schmitt and his co-workers developed a quick neutron-activation method for determining several elements including bromine and chlorine in milk products and fruit (Castro and Schmitt, 1962; Schmitt and Zweig, 1962).

Radioactive isotopes have often been used in metabolism studies (for example, metabolism of trichlorphon using ^{32}P (Bull and Ridgway, 1969), and in the development of analytical methods for routine residue determinations. The technique involves the introduction of a radioactive atom into the pesticide molecule. By tracing the radioactivity emitted, the progress of the pesticide, through the metabolism pathway or analytical method, may be closely followed. The method is quite sensitive (0.1 μg per kg for pesticides) and measurement is relatively independent of the colour and the chemical and physical states of the sample to be measured. The common isotopes used are ^{3}H, ^{14}C, ^{32}P, ^{35}S, ^{36}Cl, ^{82}Br of which ^{14}C is most important (Watts, 1971; Cannizzaro *et al.*, 1970). A disadvantage is that most pesticide laboratories are not equipped to synthesise their own radioactive compounds.

Isotope dilution analysis is a means of measuring the yield of a non-quantitative process and it also enables an analysis to be performed where no quantitative isolation procedure is known. A known weight of a radioactive compound is added to an unknown mixture containing that compound. They mix so that they are chemically indistinguishable. A small amount is isolated and determined chemically and radioactively. The proportion of radioactive compound to the total gives the dilution and hence the original concentration. The method has been used by Bazzi (1966) for determining dimethoate and phenthoate.

1.5 CONFIRMATORY TECHNIQUES

It is important that the identity and level of pesticide residues determined, should be confirmed by a different method from that used in the determination. When the end determination method is thin-layer chromatography, confirmation can be obtained using alternative developing solvents or visualising agents. Often, the pesticide can be removed from the thin-layer plate with suitable solvent and injected directly on to a gas chromatograph.

Interference peaks may occur on gas-liquid chromatography traces even when the most selective detectors available are used. It is poor technique to run samples on only one column and in some cases two columns are insufficient. The diagnosis of BHC isomers on chromatograms is hindered at times by the presence of hexachlorobenzene (HCB). This compound occurs at the position of a-BHC on silicone columns and at the same retention time as γ-BHC on Apiezon but all three compounds are resolved on cyano-silicone. When dieldrin and p,p'-DDE occur together in a sample extract, there is no separation on the silicone column. Both the Apiezon and cyano-silicone columns will resolve the mixture but with the latter the order of emergence is reversed, i.e. p,p'-DDE appears before dieldrin.

A simple confirmation technique for gas-liquid chromatography is afforded by thin-layer chromatography when the residues present are large enough. An aliquot of the cleaned-up extract, calculated to contain a suitable amount of the pesticides, can be evaporated to dryness, dissolved in suitable solvent and run on a thin-layer plate together with the appropriate standards. The identity of the residues can be confirmed, and the levels substantiated within the limited quantitative nature of thin-layer chromatography.

For a few pesticides, gas-liquid chromatography results using one detector can be confirmed by using another type of detector. Thus, a microcoulometric detector can be used with an electron capture detector for organochlorine compounds and the thermionic and flame photometric detector for the organophosphorus compounds.

Nuclear magnetic resonance spectroscopy is one of the most important tools for the structural elucidation of organic molecules, and finds a place in pesticide analysis. It is extremely useful for structural elucidation of pesticide metabolites and degradation products, for structural confirmation of new pesticides, or for product analysis of existing ones. It has also been used for the confirmation and estimation of relative concentrations of p,p'-DDT and p,p'-DDE isolated from adipose and liver tissue-samples. Because nuclear magnetic resonance spectroscopy has low sensitivity and pesticide residues encountered are of low concentration, at present it has only limited use in residue work. Reference spectra of pesticides have been published and will be useful for identity confirmation. Keith *et al.* (1968) presented a description of the 100-Mhz nuclear magnetic resonance spectra of 40 organophosphorus pesticides. The more interesting and unusual spectral features are discussed, and reproductions of some of the spectra have been included. Keith *et al.* (1969) published reference spectra for commercially available pesticides having the diphenylmethane or substituted diphenylmethane skeleton. The N.M.R. spectrometer was used quantitatively to determine the percentage of each isomer present in a mixture. The mixtures were technical DDT and a mixture of 'Bulan' and 'Prolan'. Keith and Alford (1970a) extended their work by the publication of the N.M.R. spectra of compounds of the carbamate pesticides. They also reviewed the application of N.M.R. spectroscopy to pesticide analysis (1970b). Residues of p,p'-DDT and p,p'-DDE have been isolated from liver and adipose tissue and subjected to N.M.R. and gas liquid chromatography using electron capture and conductivity detectors. The results show that N.M.R. can be used to confirm the presence and to ascertain (semiquantitatively) the relative composition of pesticide residues isolated from tissue samples. In preparation for the application of photo-electron spectrometry to pesticide analysis, Baker *et al.* (1970) examined fivemembered aromatic heterocyclic and related molecules. This technique, which measures the binding energies of electrons in molecules, can be used to give valuable structural information.

Mass spectrometry, because of its high sensitivity, is a useful technique in the identification of pesticides isolated from residues. It is widely used for the analysis of individual pesticides and metabolites isolated by conventional separation techniques such as thin-layer chromatography, gas-liquid chromatography, etc. When mass spectrometry is coupled to gas chromatography it enables positive identification of the components of mixtures, without prior separation and at a sensitivity compatible with the low level of residues encountered. Under ideal conditions, quantities as small as 10 nanograms will produce a useful mass spectrum. Using an extensive extraction and clean-up procedure adapted from existing analytical methods, Biros and Walker (1970) separated pesticide residues from human adipose and liver tissues. Identification and quantitative and qualitative determination of the pesticide residues were carried out by a combined gas chromatograph (with electron-capture detection) and mass spectrometer. The pesticides found were β-BHC, γ-BHC, heptachlor epoxide, dieldrin, p,p'-DDE, p,p'-TDE, o,p'-DDT, and p,p'-DDT. The authors discuss the advantages and limitations, and suggest modifications for improving the combined technique.

One very important use of mass spectrometry is in the confirmation of pesticides in the presence of polychlorinated biphenyls (PCB's). It is

One of the methods of determining certain organophosphorus and carbamate pesticides is an enzymatic method based on the inhibition of cholinesterase. When the enzyme is incubated at pH 7.4 with a thiocholine ester, enzymatic hydrolysis takes place, yielding thiocholine. This is reacted with 5,5-dithiobis-(2-nitrobenzoic acid) to give a yellow anion of 5-thio-2-nitrobenzoic acid which can be measured colorimetrically at 420 nm. In the presence of organophosphorus pesticides, the cholinesterase is partially inactivated, resulting in decreased hydrolysis of the ester. This inhibition of the enzyme can be measured and readily used in an auto-mated determination of organophosphorus pesticides. For crop samples, manual clean-up to remove interfering non-pesticidal phosphorus compounds is required prior to the automated technique. The samples can then be wet-digested automatically, and the resulting orthophosphate determined by the automated measurement of the phosphomolybdenum complex (Ott and Gunther, 1968). The combustion products have also been adapted to an automated procedure (Ott and Gunther, 1966). Ott (1968) extended the determination to a dual system. The stream was split and the organically bound phosphorus determined simultaneously by a colorimetric method, and by cholinesterase inhibition. This method used in a screening programme offers greater validity than either system alone, with little extra effort required; however, other evidence regarding the identity of the residue will also be required.

For the determination of organophosphorus insecticides and carba-mates, Voss (1969) also used an automated cholinesterase-inhibition method. He found that, by changing the type of cholinesterase, sensitivity could be improved and up to 40 samples an hour could be determined with a precision equal to that of manually-performed chemical methods. Many of the thiolo and thionophosphorus pesticides are converted to much more potent cholinesterase inhibitors by selective oxidation of the sulphur atom. An automated screening procedure of inhibition, before and after such oxidation, is useful in giving total residue information. Leegwater and van Gend (1968) used an automated enzymic procedure for the detection and determination of organophosphates, and carried out differential analyses on samples before and after oxidation with bromine water. The method was applied to the screening of lettuce without requiring clean-up of the extracts.

A review article of the automated analysis of pesticides and screening techniques has been published by Gunther and Ott (1966). As an example, the completely automated determination of biphenyl in citrus rind was given. The rind samples were automatically homogenised in water and then steam distilled to liberate oils, waxes and biphenyl. The steam volatiles were trapped in cyclohexane, and the oils and waxes extracted into concentrated sulphuric acid and discarded. The isolated biphenyl was then determined at a wavelength of 246 nm. This automated method takes about ten minutes compared with two hours for the normal method.

The Technicon Auto Analyser system has been used by Friestad (1967) for the determination of linuron in soils. The linuron is hydrolysed to 3,4-dichloroaniline and evaluated on the autoanalyser by diazotising and coupling with *N*-(1-naphthyl)ethylenediamine, the resultant diazo dye being readily determined spectrophotometrically at 550 nm.

Recent progress in the application of computers to determine low levels of pesticide has resulted in a reduction in the time required for calculations. Modern gas chromatographic systems can be readily adapted to computers because the nature of the gas chromatographic signals permit the insertion of digitisers between the chromatograph and the computer. A system has been devised by Jennings *et al.* (1970), who used a magnetic tape analogue to digital conversion, in which a tape playback supplies data to the digital integrator interfaced with the computer. The recording device receives the analogue signal, converts it to a digital signal and impresses it on magnetic tape. The magnetic tape is replayed on the playback system to yield results through the integrator computer system.

1.7 FUMIGANTS AND FUNGICIDES

1.7.1 Fumigants

For the protection of stored food, particularly cereals, against loss or deterioration due to insect damage, two types of protectants are used. The first is the use of insecticides such as dichlorvos, which are volatile and leave only residual amounts of pesticide. The other method is fumigation, which is used to control an infestation already built up. Amounts of fumigants taken up initially and during exposure depend on a number of factors, including the physical structure, moisture and oil content of the sorbent, temperature and also physiochemical characteristics of the fumigant. The most commonly used fumigant is methyl bromide and this can give the following residues: inorganic bromide, unchanged methyl bromide and reaction products such as methylated compounds.

A method for the determination of multiple residues of organic fumigants in grain involves boiling the whole or ground grain in an acid medium. The volatile fumigants are then dried and collected in a solvent medium. Aliquots are injected on to a gas-liquid chromatography column with electron-capture detection (Malone, 1970). Heuser and Scudamore (1969) investigated the use of polar and non-polar columns and two solvent systems with electron-capture and β-ionisation detection. They found the most satisfactory combination for a variety of fumigants. A flame-ionisation detector was operated in series with each of these detectors and solvent extracts were run on each column. In this way, each of the compounds of interest could be determined with high sensitivity by one or more of the detectors, without solvent interference and with some confirmation of identity.

Bielorai and Alumot (1966) have studied residues of carbon disulphide, carbon tetrachloride, chloroform and trichlorethylene in cereals. The samples were steam distilled in the presence of toluene, and an aliquot of the toluene layer examined by gas-liquid chromatography. The column used was 10% DC-710 on 80/100 chromosorb W with electron-capture detection. Scudamore and Heuser (1970) determined residual methyl bromide in fumigated samples of wheat, wheat flour, maize, sultanas, groundnuts, groundnut and cottonseed cakes and cocoa beans. They also investigated the effect of extraction at low temperatures ($0°-2°C$) to inhibit breakdown of methyl bromide during recovery.

Ben-Yehoshua and Krinsky (1968) determined residues of ethylene oxide, ethylene glycol, diethylene glycol and ethylene chlorohydrin in acetone extracts of treated date fruit by gas chromatography. Less than 1 ppm of each of these materials was determined using columns of Porapak R and polypropylene glycol. Manchon and Buquet (1970) determined ethylene chlorohydrin residues in bread treated with ethylene oxide. Residues were extracted and determined by gas-liquid chromatography, using a silicone oil on Diatoport S column and flame ionisation detection. For the determination of chlorohydrins in aqueous or solvent extracts, Pagington (1968) used a detector which is only sensitive to halogenated compounds. The detector, which was made from an Ozatron type J detector element from an A.E.I. leak detector type H.A., was used to determine chlorohydrins at the 0.2 ppm level in aqueous solutions. An X-ray fluorescence method for the determination of bromide residues in cereals and other products has been reported by Getzendaner et al., (1968). Their method determines the bromide ions and therefore cannot differentiate between those occurring naturally and those resulting from fumigation.

Another fumigant commonly used for the protection of grain is phosphine. There are methods for employing gas-liquid chromatography for the determination of phosphine in air, but these techniques are not readily adaptable to the determination of residues in foodstuffs. Dumas (1964) used a thermistor thermal conductivity detector, with a column of Apiezon L on firebrick, for concentrations in air from 0.5-10 mg per litre. He has also used the thermionic detector with a column of Apiezon L on chromosorb W (Dumas, 1969). Berck (1965) used a thermal conductivity detector with a column of 10% S.E. 30 on Diatoport S. Samples of phosphine dissolved in 1-butanol were injected, and the peaks obtained resulted from a combination of air and phosphine. The peak area for air was small and constant, and could be allowed for (Wainman and Taylor, 1967). Heuser and Wainman (1968) obtained good separation from air and carbon dioxide using a Porapak Q column but concluded that a micro cross-section detector and micro katharometer detector used lacked sensitivity. The flame ionisation detector used was sensitive only to the higher range of concentrations and behaved erratically from time to time. The phosphorus detector (with a caesium bromide tip) was highly sensitive to phosphine but linear only over a limited range. A β-ionisation detector was used for the determination of concentrations normally encountered in fumigation work. Standard mixtures of phosphine in argon were used to calibrate the instrument daily because the response varied from day to day. Berck and Gunther (1970) recently investigated three sensitive detectors for the determination of phosphine by gas-liquid chromatography. The microcoulometric detector, using a carrier gas of nitrogen, was useful in the 5-500 ng range. The thermionic detector with a caesium bromide tip was useful for amounts as low as 20 pg, but it was non-linear above 400 pg. The flame photometric detector gave the best results and a linear relationship between peak height and picograms of phosphine was obtained in the range 2-600 pg.

Recently a bioassay method for determining fumigant concentrations in the air has been published (Muthu *et al.,* 1971). The principle of the method is to aspirate samples from fumigated spaces through tubes containing insects, noting the knock-down times. The established concentration-time product divided by the knock-down time will give the effective concentration of fumigant. The fumigant concentration values obtained by bioassay compared favourably with the results obtained by chemical analysis in field trials using methyl bromide, ethylene oxide and phosphine.

1.7.2 Fungicides
These compounds can be split into two types according to their mode of action: the systemic and the non-systemic. The systemic fungicides include benzimidazole, 1-4 oxathiin and pyrimidine derivatives. The non-systemic group include phthalimide and reduced phthalimide derivatives, pentachloronitrobenzene, dithiocarbamates and the organometallic fungicides. The former types of compound are of recent introduction and methods for their detection and determination are still being investigated. Fungicides have different chemical structures and there is no specific approach or method for their detection and determination. Many fungicidal compounds are organometallic, such as the dithiocarbamates, the organotins and organomercurials. The chloro and nitro-aromatic class of fungicides have recently been investigated by low temperature fluorescence and phosphorescence (Zander and Hutzinger, 1971). It was concluded that five of the fungicides investigated, including dicloran, anilazine, tetrachloroisophthalonitrile, chloroneb and 1-chloro-2, 4-dinitronapthalene, could be determined by phosphorescent measurements at low detection limits.

The detection of quintozene at the residue level is widely documented (Zweig, 1964). Three basic methods are principally in use: a colorimetric method (Ackermann *et al.,* 1958), a polarographic method (Klein and Gajan, 1961) and gas chromatographic methods (Klein and Gajan, 1961; Gorbach and Wagner, 1967; Methratta *et al.,* 1967; Kuchar *et al.,* 1969). The development of the electron-capture detector has made possible the detection of nanogram amounts of quintozene. Baker and Flaherty (in press) have developed a gas-liquid chromatographic method involving simple extraction and clean-up procedures which can be applied to vegetables and fruits. In addition, a confirmatory test for quintozene is proposed, involving reduction with lithium aluminium hydroxide in ether, to produce pentachloroaniline with a considerably increased retention time on all the columns used. Interest has been shown in the detection and determination of captan and other structurally-related fungicides (Kittleson, 1952; Wagner *et al.,* 1956; Burchfield and Schechtman, 1958; Kilgore *et al.,* 1967; Beveneue and Ogata, 1968; Kilgore and White, 1967) and a method has been published for the determination of captan, folpet and difolatan in crops (Pomerantz *et al.,* 1970). The fungicides were extracted from the crops with acetonitrile, partitioned into methylene chloride-petroleum ether and cleaned up on Florisil. The determination was then made after injection on to a QF-1 or XE60 column with electron-capture detection.

Since many of the fungicides do not lend themselves to detection by gas-liquid chromatography, the tendency is to use spectrometric methods for their detection and determination.

An important class of organometallic fungicides are salts of derivatives of dithiocarbamic acid, the metals involved being iron, zinc, manganese and sodium. At present, the following methods are used for the analysis of residues of these compounds:

(1) The dithiocarbamate is decomposed by hot acid and the resulting evolved carbon disulphide is either: (a) absorbed in alcoholic potassium hydroxide and determined by iodometric titration (Callan and Strafford, 1924), (b) absorbed in copper acetate/diethyl or dimethylamine or diethanolamine reagent and the resulting coloured complex measured spectrophotometrically (Viles, 1946; Dickenson, 1946; Lowen, 1951; Clarke *et al.,* 1951; Pease, 1957; Cullen, 1964; Gordon *et al.,* 1967; Keppel, 1969, 1971), or (c) determined gas chromatographically (McLeod and McCully, 1969).

(2) Infrared analysis of pesticides in the solid state (Susi and Rector, 1958).

(3) Conventional and oscillographic polarography (Nangniot, 1960, 1966; Vogeler, 1967).

(4) Atomic absorption spectroscopy (Gudzinowicz and Luciano, 1966).

(5) Thin-layer chromatography (Hylin, 1966).

Two organotin compounds, widely used commercially as fungicides, are triphenyltin hydroxide and triphenyltin acetate. Methods have been published for their determination at the residue level either as the triphenyl tin dithizone complex (Hardon *et al.,* 1960; Lloyd *et al.,* 1962) or, after extraction and wet oxidation, the resulting inorganic tin determined by polarography, atomic absorption or colorimetry (Thomas and Tann, 1971).

The organomercury compounds used in agriculture can be divided chemically into three groups: alkylmercury, alkoxyalkylmercury and aryl-mercury. Residues of organomercurials can be determined by two methods: analysis for the total mercury or the intact organomercurial. For the determination of total mercury, the sample is generally oxidised either by a sulphuric and nitric acid mixture or by acid potassium permanganate. The ionic mercury is then reduced to the metal, which is vaporised and driven into a flameless atomic absorption spectrophotometer or a mercury vapour detector (Rathje, 1969; Omang, 1971; Lindstedt, 1970; Munns and Holland, 1971; Thorpe, 1971; Lindstedt and Skare, 1971; Jeffus *et al.,* 1970).

In 1961, Brodersen and Schlenker (1961) applied gas chromatography to the analysis of alkylmercury compounds but the separation and sensitivity was poor and not suitable for residue work. Kitamura *et al.* (1966a, b) improved the sensitivity and achieved better separation by using polar columns and both flame ionisation and electron-capture detectors. Methylmercury was identified and determined in human hair and fish, and there were also phenylmercury compounds in various samples. Westoo (1966, 1967, 1968) identified and determined methylmercury in fish, egg, meat, liver and various kinds of biological materials by gas chromatography. The organomercury was converted to the chloride or bromide by treatment with hydrochloric or hydrobromic acids and then extracted with toluene. Clean-up was by conversion of the mercurial into a water soluble form such as the hydroxide, sulphate or cysteine derivative, followed by acidification and back extraction into benzene. This method was suitable for the detection of methylmercury compounds but unsuitable for alkoxyalkyl compounds. Sumino (1968) established a method for the determination of alkyl- and phenylmercury compounds, which was used for the analysis of human organs after death from methylmercury due to the consumption of shellfish and fish. Various kinds of column packings were used and several methods of extraction of residues from biological material were employed. Electron-capture detection with DEGS and 1,4-BDS stationary phases, for alkylmercury analysis, and DEGS, for phenylmercury analysis, were the most successful. Tatton and Wagstaffe (1969) developed a method for the extraction and clean up of all the organomercurials using a slightly alkaline solution of cysteine hydro-chloride in propan-2-ol. The organomercurials were gas-liquid chromato-graphed as the dithizonates, the method being applicable to potatoes, tomatoes and apples. The separation of organomercurials by thin-layer chromatography has been dealt with in the section on end methods of determination.

Sensitive methods have been developed for the determination of benomyl in plant and animal tissue and in soil. Benomyl is isolated, purified and converted to 2-aminobenzimidazole and determined by direct fluorometric measurement or colorimetric measurement after bromination (Pease and Gardiner, 1969).

1.8   PROBLEMS IN RESIDUE ANALYSIS

The residue analyst's main task is the identification and estimation of pesticide residues and this presents many problems

1.8.1   Loss of the pesticide due to breakdown or volatilization

The complexity of this problem can be gauged by referring to the diagram showing the degradation of malathion (Fig. 1.1). Breakdown of such organophosphorus pesticides often occurs within a few days or hours of application. For example, the P=S grouping can be readily oxidised to P=O, and sulphide groupings can be oxidised to sulphoxides and sulphones. Therefore, a sample initially treated with malathion may contain all the metabolites shown in Fig. 1.1 and it is necessary to analyse for them as well. As the polarity of the metabolites differs from those of the parent material, extraction and clean-up may have to be modified to avoid losses during these procedures.

1.8.2   Presence of interfering compounds

Difficulties arise due to the presence in the samples of compounds whose analytical behaviour is similar to that of pesticides and may lead to consequent erroneous conclusions. One of the problems in the analysis of material for organochlorine pesticides has been the interference from a group of compounds known as polychlorinated biphenyls (PCB's). These compounds have similar structures and are obtained on a large scale by free radical reaction between benzene and chlorine. Fractional distillation of the products gives fractions which are graded according to carbon atom content and percentage chlorine present. These fractions vary from a pale yellow mobile oil with a low chlorine content, to a yellow brittle resin with a high chlorine content. The PCB compounds have been widely used in industry as lubricants, heat transfer media, and insulators, and they have been added to paints, varnishes, synthetic resins, etc., to improve the product resistance to chemicals, water, etc. Due to this they appear to be as widely distributed as organochlorine pesticides and because they are resistant to oxidation, and to both acidic and basic hydrolysis, they tend to persist in the environment.

The presence of the multicomponent PCB's in a sample examined for pesticide residues can interfere with, or prohibit, identification and measurement of some chlorinated pesticides. When PCB's are present, they give a pattern of peaks on a gas-liquid chromatogram with retention times similar to those of dieldrin, DDT, DDE, aldrin and heptachlor expoxide (Fig. 1.2). Methods for the separation of PCB's from organochlorine pesticides generally involve separation on a silica-gel column (although this does not separate p,p' DDE and PCB's). Armour and Burke (1970), using a column of silicic acid with controlled activity, separated PCB's from DDT and its analogues, although aldrin was eluted with the PCB's. This work

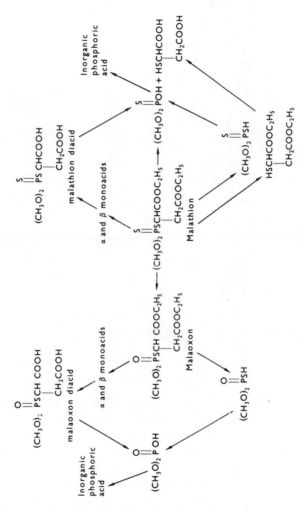

FIG. 1.1.   The degradation of malathion.

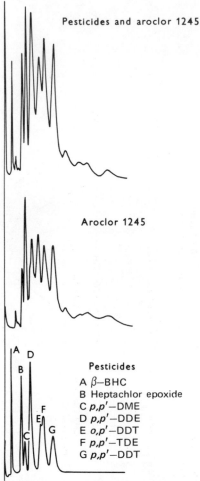

FIG. 1.2.   Gas chromatograms showing interference of pesticides by
P.C.B's on a SE.30 column using electron capture detection.

was later extended to chlorinated naphthalenes, (Armour and Burke,
1971), which may also interfere with the gas-liquid chromatographic
determination of several organochlorine pesticides. Gas chromatography
combined with mass spectrometry has been used for separating and
identifying the PCB's in ecological samples. The technique is discussed in
greater detail in the section on confirmatory techniques. Recently, the
individual chlorobiphenyls have been synthesised (Hutzinger *et al.*, 1971)
and their retention times and the response of an electron-capture detector
to them have been reported (Zito *et al.*, 1971), which aids positive
identification of these compounds. In addition to gas chromatography,
reversed phase partition thin-layer chromatography has been used success-

fully by de Vos and Peet (1971) for the separation of the components of a PCB mixture.

1.8.3 Solvent and reagent contamination

Solvents or reagents used in analysis may contain contaminants, such as di-n-butyl phthalate (commonly used as a plasticiser), that could be confused with BHC and aldrin in gas-liquid chromatography with electron-capture detection. Organic solvents when in contact with polythene can extract a contaminant, which reacts with the common thin-layer chromatographic reagent silver nitrate, giving a spot at an R_f value close to that of p,p'-DDE and having similar retention times to o,p'-DDE and p,p'-DDE. Sulphur and sulphur-containing compounds may be present in solvents, column materials (e.g. carbon, cellulose) (McKinley and McCully, 1967) or in substrates such as onion, cabbage, turnips, etc. (Oaks *et al.,* 1964; Aerograph Research Notes, 1964). Elemental sulphur can give rise to a 'pseudo aldrin' peak when using gas-liquid chromatography.

1.8.4 Positive identification

One of the more difficult problems in pesticide residue analysis is the positive identification of the pesticides present in a sample. In general, chromatographic techniques alone can rarely give absolute identification of a pesticide. Dieldrin and 'photo-dieldrin' have similar retention times irrespective of the gas-liquid chromatographic column used, and correspondingly close R_f values with a number of thin-layer solvent systems. Dieldrin and p,p'-DDE is another example of two pesticides with similar retention times on certain stationary phrases. There are many other examples of pesticides with similar retention times, as demonstrated in retention data given by Burke and Holswade (1964, 1966).

It must therefore be emphasised that confirmation of the identity of an unknown residue is essential before taking action on public health or legal grounds. This need for a full understanding of results is particularly important, because it is easy for alternative and misleading deductions to be made from the simpler non-specific residues analysis procedures.

REFERENCES

D. C. Abbott, K. W. Blake, K. R. Tarrant and J. Thomson, *J. Chromat.,* **30**, 136 (1967)

D. C. Abbott, N. T. Crosby and J. Thomson, Proc. SAC Conference Nottingham 1965, p. 121, W. Heffer Son Ltd (1965)

D. C. Abbott and H. Egan, *Analyst,* **92**, 475 (1967)

D. C. Abbott, H. Egan, E. W. Hammond and J. Thomson, *Analyst,* **89**, 480 (1964)

D. C. Abbott, R. B. Harrison, J. O'G. Tatton and J. Thomson, *Nature* (Lond.), **211**, 259 (1966)

D. C. Abbott, J. O'G. Tatton and N. F. Wood, *J. Chromat.*, **42**, 83 (1969)
D. C. Abbott and J. Thomson, *Chemy Ind.*, 481 (1964a)
D. C. Abbott and J. Thomson, *Analyst*, **89**, 613 (1964b)
D. C. Abbott and J. Thomson, *Residue Rev.*, **11**, 1 (1965)
H. Ackermann, *J. Chromat.*, **36**, 309 (1968)
H. Ackermann, *J. Chromat.*, **44**, 414 (1969)
H. J. Ackermann, H. A. Baltrush, H. H. Berges, D. O. Brookover and B. B. Brown, *J. agric. Fd. Chem.*, **6**, 747 (1958)
Aerograph Research Notes, Wilkins Instrument and Research, Summer Issue (1964)
P. T. Allen, Pesticides, Plant Growth Regulators and Food Additives, vol. 5, p. 67. Academic Press (1967)
H. G. Applegate and G. Chittwood, *Bull. envir. Contam. Toxicol.*, **3**, 211 (1968)
R. J. Argauer, H. Shimanuki and C. Alvarez, Abstracts, 158th National Meeting of the American Chem. Soc., New York, Sept. 1969, Pesticide Chemicals Division
J. A. Armour and J. A. Burke, *J. Ass. off. analyt. Chem.*, **53**, 761 (1970)
J. A. Armour and J. A. Burke, *J. Ass. off. analyt. Chem.*, **54**, 175 (1971)
J. Askew, J. H. Ruzicka and B. B. Wheals, *Analyst*, **94**, 275 (1969)
C. A. Bache and D. J. Lisk, *Analyt. Chem.*, **37**, 1477 (1965)
C. A. Bache and D. J. Lisk, *Analyt. Chem.*, **38**, 783 (1966a)
C. A. Bache and D. J. Lisk, *Analyt. Chem.*, **38**, 1757 (1966b)
C. A. Bache and D. J. Lisk, *Analyt. Chem.*, **39**, 786 (1967a)
C. A. Bache and D. J. Lisk, *J. Ass. off. analyt. Chem.*, **50**, 1246 (1967b)
C. A. Bache and D. J. Lisk, *J. gas Chromat.*, **6**, 301 (1968)
G. E. Bagley, W. L. Reichel and E. Cromartie, *J. Ass. off. analyt. Chem.*, **53**, 251 (1970)
A. D. Baker, D. Betteridge, N. R. Kemp and R. E. Kirby, *Analyt. Chem.*, **42**, 1064 (1970)
P. B. Baker and B. Flaherty (in press)
J. A. R. Bates, *Analyst*, **90**, 453 (1965)
B. Bazzi, Radioisotopes in the detection of pesticide residues, p. 27. International Atomic Energy Agency, Vienna (1966)
T. H. Beasley and H. W. Ziegler, *J. Ass. off. analyt. Chem.*, **53**, 1010 (1970)
H. Beckman and W. O. Gauer, *Residue Rev.*, **18**, 1 (1967)
W. R. Benson and J. N. Damico, *J. Ass. off. analyt. Chem.*, **51**, 347 (1968)
S. Ben-Yehoshua and P. Krinsky, *J. gas Chromat.*, **6**, 350 (1968)
B. Berck, *J. agric. Fd. Chem.*, **13**, 373 (1965)
B. Berck and F. Gunther, *J. agric. Fd. Chem.*, **18**, 148 (1970)
M. Beroza and M. C. Bowman, *Envir. Sci. Technol.*, **2**, 450 (1968)
A. Bevenue and J. N. Ogata, *J. Chromat.*, **36**, 529 (1968)
R. Bielorai and E. Alumot, *J. agric. Fd. Chem.*, **14**, 622 (1966)
F. J. Biros and A. C. Walker, *J. agric. Fd. Chem.*, **18**, 425 (1970)
R. L. Bogner, Radioisotopes in the detection of pesticide residues, p. 67. International Atomic Energy Agency, Vienna (1966)
K. J. Bombaugh, R. F. Levangu, R. N. King and L. Abraham, *J. Chromat. Sci.*, **8**, 657 (1970)
M. D. Booth, M. J. D. Brand and B. Fleet, *Talanta*, **17**, 1059 (1970)

M. D. Booth and B. Fleet, *Talanta,* 17, 491 (1970a)
M. D. Booth and B. Fleet, *Analyt. Chem.,* 42, 823 (1970b)
M. Bowman and M. Beroza, *Residue Rev.,* 17, 1 (1966a)
M. Bowman and M. Beroza, *Residue Rev.,* 17, 23 (1966b)
M. C. Bowman and M. Beroza, *Analyt. Chem.,* 40, 1448 (1968)
M. C. Bowman and M. Beroza, *J. Ass. off. agric. Chem.,* 48, 943 (1965)
M. C. Bowman and M. Beroza, *J. Chromat. Sci.,* 7, 484 (1969)
M. J. D. Brand and B. Fleet, *Analyst,* 95, 905 (1970a)
M. J. D. Brand and B. Fleet, *Analyst,* 95, 1023 (1970b)
V. V. Brazhnikov, M. V. Gurev and K. I. Sakodynsky, *Chromat. Rev.,* 12, 1 (1970)
H. Brockmann and H. Schodder, *Chem. Ber.,* 74, 73 (1941)
K. Brodersen and U. Schlenker, *Z. analyt. Chem.,* 182, 421 (1961)
S. S. Brody and J. E. Chaney, *J. gas Chromat.,* 4, 42 (1966)
D. L. Bull and R. L. Ridgway, *J. agric. Fd. Chem.,* 17, 837 (1969)
H. P. Burchfield and J. Schechtman, *Contrib. Boyce Thomson Inst.,* 19, 411 (1958)
H. P. Burchfield and E. E. Storrs, *Contrib. Boyce Thomson Inst.,* 17, 333 (1953)
J. Burke and W. Holswade, *J. Ass. off. agric. Chem.,* 47, 845 (1964)
J. Burke and W. Holswade, *J. Ass. off. analyt. Chem.,* 49, 374 (1966)
H. Burtchell and H. E. Boyle, Preprint, 14th Meeting Am. Chem. Soc. Div. of Water and Waste Chemistry, Philadelphia, Pennsylvania, April 1964, vol. 4, p. 134
A. Calderbank, *Residue Rev.,* 12, 14 (1966)
A. Calderbank, C. B. Morgan and S. H. Yuen, *Analyst,* 86, 589 (1961)
A. Calderbank and S. H. Yuen, *Analyst,* 90, 99 (1965)
T. Callan and N. Strafford, *J. Soc. chem. Ind.,* 43, 8T (1924)
R. D. Cannizzaro, T. E. Cullen and R. T. Murphy, *J. agric. Fd. Chem.,* 18, 728 (1970)
P. E. Caplan, D. Culver and W. C. Thielen, *A.M.A. Arch. ind. Health,* 14, 326 (1956)
C. E. Castro and R. A. Schmitt, *J. agric. Fd. Chem.,* 10, 236 (1962)
S. J. Clark, *Residue Reviews,* 5, 32 (1964)
D. G. Clarke, H. Baum, E. L. Stanley and W. F. Hester, *Analyt. Chem.,* 23, 1842 (1951)
D. E. Coffin, *J. Ass. off. analyt. Chem.,* 49, 1018 (1966)
F. Čoha and R. Kljajić, *J. Chromat.,* 40, 304 (1969)
I. C. Cohen and B. B. Wheals, *J. Chromat.,* 43, 233 (1969)
I. C. Cohen, J. Norcup, J. H. A. Ruzicka and B. B. Wheals, *J. Chromat.,* 44, 251 (1969)
D. M. Coulson, *J. gas. Chromat.,* 3, 134 (1965)
J. Crossley, *J. chromat. Sci.,* 8, 426 (1970)
T. E. Cullen, *Analyt. Chem.,* 36, 221 (1964)
W. E. Dale and C. C. Hughes, *J. gas Chromat.,* 6, 603 (1968)
J. N. Damico, *J. Ass. off. analyt. Chem.,* 49, 1027 (1966)
J. N. Damico, R. P. Barron and J. M. Ruth, *Org. Mass Spectrom.,* 1, 331 (1968)
J. N. Damico and W. R. Benson, *J. Ass. off. agric. Chem.,* 48, 344 (1965)
D. Dickenson, *Analyst,* 71, 327 (1946)

K. P. Dimick and H. Hartmann, *Residue Rev.*, **4**, 150 (1963)
H. Draegerwerk and W. Draeger, W. German Patent 1, 133, 918 (1962)
T. Dumas, *J. agric. Fd. Chem.*, **12**, 257 (1964)
T. Dumas, *J. agric Fd. Chem.*, **17**, 1164 (1969)
D. Eberle, D. Naumann and A. Wuethrich, *J. Chromat.*, **45**, 351 (1969)
M. Eidelman, *J. Ass. off. agric. Chem.*, **45**, 672 (1962)
M. Eidelman, *J. Ass. off. agric. Chem.*, **46**, 182 (1963)
M. A. El-Dib, *J. Ass. off. analyt. Chem.*, **53**, 756 (1970)
C. E. Eliakis, A. S. Coutselinis and E. C. Eliakis, *Analyst*, **93**, 368 (1968)
K. Erne, *Acta vet. scand.*, **7**, 77 (1966)
G. E. Ernst and F. Schuring, *J. Chromat.*, **49**, 325 (1970)
N. V. Fehringer and J. D. Ogger, *J. Chromat.*, **25**, 95 (1966)
M. Felton, *J. chromat. Sci.*, **7**, 13 (1969)
V. H. Freed and R. E. Hughes, *Weeds*, **7**, 364 (1959)
R. W. Frei and J. R. Duffy, *Mikrochim. Acta*, 480 (1969)
R. W. Frei and C. D. Freemann, *Mikrochim. Acta*, 1214 (1968)
R. W. Frei and N. S. Nomura, *Mikrochim. Acta*, 565 (1968)
H. O. Friestad, *Bull. envir. Contam. Toxicol.*, **2**, 236 (1967)
R. J. Gajan, *Residue Reviews*, **5**, 80; **6**, 75 (1964)
R. J. Gajan, *J. Ass. off. analyt. Chem.*, **52**, 811 (1969)
L. K. Gaston, *Residue Rev.*, **5**, 21 (1964)
H. Geissbühler and D. Gross, *J. Chromat.*, **27**, 296 (1967)
M. E. Getz, *Residue Rev.*, **2**, 9 (1963)
M. E. Getz, *J. gas. Chromat.*, **5**, 377 (1967)
M. E. Getz and H. G. Wheeler, *J. Ass. off. analyt. Chem.*, **51**, 1101 (1968)
M. E. Getzendaner, A. E. Doty, E. L. McLaughlin and D. L. Lindgren, *J. agric. Fd. Chem.*, **16**, 265 (1968)
L. Giuffrida, *J. Ass. off. agric. Chem.*, **47**, 293 (1964)
S. Gorbach and U. Wagner, *J. agric. Fd. Chem.*, **15**, 654 (1967)
C. F. Gordon, R. J. Schuckert and W. E. Bornak, *J. Ass. off. analyt. Chem.*, **50**, 1102 (1967)
O. W. Grussendorf, A. J. McGinnis and J. Solomon, *J. Ass. off. analyt. Chem.*, **53**, 1048 (1970)
B. J. Gudzinowicz and V. J. Luciano, *J. Ass. off. analyt. Chem.*, **49**, 1 (1966)
F. A. Gunther and D. E. Ott, *Residue Rev.*, **14**, 12 (1966)
E. O. Haenni, J. W. Howard and F. L. Joe, *J. Ass. off. agric. Chem.*, **45**, 67 (1962)
D. J. Hamilton and B. W. Simpson, *J. Chromat.*, **39**, 186 (1969)
H. J. Hardon, H. Brunink and R. W. Van der Pol, *Analyst*, **85**, 847 (1960)
C. H. Hartmann, *Bull envir. Contam. Toxicol.*, **1**, 159 (1966)
F. E. Hearth, D. E. Ott and F. A. Gunther, *J. Ass. off analyt. Chem.*, **51**, 690 (1968)
G. Hesse, I. Daniel and G. Wohlleben, *Angew. Chem.*, **64**, 103 (1952)
S. G. Heuser and K. A. Scudamore, *J. Sci. Fd. Agric.*, **20**, 566 (1969)
S. G. Heuser and H. E. Wainman, *Pest Infestation Research*, 54 (1968)
R. C. Hirt and J. B. Gisclard, *Analyt. Chem.*, **23**, 185 (1951)
A. L. Homans and A. Fuchs, *J. Chromat.*, **51**, 327 (1970)
D. F. Horler, *J. Sci. Fd. Agric.*, **19**, 229 (1968)
I. Hornstein and W. N. Sullivan, *Analyt. Chem.*, **25**, 496 (1953)

J. J. Huang, H. C. Hsiu, T. B. Shih, U. T. Chou, K. T. Wang and C. T. Cheng, *J. Pharm. Sci.*, **57**, 1620 (1968)

O. Hutzinger, W. D. Jamieson and V. Zitko, *Nature*, **226**, 664 (1970)

O. Hutzinger, S. Safe and V. Zitko, *Bull. envir. Contam. Toxicol.*, **6**, 160 (1971)

J. W. Hylin, *Bull. envir. Contam. Toxicol.*, **1**, 76 (1966)

M. T. Jeffus, J. S. Elkins and C. T. Kenner, *J. Ass. off. analyt. Chem.*, **53**, 1172 (1970)

Z. Jegier, *Arch. envir. Health*, **8**, 565 (1964a)

Z. Jegier, *Arch. envir. Health*, **8**, 670 (1964b)

R. W. Jennings, D. L. Jue and J. E. Suggs, *Chromatographia*, **3**, 353 (1970)

R. E. Johnsen and R. I. Starr, *J. econ. Ent.*, **60 (6)**, 1679 (1967)

R. E. Johnsen and R. I. Starr, *J. econ. Ent.*, **63**, 165 (1970)

L. A. Jones and J. A. Riddick, *Analyt. Chem.*, **24**, 569 (1952)

L. H. Keith and A. L. Alford, *J. Ass. off. analyt. Chem.*, **53**, 157 (1970a)

L. H. Keith and A. L. Alford, *J. Ass. off. analyt. Chem.*, **53**, 1018 (1970b)

L. H. Keith, A. L. Alford and A. W. Garrison, *J. Ass. off. analyt. Chem.*, **52**, 1074 (1969)

L. H. Keith, A. W. Garrison and A. L. Alford, *J. Ass. off. analyt. Chem.*, **51**, 1063 (1968)

G. E. Keppel, *J. Ass. off. analyt. Chem.*, **52**, 162 (1969)

G. E. Keppel, *J. Ass. off. analyt. Chem.*, **54**, 528 (1971)

W. W. Kilgore and E. R. White, *J. agric. Fd. Chem.*, **15**, 1118 (1967)

W. W. Kilgore, W. Winterlin and R. White, *J. agric. Fd. Chem.*, **15**, 1035 (1967)

J. J. S. Kim and C. W. Wilson, *J. agric. Fd. Chem.*, **14**, 615 (1966)

J. J. Kirkland, *Analyt. Chem.*, **41**, 218 (1969a)

J. J. Kirkland, *J. Chromat. Sci.*, **7**, 7 (1969b)

S. Kitamura, K. Sumino and K. Hayakawa, *Medicine and Biology*, **72**, 274 (1966a)

S. Kitamura, K. Sumino and K. Hayakawa, *Medicine and Biology*, **73**, 276 (1966b)

A. R. Kittleson, *Analyt. Chem.*, **24**, 1173 (1952)

A. K. Klein and R. Gajan, *J. Ass. off. agric. Chem.*, **44**, 712 (1961)

J. G. Koen and J. F. K. Huber, *Analytica chim. Acta*, **51**, 303 (1970)

J. G. Koen, J. F. K. Huber, H. Poppe and G. Den Boef, *J. chromat. Sci.*, **8**, 192 (1970)

A. C. Koutselinis, G. Dimopoulos and Z. I. Smirnakis, *Medicine Sci. Law*, **10**, 178 (1970)

E. J. Kuchar, F. O. Geenty, W. P. Griffith and R. J. Thomas, *J. agric. Fd. Chem.*, **17**, 1237 (1969)

S. Lakota and W. A. Aue, *J. Chromat.*, **44**, 472 (1969)

S. M. Lambert and P. E. Porter, *Analyt. Chem.*, **36**, 99 (1964)

J. R. Lane, *J. agric. Fd. Chem.*, **18**, 409 (1970)

D. C. Leegwater and H. W. Van Gend, *J. Sci. Fd. Agric.*, **19**, 513 (1968)

V. Leoni and G. Puccetti, *J. Chromat.*, **43**, 388 (1969)

G. Lindstedt, *Analyst*, **95**, 264 (1970)

G. Lindstedt and I. Skare, *Analyst*, **96**, 223 (1971)

G. A. Lloyd, C. Otaci and F. J. Last, *J. Sci. Fd. Agric.*, **13**, 353 (1962)

J. E. Lovelock, *Analyt. Chem.*, **33**, 162 (1961a)

J. E. Lovelock, *Nature* (Lond.), **189**, 729 (1961b)

J. E. Lovelock, *Analyt. Chem.*, **35**, 474 (1963)

J. E. Lovelock and S. R. Lipsky, *J. Amer. Chem. Soc.*, **82**, 431 (1960)

R. E. Lovins, *J. agric. Fd. Chem.*, **17**, 663 (1969)

W. K. Lowen, *Analyt. Chem.*, **23**, 1846 (1951)

D. MacDougall, *Residue Rev.*, **1**, 24 (1962)

D. MacDougall, *Residue Rev.*, **5**, 119 (1964)

J. C. Maitlen, L. M. McDonough and M. Beroza, *J. agric. Fd. Chem.*, **16**, 549 (1968)

B. Malone, *J. Ass. off. analyt. Chem.*, **53**, 742 (1970)

P. Manchon and A. Buquet, *Fd. Cosmet. Toxicol.*, **8**, 9 (1970)

A. M. Mattson, V. A. Sedlak and J. W. Miles, Determination of DDVP in Air, presented at 138th meeting Am. Chem. Soc., New York City, Sept. 13th (1960)

M. J. de Faubert Maunder, H. Egan, E. W. Godley, E. W. Hammond, J. Roburn and J. Thomson, *Analyst*, **89**, 168 (1964)

A. J. McCormack, S. C. Tong and W. D. Cooke, *Analyt. Chem.*, **37**, 1470 (1965)

W. P. McKinley and S. A. Magarvey, *J. Ass. off. agric. Chem.*, **43**, 717 (1960)

W. P. McKinley and K. A. McCully, *Canadian Food Industries*, Part I, **38**, 47 (1967)

W. P. McKinley and S. I. Read, *J. Ass. off agric. Chem.*, **45**, 467 (1962)

H. A. McLeod and K. A. McCully, *J. Ass. off. analyt. Chem.*, **52**, 1226 (1969)

H. A. McLeod, C. Mendoza, P. Wales and W. P. McKinley, *J. Ass. off. analyt. Chem.*, **50**, 1216 (1967)

C. E. Mendoza and J. B. Shields, *J. Chromat.*, **50**, 92 (1970)

C. E. Mendoza, P. J. Wales, D. L. Grant and K. A. McCully, *J. agric. Fd. Chem.*, **17**, 1196 (1969)

C. E. Mendoza, P. J. Wales, H. A. McLeod and W. P. McKinley, *Analyst*, **93**, 34 (1968a)

C. E. Mendoza, P. J. Wales, H. A. McLeod and W. P. McKinley, *Analyst*, **93**, 173 (1968b)

R. Mestres and F. Barthes, *Bull. envir. Contam. Toxicol.*, **1**, 245 (1966)

T. P. Methratta, R. W. Montagna and W. P. Griffith, *J. agric. Fd. Chem.*, **15**, 648 (1967)

J. W. Miles, L. E. Fetzer and G. W. Pearce, *Envir. Sci. Technol.*, **4**, 425 (1970)

P. A. Mills, *J. Ass. off. agric. Chem.*, **42**, 734 (1959)

P. A. Mills, *J. Ass. off. analyt. Chem.*, **51**, 29 (1968)

L. C. Mitchell, *J. Ass. off. analyt. Chem.*, **49**, 1163 (1966)

H. A. Moye, *Analyt. Chem.*, **39**, 1441 (1967)

H. A. Moye and J. D. Winefordner, *J. agric. Fd. Chem.*, **13**, 516 (1965)

R. K. Munns and D. C. Holland, *J. Ass. off. analyt. Chem.*, **54**, 202 (1971)

M. Muthu, H. R. Gundu Rao and S. K. Majumder, *International Pest Control*, **13**, 11 (1971)

K. Nagasawa, H. Yoshidome and F. Kamata, *J. Chromat.*, **52**, 453 (1970)

P. Nangniot, *Bull. Inst. agron. stns. Rech. Gembloux*, **28**, 365 (1960)

P. Nangniot, *Meded. Landbhoogesch. Gent*, **31**, 447 (1966)

D. M. Oaks, H. Hartmann and K. P. Dimick, *Analyt. Chem.*, **36**, 1560 (1964)

S. H. Omang, *Analytica chim. Acta*, **53**, 415 (1971)

D. E. Ott, *J. agric. Fd. Chem.*, **16**, 874 (1968)

D. E. Ott and F. A. Gunther, *Bull. envir. Contam. Toxicol.*, **1**, 90 (1966)

D. E. Ott and F. A. Gunther, *J. Ass. off. analyt. Chem.*, **51**, 697 (1968)

J. S. Pagington, *J. Chromat.*, **36**, 528 (1968)

G. Patchett, *J. chromat. Sci.*, **8**, 155 (1970)

H. L. Pease, *J. Ass. off. agric. Chem.*, **40**, 1113 (1957)

H. L. Pease and J. A. Gardiner, *J. agric. Fd. Chem.*, **17**, 267 (1969)

U. J. Peters and J. A. Schmidt, *F. and M. Res. Bull.*, **5**, 1 (1964)

I. H. Pomerantz, L. J. Miller and G. Kava, *J. Ass. off. analyt. Chem.*, **53**, 154 (1970)

M. T. H. Ragab, *Lab. Pract.*, **17**, 1342 (1968)

M. Ramasamy, *Analyst*, **94**, 1075 (1969)

A. O. Rathje, *Am. ind. Hyg. Ass. J.*, **30**, 126 (1969)

M. Riva and A. Carisano, *J. Chromat.*, **42**, 464 (1969)

A. A. Rosen and F. M. Middleton, *Analyt. Chem.*, **31**, 1729 (1959)

J. H. Ruzicka, J. Thomson, B. B. Wheals and N. F. Wood, *J. Chromat.*, **34**, 14 (1968)

W. W. Sans, *J. agric. Fd. Chem.*, **15**, 192 (1967)

R. A. Schmitt and G. Zweig, *J. agric. Fd. Chem.*, **10**, 481 (1962)

J. E. Scoggins and C. H. Fitzgerald, *J. agric. Fd. Chem.*, **17**, 157 (1969)

M. Scolnick, *J. chromat. Sci.*, **8**, 462 (1970)

K. A. Scudamore and S. G. Heuser, *Pestic. Sci.*, **1**, 14 (1970)

Shell Development Co., Divisions of Shell Oil Co. Modesto, California, Analytical Method MMS-8/59, Nov. 19th (1959)

G. R. Simpson and A. Beck, *Arch. envir. Health*, **11**, 784 (1965)

C. W. Stanley, *J. agric. Fd. Chem.*, **14**, 321 (1966)

C. W. Stanley, J. E. Barney, M. R. Helton and A. R. Yobs, *Envir. Sci. Technol.*, **5**, 430 (1971)

C. W. Stanley and J. I. Morrison, *J. Chromat.*, **40**, 289 (1969)

R. K. Stevens, *J. Ass. off. analyt. Chem.*, **50**, 1236 (1967)

P. B. Stockwell, W. Bunting, R. Sawyer. R. Stock (editor), Gas Chromatography, p. 204. Institute of Petroleum, London (1970)

R. W. Storherr and R. R. Watts, *J. Ass. off. agric. Chem.*, **48**, 1154 (1965)

K. Sumino, *Kobe J. med. Sci.*, **14**, 115, 131 (1968)

H. Susi and H. E. Rector, *Analyt. Chem.*, **30**, 1933 (1958)

A. Szokolay and A. Madarič, *J. Chromat.*, **42**, 509 (1969)

E. C. Tabor, *Trans, N. Y. Acad. Sci.*, **28**, 569 (1966)

R. Takeshita, H. A. Akagi, M. Fujita and Y. Sakagami, *J. Chromat.*, **51**, 283 (1970)

J. O'G. Tatton and P. Wagstaffe, *J. Chromat.*, **44**, 284 (1969)

I. S. Taylor, *J. Chromat.*, **52**, 141 (1970)

B. Thomas and H. L. Tann, *Pestic. Sci.*, **2**, 45 (1971)

W. W. Thornburg, *Residue Rev.*, **14**, 1 (1966)

V. A. Thorpe, *J. Ass. off. analyt. Chem.*, **54**, 206 (1971)

S. S. C. Tong, W. H. Gutenmann and D. J. Lisk, *Analyt. Chem.*, **41**, 1872 (1969)

U.S. Dept. Health, Education and Welfare, Public Health Service Publ. No. 1241. Washington, D.C. (1964)

B. Versino and G. Rossi, *Chromatographia*, **4**, 331 (1971)

F. J. Viles, *J. industr. Hyg. Toxicol.*, **22**, 188 (1946)

D. C. Villeneuve, A. G. Butterfield, D. L. Grant and K. A. McCully, *J. Chromat.*, **48**, 567 (1970)

K. Vogeler, *Pflanzenschutz Nachr. Bayer*, **20**, 525 (1967)

R. H. de Vos and E. W. Peet, *Bull. envir. Contam. Toxicol.*, **6**, 164 (1971)

G. Voss, *J. Ass. off. analyt. Chem.*, **52**, 1027 (1969)

J. Wagner, V. Wallace and J. M. Lawrence, *J. agric. Fd. Chem.*, **4**, 1035 (1956)

H. E. Wainman and R. W. Taylor, *Pest. Infestation Research*, 51 (1967)

K. T. Wang and U. T. Chou, *J. Chromat.*, **42**, 416 (1969)

R. R. Watts, *Residue Reviews*, **18**, 105 (1967)

R. R. Watts, *J. Ass. off. analyt. Chem.*, **54**, 953 (1971)

R. R. Watts and R. W. Storherr, *J. Ass. analyt. Chem.*, **52**, 513 (1969)

G. Westoo, *Acta chem. scand.*, **20**, 2131 (1966)

G. Westoo, *Acta chem. scand.*, **21**, 1790 (1967)

G. Westoo, *Acta chem. scand.*, **22**, 2277 (1968)

G. H. Willis, J. F. Parr, R. I. Papendick and S. Smith, *Pesticides Monit. J.*, **3**, 172 (1969)

W. Winterlin, G. Walker and H. Frank, *J. agric. Fd. Chem.*, **16**, 808 (1968)

H. R. Wolfe, J. F. Armstrong and W. F. Durham, *Arch. envir. Health*, **13**, 340 (1966)

H. R. Wolfe, W. F. Durham and J. F. Armstrong, *Arch. Envir. Health*, **14**, 622 (1967)

N. F. Wood, *Analyst*, **94**, 399 (1969)

G. Yip and S. F. Howard, *J. Ass. off. analyt. Chem.*, **49**, 1166 (1966)

M. Zander and O. Hutzinger, *Bull. envir. Contam. Toxicol.*, **5**, 565 (1970)

P. Zink, *Arch. Tox.*, **25**, 1 (1969)

V. Zitko, O. Hutzinger and S. Safe, *Bull. envir. Contam. Toxicol.*, **6**, 160 (1971)

G. Zweig, Analytical Methods for Pesticides, vol. 3, p. 127. Academic Press, London (1964)

Chapter 2

Pesticide Residues in Plants

D. G. Finlayson* and H. R. MacCarthy*

*Agriculture Canada Research Station,
Vancouver, B.C., Canada*

2.1 INTRODUCTION

Much of the recent literature on various aspects of pesticide[1] residues associated with plants has been thoroughly and critically reviewed. Amongst the outstanding reviews are those on persistent pesticides by Robinson (1970); on pesticides in people by Wassermann *et al.* (1967); on dipyridylium herbicides by Akhavein and Linscott (1968); on translocation of fungicides by Wain and Carter (1967); on mercury compounds by Smart (1968) and by Caldwell (1971); on systemic pesticides in woody plants by Riley *et al.* (1965); on residues and the cuticula of leaves by Linskens *et al.* (1965); on molluscicides by Strufe (1968); and on forage crops by George *et al.* (1967a). Research literature on residues in food, feed, and fibre plants and in tobacco, is scanned here from 1964, continuing our earlier review (Finlayson and MacCarthy, 1965), and emphasising recent contributions. We have confined our study largely to work on accepted compounds used in commercial production of crops, but have perforce ignored much good work on weeds and ornamentals.

* Contribution No. 226, Research Station, Canada Agriculture, 6660 N.W. Marine Drive, Vancouver 8, Canada.
[1] Common names of pesticides are used throughout the text: where common names have not been assigned, registered names are used; all are identified chemically in Table 2.7.

We define some common terms as follows: a deposit is the material laid down by the initial treatment (Ebling, 1963), but a residue is the material remaining after an interval of weathering or degradation (Gunther and Blinn, 1955). Translocation is the movement of solutes from one part of a plant to another in the vascular system; lateral, cell-to-cell movement is diffusion. Systemic pesticides (Metcalfe, 1957) are those having the ability to penetrate into the plant through roots, leaves, or stems, with enough water solubility to move into and by way of the vascular system, and enough stability or conversion to metabolites toxic enough to maintain the required degree of insecticidal action for a reasonable period (Ripper, 1957), when applied in realistic amounts (Reynolds, 1958).

2.2 SOURCES OF RESIDUES

Unintentional residues of pesticides in foods of plant origin, whether demonstrably harmful or not, are universally condemned. Their significance lies not merely in aesthetics or even in their direct toxicity, but also in their still unknown long-term effects, notably the biological concentration of certain compounds which are trapped by lipids and stored in the fatty tissues of consuming organisms.

Residues may originate from accidental or incidental contamination, as from spillage, from pesticides in adhering particles of treated soil, by volatilization of pesticides from treated soil, or from dusts or sprays drifted by wind. In a recent example, Watters and Grussendorf (1969) showed experimentally that organochlorine insecticide deposits on the surfaces of storage buildings might be a source of residues in grains, directly and even by diffusion. However, residues result most often from deliberate treatments.

External treatments are those applied to growing plants or to harvested or processed plant products. The residues that result are nearly always unintentional and are confined to a comparatively few persistent and stable compounds, which occur in all three major groups of insecticides. Efficient and sensitive analytical methods have shown where the dangers lie.

Internal treatments are systemic. They are applied to living plants, sometimes directly to the surfaces of the plants from which they are absorbed, diffused, and translocated. Translocation and distribution of compounds absorbed through leaves tends to be inefficient, because new and apical growth is not protected from pests, but the method has the advantage of avoiding accumulation of toxins in fruits. Most often, systemics are applied to the soil, from where they are absorbed by the roots and readily translocated upwards. Slow absorption of nearly insoluble or slightly soluble compounds, such as disulfoton and aldicarb, leads to efficient translocation and distribution of the effective toxins throughout the plant for comparatively long periods. Systemics are regarded by the public with interest and suspicion: interest because they promise freedom from the persistent organochlorine compounds; suspicion because all residues are mistrusted and because the metabolites sometimes exceed the parent materials in mammalian toxicity.

A few organochlorine and organocarbamate insecticides are feebly systemic. These are absorbed by roots or leaves, metabolised, and translocated in non-insecticidal amounts, to leave persistent and undesirable residues at harvest. Crops having seeds with a high oil content may concentrate these small amounts in the oil (Duggan, 1968) to as much as one-tenth of the original content in the soil. Forage and fodder-carrying residues, even in parts per billion*, may represent a danger when fed to livestock, in which the chemicals reappear concentrated in body and milk fats.

Menzie (1969) has classified the possible biotransformations of pesticides in plants and soils. These include: oxidation, dehydrogenation and dehydrohalogenation, reduction, conjugation, hydrolytic reactions, exchange reactions, and isomerisation. Table 2.1 contains examples of

TABLE 2.1. Metabolic and other degradation products from pesticide residues in plants and soil. (Compiled from Gunther, 1970 and Menzie, 1969.)

Pesticide	Substrate		Product
	Plants	Soil	
aldicarb	+	+	sulfoxide, sulfone, oxime
aldrin	+	+	dieldrin
amitrole	+	+	several
captan	+		thiophosgene
carbaryl	+	+	a-naphthol
carbofuran	+		3-hydroxy, 3-keto, others
DDT	+	+	DDE and others
disulfoton	+		sulfoxide, sulfone
endosulfan	+		sulfate, others
fensulfothion	+	+	sulfone, oxygen analog
heptachlor	+	+	epoxide
hexachlorocyclohexane	+	+	pentachlorocyclohexene
linuron	+		3,4-dichloroaniline
pentachloronitrobenzene	+	+	pentachloroaniline
phenoxyacetic comp's	+		several
phorate	+	+	sulfoxide, sulfone
trifluralin	+	+	several
vegadex	+	+	lactic acid
Zectran	⌐		several

* parts per 10^9

fungicides, herbicides and insecticides which are subject to biotransformation, the substrates, and products.

Since 1945, the use of pesticides has accelerated rapidly. Their continued use is essential to meet the food requirements of an increasing population. In the years immediately following 1945, broad-spectrum, persistent compounds were considered ideal, but these characteristics were to limit their future use-patterns.

Pesticide sales in the United States increased from 634 million pounds in 1962 to 960 million in 1968 with a slight decline in 1969. The trends in pesticide sales from 1962 through 1969 are presented graphically in Fig. 2.1. It can be seen that after the appearance of the book 'Silent Spring' by

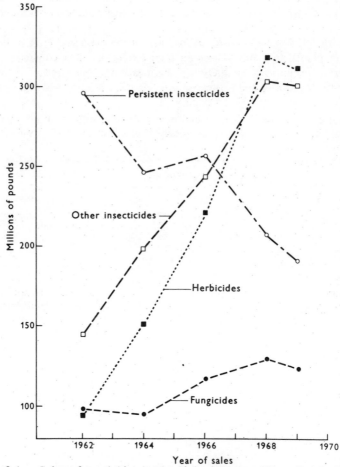

FIG. 2.1.   Sales of pesticides in the United States. (Compiled from U.S. Tariff Commission Publications, 'Synthetic Organic Chemicals', 1962, 1964, 1966, 1968, and 1969.)

Carson in 1962, the decline in the use of persistent insecticides coincided with a rise in the use of less persistent but more immediately toxic compounds.

From the 30 million pounds of DDT produced in 1945 (Martin, 1959) to the 300 million pounds of persistent insecticides sold in 1962 by the U.S.A. alone, it is not difficult to calculate a figure in excess of three thousand million pounds of persistent insecticides used during the 20-year period, 1950 to 1970. Dr. George Woodwell estimates that there are 10^9 pounds of DDT alone now circulating in the biosphere (Niering, 1968).

As early as 1950 (Starnes), BHC and lindane residues were reported in potato tubers. Reynolds et al. (1953, 1954) found significant quantities of BHC in peanuts and soil from land treated the previous year. Gannon and Biggar demonstrated the conversion of aldrin and heptachlor to their epoxides in soil (1958a) and on plants (1958b, c). The conversion of aldrin to dieldrin in plants was reported almost simultaneously (Glasser et al., 1958). By 1964, Wilkinson et al. had shown, both by bioassay and chemical analysis, that silt loam, treated nine years previously with normal applications of aldrin and heptachlor, still contained measurable amounts of their epoxides. Thus, the persistence of these broad-spectrum insecticides in the soil and their movement into plants was demonstrated early in their histories. As recently as 1964, 34.2 million pounds of toxaphene, 31.8 million pounds of DDT and 11.1 million pounds of aldrin were used on crops by farmers in the U.S.A. Of these amounts, 79% of the toxaphene and 74% of the DDT were used to combat insects attacking cotton. Of the aldrin, 96% was used for protection of corn (Agr. Econ. Rept. No. 131, 1968).

Considering the amounts of organochlorine insecticides used since 1950 for pest control on crops, it was no surprise when surveys of highly productive crop land in Ontario revealed residues of more than 130 ppm of DDT and related products, in the top two inches of sandy loam soil under trees in an orchard, and an average of 2.5 ppm of aldrin, 2.0 ppm of dieldrin and 8.0 ppm of endrin in an 8-inch profile of muck soil used for vegetable production (Harris and Sans, 1969a). In the U.S.A., a survey to determine residues in soils from areas of regular, limited and no pesticide usage, revealed a wide variety of residues in soils under regular use, DDT and dieldrin in those under limited use, and DDT in one area where it had not been used (Stevens et al., 1970). In West Germany, during investigations to determine the DDT- and lindane-contamination of intensively used agricultural and horticultural soils, samples from 222 sites in 13 counties revealed considerable contamination with DDT. The results showed that only 4.6% of the soils were free from DDT, while 10.9% contained more than 2 ppm of DDT. Approximately 16% of the samples contained lindane, in concentrations up to 0.5 ppm (Heinisch et al., 1968). In Washington State, residues of DDT and related compounds were found in alfalfa plants grown in land nine years after it had been used as an orchard and sprayed with DDT for orchard pests (Butler et al., 1970). Many instances of soils retaining residues from previous treatments are available in the literature (see chapter 11). Some of these have been selected and compiled into Table 2.2, to show how residues can and do occur in crops grown in soil treated from one to ten years previously.

2.3   FACTORS INFLUENCING RESIDUES

The factors influencing the use of pesticides, and hence the presence of residues, may be artificial or natural. Of the former, simple economics is the chief factor. The remarkable effects of pesticides and chemical fertilizers on the production of food and fibre in the U.S.A., during the last 30 years, may be gauged from Table 2.3. In Canada also, agricultural production has grown at almost 3% annually for the last 20 years. But it is not merely because it is profitable to use them, that pesticides will continue to be used. That fact is that the world has passed the point of no return and cannot afford not to use them. Demand for food continually increases but there are no new lands to colonize and cities everywhere spread themselves over the best acreage; thus, more food must come from the same or shrinking areas (Collingwood, 1965), or from marginal land. The world has a limited area of productive land from which must come the food for an inexorably growing population. Parallel with this growth are changing trends in food demand; as nations develop, they demand high protein foods such as wheat or meat, and discard their former dietary habits (Sci. Counc. Canada, 1971). To increase their own agricultural production, developing countries continue to use many of the most persistent organochlorine compounds because they are effective, long-lasting, and cheap (Reed, 1967; Appert, 1967; Ingram, 1967a and b; Materu and Hopkinson, 1969; Robertson, 1969; Lazarevic, 1970). Residues of several kinds are thus likely to remain as acute environmental problems for the forseeable future (Butler et al., 1970).

The choice of type, longevity and formulation of an insecticide is based on such factors as the pests it is to be used on, the skill and knowledge of the applicators, the degree of isolation and the climate. In general, systemic poisons within plants are used against small pests such as aphids, leafhoppers, scales, thrips and mites that suck large volumes of sap. Leaf-eating forms such as caterpillars, except when very small, do not usually take in enough plant sap to obtain a fatal dose of systemic poison. For these and for most soil pests, contact and stomach poisons must be used.

Since the movement of pesticides into plants begins with penetration of the root or leaf, formulations and methods of application greatly influence residues in plants (Ridgway et al., 1965; Dorough and Randolph, 1967; Ishiguro and Saito, 1970). Once the pesticide has been applied, the site of the deposit on the plant is another factor influencing penetration.

Natural factors are more subtle in their effect on residues in plants. They include the plants, soil, weather, penetration, absorption, trans-

TABLE 2.2. Pesticides in soil and movement into plants.

Pesticide	(Kg/ha)	Crop periods after final application	Crop	Residue	Amount (ppm)	Reference
aldrin	28.0	5	cucumber	dieldrin	0.116	Lichtenstein et al. (1965b)
	5.0×2	3	carrot	aldrin/dieldrin	0.2	Bro-Rasmussen et al. (1966)
	5.6	3	soybean	aldrin/dieldrin	0.044	Bruce and Decker (1966)
	1.12×2	1	alfalfa	dieldrin	0.015	Moubry et al. (1967)
	22.4	3	pumpkin	dieldrin	0.112	Bruce et al. (1967)
	8.4	3	sugarbeet	aldrin/dieldrin	0.03	Onsager et al. (1970)
chlordane	11.2	3		chlordane	0.12	Popov and Donev (1970)
DDT	15.0	3		DDT	0.5	Onsager et al. (1970)
	16.8	3		DDE; o,p'; p,p'-DDT	0.11	Butler et al. (1970)
	(old orchard to 1959)	9	alfalfa	DDE	0.017	
				o,p'-DDT	0.030	
				p,p'-DDT	0.054	
dieldrin	9.4	3	bean	dieldrin	0.5	Popov and Donev (1970)
heptachlor	5.6	3	soybean	hept/epoxide	0.038	Bruce and Decker (1966)
	22.4	3	pumpkin	hept/epoxide	0.036	Bruce et al. (1967)
	6.6	current	rutabaga	heptachlor	0.040	Saha and Stewart (1967)
				hept/epoxide	0.012	
				γ-chlordane	0.008	
	1.12	3	alfalfa	hept/epoxide	0.111	Moubry et al. (1967)
	28.0	10	carrot	hept/epoxide	0.223	Lichtenstein et al. (1970)
			radish		0.130	
			cucumber		0.068	
lindane	2.8	3	maize	lindane	0.6	Popov and Donev (1970)
methanearsenate	9.0×4	current	cottonseed	arsenic	5.2	Johnson and Hiltbold (1969)
			soybean		4.5	
			sorghum		3.3	
			corn		2.4	
organochlorines	several	1	sugarbeet	dieldrin	0.07	Harris and Sans (1969b)
			carrots		0.04	
picloram	1.12	1	grass	picloram	12.0	Getzendaner et al. (1969)

location and diffusion. The most important factor is the plant itself and the use to which it is put. Carrots and garden beets, at one extreme, appear to have an affinity for many insecticides, with an unusual ability to pick them up from soil and store them in the roots as original material or as metabolites. Residues in cotton, at the other extreme, are of little direct significance, except those that survive the processing of cottonseed oil, but persistent pesticides remaining in soil and trash after harvest may be of great importance in a subsequent food crop. Recent studies have shown that not only the species of plant (Harris and Sans, 1967; Stoller, 1970), but also the variety (Lichtenstein et al., 1965a; Stewart and Ross, 1967; Hulpke and Schuphan, 1970) and the stage of the plant (Hermanson et al., 1970) have a bearing on the residues found. Hartisch et al. (1969) found that crops such as rape, mustard, and carrots, which contain lipophilic constituents, were able to absorb DDT even from the soil.

TABLE 2.3. Yield per acre of eight major crops in the United States, 1938-1968.

Crop	1938	1948	1958	1968	30-year % gain
Barley, bu/A	24.2	26.5	32.3	43.7	80.5
Corn, bu/A	27.7	42.5	51.6	78.5	183.4
Cotton, lb/A	235.8	311.3	466.0	511.0	116.7
Potatoes, cwt/A	74.4	136.3	186.9	213.0	186.3
Rice, cwt/A	21.9	21.2	31.6	44.7	104.1
Sorghum, bu/A	14.3	18.0	35.2	52.9	269.9
Soybeans, bu/A	20.4	21.3	24.2	26.6	30.4
Wheat, bu/A	13.3	17.9	27.5	28.4	116.5

Compiled from Agricultural statistics U.S.D.A. (Crafts, 1970).

The type of soil is influential, because pesticides are held more tightly and are thus inactivated and less available to the plant roots or the pests as the organic content of the soil increases. Other factors affect root penetration; for example, the physical properties of soil have great influence on pesticide movement into plants. Wheeler et al. (1967) showed that wheat, orchard grass, alfalfa and corn plants grown in sand, contained from two to six times more dieldrin than those grown in soil. Harris (1969b) found that carrots from mineral soil with small amounts of dieldrin had higher residues of dieldrin than those from organic soils with greater amounts of dieldrin. Beestman et al. (1969) concluded that the organic content of soil was inversely related to the concentration of dieldrin in the roots and shoots of corn, and was directly related to the concentration of dieldrin recovered. Soil pH, temperature, organic matter and clay content affected the uptake of the labelled herbicide ^{14}C-diuron, by oat seedlings (Nash, 1968b). Soil moisture at field capacity induced the highest insecticidal uptake by plants; a gradual reduction occurred as the moisture level approached either the wilting-point or water-logging (Naoum, 1965). Ridgway et al. (1965), Chisholm and Specht (1967), Graham-Bryce and Etheridge (1967), and Etheridge and Graham-Bryce (1970), have all reported on the effects of moisture on absorption of pesticides from the soil. Burt et al. (1965) compared the absorption of four systemic insecticides and concluded that in moist soil all were equally effective. However, in dry soil the volatility of the insecticides determined whether the roots were capable of absorbing a toxic dose.

Rainfall produces its greatest effect on pesticide deposits within 24 hours of application. The mechanical action of the rain-drops affects dusts and suspensions more than it does emulsions. Radiation also reduces deposits, especially radiation near and above the violet end of the spectrum, but suspensions are more persistent than emulsions (Linskens et al., 1965). Some pesticide is usually lost by photolysis or photodecomposition.

The structure of the outer layer of plants has been reviewed by Frey-Wyssling (1948), van Overbeek (1956), Crafts and Foy (1962) and Linskens et al. (1965). Various pathways for penetration have been summarised by Currier and Dybing (1959) and Mitchell et al. (1960). Most commonly these pathways are: walls of root hairs or epidermal cells of roots; stomata and the cuticle of cells in the spongy mesophyll; lenticels or cracks in the cuticle and periderm, giving entry to cells of the phellogen; and plasmodesmata which are the protoplasmic threads between protoplasts of adjoining cells. In 1970, Hull reviewed extensively the relationship of leaf structure with the absorption of pesticides and other compounds. Franke (1967) has also reviewed the mechanisms of foliar penetration by solutions, and more recently (1970) discussed his theory of entry through ectodesmata in the outer walls of epidermal cells.

Sargent (1965) reviewed the penetration of growth regulators into leaves, and defined penetration as that part of the absorption process by which applied substances move through the outermost plant surfaces. Absorption is the movement of a substance through the outer, non-living layers of a cell and into the plasma membrane. With pesticides, the amount of absorption depends largely on their molecular polarity. On foliage, polar, aqueous compounds do not penetrate waxy, hydrophobic layers so readily as non-polar, lipoid and lipophilic compounds. Recent investigations on cuticular components by Baker and Bukovac (1971) showed that absorption of 2,4-D by the membrane was inversely related to the amount of cuticular wax. Thrower et al. (1965) found that cuticular wax on the leaves interfered with the penetration of the herbicide diquat. Persistence of residues on the surface of leaves, and their penetration are enhanced by increased humidity (Clor et al., 1962; Hopkins, 1967).

Penetration of leaves and the resultant residues tend to be greater with ultra-low-volume applications at high concentration, than with water emulsions at high dilution (Dorough and Randolph, 1967). Recently, residues of insecticides in plants have been attributed to penetration of leaves by the vapour phase of compounds with comparatively low vapour pressures (Whitacre and Ware, 1967; Hulpke, 1970). Nevertheless, pesticides with high vapour pressure disappear quickly from the surfaces of living leaves, especially in hot weather. The water solubilities of selected agricultural pesticides absorbed by, and resulting in residues in plants, together with their vapour pressures are shown in Table 2.4.

Although water-solubility is all-important in the translocation of pesticides, in recent years, advanced techniques and sophisticated equipment have revealed residues of pesticides that are practically water-insoluble. Penetration and translocation of these compounds in plants do occur, although not in amounts toxic to animals. Nash and Beall (1970) concluded that the major source of DDT residues in soybean plants was vapour from contaminated soil surfaces. By contrast, residues of the cyclodiene group of organochlorine insecticides result primarily from root uptake and translocation.

By far the most important site of absorption is the roots. This is the logical point of entry since absorption of water and mineral salts is their function. Linquist et al. (1961) compared the various absorptive regions and concluded that the most active site is 20 to 40 mm above the root cap in the zone of root hairs. Roots absorb lipid-soluble pesticides much less readily than water-soluble ones (Mitchell et al., 1960).

Translocation, sometimes called transport or conduction, may be upward or acropetally, downward or basipetally, or laterally. Robertson and Kirkwood (1969 and 1970) have reviewed the mechanisms and factors influencing absorption, translocation, metabolism and biochemical inhibition of phenoxy-acid herbicide compounds applied to foliage. They concluded that after penetration, these compounds may move in the free space of the cell walls or to the phloem by way of the cytoplasmic connections of the symplast. Humidity influences the rate and direction of translocation. Basler et al. (1970) injected ^{14}C labelled 2,4,5-T into bean seedlings and found that high humidity enhanced the basipetal transport of the herbicide in the phloem, whereas low humidity contributed to acropetal transport in the xylem. Their results confirm the conclusions of others. Translocation from the roots in the sap stream of the xylem is the normal pathway for systemic pesticides. From application to seed or soil, toxic amounts penetrate the roots and are carried to the aerial parts of plants. However, non-toxic amounts may also be carried and this transportation is not restricted to systemic compounds. Miles et al. (1967) detected diazinon in cabbage and tobacco plants up to eight weeks after transplanting. The insecticide had been applied to the roots in the transplanting water, but was absorbed only during restoration of turgidity. With recently developed techniques, and by use of compounds with radioactive labels, residues of persistent non-systemic pesticides have been detected in most parts of plants.

Metabolism within the plant may also determine the persistence of the residue. Metabolites of systemic insecticides, for example the sulfoxide of

TABLE 2.4. Solubility in water and vapour pressure of selected agricultural pesticides at various temperatures.

Pesticide	Solubility*		Vapour pressure*	
	ppm	temp°C	mm	temp°C
aldicarb	6,000	25†	0.05	25†
aldrin	0.27	30‡	6.0×10^{-6}	25‡
amitrole	28 g/100 ml	25	—	
captan	<0.5		—	
carbaryl	40	40	<0.005	26
carbofuran	700	25	—	
chlordane	insoluble		1×10^{-5}	25
chlorfenvinphos	sparingly soluble		1.7×10^{-7}	25
D-D	2,800§		31.3	20
DDT	practically insoluble		1.9×10^{-7}	20
diazinon	40	room	1.4×10^{-4}	20
dieldrin	0.18	30‡	7.78×10^{-7}	25
disulfoton	25		18	20
endosulfan	insoluble		0	75
endrin	0.23	25‡	2×10^{-7}	25
ethion	2§		—	
fensulfothion	slightly soluble		0.01	138-141
fonofos	13	21	0.1	130§
heptachlor	0.056	30‡	4×10^{-4}	25
hexachlorobenzene	practically insoluble		1.089×10^{-5}	20
lindane	10	20§	9.4×10^{-6}	20
linuron	75	25	1.1×10^{-5}	24
parathion	24	25	3.78×10^{-5}	20
pentachloronitro-benzene	practically insoluble		11.3×10^{-5}	25
phorate	50		0.8	118-120
sulfallate	100	25	2.2×10^{-3}	20
tetradifon	200	50	—	
thionazin	1,140	24.8	3×10^{-3}	30
toxaphene	3	room	0.2 to 0.4	25
trichloronate	25	25§	0.01	108
trifluralin	24	27	1.99×10^{-4}	29.5
Zectran	100	25	0.1	139

* From Spencer (1968).
† Chemical Company technical information brochures.
‡ Park and Bruce (1968).
§ Gunther *et al.* (1968).

aldicarb, may persist longer and be considerably more effective than the parent compound (Tamaki *et al.,* 1969). Stoller (1970) found differential absorption and translocation in several species of plants of a ^{14}C labelled herbicide, chloramben. He postulated that the mechanism in wheat, a tolerant species, is the conversion of ^{14}C labelled chloramben in the roots to a non-transportable compound, whereas only free chloramben is translocated.

Movement of pesticides acropetally and basipetally out of plants, although rare, does occur. Gunner *et al.* (1966) applied labelled diazinon to the leaves of bean plants and recovered an active compound from the exudate of the roots. Picloram was recovered from root exudates of ash and maple by Reid and Hurtt (1970). Movement out of plants acropetally was recorded as early as 1952 when Park and Heath found schradan in guttation droplets on leaves of wheat, strawberries and Brussels sprouts. Stoller (1970) found ^{14}C labelled chloramben deposited by guttation on leaves of barnyard grass. Of possibly more consequence was the recent finding of Lord *et al.* (1968) that amounts of dimethoate, toxic to bees and *Drosophila melanogaster* could be translocated from the soil to the nectaries of beans.

2.4 AMOUNTS OF RESIDUES

Pesticides range in toxicity to man, from much lower than that of table salt to much higher than that of strychnine. On the assumption that nowadays some chemical residues in crop plants are virtually inevitable, government agencies in developed countries have set legal limits or tolerances to these. The permissible amounts agree between countries within fairly narrow limits, as indeed they must between trading partners. The basis on which they are set is usually some minor variant of the following: the maximum amount that should exist when a chemical has been properly used, or not more than 1/100th of the minimum dose that will produce any physiological effect on the most sensitive of two or more species of laboratory animals such as standard white rats, white mice, guinea pigs, or rabbits, for two to four lifetimes. Typically, in Canada these limits are not set by the Department of Agriculture but by the Department of Health and Welfare. Where it is feasible to grow a crop with little or no residues, the tolerances are set at these levels.

Pesticides available to plants in the soil may be from an application made immediately before planting or from deposits in previous years which have left a residue in the soil. Figure 2.2 diagrams the progression of a soil or plant deposit from its initial contact with the plant to utilisation of the crop. The residue may be parent pesticide (P), metabolites or degradation products (M), or both. From application to harvest, the absorbed pesticide is subjected to further degradation and metabolism. At harvest, residues in the edible parts of plants must be less than the established residue tolerance if the plant is to be used for food or fodder.

Harvest is the first point (T_1, Fig. 2.2) for enforcement of residue tolerances. After harvest, various treatments of the crops may follow, such as drying, ensilaging, and chilling. During this period, further metabolism or degradation often reduces residues to negligible amounts. However, in fodder crops, the preparation of hay may concentrate residues simply by removal of water. Edible portions of plants are usually processed in minimal time from harvest. Processing includes such treatments as canning or freezing, during which residues are often reduced or eliminated. The processed product offers the second point (T_2, Fig. 2.2) for enforcement of residue tolerances. Several countries take advantage of both points T_1 and T_2, by taking random samples of fresh and processed foods from food markets. The residues in these samples indicate the amounts and nature of the residues that populations are consuming in their daily diet. Storage or cooking of processed plants may further lower any residues to the terminal level.

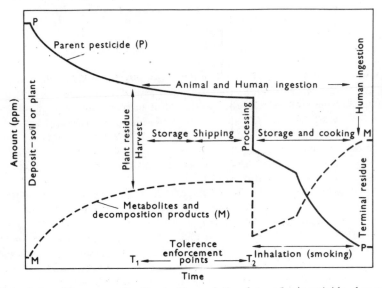

FIG. 2.2.   Diagrammatic illustration of the fate of a pesticide deposit. (Modified from Report of the Panel, Nuclear Techniques for Studying Pesticide Residue Problems. International Atomic Energy Agency, Vienna, 1970.)

Tobacco, although it generally contains residues of pesticides, is not normally subjected to tolerance levels. Residues which are present in tobacco are inhaled in the mainstream smoke unless they are destroyed by pyrolysis. In studies to determine the effects of pyrolysis, Chopra and Osborne (1971) found that when p,p'-DDT was pyrolyzed in a nitrogen atmosphere at $900°C$, which is the temperature of the on-puff burning zone of a cigarette, the products isolated and identified included p,p'-DDT, p,p'-DDE and p,p'-TDE.

Much research is conducted annually to determine the efficacy of pesticides in crops and the resultant residues at harvest. Although processes such as extracting sugar from beets or oil from seeds, may and often do remove the residue from the plant portion consumed by humans, the fraction remaining of the beet pulp, mint hay, peanut hay or citrus peel, is often fed to livestock. This fraction sometimes contains undesirable residues.

Table 2.5 presents selected records of residues in crops grown for food and fodder and in tobacco. The period from final application to harvest ranges from 8 days for tobacco, which left residues of 128.7 ppm of DDT and its metabolites (Tappen et al., 1967), to 174 days for soybean where the content of heptachlor epoxide in the oil was 0.38 and 0.81 ppm (Dorough et al., 1969). The residues shown in the table result from experimental work and in many instances, are well above established tolerance levels.

2.5   HAZARDS FROM RESIDUES

In spite of nearly 30 years of worldwide experience with DDT, and upwards of 20 with lindane, dieldrin, heptachlor and other cyclodiene organochlorine compounds, the hazards to man from these most common residues are still not completely understood. The consensus appears to be that man and most other mammals are able to cope fairly effectively with residues consumed in food, but that many birds and many cold-blooded animals are less able, and some are unable, to do so. At greatest risk, are predators at, or near, the tops of food pyramids. These take in the accumulated residues held in the fatty tissues of their prey, which they tend to store for themselves, with some demonstrably deleterious effects. Considerable ecological damage appears to have been done, mostly by the long-lasting organochlorines mentioned, which are stable enough to be transported in air currents or on air- or water-borne particles (Tarrant and Tatton, 1968). DDT is the most pervasive; it was recently established that even benthic whales of two species were carrying up to 6.0 ppm in their blubber (Wolman and Wilson, 1970).

In man, DDT and its metabolites ingested as small residues over a long period, may build up to surprising levels. Irrefutable evidence of damage to humans, caused by residues of DDT picked up from commercial foods, appears still to be lacking. However, there are clear indications that DDT will pass the placental barrier and appear in newborn children (Wassermann et al., 1967). How widespread it has been recently (1955-1966) in the general population, may be judged from Chapter 8 and Table 2.6, which is condensed from Wassermann et al. (1967). Similar data for DDT and heptachlor, covering some of these and other countries, were superimposed on maps of Europe and North America by Robinson (1970). Following bans or restrictions on the use of DDT in many western countries, some of the levels in Table 2.6 may now be lower. The daily intake of DDT in Canada (1969), in the U.S.A. (1968), and in the U.K. (1967) were 18, 55 and 44 μg per person per day. These levels were 5%, 16%, and 13%, respectively, of the acceptable daily intake as laid down by the W.H.O. The intake of dieldrin was 51%, 57%, and 94%, respectively, of the acceptable intake (Smith, 1971). It is quite evident from the literature, that DDT and the cyclodiene organochlorine insecticides are still being used in large amounts in underdeveloped countries, not only against pests of public health but also against pests of agricultural food crops.

Tobacco remains as an exception, for it is neither a food nor a drug and hence is exempt from pesticide tolerances, at least in the United States. Recent studies (Guthrie and Sheets, 1970) have shown that residues of 20 to 30 ppm of DDT-TDE are present in commercial cigarettes. They state, 'The average daily intake of total DDT components by the one-pack-per-day smoker through the mainstream smoke nearly equals the daily intake in food'. Residues in cigarettes are not confined to organochlorine insecticides. The herbicides metobromuron and 'Molipan' applied to the soil were translocated to the leaves of tobacco and delivered into the mainstream smoke. Retention of these residues by cellulose and cellulose acetate filters was low, of the same magnitude as the retention of nicotine (Corbaz *et al.,* 1969). Analysis of commercial brands of cigarettes from England, Germany, Japan and the U.S.A. (Maruyama *et al.,* 1970) revealed a mercury content of 0.70 ppm in Japanese cigarettes, the others averaged 0.16 ppm. Copper ranged from 6.9 to 38.0 ppm and arsenic from 0.37 to 1.9 ppm. The high rate of mercury was attributed to the use of mercury pesticides on tobacco farms in Japan. West Germany initiated legislation in 1966 which extended pesticide tolerances to tobacco. The German tobacco industry and countries exporting cigarettes to Germany were given six years in which to comply with the established tolerances of 0.1 ppm of DDT-TDE or 0.1 ppm of maleic hydrazide (Guthrie and Sheets, 1970).

The organophosphorus compounds have different characteristics. They fall into three well-defined chemical subgroups: aliphatics, having open chain molecules; aryls, having one or more phenyl rings in the molecules; and heterocyclics, having oxygen, nitrogen, or phosphorus atoms included in the rings.

Aliphatics are often partly or largely water-soluble, and the group includes some useful plant systemics. A few are slightly phytotoxic. Their mammalian toxicities range from the highest known to others so low that they are used as systemics for livestock. Generally, they break down completely and in a comparatively short time. Aryls and heterocyclics are soluble in organic solvents but only slightly soluble in water, and also have a wide range of toxicity. Some are persistent and stable enough to use in soil against insects, nematodes, and symphylids. Nevertheless, all of the

TABLE 2.5.   Pesticide residues in crops after various intervals from last application to harvest.

Pesticide	Rate	Interval (days)	Crop	Residue Compound	Residue ppm	Reference
aldicarb	113.4g/tree	143	apples	sulfoxide	0.83	Maitlen et al. (1970)
				sulfone	0.14	
	1.6g/m²	48	grapes	—*	2.00	Purokoski (1969)
	40.0g/tree	100	orange peel	sulfoxide	12.75	Tashiro et al. (1969)
				sulfone	2.24	
			orange pulp	sulfoxide	2.63	
				sulfone	0.58	
	2.24kg/ha	126	sugarbeet	sulfoxide	0.58	Maitlen et al. (1970)
			foliage	sulfone	0.86	
aldrin	10g/100g seed dressing	—	carrot	aldrin	0.41	Hulpke and Schuphan (1970)
azinphosmethyl	1.12kg/ha	14	sorghum	—	0.99	Dorough et al. (1966)
	3.0ml/tree	21	apple	—	0.75	Purokoski (1970)
calomel	3.6g/10m	84	rutabaga	mercury	1.6	Purokoski (1967)
carbofuran	1.12kg/ha	21	alfalfa	3-OH	1.26	Shaw et al. (1969)
chlorfenvinphos	5.6kg/ha	50	carrot	—	0.40	Finlayson (unpublished)
dalapon	2.24kg/ha	84	alfalfa	—	6.00	Scholl and Williams (1969)
			trefoil	—	6.00	
DDT	1.12kg/ha	24	mint hay	DDT⁺	4.21	Ballee et al. (1970)
			mint oil	DDT⁺	1.54	
	4.48kg/ha	8	tobacco	—	128.70	Tappan et al. (1967)
	2.24kg/ha	90	tobacco	p,p'-DDE	0.09	Reed and Priester (1969)
				o,p'-DDT	0.18	
				p,p'-DDT	0.29	
			soybean	o,p'-DDT	0.08	

	Rate	Days	Crop	Metabolite	Residue	Reference
						...and Novak (1967)
dieldrin	12.32 kg/ha	70	carrot	—	0.30	Finlayson et al. (1968)
	1.0 kg/ha	70	peanut hay	—	0.13	Thompson et al. (1970)
		112	peanut	—	1.27	
		112	peanut shell	—	3.97	
dimethoate	1.12 kg/ha	70	grass (pasture)	—	0.56	Beck et al. (1966)
	1.12 kg/ha	14	sorghum	—	0.64	Dorough et al. (1966)
disulfoton	2.24 kg/ha	49	spinach	—	0.27	Menzer and Ditman (1968)
	1.68 kg/ha	77	potato	—	0.29	Brewerton and Close (1967)
endrin	1.4 l/ha	31	tobacco	I+II	2.10	Joshi and Ramaprased (1969)
endosulfan	2.24 kg/ha	8	tobacco	I+II	33.10	Tappan et al. (1967)
	8 ml/bush	54	black currant	sulfate	1.31	Purokoski (1969)
					0.40	
fensulfothion	6.72 kg/ha	56	carrot	—	0.32	Finlayson et al. (1970)
fonofos	2.2 kg/ha	at seeding	carrot	—	0.37	Suett and Wheatley (1970)
heptachlor	3.36 kg/ha	174	soybean oil	epoxide	0.38	Dorough et al. (1969)
	6.72 kg/ha	174	soybean oil	epoxide	0.81	
lindane	90 g/kg seed	28	radish	—	0.80	Purokoski (1967)
		51	turnip	—	0.50	
parathion ethyl	0.56 kg/ha	15	alfalfa	—	2.61	Waldron and Goleman (1969)
parathion methyl	0.56 kg/ha	15	alfalfa	—	0.60	
plictran	30 g/tree	28	apples	tin	0.29	Trombetti and Maini (1970)
			pears	tin	0.07	
propoxur	2.24 kg/ha	86	potato	—	0.18	George et al. (1967b)
	2.86 kg/ha	55	sugarbeet-foliage	—		
			black currant	—	0.43	
thionazin	40 ml/bush	64	black currant	—	0.55	Purokoski (1970)
	6.72 kg/ha	56	carrot	—	0.33	Finlayson et al. (1970)
trichloronate	20 g/kg seed	47	radish	—	0.55	Purokoski (1970)
	9 g/100 m	122	carrot	—	0.82	

* A dash = residues not stated, assumed to be parent compound.

group are short-lived by comparison and they break down completely, in some cases within hours.

There is often confusion in the public mind between the extreme hazards of handling and applying organophosphorus compounds, and the hazards of their residues; any residues from materials of such terrifying toxicity are thought to be correspondingly dangerous. Indeed, some of these chemicals are so dangerous to apply that their value is impaired or they cannot be used in some circumstances; for instance, in the tropics it is often physically impossible for workers to wear, even for short periods, the heavy, natural rubber coat, hat, boots, gloves and mask that are essential for safety in applying the very toxic compounds. The few residues found in extensive surveys of foods, however, have been extremely low (Abbott et al., 1970; Corneliussen, 1970).

TABLE 2.6. Concentration of DDT and its metabolite DDE in the body fat of the general population of various countries, 1955-1966. Condensed from Wassermann et al., 1967.

Country	Years	Sample size	DDT plus DDE (ppm)
W. Germany	1958-59	60	2.3
England	1961-64	231	2.9
Canada	1959-60	62	4.9
France	1961	10	5.2
Czechoslovakia	1963-64	229	9.2
U.S.A.	1954-64	575	12.0
Hungary	1960	50	12.4
Israel	1963-66	458	12.7
India	1964	102	24.9

There has been a sharp increase in the use of organophosphorus compounds following the restriction or banning of persistent organochlorines. Smith (1970) pointed out and discussed the troubles associated with the change, such as: outbreaks of secondary pests resulting from destruction of their natural enemies; the undesirable side-effects on non-target organisms including parasites and predators, bees and other pollinators, and on fish, birds and other wildlife; the direct hazards to field workers, and ground and air applicators from the pesticides; and the need for frequent repetition of the treatments, which inevitably leads to increases in resistance. Nevertheless, although the transition from organochlorine chemicals is costly and extraordinarily difficult (Ware et al., 1970), it seems clear that there has been a reduction in their residues in foods of plant origin and further such declines may be expected.

A somewhat disquieting finding was that the aliphatic organophosphorus systemics, dimethoate and phorate, applied at a very high rate to soil, could be transported from roots so efficiently as to reappear in the nectar of fuchsia and nasturtium, and in the nectaries of field beans (Lord et al., 1968). Dimethoate was found in amounts toxic to bees and fruit flies but phorate was not. The toxicity was lost after about 20 days.

With the increased usage of organophosphorus insecticides and especially of herbicides, to reduce labour costs, residual amounts of either in the soil may sometimes result in unexpected crop losses. Nash (1968a) found that when the herbicide diuron was combined with the insecticides disulfoton or phorate and applied to the soil, synergistic phytotoxicity resulted.

The organocarbamates are a comparatively small group of compounds, but of growing importance. They bridge the gap between the persistent organochlorines and the short-lived but excessively toxic organophosphorus chemicals. Their wide adoption is delayed because they all are difficult to analyse with precision and a few are hardly less toxic than their predecessors. The group includes highly effective systemics. The most widely used carbamate is carbaryl, a contact insecticide that is also feebly systemic in fruit trees by diffusion through the bark. In the extensive study of pesticide residues in ready-to-eat foods, conducted by the Food and Drug Administration of the U.S.A., carbaryl residues are listed (Corneliussen, 1970) as occurring once at each of only two centres during 1968-1969, in trace amounts in legume vegetables, and once in fruit in another city. Since carbaryl is used very widely the conclusion is that the hazards from its residues are negligible (Mann and Chopra, 1969). A number of organocarbamates are still in the experimental and testing stage.

Synthetic organic pesticides based on tin are of recent origin and are mostly still experimental. The group includes fungicides, chemosterilants and at least one insect repellent but not, so far, systemics. External residues on plants appear to be easy to avoid, at least from the effective miticide 'Plictran' currently in commercial use (Allison et al., 1968; Trombetti et al., 1971).

A number of other miticides in use are derived from such chemical groups as the dinitros, carbonates, sulfites, sulfides, sulfates, sulfones, and sulfonates. None is systemic or has left troublesome residues.

The old-fashioned pesticides, compounds of such inorganics as lead, sulphur, mercury, fluorine, sodium, calcium, arsenic or copper, although dangerously toxic and often highly persistent in soil, do not normally penetrate into or translocate within plants. A rare exception is the element selenium, which occurs naturally in some soils, in amounts large enough to kill sucking insects and to make the plants poisonous to livestock or humans.

2.6   REMOVAL OF RESIDUES

The extensive use of pesticides in crop production has not only contributed to the increase in yield per unit area and quality of produce, but has also helped to keep the cost of food well below the corresponding increases of other commodities. However, insecticides, and in particular organochlorines, are of great concern because they accumulate in humans, concentrate in food chains and may resist biodegradation. Their over-use has led to the development of tolerant pests, to the reduction of parasites and predators, and to the decrease of pollinators. Because of these characteristics, research towards less persistent pesticides has produced compounds that are more toxic, but less persistent, more costly, but often less effective. This has resulted in higher dosage rates and more applications, and has increased the problem of residues in harvested plants.

Residues in soil from crops grown previously are sometimes a serious problem in current crops. Residues may occur in crops as long as 10 years after the application of the pesticides (Table 2.2). If the history of the land usage is known, selection of crops other than those known to absorb and translocate soil residues is the first approach. Carbon applied to soil containing residues of the cyclodiene insecticides decreases the amount of translocated insecticide (Ahrens and Kring, 1968, Lichtenstein et al., 1968). The latter group of workers showed that residues of aldrin and its metabolite dieldrin, in radishes, carrots, and potatoes were reduced 66%, 53% and 71% respectively by the addition of carbon to the soil at 2,000 ppm.

Photoisomerisation and the effects of several fractions of the light spectrum have been reported effective in reducing residues of cyclodiene insecticides on plants. Ivie and Casida (1970) found that applications of low concentrations of rotenone catalysed the process of photoisomerisation. This suggests the possibility of controlling residues by the use of pesticide-photosensitiser combinations. Archer (1969) found that residues of DDT on alfalfa that had been air-dried in darkness, in sunlight and under ultra-violet lamps reacted quite differently. In darkness, DDT remained as the parent compound, but 49% of the original amount was lost during drying. Under ultra-violet light, DDD was formed on the fourth day, no change occurred with DDE, and 67% of the DDT was lost. In sunlight, DDD was present by the sixth day, there was no change in DDE and 72% was lost. These findings are in agreement with those of residues on tobacco. In the curing process, at a relatively low temperature (65.6°C) the organophosphorus and organocarbamate residues were reduced by 75-95%. The losses of organochlorine residues were 40-50% or similar to those for DDT on alfalfa in darkness (Guthrie and Sheets, 1970). Joshi and Ramaprasad (1969) concluded that organophosphorus residues decomposed during flue-curing but that those of endrin persisted.

The usual method for limiting residues in agricultural produce is to specify, in the registration for usage, a time interval in days which must elapse between the last application and harvest, to allow degradation of the pesticide. This time interval may be long, for example 90 days for preplant soil treatments with disulfoton in potatoes, or short, one day for malathion on blueberry (Compendium on Registered Uses of Pesticides in

Canada, 1968). Included with the instructions for use, and the time interval is a statement of the allowable residue.

In spite of these statements and the instructions, misuse of pesticides often occurs and residues above tolerance are sometimes found in harvested crops. Storage and processing are both means of reducing residues. Kilgore and Windham (1970) analysed malathion residues on broccoli 1, 2, 3 and 4 days after treatment. They found that cooking for five minutes reduced the residues 9, 34, 8 and 7% respectively. When they blanched, froze and stored the broccoli at −9°C for six months the residues were reduced by 45, 69, 75 and 77%. Another example is provided by Fahey et al. (1969) who treated tomatoes with Gardona at 1.12 kg/ha at 10 days and again 24 hours before harvest. Residues were: in unwashed tomatoes, 1.22 ppm; in washed, 0.64 ppm; in juiced or canned, 0.03 ppm. The tomato pomace contained 1.68 ppm. When Fahey et al. (1970) treated cherries, peaches and pears with Gardona at the recommended dosage and at twice the recommended dosage the residues were respectively: cherries, 4.3 and 9.4 ppm; peaches, 6.5 and 15.0 ppm; and pears, 5.6 and 10.4 ppm. Plumped and pitted cherries contained 0.95 and 2.11 ppm, and when frozen they had 0.81 and 1.88 ppm after treatment at the same two rates. When the peaches and pears were peeled they had 0.05 and 0.26 ppm of Gardona. Canning reduced the terminal residue to 0.02 and 0.2 ppm.

Root crops, in direct contact with pesticides in the soil, absorb and hold most of the residue in the peel. Suett and Wheatley (1970) found 6.4 ppm of fonofos in the peel, compared with only 0.37 in the whole carrot. Williams and Finlayson (unpublished) determined residues of fensulfothion in pulp and peel for various parts of carrots. In the top 3 cm of a carrot the peel contained 10.54 ppm, the pulp 0.51 and the whole fraction 0.93 ppm of fensulfothion and 2.86, 0.07, and 0.22 ppm respectively of the sulfone. In the next 3-cm fraction the peel had 1.97 and 0.90 ppm of fensulfothion and the sulfone respectively, the pulp 0.07 and 0.01 ppm and the whole fraction 0.25 and 0.08 ppm. The residues continued to decrease progressively to the tip of the carrot. Saha and Stewart (1967) found 0.15 ppm of heptachlor and 0.06 ppm of its epoxide in rutabaga peel, compared with only 0.002 ppm of the heptachlor epoxide in the pulp. Boiling the peel reduced the residue approximately 50%. Contrary to this, Hulpke (1969) found that, for carrots that neither cooking nor storage at room temperature till they rotted, reduced the content of ^{14}C labelled aldrin in the periderm. Jacobs et al. (1970) found 0.5 ppm of arsenic in the flesh of potato tubers but up to 84 ppm in the peel. They recommended that the use of arsenic as a potato vine defoliant, be discontinued.

In citrus production, Gunther (1969) reviewed residue legislation, tolerances in several countries, insecticide residues in juice and citrus pulp cattle-feed, and washing for residue removal. He reported removal of residues from the fruits ranging from traces to almost 100%. Residues in the juice ranged from less than 0.1 ppm for many insecticides to 2.0 ppm of parathion and about 5.0 ppm of DDT. In dried pulp for cattle-feed, the residues ranged from 0.8 ppm of tetradifon to 5.6 ppm of azinphos-methyl. In a second review, Jepson and Gunther (1970) reported on acaracide residues in citrus foliage and fruits and their biological significance.

Studies on the persistence of residues of fenitrothion, malathion, and methyl-parathion (Rippel *et al.,* 1970), revealed that degradation of organophosphorus compounds, such as parathion, occurred in citrus juice after packaging. The decrease was brought about by the chemical reduction of tin which lined the containers. The degradation was substantially lower in containers lined with a protective layer and negligible in glass containers. They concluded that elemental tin caused greater reduction than bivalent tin.

As with root vegetables and fruit, residues in other vegetables can be removed both by commercial and home preparative methods. Lamb *et al.,* (1968) found that washing and blanching spinach removed more than 50% of the residues of *p,p'*-DDT, *o,p'*-DDT and carbaryl, and about 45% of the residues of parathion present on unwashed spinach. Similar reductions were obtained when tomatoes treated with DDT, malathion and carbaryl were washed, peeled, and canned (Farrow *et al.,* 1968); when green beans were washed, blanched, canned or frozen (Elkins *et al.,* 1968); and when broccoli with parathion and carbaryl residues were washed, blanched and frozen (Farrow *et al.,* 1969).

The processing of sugar beets also presents special residue problems. Sugar beets have become a major source of sugar and since they must be treated with insecticides, undesirable residues in the sugar and by-products must occur. In 1963, the U.S.A. produced 2.7 million tons of wet beet pulp, 215,000 tons of dried beet pulp and 870,000 tons of dried pulp plus molasses. Beet pulp is often used in the diet of lactating and meat-producing livestock. In 1965, Walker *et al.* reported on the fate in sugar beet of aldrin, dieldrin and endrin applied as a preseeding application to the soil, and dieldrin as a seed treatment. Raw sugar beet treated with aldrin contained 0.018 ppm aldrin and 0.17 ppm dieldrin; those treated with dieldrin, 0.095 ppm dieldrin, and those with endrin 0.24 ppm endrin. In the dried pulp, aldrin treatment produced 0.03 ppm aldrin and 1.71 ppm dieldrin, dieldrin treatment 1.49 ppm and the endrin treatment 1.62 ppm. In the first carbonation, juice residues were only from 0.001 to 0.004 ppm.

2.7   THE FUTURE

The ability of chemical pesticides to concentrate in biological tissues, with the continuing risk of long-term damage to populations of men, and domestic and wild creatures, means that fewer chemicals in smaller amounts must be used, and that in time lower standards for produce must be accepted. A continuing programme of education is needed.

There may be more unpleasant surprises from residues of heavy metals in plants, such as the realisation of the ubiquity of mercury (Andersson, 1967). Upon the natural distribution of mercury is super-imposed massive contamination of the environment with this element resulting from industrial wastes and from unexpected impurities in bulk reagents such as sulphuric acid and hence from synthetic fertilisers. Internal residues of mercury in plants are not yet a problem, but it is clear that the use of mercury in fungicides must be reduced or ended. It is now realised that mercury can be converted to an organic and more deadly form and then concentrated by micro-organisms.

A new and hopeful sign is the widespread recent recognition of the role of natural pest control measures, and the realisation of their value. It is more generally accepted today than in the past that insect pests cannot be exterminated, even if this were desirable. Eventually, we must come to terms with pest problems.

As a workable concept, integrated pest control is already well entrenched and in process of refinement. This is the antithesis of prophylactic or preventative control, in which chemical treatments are applied by the calendar whether there is need or not. In integrated control, dosages and applications are minimal and are aimed at reinforcing, or at least avoiding inhibiting parasites, predators, or other natural control agents. It follows that the risk of internal and external residue of chemicals in plants is greatly reduced under such a regime. The present lack is in intimate knowledge of the interactions between the pests and their control agents, on which the regimes must be based.

Most chemosterilants are carcinogens. They may have a role to play, but even when they are confined to traps and the victims are lured into contact by sex attractants and other lures, the problem remains of releasing into the environment numerous insects carrying carcinogens. Insectivorous birds would clearly be at risk.

The root of the residue problem lies in the need to overproduce food, and thus to abuse the land, in order to feed the world population. Even with about one third of the earth's land area already filled or in pasture, food production per capita is declining steadily. Unless the population increase is arrested before long, it may be impossible to maintain good standards of residue levels in underdeveloped countries or to adhere to the present meticulous standards set and maintained by the developed countries.

TABLE 2.7. Chemical name, type and use of pesticides cited in the text.

Pesticide	Type*	Acar	Fung	Herb	Ins	Nem	Chemical Name†
aldicarb	(1)	+			+	+	2-methyl-2-(methylthio) propionaldehyde O-(methylcarbamoyl) oxime
aldrin	(2)				+	+	1,2,3,4,10,10-hexachloro-1,4,4a,5,8,8a-hexahydro-1,4-*endo-exo*-5,8-dimethanonaphthalene
amitrole	(4)			+			3-amino-1,2,4-triazole
azinphosmethyl	(3)	+			+		O,O-dimethyl S(4-oxo-1,2,3-benzotriazin-3(4H)-ylmethyl) phosphorodithioate
benzene hexachloride	(2)				+		1,2,3,4,5,6-hexachlorocyclohexane, mixed isomers and a specified percentage of gamma
calomel	(4)		+				mercurous chloride
carbaryl	(1)				+		1-naphthyl methylcarbamate
carbofuran	(1)				+		2,3-dihydro-2,2-dimethyl-7-benzofuranyl methylcarbamate
captan	(4)		+				N-trichloromethylmercapto-4-cyclohexene-1,2-dicarboximide
chloramben	(4)			+			3-amino-2,5-dichlorobenzoic acid
chlordane	(2)				+		1,2,4,5,6,7,8,8-octachloro-3a,4,7,7a-tetrahydro-4,7-methanoindane
chlorfenvinphos	(3)				+		2-chloro-1-(2,4-dichlorophenyl)vinyl diethyl phosphate
2,4-D	(4)			+			2,4-dichlorophenoxyacetic acid
dalapon	(4)			+			a a-dichloropropionic acid
D-D	(4)					+	1,3-dichloropropene and 1,2-dichloropropane mixture
DDE	(2)						metabolite of DDT, 1,1-dichloro-2,2-bis(p,-chlorophenyl)ethylene
DDT	(2)				+		1,1,1-trichloro-2,2-bis(p,-chlorophenyl)ethane
diazinon	(3)	+			+		O,O-diethyl O-(2-isopropyl-4-methyl-6-pyrimidyl)phosphorothioate
dieldrin	(2)				+		1,2,3,4,10,10-hexachloro-6,7-epoxy-1,4,4a,5,6,7,8,8a-octahydro-1,4-*endo-exo*-5,8-dimethanonaphthalene
dimethoate	(3)	+			+		O,O-dimethyl S-(N-methylcarbamoylmethyl) phosphorodithioate
diquat	(4)			+			9,10-dihydro-8a,10a-diazoniaphenonthrene (1,1'-ethylene-2,2'-bipyridylium) dibromide
disulfoton	(3)				+		O,O-diethyl S-2[(ethylthio)ethyl] phosphorodithioate
diuron	(4)			+			3-(3,4-dichlorophenyl)-1,1-dimethylurea
endosulfan	(2)	+			+		6,7,8,9,10,10-hexachloro-1,5,5a,6,9,9a-hexahydro-6,9-methano-2,4,3-benzodioxathiepin 3-oxide
endrin	(2)				+		1,2,3,4,10,10-hexachloro-6,7-epoxy-1,4,4a,5,6,7,8,8a-octahydro-1,4-*endo*-

O-ethyl S-phenyl ethylphosphonodithioate

fonofos	(3)	O-ethyl S-phenyl ethylphosphonodithioate	+	
'Gardona'	(3)	2-chloro-1-(2,4,5-trichlorophenyl)vinyl dimethyl phosphate	+	
heptachlor	(2)	1,4,5,6,7,8,8-heptachloro-3a,4,7,7a-tetrahydro-4,7-methanoindene	+	
hexachloro-cyclohexane	(2)	1,2,3,4,5,6-hexachlorocyclohexane mixed isomers and specified percentage of gamma	+	
lindane	(2)	1,2,3,4,5,6-hexachlorocyclohexane gamma isomer	+	
linuron	(4)	3-(3,4-dichlorophenyl)-1-methoxy-1-methylurea		+
maleic hydrazide	(4)	6-hydroxy-3(2H)-pyridazinone		+
methyl parathion	(3)	O,O-dimethyl O-p-nitrophenyl phosphorothioate	+	
'Molipan'	(4)	(monolinuron & linuron)		+
monolinuron	(4)	3-(4-chlorophenyl)-1-methoxy-1-methylurea		+
parathion ethyl	(3)	O,O-diethyl O-p-nitrophenyl phosphorothioate	−	
'Patoran'	(4)	3-(p-bromophenyl)-1-methyl-1-methoxyurea		+
phenitrothion	(3)	see fenitrothion	+	
phorate	(3)	O,O-diethyl-S-[(ethylthio)methyl] phosphorodithioate	+	
picloram‡	(4)	4-amino-3,5,6-trichloropicolinic acid		+
'Plictran'	(4)	tricyclohexylhydroxytin		+
propoxur	(1)	o-isopropoxyphenyl methylcarbamate	+	
schradan§	(3)	bis-N,N,N',N'-tetramethylphosphorodiamidic anhydride	+	
sulfallate	(4)	2-chloroallyl N,N-diethyldithiocarbamate		+
2,4,5-T	(4)	2,4,5-trichlorophenoxyacetic acid		+
TDE	(1)	1,1-dichloro-2,2-bis(p-chlorophenyl)ethane	+	
tetradifon	(4)	p-chlorophenyl 2,4,5-trichlorophenyl sulfone		+
thioazine	(3)	O,O-diethyl O-2-pyrazinyl phosphorothioate	+	+
Tordon	(4)	see picloram	+	
toxaphene	(1)	chlorinated camphene containing 67-69% chlorine	+	
trichloronate	(3)	O-ethyl O-2,4,4-trichlorophenyl ethylphosphonothioate	+	
trifluralin	(4)	-trifluoro-2,6-dinitro-N,N-dipropyl-p-toluidine		+
'Zectran'	(1)	4-dimethylamino-3,5-xylyl methylcarbamate	+	

* Pesticides are categorised by their use and chemical structure as follows: insecticides, organocarbamates (1), organo-chlorines (2), organophosphorus (3) and others (4).

† Chemical names of insecticides are in accordance with Kenaga and Allison (1969), herbicides and fungicides from Spencer (1968).

‡ From Menzie (1969).   § From Spencer (1968).

REFERENCES

D. C. Abbott, S. Crisp, K. R. Tarrant and J. O'G. Tatton, *Pestic. Sci.*, 1, 10 (1970)

Agricultural Economic Report No. 131, Quantities of Pesticides Used by Farmers in 1964. Economic Research Service, U.S.D.A. Washington (1968)

J. F. Ahrens and J. B. Kring, *J. Econ. Entomol.*, 61, 1540 (1968)

A. A. Akhavein and D. L. Linscott, *Residue Rev.*, 23, 97 (1968)

W. E. Allison, A. E. Doty, J. L. Hardy, E. E. Kenaga and W. K. Whitney, *J. Econ. Entomol.*, 61, 1254 (1968)

A. Andersson, *Grundförbättring*, 20, 95 (1967)

J. Appert, *Docum. Inst. Rech. Agron. Madagascar*, 113, (1967)

T. E. Archer, *J. Dairy Sci.*, 52, 1806 (1969)

E. A. Baker and M. J. Bukovac, *Ann. Appl. Biol.*, 67, 243 (1971)

D. L. Ballee, G. E. Gould and J. E. Fahey, *J. Econ. Entomol.*, 63, 1658 (1970)

E. Basler, F. W. Slife and J. W. Long, *Weed Sci.*, 18, 396 (1970)

E. W. Beck, L. H. Dawsey, D. W. Woodham and D. B. Leuck, *J. Econ. Entomol.*, 59, 78 (1966)

G. B. Beestman, D. R. Keeney and G. Chesters, *Agron. J.*, 61, 247 (1969)

H. V. Brewerton and R. C. Close, *N.Z. J. Agric. Res.*, 10, 272 (1967)

F. Bro-Ramussen, K. Boldum-Clausen, J. Jørgensen and T. Thygesen, *Tidsskrift for Planteavl.*, 70, 232 (1966)

W. N. Bruce and G. C. Decker, *J. Agr. Food Chem.*, 14, 395 (1966)

W. N. Bruce, G. C. Decker and W. H. Luckmann, *J. Econ. Entomol.*, 60, 707 (1967)

P. E. Burt, R. Bardner and P. Etheridge, *Ann. Appl. Biol.*, 56, 411 (1965)

L. I. Butler, B. J. Landis and L. M. McDonough, Circular #522, Washington Agr. Exp. Sta., College of Agr., Wash. State U. (1970)

R. L. Caldwell, *Phytopath. News*, 5(4), 2 (1971)

R. Carson, *Silent Spring* (Houghton Mifflin Co., 1962)

D. Chisholm and H. B. Specht, *Can. J. Plant. Sci.*, 47, 175 (1967)

N. M. Chopra and N. B. Osborne, *Analytical Chem.*, 43, 849 (1971)

M. A. Clor, A. S. Crafts and S. Yamaguchi, *Plant Physiol.*, 37, 609 (1962)

C. A. Collingwood, *Ann. Appl. Biol.*, 55, 335 (1965)

Compendium on Registered Uses of Pesticides in Canada, Cat. A41-16 Queen's Printer, Ottawa (1968)

R. Corbaz, A. Artho, P. Ceschini, M. Haeusermann and J. C. Plantefeve, *Beitr. Tabakforsch.*, 5, 80 (1969)

P. E. Corneliussen, *Pesticides Monit. J.*, 4, 89 (1970)

A. S. Crafts, *Pure Appl. Chem.*, 21, 295 (1970)

A. S. Crafts and C. L. Foy, *Residue Rev.*, 1, 112 (1962)

H. B. Currier and C. D. Dybing, *Weeds*, 7, 195 (1959)

H. W. Dorough and N. M. Randolph, *Bull. Environ. Contam. Toxic.*, 2, 340 (1967)

H. W. Dorough, N. M. Randolph and G. L. Teetes, Texas Agr. Expt. Sta. Bull. MP-936 (1969)

H. W. Dorough, N. M. Randolph and G. H. Wimbish, *Bull. Environ. Contam. Toxic.*, 1, 46 (1966)

R. E. Duggan, *Pesticides Monit. J.*, 1, 3 (1968)

W. Ebeling, *Residue Rev.*, **3**, 35 (1963)

D. O. Eberle and D. Novak, *J.A.O.A.C.*, **52**, 1067 (1969)

E. R. Elkins, F. C. Lamb, R. P. Farrow, R. W. Cook, M. Kawai and J. R. Kimball, *J. Agr. Food Chem.*, **16** 962 (1968)

P. Etheridge and I. J. Graham-Bryce, *Ann. Appl. Biol.*, **65**, 15 (1970)

J. E. Fahey, G. E. Gould and P. E. Nelson, *J. Agr. Food Chem.*, **17**, 1204 (1969)

J. E. Fahey, P. E. Nelson and D. L. Ballee, *J. Agr. Food Chem.*, **18**, 266 (1970)

R. P. Farrow, F. C. Lamb, R. W. Cook, J. R. Kimball and E. R. Elkins, *J. Agr. Food Chem.*, **16**, 65 (1968)

R. P. Farrow, F. C. Lamb, E. R. Elkins, R. W. Cook, M. Kawai and A. Cortes, *J. Agr. Food Chem.*, **17**, 75 (1969)

D. G. Finlayson and H. R. MacCarthy, *Residue Rev.*, **9**, 114 (1965)

D. G. Finlayson, I. H. Williams and H. G. Fulton, *J. Econ. Entomol.*, **61**, 1174 (1968)

D. G. Finlayson, H. G. Fulton, R. Kore and I. H. Williams, *J. Econ. Entomol.*, **63**, 1304 (1970)

W. Franke, *Ann. Rev. Plant Physiol.*, **18**, 281 (1967)

W. Franke, *Pestic. Sci.*, **1**, 164 (1970)

A. Frey-Wyssling, Sub-microscopic Morphology of Protoplasm and its derivatives. Elsevier Publ. Co. Inc., New York (1948)

N. Gannon and J. H. Biggar, *J. Econ. Entomol.*, **51**, 1 (1958a)

N. Gannon and J. H. Biggar, *J. Econ. Entomol.*, **51**, 3 (1958b)

N. Gannon and J. H. Biggar, *J. Econ. Entomol.*, **51**, 8 (1958c)

D. A. George, L. I. Butler, J. C. Maitlen, H. W. Rusk and K. C. Walker, *U.S.D.A., A.R.S.*, **33-112**, (1967a)

D. A. George, H. W. Rusk, D. M. Powell and B. J. Landis, *J. Econ. Entomol.*, **60**, 82 (1967b)

M. E. Getzendaner, J. L. Herman and B. Van Giessen, *J. Agr. Food Chem.*, **17**, 1251 (1969)

R. F. Glasser, R. G. Blenk, J. E. Dewey, B. D. Hilton and M. H. J. Weiden, *J. Econ. Entomol.*, **51**, 337 (1958)

I. J. Graham-Bryce and P. Etheridge, Proc. 4th Brit. Insecticide and Fungicide Conf., 336 (1967)

H. D. Gunner, B. M. Zukerman, R. W. Walker, C. W. Miller, K. W. Deubert and R. E. Longley, *Plant Soil*, **25**, 249 (1966)

F. A. Gunther, *Residue Rev.*, **28**, 1 (1969)

F. A. Gunther, *Pure Appl. Chem.*, **21**, 355 (1970)

F. A. Gunther and R. C. Blinn, Analysis of Insecticides and Acarcides, p. 76. Interscience, New York and London (1955)

F. A. Gunther, W. E. Westlake and P. S. Jaglan, *Residue Rev.*, **20**, 1 (1968)

F. E. Guthrie and T. J. Sheets, *Tob. Sci.*, **xiv**, 44 (1970)

C. R. Harris, *Proc. Entomol. Soc. Ontario*, **100**, 14 (1969)

C. R. Harris and W. W. Sans, *J. Agr. Food Chem.*, **15**, 861 (1967)

C. R. Harris and W. W. Sans, *Proc. Entomol. Soc. Ontario*, **100**, 156 (1969a)

C. R. Harris and W. W. Sans, *Pesticides Monit. J.*, **3**, 182 (1969b)

J. Hartisch, H. Beitz and E. Heinisch, *Nachrichtenbl. Deut. Pflanzenschutzdienst.* (Berlin) **NF 23**, 101 (1969)

84   D. G. Finlayson and H. R. MacCarthy

E. Heinisch, H. Beitz and J. Hartisch, *Nachrichtenbl. Deut. Pflanzenschutz-dienst.* (Berlin) **NF 22**, 61 (1968)
H. P. Hermanson, L. D. Anderson and F. A. Gunther, *J. Econ. Entomol.,* **63**, 1651 (1970)
T. L. Hopkins, *J. Econ. Entomol.,* **60**, 1167 (1967)
H. M. Hull, *Residue Rev.,* **31**, 1 (1970)
H. Hulpke, *Qual. Plant. Mater. Veg.,* **xviii**, 331 (1969)
H. Hulpke, *Qual. Plant. Mater. Veg.,* **xix**, 333 (1970)
H. Hulpke and W. Schuphan, *Qual. Plant. Mater. Veg.,* **xix**, 347 (1970)
W. R. Ingram, *E. Afr. Agric. For. J.,* **33**, 206 (1967a)
W. R. Ingram, *Cott. Grow. Rev.,* **44**, 203 (1967b)
T. Ishiguro and T. Saito, *Botyu-Kagaku,* **35**, 1 (1970)
G. W. Ivie and J. E. Casida, *Science,* **167**, 1620 (1970)
L. W. Jacobs, D. R. Keeney and L. M. Walsh, *Agron. J.,* **62**, 588 (1970)
L. R. Jeppson and F. A. Gunther, *Residue Rev.,* **33**, 101 (1970)
L. R. Johnson and A. E. Hiltbold, *Soil Sci. Soc. Amer. Proc.,* **33**, 279 (1969)
B. G. Joshi and G. Ramaprased, *Pesticides,* **3(9)**, 17 (1969)
E. E. Kenaga and W. E. Allison, *Bull. Entomol. Soc. Amer.,* **15**, 85 (1969)
L. Kilgore and F. Windham, *J. Agr. Food Chem.,* **18**, 162 (1970)
F. C. Lamb, R. P. Farrow, E. R. Elkins, J. R. Kimball and R. W. Cook, *J. Agr. Food Chem.,* **16**, 967 (1968)
B. M. Lazarevic, *J. Econ. Entomol.,* **63**, 629 (1970)
E. P. Lichtenstein, T. W. Fuhremann and K. R. Schulz, *J. Agr. Food Chem.,* **16**, 348 (1968)
E. P. Lichtenstein, G. R. Myrdal and K. R. Schulz, *J. Agr. Food Chem.,* **13**, 126 (1965a)
E. P. Lichtenstein, K. R. Schulz, R. F. Skrentny and P. A. Stitt, *J. Econ. Entomol.,* **58**, 742 (1965b)
E. P. Lichtenstein, K. R. Schulz, T. W. Fuhremann and T. T. Liang, *J. Agr. Food Chem.,* **18**, 100 (1970)
D. A. Lindquist, J. Hacskaylo and T. B. Davich, *Bot. Gaz.,* **123**, 137 (1961)
H. F. Linskens, W. Heinen and A. L. Stoffers, *Residue Rev.,* **8**, 136 (1965)
K. A. Lord, M. A. May, and J. H. Stevenson, *Ann. Appl. Biol.,* **61**, 19 (1968)
J. C. Maitlen, L. M. McDonough, F. Dean, B. A. Butt and B. J. Landis, *U.S.D.A., A.R.S.,* **33-135** (1970)
G. S. Mann and S. L. Chopra, *Pesticides Monit. J.,* **2**, 163 (1969)
H. Martin, The Scientific Principles of Crop Protection, p. 202. Edward Arnold (Publishers) Ltd., London (1959)
Y. Maruyama, K. Komiya and T. Manri, *Radioisotopes,* **19**, 250 (1970)
M. E. A. Materu and D. Hopkinson, *E. Afr. Agric. For. J.,* **35**, 78 (1969)
R. E. Menzer and L. P. Ditman, *J. Econ. Entomol.,* **61**, 225 (1968)
C. M. Menzie, Metabolism of Pesticides. Special Scientific Report-Wildlife No. 127, Washington (1969)
R. L. Metcalf, Proc. 2nd Internat. Plant Prot. Conf., 129 (1957)
J. R. W. Miles, G. F. Manson, W. W. Sans and H. D. Niemczyk, *Can. J. Plant Sci.,* **47**, 187 (1967)

J. W. Mitchell, B. C. Smale and R. L. Metcalf, Advances in Pest Control Research, iii, 359 (1960)

R. S. Moubry, G. R. Myrdal and H. P. Jensen, *Pesticides Monit. J.*, **1**, 13 (1967)

A. N. A. W. Naoum, *Diss. Abstr.*, **25**, 6860 (1965)

R. G. Nash, *Weed Sci.*, **18**, 74 (1968a)

R. G. Nash, *Agron. J.*, **60**, 177 (1968b)

R. G. Nash and M. L. Beall, Jr., *Science*, **168**, 1109 (1970)

W. A. Niering, *BioScience*, **18** 869 (1968)

J. A. Onsager, H. W. Rusk and L. I. Butler, *J. Econ. Entomol.*, **63**, 1143 (1970)

J. R. Pardue, E. A. Hansen, R. P. Barron and J. T. Chen, *J. Agr. Food Chem.*, **18**, 405 (1970)

K. S. Park and W. N. Bruce, *J. Econ. Entomol.*, **61**, 770 (1968)

O. Park and D. F. Heath, 111e Congrès Intern. Phytopharmacie, 742 (1952)

P. Popov and L. Donev, *Acta Agron. Acad. Sci. Hung.*, **19**, 89 (1970)

P. Purokoski, Investigations on Pesticide Residues 1963-1966. *Publ. State Inst. Agric. Chem.*, **1**, Helsinki (1967)

P. Purokoski, Investigations on Pesticide Residues 1968. *Publ. State Inst. Agric. Chem.*, **4**, Helsinki (1969)

P. Purokoski, Investigations on Pesticide Residues 1969. *Publ. State Inst. Agric. Chem.*, **5**, Helsinki (1970)

J. K. Reed and L. E. Priester, *Pesticides Monit. J.*, **3**, 87 (1969)

W. Reed, *Agric. Vet. Chem.*, **8**, 8 (1967)

C. P. P. Reid and W. Hurtt, *Nature*, **225**, 29 (1970)

H. T. Reynolds, Advances in Pest Control Research, ii, 135 (1958)

H. Reynolds, G. L. Gilpin, and I. Hornstein, *J. Agric. Food Chem.*, **1**, 772 (1953)

H. Reynolds, G. L. Gilpin and I. Hornstein, *U.S. Dept. Agr. Circ.*, #952 (1954)

R. L. Ridgway, D. A. Lindquist and D. L. Bull, *J. Econ. Entomol.*, **58**, 349 (1965)

G. B. Riley, D. M. Norris, N. E. Johnson, J. H. Rediske, D. A. Lindquist and H. E. Thompson, *Bull. Entomol. Soc. Amer.*, **11**, 187 (1965)

A. Rippel, J. Kovac and A. Szokolay, *Nahrung*, **14**, 223 (1970)

W. E. Ripper, Advances in Pest Control Research, i, 305 (1957)

I. A. D. Robertson, *E. Afr. Agric. For. J.*, **35**, 181 (1969)

M. M. Robertson and R. C. Kirkwood, *Weed Res.*, **9**, 224 (1969)

M. M. Robertson and R. C. Kirkwood, *Weed Res.*, **10**, 94 (1970)

J. Robinson, *Ann. Rev. Pharmacology*, **10**, 353 (1970)

J. G. Saha and W. W. A. Stewart, *Can. J. Plant Sci.*, **47**, 79 (1967)

J. A. Sargent, *Ann. Rev. Plant Physiol.*, **16**, 1 (1965)

J. M. Scholl and C. S. Williams, *Down to Earth*, **25(3)**, 15 (1969)

Science Council of Canada, Report No. 12, Two Blades of Grass: the Challenge Facing Agriculture. SS22-1970/12, Ottawa (1971).

F. R. Shaw, D. Miller, M. C. Miller and C. P. S. Yadava, *J. Econ. Entomol.*, **62**, 953 (1969)

N. A. Smart, *Residue Rev.*, **23**, 1 (1968)

D. C. Smith, *Pestic. Sci.*, **2**, 42 (1971)

R. F. Smith. R. L. Rabb and F. E. Guthrie (Editors), Concepts of Pest Management, p. 103. Conf. Proc. North Carolina State Univ. (1970)

E. Y. Spencer, Guide to Chemicals used in Crop Protection. Publ. #1093 (Fifth ed.) Cat. No. A43-1093 Can. Dept. Agr. Queen's Printer, Ottawa (1968)

O. Starnes, *J. Econ. Entomol.*, **50**, 338 (1950)

L. J. Stevens, C. W. Collier and D. W. Woodham, *Pesticides Monit. J.*, **4**, 145 (1970)

D. K. R. Stewart and R. G. Ross, *Can. J. Plant Sci.*, **47**, 169 (1967)

E. W. Stoller, *Plant Physiol.*, **46**, 732 (1970)

R. Strufe, *Residue Rev.*, **24**, 79 (1968)

D. L. Suett and G. A. Wheatley, Ann. Report (1969) Nat. Veg. Res. Sta., Wellesbourne, 95 (1970)

G. Tamaki, J. E. Halfhill and J. C. Maitlen, *J. Econ. Entomol.*, **62**, 678 (1969)

W. B. Tappan, C. H. Van Middelem and H. A. Moye, *J. Econ. Entomol.*, **60**, 765 (1967)

K. R. Tarrant and J. O'G. Tatton, *Nature*, **219**, 725 (1968)

H. Tashiro, D. L. Chambers, J. G. Shaw, J. B. Beavers and J. C. Maitlen, *J. Econ. Entomol.*, **62**, 443 (1969)

N. P. Thompson, W. B. Wheeler and A. J. Norden, *J. Agr. Food Chem.*, **18**, 862 (1970)

S. L. Thrower, N. D. Hallam and L. B. Thrower, *Ann. Appl. Biol.*, **55**, 253 (1965)

G. Trombetti and P. Maini, *Pestic. Sci.*, **1**, 144 (1970)

G. Trombetti, P. Maini, B. Caumo and A. Kovacs, *Pestic. Sci.*, **2**, 129 (1971)

J. Van Overbeek, *Ann. Rev. Plant Physiol.*, **7**, 355 (1956)

R. L. Wain. G. A. Carter and D. C. Torgeson (Editors), Fungicides, an Advanced Treatise, vol. 1, p. 561. Academic Press, New York and London (1967)

A. C. Waldron and D. L. Goleman, *J. Agr. Food Chem.*, **17**, 1066 (1969)

K. C. Walker, J. C. Maitlen, J. A. Onsager, D. M. Powell, L. I. Butler, A. E. Goodban and R. M. McCready, *U.S.D.A., A.R.S.*, **33-107** (1965)

G. W. Ware, B. J. Estesen, C. D. Jahn and W. P. Cahill, *Pesticides Monit. J.*, **4**, 21 (1970)

M. Wassermann, D. Wassermann, L. Zellermayer and M. Gon, *Pesticides Monit. J.*, **1(2)**, 15 (1967)

F. L. Watters and O. W. Grussendorf, *J. Econ. Entomol.*, **62**, 1101 (1969)

W. B. Wheeler, D. E. II. Freai, R. O. Mumma, R. H. Hamilton and R. C. Cotner, *J. Agr. Food Chem.*, **15**, 231 (1967)

D. M. Whitacre and G. W. Ware, *J. Agr. Food Chem.*, **15**, 492 (1967)

A. T. S. Wilkinson, D. G. Finlayson and H. V. Morley, *Science*, **143**, 681 (1964)

A. Wolman and A. J. Wilson, Jr., *Pesticides Monit. J.*, **4**, 8 (1970)

Chapter 3

Pesticide Residues in Soil Invertebrates

A. R. Thompson

National Vegetable Research Station,
Wellesbourne, England

3.1 INTRODUCTION

The broad spectrum activity of some pesticides was realised at least 20 years ago (Pickett, 1949) and, since then, many studies have been done with elementary ecological techniques on the effects of pesticides on species of invertebrates against which they were not directed in soil (Edwards and Thompson, 1973). Most described changes in numbers of invertebrates in treated soil but, after awareness of the need for knowledge of the distribution of pesticides in the environment became more acute about ten years ago, data on chemical analyses of residues in invertebrates have been published. Although the term 'pesticide' includes insecticides, herbicides, fungicides, molluscicides, nematicides, rodenticides and other biocides, residue work has been restricted almost entirely to insecticides, with a few studies on herbicides. Research on this aspect of pesticide ecology has not developed as fast as the production of new compounds, and although there are considerable data on residues of the persistent organochlorine insecticides, the uptake of more recent and less persistent organophosphorus and carbamate compounds into invertebrates has scarcely been investigated. The more important papers published on the subject are listed in Table 3.1 and have been grouped by the origin of the samples and the invertebrates that were analysed. The first work was by Barker (1958) in stands of elm trees, but the great majority of studies have included earthworms from arable and orchard soils. There have also been reports on beetles, almost all from arable land, and molluscs.

3.2   RESIDUES OF INSECTICIDES IN SOIL INVERTEBRATES

Many orchards and forests have similar schedules of insecticide treatments applied to them each year. The soil receives minimal or no cultivation and, although residues tend to persist less when not incorporated into soil (Edwards, 1966), foliar treatments rapidly reach the soil surface and establish a reservoir of residues in the surface layers. Leaves of deciduous trees also contribute to this reservoir. For example, a leaf sample taken from those that fell from treated elm trees contained up to 167.5 ppm of DDT residues (Boykins, 1966). The largest residues of DDT in soil have usually been found in orchards, especially in the top few inches (Harris *et al.*, 1966; Harris and Sans, 1969).

Residues of insecticides in arable land are distributed to the depth of cultivation however, and, because fields tend to have different crops including grasses in successive years, insecticide treatments vary greatly. Regular cultivation tends to lessen residues (Lichtenstein *et al.*, 1971) and their profile in arable land is very different from, and more diffuse than, that in orchards and forests (Harris and Sans, 1969).

Because of these differences between the three environments, residues in invertebrates living in them will be considered with reference to the 'crops'. It is, nevertheless, often difficult to compare analyses of samples from similar situations because investigators used different techniques but it is possible to obtain some idea of the amounts of residues present. Firstly, however, it is pertinent to discuss some elementary features of methods involved in this type of work in order to qualify some results that are quoted in this chapter.

Most estimates of residues refer to single samples taken at arbitrary intervals after an insecticide was applied, although some refer to more samples taken over periods of time that sometimes extend to years. Amounts of residues in some soil invertebrates seem to be directly related to residues in the environment (Davis, 1968; Wheatley and Hardman, 1968; Davis, 1971; Edwards and Thompson, 1973), and it is therefore possible only to contrast rather than compare analyses of samples taken soon after soil has been treated, and when residues are almost maximal, with those taken, for example, before an annual treatment is applied, when soil residues are almost minimal. Residues in soil or animal tissues are the result of a dynamic and constantly-changing balance, and results of single analyses taken at an arbitrary time after treatment are apt to distort this concept. The data in Tables 3.2 to 3.8 are therefore of considerable interest, but do not present such a complete picture as experiments which used frequent samples of tissues and soil as discussed later (p. 106) (e.g. Boykins, 1966; Dempster, 1968b; Davis and French, 1969; U.S.D.I., 1969; Edwards *et al.*, 1971).

Before residues are extracted from soil invertebrates for analysis, soil particles must be washed from the outside of the animals to minimise contamination. It is more difficult to remove particles that sometimes contain residues from the alimentary canals of invertebrates that ingest soil, without also disturbing the physiological state of the animals. Some workers have cleared the guts of earthworms by keeping the live animals moist in glass beakers or on damp filter paper in petri dishes for a few days before residues were extracted (e.g. Edwards *et al.*, 1968), and Davis (1971) transferred earthworms to untreated soil for a few days so that the alimentary canal would contain minimal residues. However, such methods involve a delay in analysis during which residues may be broken down, and it would seem preferable to analyse whole animals and allow for residues in the gut, especially as the entire animal, with the soil it contains is the focal point of food-chain studies. Davis (1966a, 1971) calculated that 17% of the live weights of individuals of *Lumbricus terrestris* consisted of the gut contents and also that evacuation of the gut reduced the dieldrin concentration of specimens by up to 20%. Frequently however, mention is not made in published reports of the state of analysed samples and it is probably wise to assume that most analyses presented in later Tables are of individuals plus associated soil, unless otherwise mentioned in the text.

Moriarty (1969) summarised essential features in the detection and estimation of synthetic insecticides and only salient points need to be developed here. It is often inconvenient to extract residues from tissues as soon as animals have been sampled, and deep freeze storage has frequently been used in the interim period. It is now evident that techniques should be carefully assessed before analyses of some insecticide components can be relied upon. Jefferies and Walker (1966) reported that the concentration of the DDT metabolite p,p'-TDE (Fig. 3.1) in avian livers increased with the period of cold storage and concluded that it was formed by the post-mortem reductive dechlorination of p,p'-DDT, possibly by bacteria. French and Jefferies (1969) showed further that, although o,p'-DDT in technical DDT is isomerised to p,p'-DDT and then to p,p'-DDE in living avain tissue, it is metabolised to o,p'-TDE in anaerobic conditions that exist after death. Tables in this paper have been compiled from published data, all of which have been assumed to be valid. Some earlier analyses, especially of breakdown products, should be confirmed before reliance can be placed upon them in view of such more recently published information.

TABLE 3.1. Studies on residues of insecticides in soil invertebrates.

Insecticide	Site			Invertebrates			Reference
	Arable	Orchard	Forest	Earth worms	Beetles	Slugs/snails	
Aldrin	*			*			Davis, 1966b, 1968; Davis and Harrison, 1966
	*			*			Wheatley and Hardman, 1968
	*			*			Gish, 1970
	*			*	*	*	Korschgen, 1970
BHC	*			*			Davis, 1966b, 1968; Davis and Harrison, 1966
	*	*		*			Wheatley and Hardman, 1968
		*				*	Cramp and Conder, 1965
		*				*	Cramp and Olney, 1967
Chlordane	*			*			Gish, 1970
Chlorfenvinphos	*			*			Edwards et al., 1968
DDT				*			Barker, 1958
				*			Hunt, 1965
			*	*			Boykins, 1966
			*	*			Hunt and Sacho, 1969
			*	*			Dimond et al., 1970
			*	*			Doane, 1962
	*			*			Cramp et al., 1964
	*			*		*	U.S.D.I., 1966
	*			*	*		El Sayed et al., 1967
	*			*	*		Wheatley and Hardman, 1968
	*			*	*	*	Davis and French, 1969

	Dieldrin	Endrin	Heptachlor epoxide	Methoxychlor	Reference
					Stringer and Pickard, 1965
					Cramp and Conder, 1965
					Stringer, 1966
					Cramp and Olney, 1967
					Collett and Harrison, 1968
					Pickard, 1968
					Stringer and Pickard, 1969
					Bailey et al., 1970
					Stringer et al., 1970
					Davis, 1966b, 1968; Davis and Harrison, 1966
					Luckman and Decker, 1960
					Cramp et al., 1964
					Raw, 1964
					Cramp and Conder, 1965
					U.S.D.I., 1966
					Wheatley and Hardman, 1968
					Davis and French, 1969
					Gish, 1970
					Korschgen, 1970
					Cramp and Conder, 1965
					U.S.D.I., 1966
					Gish, 1970
					DeWitt and George, 1960
					Smith and Glasgow, 1965
					Gish, 1970
					Hunt and Sacho, 1969

The use of gas-liquid chromatography enabled smaller residues to be detected more accurately, and colorimetric methods have gradually been phased out (see Chapter 13). Thin-layer chromatography remains a reliable method of confirming some residues, e.g. TDE. Within the last five years however, after polychlorinated biphenyls (PCB's) were reported in wildlife from Sweden (Anon, 1966), chemists have been concerned about possible interference between PCB's and organochlorine insecticides in gas-liquid chromatography. PCB's are synthetic materials which, although similar in structure and properties to the DDT insecticide group, are not used as insecticides. If present in samples, they are carried through the usual pesticide extraction and screening procedures, and because they possess electron-absorbing properties, will interfere with GLC electron capture analysis of organochlorine insecticides (Reynolds, 1969) (Chapter 13). PCB's were detected in an eagle captured in 1944 (Anon,

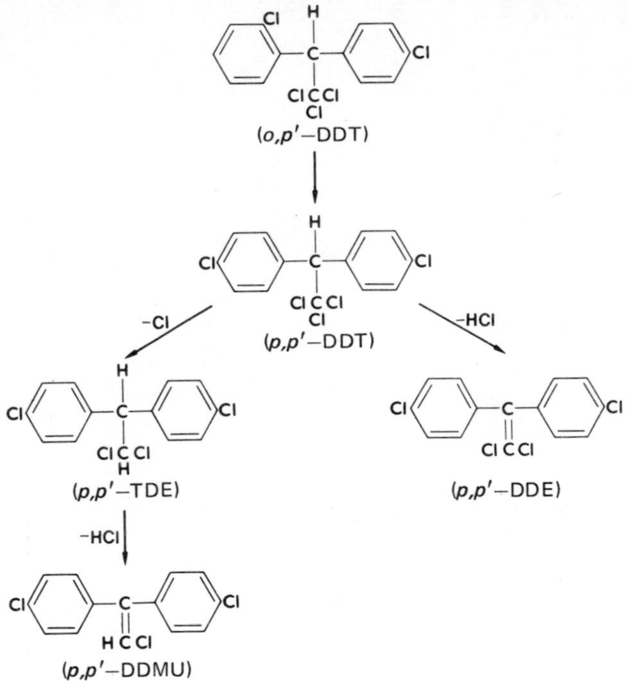

p,p'–DDT	:1,1,1–trichloro–2,2–bis(p–chlorophenyl)ethane
o,p'–DDT	:1,1,1–trichloro–2–(o–chlorophenyl)–2–(p–chlorophenyl)ethane
p,p'–DDE	:1,1–dichloro–2,2–bis(p–chlorophenyl)ethylene
p,p'–TDE	:1,1–dichloro–2,2–bis(p–chlorophenyl)ethane
p,p'–DDMU	:1–chloro–2,2–bis(p–chlorophenyl)ethylene

FIG. 3.1.   Suggested pathways of DDT metabolism in some soil invertebrates.

1966) and it is possible that they have, unknowingly, been present in samples of soil invertebrates and have been mis-identified. There have not yet been any positive identifications of PCB's in soil invertebrates and Davis (1968) stated specifically that no residues of PCB's were present in any of his soil or animal samples, but most authors have not commented on the possibility of their occurrence. Further work should be done to determine whether PCB's, which are known to be toxic (Holmes et al., 1967), occur in significant amounts in invertebrates in soil.

Accuracy of published analyses is related, not only to the correct identification of residues, but also to the expression of the amounts present. Although some papers mention the efficiency of methods of extraction and detection, i.e. the percentage recovery of treated controls (e.g. Stringer and Pickard, 1964, 1965; Edwards et al., 1968; Wheatley and Hardman, 1968; Gish, 1970), many do not, and this deficiency complicates the assessment of residues actually present in samples. Because many studies included analyses of only a few samples collected arbitrarily, it has also not been possible to refer to a range of residue values, or the standard errors of mean amounts. It is therefore not improbable, that some analyses quoted are extreme examples and that the general level of residues may be different from that suggested by atypical samples.

Residues are generally expressed in terms of either the original fresh/ live weight or an air/oven dry weight of the sample. To overcome great variability in moisture content at any one site, residues in soils are usually given as a proportion of the dry weight of the soil (e.g. Stringer and Pickard, 1964, 1965; Wheatley and Hardman, 1968; Davis and French, 1969; Hunt and Sacho, 1969; Dimond et al., 1970; Gish, 1970). Moisture content of insects and other soil invertebrates can be expected to be considerably more uniform than that of soils, and this is presumably one reason why residues in them have frequently been expressed in terms of the fresh weight of individuals (e.g. Barker, 1958; Stringer and Pickard, 1964, 1965; Davis, 1968; Wheatley and Hardman, 1968; Hunt and Sacho, 1969; Dimond et al., 1970; Korschgen, 1970). Also, it is difficult both to dry, and analyse residues in, single specimens of animals such as earthworms, without destroying or otherwise altering residues that may be present (Davis, 1971). Estimates of moisture content of independent samples vary with the condition of the gut contents (Davis, 1971) and it has therefore often been the practice to base analyses on the fresh weight of the samples. Gish (1970) considered however, that the dry weight was less variable than the fresh weight and therefore less subject to measuring errors under the different circumstances in which samples were obtained. This argument is favoured by the fact that, although earthworms normally contain between 80 and 90% of water in their tissues, they can survive losing up to 70% with no ill effects (Roots, 1956). Such possible variations could be overcome if residues were standardised on a dry weight basis in animals as well as in soils.

Tables 3.2-3.8 include storage ratios of insecticides, i.e. the ratio of

TABLE 3.2.   Residues (ppm) of insecticides in earthworms.
(i) DDT and metabolites

Species	Residues as ppm fresh (F) or dry (D) wt.	p,p' DDT	o,p' DDT	p,p' TDE	p,p' DDE	Total DDT
Lumbricis terrestris	F	33.0			24.0	57.0
L. rubellus	,,	145.0			59.0	204.0
Octolasium lacteum	,,	82.0			22.0	104.0
Helodrilus zeteki	,,	164.0			33.0	197.0
H. caliginosa	,,	39.0			14.0	53.0
Earthworms						43.0
Earthworms		0.1			0.2	0.3
L. terrestris	Worm: F: soil: D					7.7
Other spp.	,, ,,					8.1
L. terrestris	D ,,					140.6
Earthworms				0.91	0.26	1.17
,,				2.4	2.3	4.7
,,	F					67.7
,,	,,					63.6
L. terrestris	,,					62.9
Allolobophora caliginosa	,,					64.8
Earthworms	F	21.1	1.3	2.2	4.3	28.9
,,	,,	2.1	0.2	0.2	0.9	3.4
,,	Worm: F: soil: D	6.1		3.8	1.2	
,,	,, ,,	1.6		0.2	0.5	
,,	,, ,,	1.8		0.5	0.4	
L. terrestris	Worm: F: soil: D	7.8	10.0		11.0	28.8
A. caliginosa	Worm: F: soil: D	0.54	0.07		0.49	1.10
A. longa	,, ,,	1.5	0.35		0.65	2.50
A. chlorotica	,, ,,	0.77	0.19		0.38	1.34
A. rosea	,, ,,	2.9	0.72		1.0	4.62
O. cyaneum	,, ,,	1.6	0.30		0.70	2.60
L. terrestris	,, ,,	0.67	0.19		0.38	1.24
	F	10.5				17.4
,,	Worm: F: soil: D	20.1	3.6	1.25	1.45	26.40
,,	Worm: F: soil: D	9.5		5.0	10.0	24.5
,,	,, ,,	11.3		2.5	4.8	18.6
Lumbricus sp.	,, ,,	16.2		2.8	3.9	22.9
A. caliginosa	Worm: F: soil: D	28.0		4.3	6.3	38.6
L. terrestris	,, ,,	4.1		1.1	2.7	7.9
Lumbricus sp.	,, ,,	10.2		1.6	3.3	15.1
A. caliginosa	,, ,,	18.0		2.8	4.7	25.5
Earthworms		5.30		1.59	1.01	7.90
,,		17.10	2.47		4.62	24.19
,,	Worm: F: soil: D				0.01	0.10
,,	,, ,,				0.03	0.16
,,	,, ,,				0.03	0.13
,,	,, ,,				0.07	0.32
,,	D		6.28	3.57	2.66	12.30
,,	F					0.13
,,	Worm: F: soil: D	6.92		3.97		11.82

p,p' DDT	o,p' DDT	p,p' TDE	p,p' DDE	Total DDT	Storage[1] ratio	Reference
		Residues in soil				
54.9			28.3	83.2	0.69	Barker, 1958
"			"	"	2.45	"
"			"	"	1.25	"
"			"	"	2.36	"
"			"	"	0.64	"
				12.5[2]	0.34	Doane, 1962
						Cramp et al., 1964
				11.4	0.7	Stringer and Pickard, 1964
					0.7	
				"	14.20	Hunt, 1965
				9.9		Cramp and Conder, 1965
						"
			90.1		0.75	Boykins, 1966
			50.1		1.27	"
			31.0		2.03	"
					2.09	"
7.4	0.7	0.5	1.2	9.8"	2.95	Davis, 1966b, 1968; Davis and
0.3	0.04	0.03	0.1	0.47	7.24	Harrison, 1966
	3.0	0.1	0.2			Cramp and Olney, 1967
1.1		0.2	0.2			"
	0.6	0.2	0.3			"
15.4		2.3	4.5	22.2	1.30	Collett and Harrison, 1968
0.63	0.14		0.17	0.94	1.17	Wheatley and Hardman, 1968
"	"		"	"	2.66	"
"	"		"	"	1.43	"
"	"		"	"	4.92	"
"	"		"	"	2.77	"
"	"		"	"	1.32	"
						Pickard, 1968
						Davis and French, 1969
	5.4	0.5	2.5	8.4	2.92	Hunt and Sacho, 1969
	6.3	0.9	2.0	9.2	2.02	"
	"	"	"	"	2.49	"
	"	"	"	"	4.20	"
	8.6	1.0	2.5	12.1	0.65	"
	"	"	"	"	1.25	"
	"	"		"	2.10	"
	1.0		6.0	7.0	1.13	Bailey et al., 1970
	6.0		6.0	12.0	2.02	"
			0.06	0.87	0.11	Dimond et al., 1970
			0.12	1.55	0.10	"
			0.12	1.20	0.11	"
			0.16	1.30	0.24	"
	0.59	0.43	0.36	1.36	9.3	Gish, 1970
				0.06	2.17	Korschgen, 1970
2.4			0.62	3.74	3.16	Stringer, et al., 1970

[1] Storage ratio = $\dfrac{\text{ppm residues in earthworm}}{\text{ppm residues in soil}}$

[2] Calculated from initial application rate.

residues in animals to the amount in the soil, to demonstrate that, in some instances, residues in animals are larger than in their immediate environment. It is arbitrary whether fresh or dry weights are used for the calculation of these ratios, but the ratio does of course become far larger for earthworms and molluscs when dry weights are used, because of the relatively large water content of these invertebrates compared with soil. When ratios are compared, it is therefore necessary to consider also the basis for their calculation.

Analyses given in Tables 3.2-3.8 are either those in original papers or means calculated from large numbers of original analyses. In almost every instance, they are estimates of residues in live individuals that were apparently healthy when captured.

3.2.1 Earthworms
These invertebrates help to establish and maintain the structure and fertility of the soil (Satchell, 1967; Edwards *et al.*, 1970) and, although they may be of comparatively little value in arable land, there is little doubt of their importance in orchards, forests and grassland where the soil receives minimal cultivation. Edwards and Thompson (1973) reviewed the considerable amount of data that has been published on the effects of insecticides on populations of earthworms and concluded that normal agricultural dosages of most insecticides do not kill many species. Some studies included details of sublethal residues of insecticides in individuals in treated soil, and Tables 3.2 to 3.5 give analyses of organochlorine insecticides that have been found in healthy earthworms from outdoor soils. Chlorfenvinphos, the only organophosphorus compound for which there is published data, is mentioned in Table 3.5.

3.2.1.1 *Forests*
DDT was used extensively in the U.S.A. to control insect vectors of Dutch elm disease and phloem necrosis, as well as other pests, and large numbers of birds were found dying in areas that had been sprayed. Barker (1958) analysed soil samples and robins from treated areas in Illinois and estimated residues of DDT and its breakdown products in earthworms sampled from soil under sprayed trees. Residues in many individuals of five species of earthworms were larger than in the surrounding soil (Table 3.2) and the author concluded that less than 100 earthworms could accumulate as much as 3 mg of DDT. He also studied the distribution of DDT in tissues of *Lumbricus terrestris* and found that, although most were in the crop and gizzard, the insecticide was also in the entire body wall of each animal. Large concentrations (92 ppm of the fresh weight) were detected in the clitellum, presumably in glandular tissue associated with the reproductive organs.

Hunt (1965) also worked on the problems of pesticide kinetics in control of Dutch elm disease, and noted certain relationships based on residues of DDT, DDE, TDE and methoxychlor in samples from Wisconsin communities. Total residues in earthworms were eight times greater than those in the surrounding soil and the three dominant groups of earthworms (adult *L. terrestris,* immature *Lumbricus* sp. and adult *Allolobophora caliginosa*) contained comparable total residues. Among elm trees sprayed with DDT, total pesticide residues (dry weight) were 9.9 ppm in soil and 140.6 ppm in earthworms (Table 3.2), giving a storage ratio of 14.2.

Although methoxychlor seemed to control Dutch elm disease as well as DDT when applied to dormant trees in the Spring (Wootten, 1962; Hafstad, *et al.*, 1966), little was used in the control programmes. However, Hunt and Sacho (1969) compared the relationships between earthworms and both DDT and methoxychlor. They collected earthworms from the Madison campus of the University of Wisconsin, where the two insecticides had been applied, and found that methoxychlor had been stored by the three taxa studied, up to a maximum of 15.8 ppm for *Lumbricus* spp. (Table 3.5). Residues of DDT, TDE and DDE were found in earthworms and soil (Table 3.2) and there were larger concentrations of the two breakdown products in the earthworms than in the soil, even when compared on a fresh weight basis.

The Michigan State University programme for control of Dutch elm disease was accompanied by a conspicuous decrease in numbers of robins on the campus between 1954 and 1957 (Mehner and Wallace, 1959). Boykins (1966) sampled individuals from under trees that had been sprayed with DDT 15 months previously, and showed that residues of DDT were 62.9 ppm in *L. terrestris* and 64.8 ppm in *A. caliginosa* (Table 3.2) when averaged for the Spring of 1964, compared with 31.0 ppm for the soil in the same period. Boykins (1966) also showed that the bark of elm trees contained large residues of DDT immediately after spraying (242.1 ppm) and these, with the DDT in the leaves that fell in the Autumn, would contribute to the residue reservoir in the surface layers of soil.

DDT has also been used extensively in forests to control spruce budworm, and Dimond *et al.* (1970) studied the distribution of the parent material and its metabolites, in the soil and fauna of a treated area (Table 3.2). Different species of worms were not segregated because of the small size of many of the samples. Residues in worms were correlated with residues in the soil, so that all samples from treated areas had larger residues than those from untreated areas, and the largest residues in earthworms were in those from soils treated two or three times. A sample from an area treated twice at 1 lb/acre (1.12 kg/ha) contained 0.636 ppm total DDT residues.

In summary, DDT and methoxychlor have been detected in earthworms sampled in forests, often in concentrations larger than in the surrounding soil. Residues, sometimes more than 200 ppm of the fresh weight of the earthworms (Barker, 1958) and which often exceeded 20 ppm (Hunt, 1965; Boykins, 1966; Hunt and Sacho, 1969), reflected the great extent to which DDT has been used.

3.2.1.3 *Arable*

Residues in worms from arable land are usually smaller than in those from orchards and forests (Davis and Harrison, 1966; Davis 1968), but more work has been done in arable land on the relationship between residues in the tissues of different species of earthworms and the soil in which they live. Raw (1964) analysed earthworms from two fields where aldrinated fertiliser had been applied a few years previously, and concluded that uptake of residues differed between species, and that, although the actual concentrations depended on the species, residues in earthworms were several times greater than in the soil.

Wheatley and Hardman (1968) obtained earthworms from arable soils where the amounts of organochlorine insecticides were known. They found that the six residue components p,p'-DDT, o,p'-DDT, p,p'-DDE, aldrine, dieldrin and γ-BHC did not occur at the same concentrations in all species (Tables 3.2, 3.3, 3.4) and that the largest residues were generally found in *Allolobophora chlorotica,* which had 0.98 ppm (of the fresh weight) aldrin, 4.6 ppm dieldrin, 2.9 ppm p,p'-DDT, 0.72 ppm o,p'-DDT and 1.0 ppm p,p'-DDE at one site. Smaller residues of dieldrin, p,p'-DDT, o,p'-DDT, p,p'-DDE and γ-BHC usually occurred in the larger species (*L. terrestris, A. longa* and *Octolaseum cyaneum)* than in the smaller species *(A. caliginosa, A. chlorotica* and *A. rosea).* Storage ratios were greater for the three smaller species for all insecticides except aldrin and dieldrin, which appeared to be in large concentrations in *O. cyaneum.* Only one sample of this last species was collected however and the result should be confirmed.

Differences between residues in earthworms can also be related to the habits of the species. Whereas small species live and feed mainly in the cultivated layers of soil where residues are usually concentrated in arable land, *L. terrestris* lives in burrows, as has already been described. In orchards, *L. terrestris* and the smaller species tend to accumulate similar amounts of residues when these are concentrated at or near the soil surface. In arable land however, residues are distributed to a greater depth and are smaller at the surface than in orchards. *L. terrestris* therefore accumulates smaller residues in arable land than in forests and orchards. The smaller species however are in more continuous contact with the residues while they move through the top few inches of soil, and they probably ingest soil freely when they feed. Their tissues consequently contain larger concentrations than those of *L. terrestris.*

However, there are situations in arable land where treated vegetation lies close to the surface of the soil and deep-living species accumulate large residues. Davis and French (1969) analysed individuals of *L. terrestris* collected from a pea field soon after the crop was harvested. The earthworms had pulled down and fed on surface vegetation, and had, within 24 days of spraying, accumulated total DDT residues of about 26 ppm (Table 3.2). Similar results were obtained by Doane (1962), who reported that four months after DDT was applied to grass plots at 2.8 lb a.i./acre (3.1

kg/ha), samples of *L. terrestris* and *A. caliginosa* contained 21 ppm DDT, and those taken from plots treated at 25 lb/acre (28 kg/ha) had 43 ppm (Table 3.2).

Wheatley and Hardman (1968) showed that residues were intermediate between the two extremes, in individuals of *A. longa* and *O. cyaneum* from arable land. The former species lives mainly in the upper twelve inches of soil, and the latter has well-defined burrows about 6 inches below the surface.

Results published by Davis (1971) indicate that species accumulate different amounts of residues according to the substrate to which the insecticide is applied. Thus *A. caliginosa*, which probably ingests more soil than *L. terrestris*, accumulated more dieldrin than *L. terrestris* when both species were kept in treated soil. When leaves treated with DDT were placed in untreated soil however, *L. terrestris*, which normally drags leaves down into the burrows, accumulated more DDT than did *A. caliginosa*.

Wheatley and Hardman (1968) demonstrated minor differences in the uptake of parent insecticides and their breakdown products. Thus, for all the species except possibly *L. terrestris*, the storage ratios were in the order dieldrin>p,p'-DDE>p,p'-DDT =o,p'DDT=γBHC>aldrin. This relationship is similar to that obtained by Gish (1970) who summarised analyses of earthworms from agricultural land in the U.S.A. and showed that the storage ratios were more or less in the order of persistence of the insecticides, i.e. DDT(10.5)>p,p'-DDE(9.2)>dieldrin (8.3)>p,p'-TDE(7.3) >aldrin(6.3)>endrin(5.6)>heptachlorepoxide(5.5)>γ-chlordane(5.0). Edwards (1966) calculated that the average time for 95% of an agricultural dosage of heptachlor to disappear is 3½ years, and this persistence is correlated with the residues in earthworms. Thus the average heptachlor epoxide content of 32 samples, each containing up to 100 earthworms, taken 6 to 12 months after fields were treated at 2 lb (1.25 kg/ha) heptachlor per acre, was 3 ppm (De Witt and George 1960). The same authors sampled other land in Louisiana that had been treated with heptachlor to eradicate the fire-ant (*Solenopsis saevissima*) and reported that earthworms contained up to 20 ppm of heptachlor epoxide in their tissues six to ten months after the insecticide was applied (Table 3.5). Similarly, Smith and Glasgow (1965) found that 70% of samples of earthworms, taken during another eradication programme, contained heptachlor epoxide, and that amounts varied from a trace to 49 ppm (Table 3.5). One sample with 49 ppm was collected 18 months after the area was treated. Most samples had less than 10 ppm, but three had more than 20 ppm. Residues did not decrease substantially over an 18-month period.

Korschgen (1970) studied residues in food chains in two cornfields that had been treated with aldrin at 1 lb/acre (1.12 kg/ha) for 16 and 15 of the previous 17 years respectively. Combined residues of aldrin+dieldrin (the epoxide of aldrin) averaged 1.49 ppm of the fresh weight of the earthworms (mainly *A. caliginosa*) (Table 3.3). The range of residues was 0.56-5.65 ppm and was about 4.8 times the average soil residues, also

TABLE 3.5. Residues (ppm) of insecticides in earthworms. (iv) Heptachlor epoxide and methoxychlor.
(v) Chlorfenvinphos

Species	Residues as ppm fresh/dry wt.	Insecticide						Reference
		Heptachlor epoxide			Methoxychlor			
		Residues (ppm) in Earthworms	Soil	Storage Ratio	Residues (ppm) in Earthworms	Soil	Storage Ratio	
Earthworms		20.0						DeWitt and George, 1960
Earthworms		3.0						Smith and Glasgow, 1965
L. terrestris	Worm: fresh: soil: dry	Trace-49.0			5.5	3.0	1.83	Hunt and Sacho, 1969
"	"				6.0	1.8	4.24	"
Lumbricus sp.	"				6.9	1.8	4.87	"
A. caliginosa	"				7.0	1.8	4.94	"
L. terrestris	"				5.3	1.1	4.82	"
Lumbricus sp.	"				15.8	1.1	14.36	"
A. caliginosa	"				14.7	1.1	13.37	"
Earthworms	Dry	0.09	0.03	3.0				Gish, 1970

Chlorfenvinphos

Species	Residues as ppm fresh/dry wt.	Residues (ppm) in Earthworms	Soil	Reference
Earthworms	Dry	0.02	<1.7	Edwards et al., 1968

expressed as a proportion of the fresh weight. Dieldrin comprised 95.5% of the 3-year average of residues and aldrin 4.5%. Luckman and Decker (1960) also reported that residues of dieldrin persisted in soil and earthworm tissues. They sampled fields that had been treated five years previously with dieldrin to control Japanese beetle (*Popillia japonica*) and, although the results were rather variable, they found 0.31 ppm dieldrin in the earthworms in 1954 (Table 3.3). One of the characteristics of organochlorine insecticides which has led to their restricted use is their persistence in soil, and it is evident that parent materials and some of their breakdown products, e.g. dieldrin, heptachlor epoxide and DDE, are equally persistent in tissues of earthworms.

Recently, Davis (1971) showed that when individuals of *L. terrestris* were maintained in the laboratory in soil treated with DDT and dieldrin, they accumulated smaller amounts of DDT than of dieldrin. Uptake is probably a species characteristic, but it may also be a preferential process depending on the insecticide.

Residues in earthworms depend not only on the habits of the different species but also on the concentration in the soil through which the animals move. Gish (1970) analysed earthworm samples from 67 fields each planted with one of 14 crops; the earthworms belonged to four genera: *Allolobophora, Lumbricus, Helodrilus* and *Diplocardia*. His data are extensive and are summarised in Tables 3.2 to 3.5 as means of all the available analyses, although these include samples from different crops, including orchards. Earthworms from cottonfields, cornfields and orchards contained the largest residues, and those from pastures the smallest, correlating with insecticide usage on these crops. Total residues of organochlorine insecticides in earthworms were 114.67 ppm of the dry weight of the tissues in a Louisiana cottonfield, 159.43 ppm in an Alabama cottonfield, 89.37 ppm in a Missouri cornfield, 20.91 and 48.52 ppm in two Maryland apple orchards with the average for 15 Louisiana pastures of 0.21 ppm.

The behaviour and availability of insecticides in soils is influenced by many factors, among the most important being the organic matter and clay contents of the soils as these particles provide surfaces on which the insecticides can be adsorbed (Edwards, 1966; Harris, 1972). Gish (1970) determined the organic matter content of each soil in his survey but it did not appear to be correlated with the ability of earthworms to accumulate residues from the soils. Davis (1971) studied the relationship between certain physical properties of the soil and the uptake of dieldrin by *A. caliginosa* under controlled conditions. He kept individuals for four weeks in five different soil types treated with the insecticide, and showed that a multiple regression using the pH and the reciprocal of the percentage organic matter accounted for 95.3% of the variance in the uptake. When only the pH was varied, the results were inconclusive, but it is probable that the physiology of the earthworms was also affected by this change in acidity of the soil. The amount of DDT taken up from the soil by *A.*

caliginosa increased as the amount of organic matter in the soil decreased (Davis, 1971). This relationship would, however, depend on the moisture content of the soil as this influences the adsorption and availability of insecticide.

Several attempts have been made to quantify the relationship between residues in soils and earthworms. Davis (1968) elaborated the data originally published by Davis and Harrison (1966) to show that residues of DDT and of aldrin+dieldrin appeared to be correlated in a similar manner with soil residues to give a linear relationship on a log-log scale. Residues in earthworms were larger than those in soils.

Wheatley and Hardman (1968) discussed the difficulties of assessing storage ratios and proposed that their data showed that the ratios decreased as the soil residue increased. They fitted their own analyses for aldrin+dieldrin, p,p'-DDT+p,p'-DDE and γ-BHC, as well as data published by Stringer and Pickard (1964) and Davis and Harrison (1966), to this curvilinear relationship. When the logarithm of the residues in earthworms was plotted against the log soil residues, the slope of the line was appreciably less than unity, with residues tending to be relatively greater in the earthworms when concentrations in the soil were small, than when larger residues were present. Storage ratios would be expected to change from about 5 to 10 fold when residue concentrations in the soil were 0.001-0.01 ppm, to less than unity when concentrations exceeded about 10 ppm in the soil. However, this relationship now appears to be inconsistent with additional data that have been published. In laboratory studies, Davis (1971) showed that, when individuals of *L. terrestris* were kept in soil treated with DDT, the storage ratio increased as the residues in the soil increased. Data from field and laboratory studies are not necessarily comparable though, especially when field data include samples composed of many species. Possible species differences in the uptake of insecticides have already been discussed.

However, when logarithms of analyses of field specimens published by Gish (1970) were plotted with the data used by Wheatley and Hardman (1968), the slope of the line approximated closely to unity (Edwards and Thompson, 1973). It may, therefore, not be necessary to hypothesise, as Wheatley and Hardman (1968) suggested that earthworms possess some sort of self-regulatory mechanism to control uptake and storage of residues.

These conclusions have been made on the basis of samples taken at arbitrary intervals after soil was treated. Experiments have been done to study the uptake of insecticides into earthworms with samples taken at frequent intervals after treatment. In one study, soil was treated with dieldrin at ½, 2 and 8 lb/acre (0.6, 1.25, 9.0 kg/ha) (U.S.D.I., 1969), and residues in soil slowly decreased from the maximum concentration when the insecticide was applied. Residues in the earthworms increased up to 180 days (6 months) after the soil was treated. On the third day they were 4.6, 9.7 and 14.6 ppm of the fresh weight of the earthworms respectively for the three treatments. After 8 months, they had decreased to only 1.0, 2.4 and 4.7 ppm but the authors gave no reason for this sudden decrease in 2 months which seem to be uncharacteristic of such a persistent compound. Uptake of DDT was studied for less time in an experiment done by Edwards *et al.* (1971) who kept earthworms in soil treated at 1 ppm of the dry weight. Samples were analysed each week and the concentrations of DDT and DDE in the earthworms increased up to the last sample that was taken 9 weeks after the soil was treated. Total residues were then 1 ppm of the fresh weight of the tissues.

There is evidence that earthworms are able to retain residues in their tissues even when they are removed from the source of contamination. Boykins (1966) sprayed individuals of *Helodrilus foetidus* directly with 12% DDT and kept them in peat moss. At various intervals (Table 3.9) earthworms were taken from the then contaminated moss and placed in uncontaminated peat moss. Analyses of weekly samples showed that the concentration of residues decreased steadily but appeared to reach a plateau, at about 87 ppm of the fresh weight of the earthworms, about three months after they were treated. However, Edwards *et al.* (1971) found that when earthworms, that had spent the whole of their lives in soil treated with DDT, were placed in uncontaminated soil, all DDT was excreted in less than 4 weeks, although DDE was still in the worms 9 weeks after they were placed in the untreated soil. Crossley *et al.* (1971) studied the intake and turnover of radioactive cesium by earthworms and reported that when *Octolasium lacteum* fed on tagged leaf litter and were then removed to non-tagged litter, radioactivity was lost in a pattern that has been termed a 'two component retention curve'. They suggested that the first rapid phase was caused by defaecation of unassimilated materials, and that the second and longer component represented biological excretion of assimilated radioactivity. Similar two-component curves were obtained with *L. terrestris* and *Eisenia hortensis*. When individuals of *O. lacteum* were fed on ^{137}Cs-tagged soil, and then removed to untreated soil, turnover of radioactivity occurred in one phase only because the radioactive cesium was bound firmly to the soil and had therefore not been assimilated by the earthworms (Crossley *et al.*, 1971). Similarly, residues of the herbicide paraquat were detected in earthworms collected from treated soil, but the residues were rapidly excreted when the worms were placed in untreated soil (Calderbank, 1968). Presumably the herbicide had been strongly bound to the soil particles.

There is not yet enough evidence to suggest that the mechanisms of uptake and turnover of radioactive materials, herbicides and insecticides in earthworms are similar. Data published by Dindal and Wurzinger (1971), and discussed later in this paper, suggest that a species of snail is able to attain a body burden equilibrium for DDT, and comparative studies on this subject are needed to determine how long residues of organochlorine insecticides can persist in invertebrates in the absence of a source of contamination.

Although organochlorine insecticides have now been largely replaced by organophosphorus and carbamate compounds, little is known of the distribution of sublethal residues of these insecticides in soil invertebrates. They are generally less persistent than organochlorine in soil, and invertebrates are exposed to only very small residues for up to one year after the insecticides are applied. Edwards *et al.* (1968) studied the effects of the organophosphorus insecticide chlorfenvinphos on soil invertebrates, and included analyses of earthworms from field plots that had been treated with the insecticide (Table 3.5). Plots were sampled only 6 months after they had been treated at 4 and 8 lb a.i./acre (4.48 and 8.96 kg/ha) however, and, at that time, earthworms contained no more than 0.02 ppm of chlorfenvinphos. When individuals of both deep-living species (*A. long* and *L. terrestris*) and shallow-living species (*A. caliginosa* and *A. rosea*) were kept in boxes of soil treated with the same insecticide, residues of the parent material did not exceed 0.02 ppm in samples taken after 4, 10 or 21 weeks.

Thompson (1970) compared the effects of nine insecticides, which included organochlorine, organophosphorus and carbamate compounds, on earthworms in pasture, and samples from treated plots were analysed for residues three weeks after the insecticides were applied (Thompson, unpublished data). Residues of DDT and its breakdown products were greatest and totalled about 128 ppm of the fresh weight of the earthworms. Like DDT, Dursban® and trichloronate had little effect on numbers of earthworms in treated plots but individuals contained appreciable residues: 9.66 ppm of Dursban® and 26.33 ppm of trichloronate. Fensulfothion, another organophosphorus insecticide, lessened numbers of earthworms by 78.8% and those that survived the treatment contained 22.77 ppm of total fensulfothion residues. One year after the plots were treated however; analyses indicated that only DDT was present in the tissues at a concentration greater than 0.01 ppm.

To conclude, concentrations of residues tend to be greatest in earthworms from forests and orchards, and smallest in those from arable land. Residues in earthworms increase as soil residues increase, and amounts in earthworms depend on the insecticide and especially on its stability in soil, and also on the species and habit of the earthworm.

3.2.2 Molluscs

Snails and slugs are not uniformly distributed within apparently suitable sites, probably because of various micro-climatological and micro-topographical features of the habitat, of which the presence or absence of moisture, calcium, shelter and food are the most important (Newell, 1967). Their aggregated distribution has limited the number of field experiments done on the effects of insecticides on their numbers (Edwards and Thompson, 1973) but reports in the literature indicate that slugs tend to be no more susceptible to insecticides than earthworms. Many species browse on plant material on or near the surface of the soil and, like those species of earthworms that live in the surface layers, are almost continually exposed to residues of persistent insecticides in treated areas.

Residues in the tissues of slugs are usually similar to, and often larger than, residues in earthworms from the same sites. Thus, Stringer and Pickard (1965) analysed specimens from the apple orchard plot at the Long Ashton Research Station and found that DDT residues in *Agriolimax reticulatus* ranged from 2.0 to 7.4 ppm, and those in *Arion hortensis* were between 2.3 and 10.1 ppm of the fresh weight of the tissues (Table 3.6). Most slugs were under the trees and there were more at the base of the trunks where the grass was longest, and these latter contained larger residues than those under the canopy of each tree. Although residues in these two species were similar, there is evidence that some species may store larger residues than others from the same site. Stringer *et al.* (1970) sampled a commercial orchard where DDT residues in the top soil were 3.74 ppm and found that individuals of *A. reticulatus* had 1.75 ppm of DDT and those of *A. hortensis* 6.53 ppm. Samples of *Milax budapestensis* however, contained 35.06 ppm (Table 3.6). The habits of these species may account for the difference. Whereas *A. reticulatus* and *A. hortensis* tend to feed on a wide variety of plants (Newell, 1967), *M. budapestensis* also ingests soil (Stephenson, 1963) and therefore has greater contact with residues.

As he showed with earthworms, Pickard (1968) demonstrated that residues in slugs fluctuate with the season and the period between treatment and sampling. Total DDT residues in *A. hortensis* and *A. reticulatus* increased from a maximum of 2.79 ppm before the annual spray was applied to the orchard plot, to a maximum of 45 ppm after the DDT had been applied. Stringer (1966) reported that, in the same orchard, DDT residues in the same species of slugs were 5 ppm in the winter and that they slowly increased after the first application of DDT in April. In July, maximum residues were 70 ppm and these then declined to about 5 ppm again in November. Total DDT residues in the grass were about 600 ppm.

Davis and Harrison (1966) obtained samples of slugs at least two months after the final application of organochlorine insecticides to two orchards, and analyses of the specimens, which were active and healthy when captured, showed that residues in the slugs (5.3-23.8 ppm of p,p'-DDT, 1.2-9.9 ppm of p,p'-TDE and 0.4-3.4 ppm of p,p'-DDE) were larger than those in the surrounding soil (1.5-13.3 ppm of p,p'-DDT, 0.3-0.6 ppm of p,p'-TDE and 0.3-2.1 ppm of p,p'-DDE), and were similar to those in earthworms from the same site (Table 3.2).

Slugs from orchards were also analysed by Cramp and Conder (1965) who found 3.3 ppm of DDE, 5.3 ppm of TDE and 0.2 ppm of BHC (Table 3.8) in a sample from an orchard in England. Another sample contained 0.63 ppm DDE and 2.81 ppm TDE (Table 3.6). Cramp and Olney (1967) published data on residues in soil and slugs from an orchard (Table 3.6). DDT residues were between 0.00 and 11.0 ppm in slugs and 0.00-3.30

ppm in the soil. Residues of TDE were as large as 3.18 ppm in slugs six weeks after the orchard was sprayed when residues in the soil were 0.15 ppm. At that time, residues of DDE in slugs were four times those in the soil.

A similar variety of organochlorine insecticides has been found in slugs and earthworms, although most of the published work concerns DDT. Thus, Cramp *et al.* (1964) found 24.7 ppm dieldrin in slugs from a field of strawberries that had been sprayed with aldrin (Table 3.7)

TABLE 3.6.   Residues (ppm) of insecticides in (a) beetles and (b) molluscs. (i) DDT and metabolites

Species	Residues as ppm fresh (F) or dry (D) wt.	p,p' DDT	o,p' DDT	p,p' TDE	p,p' DDE	Total DDT
(a) Beetles						
Various	F	0.1		0.03	1.1	
Calosoma alternous		4.34			2.93	
,,		0.00			0.15	
Harpalus pennsylvanicus		0.10			5.25	
,,		0.00			12.89	
Agonoderus lecontei		9.16			0.00	
,,		0.00			0.70	
Harpalus rufipes	F	0.5		0.10	3.6	4.2
,,	,,				0.4	0.4
Agonum dorsale	,,				0.3	0.3
Feronia melanaria	Worms: F: soil: D	0.99	0.29	<0.01	0.52	
F. madida	,, ,,	4.3	0.93	0.14	0.84	
Nebris brevicollis	,, F ,,	0.41			75.0	
Harpalus spp.	F					0.2
Poecilus chalcites						0.2
Beetle larvae			0.19	0.06	0.44	0.6
(b) Molluscs						
Slugs				5.3	3.3	
,,				2.81	0.63	
Agriolimax reticulatus	Slugs: F: soil: D					6.5
Arion hortensis	,, ,,					6.2
Slugs	D ,,	42.73				
,,	,,	19.75				
Slugs		7.88		2.64	0.46	
,,		1.06		0.64	4.64	
A. reticulatus	F	18.3	1.8	6.9	2.3	29.6
Arion spp.	,,	5.3	0.4	1.2	0.4	7.3
A. hortensis and		1.75				2.7
A. reticulatus	F	35.0				45.0
A. reticulatus	F	60.0	10.0	7.0	5.0	82.0
Slugs	D	16.84		8.29	9.90	35.0
Snails	,,		0.32	0.83	1.06	2.2
,,	,,		0.49	1.85	0.74	3.0
,,	,,		0.28	1.51	0.66	2.4
A. reticulatus	Slugs: F: soil: D					1.7
A. hortensis	,, ,,					6.5
Milax budapestensis	,, ,,					35.0

[1] Total DDT includes 0.4 ppm of p,p'-DDMU

and Cramp and Conder (1965) detected 10.3 ppm of endrin in a sample of earthworms and slugs from a blackcurrant nursery that had been treated regularly with this insecticide (Table 3.8).

In his survey in the U.S.A., Gish (1970) obtained sufficient numbers of slugs from four sites and of snails from two sites for analysis. Average residues for molluscs from these sites are given in Tables 3.6, 3.7 and 3.8. Total residues of organochlorine insecticides averaged 89.0 ppm of the dry weight of the slugs and 3.5 ppm of that of the snails including the shells. Residues in slugs were 3.7 times those in earthworms from the same sites and, at sites that had snails, residues in earthworms were 11.5

Residues in soil					Storage Ratio	Reference
p,p' DDT	o,p' DDT	p,p' TDE	p,p' DDE	Total DDT		
0.37	0.06	0.04	0.11			Davis and Harrison, 1966
						El Sayed et $al.$, 1967
						,,
						,,
						,,
						,,
7.4	0.7	0.5	1.2	9.8	0.43	Davis, 1968
0.2			0.03	0.23	1.74	,,
0.2			0.05	0.25	1.20	,,
						Davis and French, 1969
						,,
			0.06		4.33	Korschgen, 1970
			0.06		4.33	,,
	0.01		0.01	0.02	35.1	Gish, 1970
						Cramp and Conder, 1965
						,,
			20.9[2]			Stringer and Pickard, 1965
			20.9[2]			
						U.S.D.I., 1966
						,,
	2.99	0.11	0.18			Cramp and Olney, 1967
	0.62	0.17	0.23			
7.4	0.7	0.5	1.2	9.8	3.03	Davis, 1968
1.5	0.2	0.3	0.3	2.3	3.17	,,
						Pickard, 1968
						Davis and French, 1969
	2.15	1.89	1.50	5.54	6.35	Gish, 1970
	5.38	5.56	4.36	15.30	0.1	,,
	0.63	1.68	0.63	2.94	1.0	,,
	0.63	1.68	0.63	2.94	0.8	,,
2.4			0.62	3.74	0.47	Stringer et $al.$, 1970
2.4			0.62	3.74	1.75	,,
2.4			0.62	3.74	9.37	,,

Grass residues were 5.3 ppm.

TABLE 3.7. Residues (ppm) of insecticides in (a) insects and (b) molluscs. (i) aldrin and dieldrin

Species	Residues as ppm fresh/dry wt.	Insecticide						Reference
		Aldrin			Dieldrin			
		Residues (ppm) in Animal	Soil	Storage Ratio	Residues (ppm) in Animal	Soil	Storage Ratio	
(a) Insects								
Beetles	Fresh				0.2			Cramp and Conder, 1965
Cutworms	Fresh				0.29			,,
					0.17			,,
Carabid beetles	Fresh				0.01-0.2	0.03-0.7		Davis and Harrison, 1966
Harpalus rufipes	Fresh				0.05	0.15	0.33	Davis, 1968
Agonum dorsale	,,				0.10	0.04	0.25	,,
Gryllus assimilis	Fresh	0.01	0.06	0.17	0.22	0.25	0.88	Korschgen, 1970
Harpalus spp.	,,	0.11	0.06	1.83	0.99	,,	3.96	,,
Poecilus chalcites	,,	0.34	0.06	5.80	9.33	,,	37.33	,,
Beetle larvae					0.60			Gish, 1970
(b) Molluscs								
Slugs					24.7			Cramp et al., 1964
,,					2.8			,,
Slugs	Dry				0.43			U.S.D.I., 1966
,,					0.21			,,
Agriolimax reticulatus	Fresh				0.3	0.04	8.75	Davis, 1968
Slugs	Dry				5.13	0.01	513.0	Gish, 1970

TABLE 3.8. Residues (ppm) of insecticides in molluscs. (ii) Endrin and γ-BHC

		Insecticide						
		Endrin			γ-BHC			
Species	Residues as ppm fresh/dry wt.	Residues (ppm) in Mollusc	Soil	Storage Ratio	Residues (ppm) in Mollusc	Soil	Storage Ratio	Reference
Slugs		10.3			0.2			Cramp and Conder, 1965
Slugs	Dry	1.14						U.S.D.I., 1966
"	"	1.06						
Slugs	Dry	48.74	1.20	40.6				Gish, 1970
Snails	Dry	0.90	1.74	0.52				Gish, 1970

times those of snails. Slugs showed great tolerance to some insecticides and, in a sample from an apple orchard, contained 134.06 ppm of endrin. Gish (1970) also showed that slugs usually contained larger residues of DDE, DDT, TDE and dieldrin than earthworms from the same sites.

Very little work has been done on organophosphorus insecticides. Edwards (unpublished data) studied the uptake of some of these insecticides into soil invertebrates and found that after the surface of the soil had been sprayed with various insecticides at 8 lb a.i./acre (8.96 kg/ha), residues in slugs were extremely large, even after several months. Maximum residues of diazinon and chlorfenvinphos were 162 and 280 ppm respectively of the fresh weight of the samples, and slugs from plots treated with chlorfenvinphos at 20 lb a.i./acre (22.4 kg/ha) contained 450 ppm of the insecticide.

The comparatively small amount of work that has been done suggests that the uptake of insecticides into slugs is similar to that in earthworms and that residues of organochlorine and organophosphorus insecticides can be very large in individuals on treated soil.

3.2.3 Insects

Work on residues of insecticides in soil insects has been confined almost entirely to beetles which predominate on or in the soil as predators, or which are associated with decaying animal or vegetable matter. It is however, difficult to assess the effect of insecticides on carabid and other ground beetles experimentally in field conditions, because either the treated and untreated plots must be very large or there should be physical barriers around them to prevent migration between plots by these active insects. Unfortunately, most of the data that will be considered (Tables 3.6 and 3.7) has been obtained from samples taken in an arbitrary manner and often as part of larger programmes to monitor residues in particular environments.

Individuals of three species of carabid beetles (*Harpalus caliginosus, H. pennsylvanicus* and *Poecilus chalcites*) were taken from two cornfields, in Missouri that had been treated with aldrin for at least 15 years previously, and were analysed for residues of aldrin and dieldrin (Korschgen, 1970) (Table 3.7). The average of the combined aldrin+dieldrin residues for the three years of the study was 1.1 ppm of the fresh weight of the beetles for the species of *Harpalus,* and this was 3.5 times the soil residue. The average residue in *P. chalcites* was however, 9.67 ppm, i.e. 31 times the soil residue and nearly 9 times the residue found in specimens of *Harpalus*. It is probable that different habits of species of carabid beetles, as of earthworms and slugs, account, at least in part, for these differences in residues. Although larvae of *Carabidae* are generally predaceous, some adults, e.g. those of *Harpalus* species, eat seeds and many remain predaceous, e.g. *Poecilus* species. Individuals of *P. chalcites* probably ingested more residues in prey that had already stored residues than *Harpalus* species ingested in plant matter (Korschgen 1970).

El Sayed *et al.* (1967) found no significant differences between residues in three species of carabid beetles although their feeding habits varied. They analysed selected insects and birds found in cotton fields where large amounts of organochlorine insecticides were applied annually (Table 3.6). *p,p'*-DDE and *p,p'*-DDT were found in individuals of *H. pennsylvanicus,* and of the two predaceous species *Calosoma alternous* and *Agonoderus lecontei.* A sample of *H. pennsylvanicus* taken several miles from any agricultural operation contained 12.89 ppm of *p,p'*-DDE, which indicates that either the species moved very considerable distances or that the insecticide was in some way accumulated in large amounts.

Beetles taken from strawberry fields that had been sprayed with aldrin for two years, contained 0.29 ppm of dieldrin (Cramp and Conder, 1965) (Table 3.7), and cutworms and other caterpillars in the same field had 0.17 and 0.06 ppm of dieldrin respectively. Cramp and Conder (1965) also detected 0.2 ppm of dieldrin in beetles from a potato field.

Davis and Harrison (1966) analysed samples of beetles from ten sites where the soil residues of organochlorine insecticides were typical of many arable soils. Samples of beetles from all sites contained both dieldrin and *p,p'*-DDE in amounts up to 0.2 and 2.2 ppm respectively (Tables 3.6 and 3.7). There was no obvious difference between residues in different groups of beetles (mainly carabids). Residues of dieldrin and *p,p'*-DDE were generally larger in earthworms than in beetles from the same site (Davis and Harrison, 1966). Davis (1968) considered that there was an overall relationship between residues in the soil and in *Harpalus rufipes* at these sites, although residues in the beetles, unlike those in earthworms and slugs, seemed to be smaller than those in the soil.

Edwards and Thompson (1973) concluded that predatory beetles were killed more readily than earthworms and slugs by normal dosages of many insecticides, and it seems probable that largest residues would be found very soon after a treatment was applied, before acquired doses became lethal. A sample of *Nebria brevicollis* collected 5-7 days after DDT had been applied at 2 lb a.i./acre (2.24 kg/ha) as a dust to a strawberry field contained 75.4 ppm of *p,p'*-DDT+*p,p'*-DDE, and the immediate effect of the dusting was a great reduction in the number of beetles caught in pitfall traps (Davis and French, 1969).

No data on residues of organophosphorus insecticides in beetles have been published. However, Edwards (unpublished data) applied 8 lb a.i./ acre (8.96 kg/ha) of several organophosphorus insecticides to plots of wheat surrounded by polythene barriers (Edwards and Thompson, 1969) and detected residues in carabid beetles that ranged from 0.01-1.33 ppm of the fresh weight of the sample for chlorfenvinphos, 0.01-0.55 ppm for diazinon and 0.01-0.28 ppm for phorate. These residues differed greatly from those found in slugs in the same plots and illustrate the different abilities of beetles and slugs to store residues.

3.3   METABOLISM OF INSECTICIDES
BY SOIL INVERTEBRATES

Chemical analyses have detected insecticides and breakdown products of parent materials in invertebrates in soil, but their relative proportions are not always the same in animals as in soil. A few experiments have been done on the metabolism of insecticides by soil invertebrates to explain these differences, and there is a considerable amount of circumstantial evidence, much of it from analyses of specimens collected in field conditions, to support the conclusions. The data will now be considered and pathways by which earthworms, slugs and beetles degrade some insecticides will be suggested.

3.3.1   DDT

The principal constituent of technical DDT is p,p'-DDT, and this is several times as toxic to most insects as the o,p' or any other of the possible isomers (Hoskins, 1964). p,p'-DDT is slowly degraded in soils to p,p'-DDE and p,p'-TDE (Fig. 3.1), often by microbes. DDE and other metabolites, excluding p,p'-TDE which is marketed as the insecticide Rhothane®, are not usually lethal but they have undesired effects on some animals (Heath et al., 1969; Ratcliffe, 1970). Under aerobic conditions, DDT is apparently converted to DDE, and under anaerobic conditions to TDE (Stenersen, 1965; Guenzi and Beard, 1967; Ko and Lockwood, 1968; Menzie, 1969). p,p'-DDT is less stable in animal tissues than in soils, and detoxifying enzymes similar to DDT-dehydrochlorinase which dehydrochlorinates p,p'-DDT and p,p'-TDE (O'Brien, 1967) have been found in a number of species (Lipke and Kearns, 1960). The enzymes have not been studied in earthworms however, and the main evidence of DDT metabolism in these animals comes from analyses of individuals, which show that there is usually a greater proportion of DDT metabolites in the animals than in the soil. On the basis of work that has been done with 'higher' animals, it is assumed that earthworms do possess mechanisms for degrading DDT and that it is not simply that they store breakdown products of DDT more than the parent isomers.

Hunt (1965) reported that individuals of Lumbricus terrestris from areas treated with DDT contained more total residues than the soil (Table 3.2) and also that the earthworms had a significantly lower percentage of DDT and a significantly higher percentage of DDE. The percentage of DDT decreased from May to June following spraying in April (Hunt, 1965). Similarly, DDE in earthworms from an apple orchard constituted a larger proportion of the total DDT residues than in soil (Table 3.2) suggesting that p,p'-DDT is broken down to DDE faster in the tissues than in the soil (Stringer et al., 1970). Analyses of other earthworms suggest that o,p'-DDT also is metabolised to DDE in earthworms. Dimond et al.

TABLE 3.9. DDT residues in *Helodrilus foetidus* (Lumbricidae) sprayed with 12% DDT.

Weeks after treatment	ppm DDT (fresh wt.)	Weeks after treatment	ppm DDT (fresh wt.)
0	290	11	92
1[1]	201	12	90
2	185	13	87
3[1]	147	14[1]	88
4	122	15	87
5	126	16	86
6	120	17	88
7	112	18[1]	90
8[1]	107	19	88
9	106	20[1]	87
10	94	21	86

From E. A. Boykins (1966).
[1] Earthworms placed in fresh untreated soil.

(1970) showed that DDE generally constituted about 20% of the total DDT residues in earthworms and less than 12% of those in soils (Table 3.2). Although o,p'-DDT and o,p'-DDE persisted the same time in soil, only traces of o,p'-DDT were detected in earthworms. Experiments have been done which suggest the way in which o,p'-DDT is metabolised to DDE in earthworms. Five individuals of *A. caliginosa* were kept in each of two samples of compost with 5.6 ppm of o,p'-DDT and p,p'-TDE respectively and were then analysed after 15 days (Davis, 1971). Mean residues in worms in the first group were 2.1 ppm o,p'-DDT+0.23 ppm p,p'-DDE. Klein, *et al.* (1964) demonstrated that isomerisation of o,p'-DDT to p,p'-DDT occurs *in vivo* when fed to rats, and in birds this is then metabolised to p,p'-DDE (French and Jefferies, 1969). *A. caliginosa* appears to metabolise o,p'-DDT in a similar manner. Analysis of the second group of worms showed 3.4 ppm of p,p'-TDE and this suggests that TDE is assimilated at a similar rate to o,p'-DDT in this species. DDMU, formed by the dehydro-chlorination of p,p'-TDE (Peterson and Robinson, 1964), was not detected but may not have persisted until the earthworms were analysed (Davis, 1971). Davis (1968) found DDMU in earthworms from orchards but the metabolite was not present in the soil.

Technical DDT contains small amounts of TDE and DDE and both metabolites have been identified in earthworms from treated areas. When exposed to pure DDT isomers, earthworms seem to produce more DDE than TDE. Davis (1971) kept individuals of *L. terrestris* in soil treated with p,p'-DDT at 1 to 64 ppm. The ratio of p,p'-DDE to p,p'-DDT after 4 weeks was 1:5; p,p'-TDE was detected in trace amounts and was about 6% of the DDT. Davis and French (1969) analysed individuals of the same species from a pea field just after the crop was harvested. Total DDT residues were about 26 ppm and most of this was still p,p'-DDT. However, the proportions of DDE and TDE were still larger than in the initial spray deposits (Table 3.10) and showed that some metabolism had occurred.

Comparisons between earthworms and slugs kept in the presence of p,p'-DDT indicated that whereas p,p'-DDE is the main metabolite in earthworms, slugs produce mainly p,p'-TDE (Davis, 1969). Samples of *A. reticulatus* and *A. hortensis* from orchards (Table 3.6) contained larger concentrations of p,p'-TDE than p,p'-DDE (Davis, 1968). Davis (Walker, 1966) applied recrystallised p,p'-DDT to carrot slices which were eaten by snails of the species *Helix aspersa*. These were then killed by freezing to $-11.5°C$ and were held at this temperature for 58 days until analysed. Snails that consumed 18.0 μg p,p'-DDT contained 1.3 μg p,p'-DDT but no p,p'-TDE. Those that consumed 61.8 μg p,p'-DDT contained 4.4 μg p,p'-TDE and 8.8 μg p,p'-DDT; p,p'-DDE was present in all snails analysed but amounts were not given by the author. It was not possible however to determine how much of the p,p'-TDE was formed *post mortem* by bacteria, as described by Jefferies and Walker (1966). When Stringer *et al.* (1970) fed DDT-^{14}C to *L. terrestris* and *A. reticulatus,* it was rapidly broken down to p,p'-DDE and p,p'-TDE. Earthworms produced more DDE than TDE, and slugs produced more TDE.

Davis and French (1969) examined residues in a series of paired samples of *A. reticulatus,* from a strawberry field that had been treated with DDT (Table 3.6). Residues of p,p'-DDT were about 18 ppm before the first DDT was applied in 1967 (Fig. 3.2). These doubled after DDT was applied in May and increased to a maximum of 60 ppm after DDT dust had been applied. A similar increase and later gradual decline was followed at smaller concentrations by o,p'-DDT, p,p'-TDE and p,p'-DDE. Residues of the two metabolites, DDE and TDE, decreased rather more slowly than those of the two isomers of DDT. These results also show that there is generally more TDE than DDE in slugs. Although appreciable amounts of o,p'-DDT were sometimes found, no o,p'-isomers of DDE or TDE were ever detected, nor was any DDMU. Total metabolites in earthworms and slugs seldom exceeded a third of the p,p'-DDT residues (Davis and French,

TABLE 3.10. The proportions (%) of components in DDT spray applied to pea field and in residues in *L. terrestris.*

	p,p'-DDT	o,p'-DDT	p,p'-DDE	p,p'-TDE
Spray	78.0	18.9	1.7	1.5
L. terrestris from treated field	76.25	13.47	5.6	4.7

Calculated from data by B. N. K. Davis and M. C. French (1969).

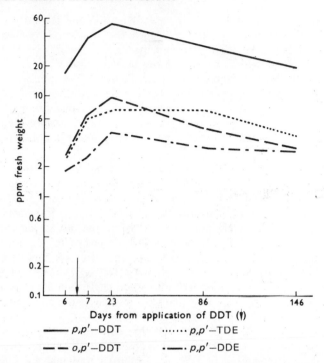

FIG. 3.2.   Seasonal changes in residues in *Agriolimax reticulatus* from a strawberry field treated with DDT. (Adapted from Davis and French, 1969.)

1969). Small amounts of DDMU have been found in *A. reticulatus* and *A. hortensis* by Stringer (1966).

Metabolism includes not only the compounds that are formed but also the rate at which they enter and leave the body.

Dindal and Wurzinger (1971) investigated the distribution of ring-labelled ^{36}Cl, *p,p'*-DDT and metabolities in the terrestrial snail *Cepaea hortensis* after a single feeding of 60 μg DDT. *Cepaea* ate the insecticide readily and most of the amount ingested was soon excreted. Some snails excreted more than half the insecticide ingested with their faeces. Residues absorbed into the body accumulated mostly in the hepatopancreas and ovotestis, so after 8 days the concentrations in these organs were 91.70 and 27.46 ppm of the fresh weight of the organs respectively. The residue load for the whole body decreased to about 11 ppm after 24 hours and remained at this level up to 8 days after treatment of the snails. They did not however, report analyses that could be compared with data on retention of residues by earthworms (Boykins, 1966; Edwards *et al.*, 1971). There were two maxima of residue concentrations in the reproductive organs and the authors suggested that these may have been caused by, firstly, the initial ingestion of the insecticide and the second by a sub-sequent redistribution of residues through the animals. This mechanism would be analogous to that proposed by Crossley *et al.* (1971) for the removal of radioactivity from earthworms in two phases.

It has already been shown that carabid beetles are frequently more susceptible to organochlorine insecticides than earthworms and slugs, possibly because they have a more complex nervous system. Experiments have been done to show that many species of carabids are able to metabolise DDT rapidly however, and in this way are able to survive on treated soil.

Davis and French (1969) studied residues in two common species, *Feronia melanaria* and *F. madida,* in a pea field. Samples contained no detectable residues of DDT components before the area was sprayed (Table 3.11). The largest concentration of DDT found in *F. melanaria* was 4 ppm, in a sample collected 13 hours after the insecticide had been applied at 2.5 lb a.i./acre (2.8 kg/ha). Proportions of the three principal compounds changed during the first 24 days after spraying from 81% *p,p'*-DDT+19% *o,p'*-DDT to 51% *p,p'*-DDT+10% *o,p'*-DDT+39% *p,p'*-DDE indicating metabolism of the DDT isomers to DDE. Samples of *F. madida* had similar residues but showed that large residues persisted in this species for more than a week (Table 3.11).

Metabolism of DDT in *F. melanaria* was also studied in the laboratory (Davis and French, 1969). Beetles were dosed topically with purified *p,p'*-DDT and kept on untreated soil for 24 hours before analysis. They contained 3.2–5.0 ppm of *p,p'*-DDT+0.21–0.36 ppm *p,p'*-DDE but no TDE was detected and only 7% of the *p,p'*-DDT applied had been metabolised to *p,p'*-DDE. Beetles from the field, comparable with a sample which

TABLE 3.11.   Residues of DDT and its metabolites in individuals of *Feronia* spp collected in a pea field sprayed with DDT at 2.5 lb ai/ac.

Species	Days after treatment	Residues (ppm fresh wt.)				
		p,p' DDT	*o,p'* DDT	*p,p'* DDE	*p,p'* TDE	Total DDT
F. melanaria	−5	—	—	—	—	—
	½	3.3	0.77	—	—	4.07
	7	0.44	0.15	0.35	0.01	0.95
	8	1.36	0.41	0.53	0.06	2.36
	9	0.99	0.29	0.52	<0.01	1.80
	24	0.31	0.06	0.24	—	0.61
F. madida	−5	—	—	—	—	—
	4	1.4	0.42	0.10	—	1.92
	8	4.3	0.93	0.84	0.14	6.21

From B. N. K. Davis and M. C. French (1969).

contained 0.99 ppm p,p'-DDT, 0.29 ppm o,p'-DDT and 0.52 ppm p,p'-DDE, lost most of their residues when kept for one week on untreated soil.

Davis and French (1969) demonstrated that some species of carabid beetles are able to metabolise DDT even more rapidly than *F. melanaria.* Individuals of *Nebria brevicollis* were collected from a strawberry field just before, and soon after, DDT was applied. Before the dust was applied, a sample of beetles contained 0.70 ppm p,p'-DDT and 38 ppm p,p'-DDE. Five to seven days after the dust was applied, a comparable sample contained 0.41 ppm p,p'-DDT and 75 ppm p,p'-DDE (Table 3.6).

Dempster (1968b) showed that *Harpalus rufipes* metabolised p,p'-DDT rapidly. Adult beetles were kept on soils that contained different concentrations of DDT and beetles that survived were frozen and kept for analysis. Those kept on soil treated at 50 ppm died within four to eight days and were the only individuals in which p,p'-DDT was detected (Table 3.12). Residues of o,p'-DDT decreased when beetles were kept on the treated soil for increasing periods of time. Those kept on soil with 20 ppm for 14 days contained 4.6 ppm o,p'-DDT, those kept for 28 days had less than 1.4 ppm while only a trace of o,p'-DDT was found in beetles kept for longer periods. DDE was eliminated very slowly but most was detected in those insects that had longest contact with DDT. Thus the concentration of DDE in beetles kept for 28 days on soil treated at 20 ppm was 16.25 ppm, and in those that were removed to untreated soil for 14 days, after 14 days on treated soil, was 11.9 ppm (Dempster, 1968b).

When soil is treated with DDT, the numbers of predatory mites in it usually decrease, and those of many species of Collembola increase after a short time (Sheals, 1956; Edwards *et al.*, 1967). Some species of Collembola are not killed by DDT residues of 50 ppm in soil (Thompson and Gore, 1972), but no work has been done on residues in specimens collected from treated soil outdoors. However, Butcher *et al.* (1969) cultured individuals of *Folsomia candida* on agar that had been treated with DDT and showed that the concentration of DDE in the medium

TABLE 3.12. DDT and its metabolites in individuals of *Harpalus rufipes* kept on treated soil.

Concn. of DDT in soil (ppm)	Days spent by beetles on soil	Residues (ppm)		
		p,p'-DDE	o,p'-DDT	p,p'-DDT
0	14	0	0	0
20	14	7.5	4.6	0
20	28	16.25	<1.4	0
20	14 (+14 on untreated)	11.9	0	0
50	6 (i.e. until dead)	7.6	5.6	4.6

From J. P. Dempster (1968b).

increased 24 hours after exposing Collembola to the agar. The concentration of DDE in macerated Collembola, that had been reared on agar treated with DDT, increased with time until the insects died.

3.3.2 Other insecticides

The best-established pathway for metabolism of cyclodiene insecticides is that of epoxidation (O'Brien, 1967). The first evidence of epoxidation in insects was reported by Gianotti *et al.* (1956) who showed qualitatively that aldrin was converted to dieldrin by American cockroaches. In 1958, it was first shown that houseflies converted heptachlor to its epoxide (Perry *et al.*, 1958). Micro-organisms convert heptachlor to heptachlor epoxide, and the epoxide to the less toxic 1-exohydroxychlordene (Miles *et al.*, 1969, 1971) but, although heptachlor epoxide has been detected in earthworms (Table 3.5), this may have been formed in soil by other organisms, and taken up later by earthworms.

Aldrin breaks down in soil to its epoxide, dieldrin (Edwards, *et al.*, 1957; Gannon and Biggar, 1958; Lichtenstein and Schultz, 1959; Tu *et al.*, 1968) and to a very limited extent soil micro-organisms are able to degrade dieldrin (Matsumura *et al.*, 1968; Tu *et al.*, 1968). Aldrin is not stored in animals as such, but is rapidly converted into dieldrin, as in beef and dairy cattle, rats and sheep (Bann *et al.*, 1956). There is only circumstantial evidence that earthworms are able to convert aldrin to dieldrin, but this is very strong. Raw (1964) mentioned dieldrin, but not aldrin, as a residue in earthworms from soil treated with aldrinated fertiliser. Davis and Harrison (1966) and Wheatley and Hardman (1968) recorded only small concentrations of aldrin in earthworms from arable soils, and residues of dieldrin were very much larger. Gish (1970) reported that dieldrin was detected in each sample of earthworms that contained aldrin. Davis and French (1969) reported analyses of *A. caliginosa* from a potato field that was treated with aldrin. Six weeks after spraying, two samples contained 0.53 and 0.77 ppm of dieldrin plus 0.12 and 0.15 ppm of aldrin respectively. Six months later, concentrations of dieldrin and aldrin had decreased by about 63% and 90% and when final samples were analysed one year after the insecticide was applied, both contained 0.09 ppm of dieldrin and no aldrin.

Beynon and Wright (1967) determined the breakdown products of an organophosphorus insecticide, chlorfenvinphos, in soils and Edwards *et al.* (1968) analysed individuals of four species of earthworms that had been kept for up to 21 weeks in treated soil for residues. Two of the breakdown products, 2,4-dichlorophenacyl chloride and 1-(2',4'-dichlorophenyl) ethan-l-ol, could not be detected when the limit of detection was near 0.05 ppm and less than 0.02 ppm of the third product, 2,4-dichloroacetophenone, was found.

The results that have been discussed suggest that soil invertebrates are able to metabolise many insecticides to the same compounds found in animals with more complex nervous and enzymic systems. Although some of the metabolites are known, the reactions by which they are produced are not certain, although the affinities of carabid beetles to other insects that have been investigated in great detail probably mean that the pathways in these animals can be more readily surmised than those in earthworms and molluscs. There is considerable scope for research into the mechanisms whereby some organochlorine, carbamate and organophosphorus insecticides readily kill earthworms while other closely related compounds have little effect (Edwards and Thompson, 1973) and are presumably themselves metabolised.

Earthworms and molluscs contribute to the long-term breakdown of the more persistent organochlorine compounds by storing the parent compounds, often in great concentrations, and metabolising them to compounds that have not yet been shown to affect them in any way.

3.4   ECOLOGICAL CONSEQUENCES OF RESIDUES OF INSECTICIDES IN SOIL INVERTEBRATES

In soil, insecticides tend to act in a cumulative, or chronic, manner so that, as animals move through treated soil, they slowly contact and acquire residues that may eventually be large enough to kill them. Because residues can reach animals in many indirect ways, chronic toxicity of insecticides cannot be accurately measured in laboratory experiments and most toxicological studies use the acute toxicity of one application. Most insecticides affect the nervous system (O'Brien, 1967) and as none is specific to a particular pest, sublethal effects may be seen in other species which may even be killed. Moriarty (1969, 1971) suggested that, as a general rule, individuals with the largest internal concentrations of insecticides are the most prone to sublethal effects and reasoned that, because the concentration in the tissues depended firstly on the rate of absorption, excretion and metabolism, and secondly on the distribution between tissues, the more persistent insecticides may cause more sublethal effects. Because of inter-specific differences, persistence alone does not necessarily imply sublethal effects (Moriarty, 1971). The converse is more certain. Some types of sublethal effects are unlikely if the insecticide disappears rapidly from the tissues. Thus, although latent toxicity, i.e. a shorter life or death at particular developmental stages, has often been associated with organochlorine insecticides, it is virtually unknown with organophosphorus compounds which are more easily metabolised and most of the literature on sublethal effects is concerned with organochlorine insecticides. Sublethal doses could have three types of ecological effect: (1) on the ability of individuals to survive; (2) on their reproductive ability; and (3) on the genetic constitution of future generations (Moriarty, 1968), but no work has been done on soil invertebrates concerning the second and third categories. Results that have been published on changes in behaviour of individuals were probably due to direct effects of the insecticides on the nervous system.

Kulash (1947) reported that, only three hours after arable land had been treated with DDT at 25 lb/acre (28 kg/ha), ground beetles were lying on their backs incapable of any co-ordinated movements. After 15 hours, many ground beetles had been unable to recover from the initial effects of the insecticide and had died. Davis (1966b) noticed numbers of large carabids lying on their backs with legs twitching after experimental plots had been sprayed with aldrin and BHC. A sample of the beetles contained 8.7 ppm dieldrin and 39 ppm γ-BHC. Although Dempster (Davis and French, 1969) observed that many diurnal species of Carabidae seek immediate shelter under foliage or burrow into the soil when exposed to insecticides, Coaker (1966) reported that sublethal residues of dieldrin caused hyperactivity in some species, including *Bembidion lampros*.

Effects on behaviour such as those noticed by Kulash (1947) and Davis (1966b) make individuals more noticeable to predators, including birds, which in turn accumulate residues when they feed on them. The incidence of residues in food chains will be considered a little later. The more subtle effects on activity of individuals, as noted by Coaker (1966), could affect the prey of predatory beetles, and Coaker (1966) considered the example in which predatory beetles were killed and numbers of their prey, in this instance cabbage root fly, increased after some insecticides had been applied. Chlorfenvinphos has been shown to increase the activity of some species of carabid beetles in experimental field plots (Edwards and Thompson, 1969) and, if the habits of the beetles are not affected in any other way, the increased activity may result in increased feeding and control of the pest species. Experiments have not been done to assess such other possible physiological effects of this insecticide.

However, Dempster (1968b) showed that residues of DDT in soil sometimes lessened the number of seeds of rye-grass *(Lolium perenne)* eaten by the carabid beetle *Harpalus rufipes*. Concentrations of less than 5 ppm DDT had no significant effect on numbers of seeds eaten, but when beetles were kept on soil with 10 ppm DDT for 12 weeks, significantly less were eaten than when residues were 5 ppm ($P = 0.02$-0.05) and 0 ppm ($P = 0.01$-0.02). No p,p'-DDT, and less than 0.02 ppm of the o,p'-isomer, was detected in beetles that had been kept on soil treated at these concentrations, although some breakdown of the isomers may have occurred after the beetles were killed and before they were analysed (Dempster, 1968b). All beetles contained p,p'-DDE, but, when beetles that had been on soil treated at 20 ppm were placed on untreated soil, their rate of feeding returned to normal again. Contact with the p,p'-isomer, and possibly the o,p'-isomer also, therefore, seems to decrease the rate of feeding. *Harpalus rufipes* is an important predator of caterpillars of *Pieris rapae* and in soils with sublethal residues of p,p'-DDT, natural control of the caterpillars could be reduced (Dempster, 1968a).

Some species of invertebrates respond to residues of DDT faster than carabid beetles. When individuals of the spider *Lycosa amentata* were placed on soil with and without DDT, and the distance moved by individuals was measured, residues of 2.5-10 ppm p,p'-DDT in the soil caused an immediate significant increase in the distance moved (Dempster, 1968a). Several hours elapsed before carabid beetles of the species *Bembidion obtusum* and *B. lampros* moved significantly greater distances.

Other insecticides affect the feeding habits of some soil invertebrates. Long and Lilly (1958, 1959) found that aldrin, dieldrin, endrin, heptachlor and lindane seed dressings tended to inhibit the feeding reaction of wireworms (*Melanotus communis*). Aldrin and lindane also repelled wireworms from treated seeds. Griffiths (1967) compared the effects of aldrin and more recent compounds on the biting ability of wireworms (*Agriotes* spp.) and found that, in soil treated with 3.7 ppm aldrin, individuals bit significantly less ($P = 0.01$) at filter paper than those in untreated soil. Wireworms did not recover the ability to bite after they had been exposed to aldrin, Bayer 38156 or thionazin, and eventually died. Insecticides that did not kill many wireworms, e.g. dichlofenthion and ethion, had much less effect than aldrin on ability to bite and there seemed to be an apparent correlation between the ability of insecticides to kill wireworms and their ability to stop them from biting (Griffiths, 1967).

Wireworms that have been in contact with insecticides may also be unable to co-ordinate their movements. When placed in beakers of untreated soil, they occasionally crawl to the surface and then burrow down again away from the light (Bardner, 1963). In soil treated with γ-BHC, individuals of *Agriotes* sp. remained on the surface and, although capable of movement, they seemed to lack the co-ordination needed to bury themselves again (Bardner, 1963; Griffiths and Bardner, 1964). Affected individuals that remain on the surface of the soil are liable to become desiccated and die, and, being conspicuous, they are more liable to be eaten by predators.

Although there is now much more concern about the quantities of insecticides used in agriculture and about their non-specific sublethal effects, there remains a conspicuous absence of data on either relationships between different species or on the effects of residues. Because it is not possible to reproduce chronic effects accurately in laboratories, it is necessary to predict sublethal effects, and this cannot be done without knowledge, firstly of the minimum amount of residues needed to induce effects, and, secondly, of the dynamic relationship between residues and tissues. The metabolism of residues has already been discussed and it is clear that much research has yet to be done to indicate whether the very small residues that are present almost universally (Edwards, 1970) do have any subtle effects on species against which insecticides, and all other pesticides, are not directed.

Pollution of the environment by radioactive wastes and fallout increases annually and it is pertinent to consider their relationship to soil invertebrates. Giles and Peterle (1964) found that, after ^{35}S-labelled malathion had been applied aerially to a watershed, measurable levels of radioactivity occurred in earthworms one year after the area was treated. Edwards (1968) irradiated individuals of many taxa in the laboratory, directly and in soil, and showed that some species of oribatid mites, which are also not directly affected by a large number of insecticides (Edwards and Thompson, 1973), were resistant to doses up to 200 kilorads. The activity of individuals and their susceptibility to radiation were well correlated and more active animals were more susceptible than those which often had saprophagous habits (Edwards, 1968). Animals from all groups tended to become more sluggish and inactive after exposure to large doses of irradiation if they were not immediately killed or paralysed. Paralysed animals never recovered, and the paralysis, which often lasted several days, was interrupted only by periodic tremors of the legs and antennae. Females of some species, especially woodlice and millipedes, tended to lay eggs just before they died but none of these were viable. Again, as has already been shown in relation to the uptake and excretion of residues from tissues, there appears to be a considerable similarity between effects of insecticides and radioactive substances, and further work with labelled compounds could possibly help to elaborate the complex relationships between pollutants and invertebrates in soil, and also among invertebrates themselves. In preliminary studies, Peredel'skii et al. (1960b) showed that wireworms and earthworms redistribute ^{45}Ca and ^{90}Sr through soil and that they also dissipate accumulated isotopes. Wireworms in pot soil containing ^{45}Ca or ^{90}Sr became only slightly radioactive and the specific radioactivity of earthworms in similar pots decreased rapidly when the worms were then transferred to pots of untreated soil, suggesting that little isotope was assimilated (Peredel'skii et al., 1960a).

Invertebrates, recently killed in soil by insecticides, contain residues and some that are not killed, e.g. earthworms and slugs, can store large residues of insecticides and their metabolites. Sublethal residues can alter the behaviour of some species and make them more conspicuous than usual to predators, and a considerable amount of work has now been done to assess the hazards presented by residues in soil invertebrates to animals such as birds and other vertebrates (e.g. Stickel et al., 1965; Jefferies and Davis, 1968). Experiments in the early 1960's in Great Britain showed that residues of organochlorine insecticides in corpses of birds found on agricultural land were large enough to have killed them (Turtle et al., 1963) and it was suspected that the residues may have originated as seed dressings. About the same time, in forests treated with DDT in the U.S.A., birds were found with tremors believed to be due to DDT poisoning and Barker (1958) calculated that earthworms were a probable source of poisoning for robins. A 50 g robin with 60 ppm of DDT contained the equivalent of eleven specimens of Lumbricus terrestris such as he analysed in his study, and a diet of this size seemed to be well within the ability of a robin. Residues are generally larger in raptorial and fish-eating birds than in herbivorous species (Edwards, 1970). Residues in soil invertebrates certainly contribute to those in birds but it is unresolved how often residues accumulated by birds in this way only are lethal. Davis (1966b) described how birds tend to concentrate on particular foods at various times, depending on their availability. For example, earthworms are an important part of the diet of rooks (*Corvus frugilegus*) and these birds feed for most of the year in pastures where residues tend to be small (Harris *et al.*, 1966) and where they are unlikely to accumulate appreciable residues. When arable land is ploughed however, rooks may feed on large numbers of earthworms that are suddenly exposed on the surface and which may contain relatively large insecticide residues. Wheatley and Hardman (1968) determined the amounts of residues of six insecticidal components in six species of earthworms from arable land and showed that total residues were between 1.1 μg per worm for *Allolobophora rosea* and 8.9 μg for *L. terrestris.* Ploughing exposes earthworms on the surface of the soil and birds may, in a short time, acquire toxic doses. Jefferies and Davis (1968) considered however, that even if a thrush could feed continuously on earthworms from an arable site as described by Davis and Harrison (1966), its dieldrin intake would be well below danger level and excretion would prevent accumulation of dangerous residues.

The hazards are increased in forests and orchards where insecticide residues in the soil, and therefore in animal tissues, are larger than in arable land. Soon after Barker (1958) analysed earthworms and robins for residues of DDT and DDE, and showed that earthworms were probably instrumental in the decline in numbers of birds in areas treated with DDT, Hunt (1965) estimated that total pesticide residues of methoxychlor and DDT and its breakdown products accumulated from 9.9 ppm in soil to 140.6 ppm in earthworms and to 443.9 ppm in the brains of adult robins but, although residues in soil invertebrates certainly contribute to such residues, the extent of their contribution can be reasonably questioned. In field conditions of stress, such as exposure to cold, nesting activity, food restriction and migration, birds mobilise fat reserves for energy, and release lipid–soluble insecticides stored there (Ecobichon and Saschenbrecker, 1969). Some studies of the effects of feeding birds on diets comprising earthworms with residues of insecticides have demonstrated that residues are transmitted from the diet to the birds, but that they are not necessarily concentrated in avian tissues (Jefferies and Davis, 1968) and the extent of the role of earthworms and other soil invertebrates in poisoning birds outdoors is clearly strongly influenced by other environmental factors.

It is difficult to assess ecological effects of pesticide residues accurately because of the many variables involved, but they depend essentially on whether a significant amount of the pesticide comes into contact with a significant proportion of the population of the species being considered (Moore, 1969). This, in turn, is influenced by the amount of the compound used and on the dispersal of its residues, as well as on the amount consumed by individuals. In Great Britain, the localised use of pesticides on fields which tend to be small has probably prevented the residues from having significant effects on total populations but, in countries where acreages are larger and often sprayed aerially, e.g. North America, effects may be more severe.

REFERENCES

Anon., 'Use of pesticides', a report of the President's Science Advisory
    Committee. The White House, May 15th (1963)
Anon., *New Scient.*, **32**, 612 (1966)
Anon., *Pesticide Information* (9) 3 (1970)
S. Bailey, P. J. Bunyan, D. M. Jennings and A. Taylor, *Pestic. Sci.*, **1**, 66
    (1970)
J. M. Bann, T. J. DeCino, N. W. Earle and Y. P. Sun, *J. Agric. Fd. Chem.*,
    4, 937 (1956)
R. Bardner, Rep. Rothamst. Exp. Sta. for 1962, 149 (1963)
R. J. Barker, *J. Wildl. Mgmt.*, **22**, 269 (1958)
K. I. Beynon and A. N. Wright, *J. Sci. Fd. Agric.*, **18**, 143 (1967)
E. A. Boykins, *Atlantic Nat.*, **21**, 18 (1966)
J. W. Butcher, E. Kirknel and M. Zabik, *Rev. Ecol. Biol. Sol.*, **6**, 291
    (1969)
A. Calderbank, R. L. Metcalf (Editor), Advances in Pest Control Research,
    8, 127 (1968)
T. H. Coaker, *Ann. appl. Biol.*, **57**, 397 (1966)
N. Collett and D. L. Harrison, *N.Z. Jl. Sci.*, **11**, 371 (1968)
J. W. Cook, Review of the persistent organochlorine pesticides, Report by
    the Advisory Committee on poisonous substances used in agriculture
    and food storage. H.M.S.O., London (1964)
S. Cramp and P. Conder, Fifth Report Joint Committee British Trust for
    Ornithology and Roy. Soc. Protection of Birds on Toxic Chemicals. 19
    pp. (1965)
S. Cramp, P. J. Conder and J. S. Ash, Fourth Report British Trust for
    Ornithology and Roy. Soc. Protection of Birds on Toxic Chemicals, in
    collaboration with Game Res. Assoc. 24 pp. (1964)
S. Cramp and P. J. S. Olney, Sixth Report Joint Committee British Trust
    for Ornithology and Roy. Soc. Protection of Birds on Toxic Chemicals.
    26 pp. (1967)
D. A. Crossley, D. E. Reichle and C. A. Edwards, *Pedobiologia*, **11**, 71
    (1971)
B. N. K. Davis, Monks Wood Exp. Sta. Rep. for 1960-1965, 55 (1966a)
B. N. K. Davis, *J. appl. Ecol.*, **3** (Suppl.), 133 (1966b)
B. N. K. Davis, *Ann. appl. Biol.*, **61**, 29 (1968)
B. N. K. Davis, Monks Wood Exp. Sta. Rep. for 1966-1968, 16 (1969)
B. N. K. Davis, *Soil Biol. Biochem.*, **3**, 221 (1971)
B. N. K. Davis and M. C. French, *Soil Biol. Biochem.*, **1**, 45 (1969)
B. N. K. Davis and R. B. Harrison, *Nature, Lond.*, **211**, 1424 (1966)
J. P. Dempster, *J. appl. Ecol.*, **5**, 463 (1968a)
J. P. Dempster, *Entomologia exp. appl.*, **11**, 51 (1968b)
J. B. DeWitt and J. L. George, U.S.D.I. Fish and Wildl. Serv. Bur. Sport
    Fisheries and Wildl. Circ., **84**, (1960)
J. B. Dimond, G. Y. Belyea, R. E. Kadunce, A. S. Getchell and J. A.
    Blease, *Can. Ent.*, **102**, 1122 (1970)

D. L. Dindal and K. H. Wurzinger, *Bull. Env. Contam. Toxicol.*, **6**, 362 (1971)

C. C. Doane, *J. econ. Ent.*, **55**, 416 (1962)

E. H. Dustman, 'Scientific aspects of pest control'. Nat. Acad. Sci., Washington. Publ. **1402**, 343 (1966)

D. J. Ecobichon and P. W. Saschenbrecker, *Toxicol. appl. Pharmacol.*, **15**, 420 (1969)

C. A. Edwards, *Residue Reviews*, **13**, 83 (1966)

C. A. Edwards, Proc. 2nd Symp. Radioecology , p. 68 (1968)

C. A. Edwards, 'Critical reviews in environmental control', vol. **1** (1), 7. Chemical Rubber Co., Cleveland, Ohio (1970)

C. A. Edwards, S. D. Beck and E. P. Lichtenstein, *J. econ. Ent.*, **50**, 622 (1957)

C. A. Edwards, E. B. Dennis and D. W. Empson, *Ann. appl. Biol.*, **60**, 11 (1967)

C. A. Edwards, J. R. Lofty, A. E. Whiting and K. A. Jeffs, Rep. Rothamst. Exp. Sta. for 1970, 193 (1971)

C. A. Edwards, D. E. Reichle and D. A. Crossley Jr., Ecological studies. Analysis and Synthesis, **1**. D. E. Reichle (Editor), Springer-Verlag, Berlin (1970)

C. A. Edwards and A. R. Thompson, Rep. Rothamst. Exp. Sta. for 1968, 216 (1969)

C. A. Edwards and A. R. Thompson, *Residue Reviews*, **45**, 1 (1973)

C. A. Edwards, A. R. Thompson and K. I. Beynon, *Rev. Ecol. Biol. Sol.*, **5**, 199 (1968)

E. I. El Sayed, J. B. Graves and F. L. Bonner, *J. Agric. Fd. Chem.*, **15**, 1014 (1967)

A. C. Frazer, Report of the research committee on toxic chemicals. A.R.C. London, 38 pp. (1964)

A. C. Frazer, Supplementary report of the research committee on toxic chemicals. A.R.C. London, 16 pp. (1965)

M. C. French and D. J. Jefferies, *Science, N.Y.*, **165**, 914 (1969)

N. Gannon and J. H. Biggar, *J. econ. Ent.*, **51**, 1 (1958)

O. Gianotti, R. L. Metcalf and R. B. Marsh, *Ann. ent. Soc. Amer.*, **49**, 588 (1956)

R. H. Giles and T. J. Peterle, U.S.D.I. Fish & Wildl. Serv. Circ., **199**, 119 (1964)

C. D. Gish, *Pestic. Mon. J.*, **3**, 241 (1970)

D. C. Griffiths, *Entomologia exp. appl.*, **10**, 171 (1967)

D. C. Griffiths and R. Bardner, *Ann. appl. Biol.*, **54**, 241 (1964)

W. D. Guenzi and W. E. Beard, *Science, N.Y.*, **156**, 1116 (1967)

G. Hafstad, J. Libby and G. Worf, Dutch elm disease manual for Wisconsin. Extension serv., coll. of Agric., Univ. of Wisconsin (1966)

C. R. Harris, *A. Rev. Ent.*, **17**, 177 (1972)

C. R. Harris and W. W. Sans, Proc. Ent. Soc. Ontario., **100**, 156 (1969)

C. R. Harris, W. W. Sans and J. R. W. Miles, *J. Agric. Fd. Chem.*, **14**, 398 (1966)

R. G. Heath, J. W. Spann and J. F. Kreitzer, *Nature, Lond.*, **224**, 47 (1969)

D. C. Holmes, J. H. Simmons and J. O'G. Tatton, *Nature, Lond.*, **216**, 227 (1967)

W. M. Hoskins, *Wld. Rev. Pest Control,* **3**, 85 (1964)

L. B. Hunt, 'The effects of pesticides on fish and wildlife'. U.S.D.I. Fish and Wildl. Serv., Wash. Circ., **226**, p. 12 (1965)

L. B. Hunt and R. J. Sacho, *J. Wildl. Mgmt.,* **33**, 336 (1969)

D. J. Jefferies and B. N. K. Davis, *J. Wildl. Mgmt.,* **32**, 441 (1968)

D. J. Jefferies and C. H. Walker, *Nature, Lond.,* **212**, 533 (1966)

A. K. Klein, E. P. Lang, P. R. Datta, J. O. Watts and J. T. Chen, *J. Assoc. Off. Agric. Chem.,* **47**, 1129 (1964)

W. H. Ko and J. L. Lockwood, *Can. J. Microbiol.,* **14**, 1069 (1968)

L. J. Korschgen, *J. Wildl. Mgmt.,* **34**, 186 (1970)

W. M. Kulash, *J. econ. Ent.,* **40**, 851 (1947)

E. P. Lichtenstein and K. R. Schulz, *J. econ. Ent.,* **52**, 289 (1959)

E. P. Lichtenstein, K. R. Schulz and T. W. Fuhremann, *Pestic. Mon. J.,* **5**, 218 (1971)

H. Lipke and C. W. Kearns. R. L. Metcalf (Editor), Advances in Pest Control Research, **3**, 253 (1960)

W. H. Long and J. H. Lilly, *J. econ. Ent.,* **51**, 291 (1958)

W. H. Long and J. H. Lilly, *J. econ. Ent.,* **52**, 509 (1959)

W. H. Luckmann and G. C. Decker, *J. econ. Ent.,* **53**, 821 (1960)

F. Matsumura, G. M. Bousch and A. Tai, *Nature, Lond.,* **219**, 965 (1968)

J. F. Mehner and G. J. Wallace, *Atlant. Nat.,* **14**, 4 (1959)

C. M. Menzie, Bur. Sport Fisheries & Wildl. Spec. Sci. Rep. Wildl. No. **127**, Washington (1969)

J. R. W. Miles, C. M. Tu and C. R. Harris, *J. econ. Ent.,* **62**, 1334 (1969)

J. R. W. Miles, C. M. Tu and C. R. Harris, *J. econ. Ent.,* **64**, 839 (1971)

N. W. Moore, Proc. 5th Br. Insectic. Fungic. Conf., (1969), 699 (1969)

F. Moriarty, *Ann. appl. Biol.,* **62**, 371 (1968)

F. Moriarty, *Biol. Rev.,* **44**, 321 (1969)

F. Moriarty, *Meded. Fak. Landbouw-Wetenschappen, Gent.,* **36**, 27 (1971)

P. F. Newell. A. Burges, F. Raw (Editors), Soil Biology, p. 413. Academic Press, London and New York (1967)

R. D. O'Brien, Insecticides: Action and Metabolism, Academic Press, London and New York, 332 pp. (1967)

A. A. Peredel'skii, I. O. Bogatyrev and N. S. Karavianskii, Soils Fertil., Harpenden (1961) **24**, 134 (1960a)

A. A. Peredel'skii, S. S. Shain and N. S. Karavianskii *et al.,* Soils Fertil., Harpenden (1961), **24**, 127 (1960b)

A. S. Perry, A. M. Mattson and A. J. Buckner, *J. econ. Ent.,* **51**, 346 (1958)

J. E. Peterson and W. H. Robison, *Toxicol. appl. Pharmacol.,* **6**, 321 (1964)

J. A. Pickard, Rep. Agr. Hort. Res. Stn. Univ. Bristol for 1967, p. 51 (1968)

A. D. Pickett, *Can. Ent.,* **81**, 67 (1949)

D. A. Ratcliffe, *J. appl. Ecol.,* **7**, 67 (1970)

F. Raw, Rep. Rothamst. exp. Sta. for 1963, 149 (1964)

L. M. Reynolds, *Bull. Environ. Contam. Toxicol.,* **4**, 128 (1969)

B. I. Roots, *J. exp. Biol.,* **33**, 29 (1956)

J. E. Satchell. A. Burges and F. Raw (Editors), Soil Biology, p. 259. Academic Press, London and New York (1967)

J. G. Sheals, *Bull. ent. Res.,* **4**, 803 (1956)

R. D. Smith and L. L. Glasgow, Proc. 17th Ann. Conf. S.E. Ass. Game & Fish Comm. p. 140 (1965)

J. H. V. Stenersen *Nature, Lond.,* **207**, 660 (1965)

J. W. Stephenson, Rep. Rothamst. Exp. Sta. for 1962, 158 (1963)

W. H. Stickel, D. W. Hayne and L. F. Stickel, *J. Wildl. Mgmt.,* **29**, 132 (1965)

A. Stringer, Rep. Agr. Hort. Res. Stn. Univ. Bristol for 1965, 51 (1966)

A. Stringer, C. H. Lyons and J. A. Pickard, Rep. Agr. Hort. Res. Stn. Univ. Bristol for 1969, 98 (1970)

A. Stringer and J. A. Pickard, Rep. Agr. Hort. Res. Stn. Univ. Bristol for 1963, 127 (1964)

A. Stringer and J. A. Pickard, Rep. Agr. Hort. Res. Stn. Univ. Bristol for 1964, 172 (1965)

A. Stringer and J. A. Pickard, Rep. Agr. Hort. Res. Stn. Univ. Bristol for 1968, 80 (1969)

A. R. Thompson, *Bull. Environ. Contam. Toxicol.,* **5**, 577 (1970)

A. R. Thompson and F. L. Gore, *J. econ. Ent.,* **65**, 1255 (1972)

C. M. Tu, J. R. W. Miles and C. R. Harris, *Life Sciences,* **7**, 311 (1968)

E. E. Turtle, A. Taylor, E. N. Wright, R. J. P. Thearle, H. Egan, W. H. Evans and N. M. Sontar, *J. Sci. Fd. Agric.,* **14**, 567 (1963)

U.S.D.I., U.S.D.I. Fish and Wildl. Serv. Bur. Sport Fisheries and Wildl. Resource Publ., **23**, 55 (1966)

U.S.D.I., U.S.D.I. Fish and Wildl. Serv. Bur. Sport Fisheries and Wildl. Resource Publ., **74**, 52 (1969)

C. H. Walker, *J. appl. Ecol.,* **3** (Suppl.), 213 (1966)

G. A. Wheatley and J. A. Hardman, *J. Sci. Fd. Agric.,* **19**, 219 (1968)

A. Wilson, Further review of certain persistent organochlorine pesticides used in Great Britain, Report by the advisory committee on pesticides and other toxic chemicals. H.M.S.O., London, 148 pp. (1969)

J. F. Wootten, Central States Forest Expt. Sta., Sta. Note 156, 3 pp. (1962)

Chapter 4

Pesticide Residues in Aquatic Invertebrates

S. R. Kerr* and W. P. Vass

*Fisheries Research Board of Canada,
Marine Ecology Laboratory, Bedford Institute,
Dartmouth, Nova Scotia*

The basic processes governing formation of pesticide residues are common to all organisms. Separate consideration of residues in aquatic invertebrates is therefore a rather arbitrary choice, but a practical one for purposes of literature review.

* Present address: Ministry of Natural Resources, Fish and Wildlife Research Branch, Maple, Ontario, Canada.

As befits a biologically and ecologically diverse group, the available literature on accumulation of pesticide residues by aquatic invertebrates is itself an heterogeneous assemblage. The literature is conveniently grouped into four broad categories: existing residue levels, sources of residue, biological concentration of residue, and residue hazards. Within each category, we have particularly emphasised weaknesses and gaps in existing knowledge. In our view, each identified deficiency represents an essential area for further research.

4.1   AMOUNTS OF RESIDUES CURRENTLY OCCURRING

Residue levels in themselves are merely static indices of dynamic processes, but as such, collation of published levels of residues detected in aquatic invertebrates serves several useful purposes. In general, a survey of observed residue levels provides a useful overview of the prevalence of pesticide contamination in aquatic habitats. Data collation also identifies the pesticides that are major contaminants, provides some measure of their relative prevalence, and may indicate the types of organism that are most likely to accumulate detectable residues.

Many investigators have detected pesticide residues in aquatic invertebrates. We were guided by several considerations in selecting representative residue levels for presentation here from an abundance of published data. Our principal aim was to provide a general overview of the chronic residue levels occurring among aquatic invertebrates in diverse habitats. We therefore, avoided data which reflect the immediate effects of direct exposure to pesticide applications. Where possible, we selected data determined for organisms sampled in control areas, or for organisms otherwise remote from immediate sources of pesticide release to the environment. For localities directly exposed to pesticide application, we have usually selected residue levels determined from samples taken one year subsequent to the most recent pesticide application. Moreover, we have often represented large series of data with a single mean or median value. Occasionally, consistency has required that we convert units of measurement, from dry weight to wet weight for example, without precise data for the purpose. In some cases, conversion of units to $\mu g/kg$ has changed the apparent significance of published values. For these reasons, our selection of data therefore does not necessarily provide a balanced summary of any particular report.

The selection criteria applied here are clearly biased towards low residue levels, and have eliminated many of the less persistent pesticides from consideration. With the exception of the polychlorinated biphenyls (PCB's), we have also excluded contaminants whose occurrence in aquatic environments is not principally a result of their use as pest control agents. For this reason, we do not consider mercurials, arsenicals, and the like, or pesticides other than organochlorine compounds.

Marine Benthic

TABLE 4.1.   Representative residue levels of ΣDDT in marine benthic organisms, mostly invertebrates. With indicated exceptions, ΣDDT = p,p'-DDT + p,p'-DDE + p,p'-DDD (and may also include o,p'-DDT, and p,p'-DDMU). Estimates of body weight to the nearest order of magnitude, and trophic level, have been assigned by us for purposes explained in the text. For explanation of criteria used in selecting and condensing data, see text.

Name of Organism		Residue µg/Kg wet weight	Range µg/Kg wet weight	Body weight (grams)	Troph Leve
Genus	Common				
Fucus	alga	2	–	100	1
Laminaria	,,	3	–	1,000	,,
Cardium	cockle	12	–	0.1	2.1
Pastella	limpet	3	–	,,	,,
Nassarius	snail	260	–	,,	,,
Thais	,,	129	–	,,	,,
Mytilus	mussel	30	5-61	10	,,
,,	,,	<26	<10-<40	,,	,,
,,	,,	24	–	,,	,,
,,	,,	43	–	,,	,,
,,	,,	21	<9-41	,,	,,
,,	,,	183	11-423	,,	,,
,,	,,	20	<4-50	,,	,,
,,	,,	43	17-429	,,	,,
,,	,,	69	12-294	,,	,,
,,	,,	20	20-30	,,	,,
,,	,,	754	198-3970	,,	,,
,,	,,	73	<10-170	,,	,,
Crassostrea	oyster	15	10-500	,,	,,
,,	,,	60	<30-710	,,	,,
,,	,,	15	<10-30	,,	,,
,,	,,	51	N.D.-150	,,	,,
Ostrea	,,	44	10-110	,,	,,
Corbicula	clam	1123	363-2280	,,	,,
Mya	,,	22	<10-140	,,	,,
Venus	quahog	50	10-230	,,	,,
Mitella	goose barnacle	27	–	,,	,,
Emerita	crab	104	39.7-7248	1	2.3
Echinus	sea urchin	50	–	10	,,
Strongylocentrolus	urchin	5	–	,,	,,
Crassostrea	giant oyster	29	–	,,	,,
,,	,, ,,	196	N.D.-1610	,,	,,
Placopecten	scallop	30	10-90	100	,,
,,	,,	106	–	,,	,,
Stichopus	sea cucumber	93	–	,,	,,
Penaeus	shrimp	160	–	1	2.5

Reference	Remarks
obinson *et al.*, 1967	DDE only
,, ,, ,,	,, ,,
,, ,, ,,	,, ,,
,, ,, ,,	,, ,,
oodwell *et al.*, 1967	
sebrough *et al.*, 1967	mean, 2 locations
nsen *et al.*, 1969	mean, 3 locations
oeman *et al.*, 1968	representative estimate, 8 locations. DDE only
obinson *et al.*, 1967	DDE only
sebrough *et al.*, 1967	mean, 4 locations
olden, 1970a	representative estimate Netherlands, DDE only
,, ,,	mean, S. Europe, 3 countries
,, ,,	representative estimate, N. Europe, 3 countries
,, ,,	representative estimate, Br. Isles, 2 countries
,, ,,	mean, Canada, 3 locations
tko, 1971	DDE only
odin, 1969	mean for 9 months, 3 locations
rague and Duffy, 1971	mean, 3 locations
tler, 1966a, 1968	
gg *et al.*, 1967	median, for 2.5 years, 6 States
rague and Duffy, 1971	mean, 2 locations
sper, 1967	representative estimate, 8 locations
rtmann (pers. comm.)	estimated mean and range for 2 years, 2 locations
odin, 1969	mean for 1 year, 2 locations
rague and Duffy, 1971	representative estimate, 3 locations
,, ,, ,,	
sebrough *et al.*, 1967	
rnett, 1971	mean excludes values near contaminant source
obinson *et al.*, 1967	DDE only
sebrough *et al.*, 1967	
,, ,, ,,	
odin, 1969	representative estimate for 2 years, 5 locations
rague and Duffy, 1971	mean, 2 locations
cher, 1971	muscle tissue
sebrough *et al.*, 1967	
oodwell *et al.*, 1967	

S. R. Kerr and W. P. Vass

TABLE 4.1. (continued)

Name of Organism		Residue µg/Kg wet weight	Range µg/Kg wet weight	Body weight (grams)	Trophc Level
Genus	Common				
Penaeus	shrimp	43	<10-140	,,	,,
Pugettia	crab	42	–	100	,,
Uca	,,	260	250-270	,,	,,
,,	,,	235	–	,,	,,
Carcinus	,,	37	–	1000	2.8
Cancer	,,	61	–	,,	,,
,,	,,	55	44-74	,,	,,
,,	,,	71	N.D.-140	,,	,,
Paralithodes	,,	2739	–	,,	,,
Homarus	lobster	24	–	,,	,,
,,	,,	40	20-80	,,	,,
Jasus	,,	–	<2-12	,,	,,
Pisaster	starfish	20	–	100	3.3
Patiria	,,	78	–	,,	,,

N.D. = None detectable

As with earlier summaries of reported residue levels (Edwards, 1970), an important *caveat* is necessary. Existing analytical methods for determination of organochlorine pesticide residues in biological material can yield quite disparate results among different laboratories. The values cited here were determined in many different laboratories, so comparisons should be made cautiously. The potential for variability due to analytical error is particularly serious, because of the exacting requirements for scrupulous sample preparation and technique when residues are present in low concentration, as is frequently the case in our summary. In our opinion, the analytic procedures described in the various reports we cite are generally good, but analytical error undoubtedly constitutes a considerable source for variability among the collated values.

4.1.1 DDT and its derivatives

DDT and its derivatives (ΣDDT) are by far the most common pesticide residues detected in aquatic invertebrates. As shown by the values set out in Tables 4.1, 4.2, and 4.3, ΣDDT has been detected in all major categories of aquatic environment, from freshwater to remote oceanic locations, and in invertebrates that differ considerably in taxonomic status, size, trophic level, behaviour, and habitat preference. The tabulated residue levels fall in the relatively low ppm x 10^3 (μg/kg) range of values. The criteria applied to selection of the data have excluded many reports of temporarily high values, and on these grounds we suggest that the tabulations tend to reflect chronic contamination levels for many of the localities sampled. If so, the

Reference	Remarks
Nimmo *et al.*, 1970	re resentative estimate, 4 States
Risebrough *et al.*, 1967	
Croker and Wilson, 1965	mean, 2 locations
Odum *et al.*, 1969	claw muscle tissue
Robinson *et al.*, 1967	DDE only
" " "	" "
Stout, 1968	representative estimate, 2 locations
Modin, 1969	mean, 6 locations
" "	1 specimen
Robinson *et al.*, 1967	DDE only
Sprague and Duffy, 1971	muscle tissue
Aucamp *et al.*, 1971	tail flesh
Risebrough *et al.*, 1967	
" " "	

data support the contention that ΣDDT has become a widely distributed, chronic contaminant of the global biosphere.

The few data for remote oceanic sites (Harvey *et al.*, personal communication; Tatton and Ruzicka, 1967) indicate lower contamination levels than usually occur in locations closer to initial sources of DDT. But with that exception, the majority of the data do not suggest any clear differences between freshwater, estuarine, and coastal marine environments. The technically advanced nations are overwhelmingly represented in the tabulations, primarily reflecting a bias due to the availability of scientific capability for residue determinations rather than any likely global pattern of residue distribution. There seems to be no reason to suppose that invertebrates from localities unrepresented in the literature are free of ΣDDT residues, but no conclusion can be reached with the existing data. Within the limits of their geographic representation, the tabulated data therefore suggest relatively uniform distribution of residues among invertebrates inhabiting the various basic types of aquatic environment adjacent to land.

Uniformity is relative, however, and the range of values tabulated with each estimate indicates that residue determinations within and among species vary considerably within and among environments. Invertebrate species can differ considerably in their rates of exposure to contaminants, and in their capacities to accumulate residues. Residue uptake by aquatic invertebrates occurs with direct exposure to water-borne contaminants and through indirect concentration via the food chain. As a preliminary

TABLE 4.2. Representative residue levels of ΣDDT in marine pelagic organisms, mostly invertebrates. See caption of Table 4.1.

Name of Organism		Residue µg/Kg weight	Range µg/Kg weight	Body weight (grams)	Tro Lev
Genus	Common				
Cladophora	alga	83	–	1×10^{-7}	1
	phytoplankton	500	70-920	,,	,,
Rhizosolenia	alga	12	–	,,	,,
	copepods	250	–	1×10^{-4}	2.
	zooplankton	3	<0.01-9.5	,,	,,
	,,	40	–	,,	,,
	microzooplankton	30	–	,,	,,
	macrozooplankton	160	–	0.01	2.
Systellaspis	crustacean	4	3.1-5.7	1.0	,,
Euphausia	shrimp	2	–	0.01	,,
,,	,,	120	95-440	,,	,,
Loligo	squid	28	–	1000	4.

Freshwater Benthic

TABLE 4.3. Representative residue levels of ΣDDT in freshwater benthic invertebrates. See caption of Table 4.1.

Name of Organism		Residue µg/Kg wet weight	Range µg/Kg wet weight	Body weight (grams)	Tro Le
Genus	Common				
Pisidium	clam	43	43-130	1	2.
Campeloma	snail	39	7-6960*	0.1	,
Lymnaea	,,	61	N.D.-1740*	,,	,
Planorbidae	ramshorn snail	5†	5-113†	1	,
Gonidea	clam	5	<2-11.3	10	,
Chironomus	midge	52	13-727*	0.001	2.
Tabanids		42	N.D.-1250*	0.001	,
Pontoporeia	amphipod	460	410-540	0.01	,
Gammarus	,,	30	29-31	0.01	,
Limnephilus	caddisfly	33	–	0.1	,
Sialis	alderfly	16	–	1	,
	crayfish	123	40-4100*	10	2.
Procambarus	,,	500	N.D.-1820*	,,	,
Cambarus	,,	9	7-2725*	,,	,
	,,	2	–	,,	,
Orconectes	,,	52†	18-375†	,,	,
Paranephrops	,,	30	10-870*	,,	,

* Range includes treated and untreated areas, or pre-treatment and post-treatm samples.
† Converted assuming dry weight = 10% wet weight.
N.D. = None detectable.

Reference	Remarks
odwell *et al.*, 1967	
⟨, 1970	estimated mean and range
⟨, 1971a	
⟨, 1971b	
ʾvey *et al.*, (pers. comm.)	oceanic, representative estimate, 4 locations
odwell *et al.*, 1967	
binson *et al.*, 1967	DDE only
,, ,, ,,	,, ,,
ʾvey *et al.*, (pers. comm.)	oceanic, mean 2 locations
ton and Ruzicka, 1967	oceanic
⟨, 1971b	estimated mean and range
ebrough *et al.*, 1967	

Reference	Remarks
deen and Duffy, 1970	controls
,, ,, ,,	,,
wn and Brown, 1970	,,
ʾrhardt *et al.*, 1970	1 year post-treatment
dsil and Johnson, 1968	representative estimate for 1.5 years
wn and Brown, 1970	controls
, ,, ,,	,,
key *et al.*, 1966	
ubry *et al.*, 1968	mean, 2 locations
,, ,, ,,	controls
,, ,, ,,	treated area
e *et al.*, 1967	mean, 3 locations, 1 year post-treatment
dges *et al.*, 1963	1 year post-treatment
nond *et al.*, 1968	controls
nnon *et al.*, 1970	
ʾrhardt *et al.*, 1970	1 year post-treatment
pkins *et al.*, 1966	mean, controls

hypothesis, we posited that residue uptake might be related to metabolic rate, which itself is related to the familiar power function of body weight. According to this oversimplified view, observed ΣDDT residue variations might be adequately described by a power function of body weight comparable to that for metabolic rate. Alternatively, if accumulation of residue through the food chain is an important concentration process in aquatic environments, then organisms existing at higher trophic levels should contain higher residue concentrations.

Published residue data are not often accompanied by supporting biological information on the organisms examined. Accordingly, for each residue value cited in Tables 4.1-4.3 we assigned our own assessment of the trophic level, together with an order of magnitude estimate of the likely body weight of the test organism. Residue determinations for aquatic plants (Tables 4.1 and 4.2) were included in the calculations to increase the range of values. The relationships between residue level and either trophic level (Figure 4.2) or body weight (Figure 4.1) are evidently weak at best. The regression coefficient of log residue concentration on log trophic level is weak ($r = .1410$), and is further weakened ($r = .0498$) by the removal of data for aquatic plants. Regression of log trophic level on log body weight is moderately strong ($r = .4458$), and similar regression of log residue level on log body weight indicates no relationship ($r = -.0142$). Repeating the latter calculation without data for aquatic plants improves the correlation slightly ($r = .1219$). Apart from the relationship between estimates for trophic level and body weight, none of the other regression coefficients are significant ($P > .10$, t-test). The numerous factors contributing to the great variability of the assembled data have obscured any relationships that may exist between either trophic level or body weight and residue concentration.

Numerous authors (e.g. Hickey *et al.*, 1966; Robinson, 1967) have suggested that residue accumulation may be influenced by trophic level, and indeed the conclusion seems inescapable for organisms, such as birds and mammals, that are likely to acquire residues principally by ingestion. The case for aquatic invertebrates is not proven, however, because explicit investigations that satisfactorily account for the effects of confounding variables, such as body size and age, have yet to be made. No conclusions can be based on the present exercise owing to the great heterogeneity of the data and the many possible sources for error that could mask potential correlations. Various further calculations are possible with the tabulated data, but we do not feel such analysis is warranted, and defer further evaluation of the processes underlying residue accumulation for later consideration.

4.1.2   Other organochlorine contaminants
Residue levels are listed in Table 4.4 for a number of organochlorine pesticides that have been detected in various aquatic invertebrates. The listed compounds have been reported less frequently than ΣDDT, and, with

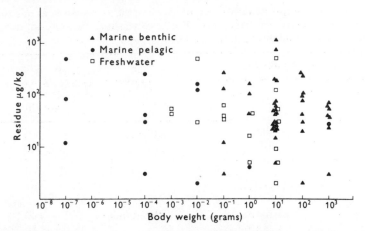

FIG. 4.1.   Relationship of ΣDDT residue concentration to body weight of organism, estimated by us to the nearest order of magnitude.

two exceptions caused by heavy local applications of aldrin/dieldrin, are generally present in lower concentrations.

Frequency of detection is only indirectly related to the true prevalence of a residue, but organochlorine pesticides have been used for many years and are detected by fundamentally similar analytical methods, thus, the citation frequencies in Table 4.4 may provide a rough measure of the relative importance of each compound as an aquatic contaminant. By this reasoning, dieldrin is second in importance to ΣDDT, a prevalence that is consistent with production statistics for pesticide manufacture (Mrak, 1969). But the rate of release of a compound to the environment is only one factor determining the prevalence of residues in aquatic organisms.

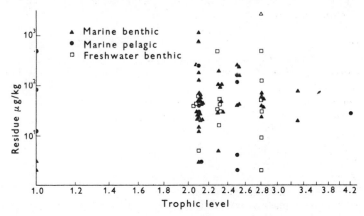

FIG. 4.2.   Relationship of ΣDDT residue concentration to our estimate of trophic level.

TABLE 4.4.   Representative residue levels of various pesticides in aquatic organisms, mostly invertebrates. For explanation of criteria used in selecting and condensing data, see text.

Name of Organism		Pesticide	Residue µg/Kg wet weight	Range µg/Kg wet weight
Genus	Common			
	scallop	Aldrin	40	—
	snail		5000	—
Crassostrea	oyster		3	N.D.-30
,,	,,		<10	<10-30
	crayfish		<1	—
	scallop	BHC-Lindane	100	—
	plankton		10	—
Crassostrea	oyster		4	N.D.-10
,,	,,		10	<10-500
Ostrea	,,		7	<10-<30
Euphausia	shrimp		5	2-7
	crayfish		1	—
Crassostrea	oyster	Chlordane	<10	<10-10
Gonidea	clam		5	<1.5-25
Fucus	alga	Dieldrin	1	—
Laminaria	,,		1	—
Cardium	cockle		18	—
Pastella	limpet		9	—
	snail		57000	—
Mytilus	mussel		1	<0.1-6
,,	,,		22	2-60
,,	,,		5	trace-57
,,	,,		7	<1-47
,,	,,		28	1-100
,,	,,		23	—
Crassostrea	oyster		4	N.D.-10
,,	,,		10	<10-30
,,	,,		17	N.D.-39
	scallop		40	—
Ostrea	oyster		13	<10-90
Corbicula	clam		18	<10-28
Gammarus	amphipod		8	3-13
	shrimp		5	—
Euphausia	,,		1	—
	microzooplankton		20	—
	macrozooplankton		160	—
	plankton		5	—
Sialis	alderfly		13	—

Reference	Remarks
Bacher, 1971	muscle tissue
Sheldon et al., 1963	heavily contaminated lake, 1 composite sample
Casper, 1967	mean, 10 samples
Bugg et al., 1967	median for 2.5 years, 6 States
Hannon et al., 1970	aldrin + dieldrin
Bacher, 1971	muscle tissue
Bailey and Hannum, 1967	
Casper, 1967	representative estimate, 10 samples
Bugg et al., 1967	median for 2.5 years, 6 States
Portmann (pers. comm.)	estimated mean and range for 2 years, 2 locations
Tatton and Ruzicka, 1967	oceanic, mean 2 samples
Hannon et al., 1970	
Bugg et al., 1967	median for 2.5 years, 6 States
Godsil and Johnson, 1968	representative estimate for 1.5 years, 1 location
Robinson et al., 1967	
,, ,, ,,	
,, ,, ,,	
,, ,, ,,	
Sheldon et al., 1963	heavily contaminated lake, 1 composite sample
Holden, 1970a	mean, Canada, 3 locations
,, ,,	mean, British Isles, 2 countries
,, ,,	representative estimate, N. Europe, 4 countries
,, ,,	representative estimate, S. Europe, 3 countries
Koeman et al., 1968	mean, 8 locations
Robinson et al., 1967	
Casper, 1967	representative estimate, 10 samples
Bugg et al., 1967	median for 2.5 years, 6 States
Modin, 1969	mean for 1 year, 2 locations
Bacher, 1971	muscle tissue
Portmann (pers. comm.)	estimated mean and range for 2 years, 2 locations
Modin, 1969	mean for 1 year, 2 locations
Moubry et al., 1968	mean, 2 locations, treated area
Bailey and Hannum, 1967	
Tatton and Ruzicka, 1967	oceanic, mean 2 samples
Robinson et al., 1967	
,, ,, ,,	
Bailey and Hannum, 1967	
Moubry et al., 1968	treated area

Table 4.4 (continued)

Name of Organism		Pesticide	Residue $\mu g/Kg$ wet weight	Range $\mu g/Kg$ wet weight
Genus	Common			
Limnephilus	caddisfly	Dieldrin	2	—
,,	,,		2	—
Homarus	lobster		24	—
Jasus	,,		<2	—
Carcinus	crab		25	—
Cancer	,,		15	—
Echinus	urchin		27	—
Gonidea	clam	Endrin	11	<0.5-90
Mytilus	mussel		91	10-360
Crassostrea	oyster		5	N.D.-20
,,	,,		<10	<10-70
Gammarus	amphipod		19	13-25
Sialis	alderfly		9	—
Limnephilus	caddisfly		3	—
Crassostrea	oyster	Heptachlor	1	N.D.-<10
,,	,,	and heptachlor	<10	—
,,	,,	epoxide	<10	—
	crayfish		<1	—
Crassostrea	oyster	Methoxychlor	<10	—
Mytilus	mussel	PCB	130	4-470
,,	,,		140	60-250
,,	,,		50	8-330
	zooplankton		217	7-450
Systellaspis	crustacean		22	8.9-35
Penaeus	shrimp		14000	—
Uca	crab		—	450-1500
Mytilus	mussel	'Telodrin®'	<1	<1-<3
Crassostrea	oyster	Camphechlor	80	<10-1000
	crayfish		<20	—

N.D.-None detectable

Various authors have suggested that the water/lipid partition coefficients of environmental contaminants (Hamelink *et al.,* 1971), or comparable coefficients related to solubility parameters (Freed, 1970), provide meaningful general indices of their affinity for accumulation in aquatic organisms. The citation frequencies in Tables 4.1-4.4 are broadly compatible with expectation from solubility criteria.

Reference	Remarks
Moubry *et al.*, 1968	control area
,, ,, ,,	treated area
Robinson *et al.*, 1967	
Aucamp *et al.*, 1971	tail flesh
Robinson *et al.*, 1967	
,, ,, ,,	
,, ,, ,,	
Godsil and Johnson, 1968	representative estimate for 1.5 years, 1 location
Koeman *et al.*, 1968	mean, 8 locations
Casper, 1967	representative estimate, 10 samples
Bugg *et al.*, 1967	median for 2.5 years, 6 States
Moubry *et al.*, 1968	mean, 2 locations, treated area
,, ,, ,,	treated area
,, ,, ,,	,, ,,
Casper, 1967	heptachlor, representative estimate, 10 samples
Bugg *et al.*, 1967	heptachlor, median for 2.5 years, 6 States
,, ,, ,,	heptachlor epoxide, median for 2.5 years, 6 States
Hannon *et al.*, 1970	heptachlor + heptachlor epoxide
Bugg *et al.*, 1967	median for 2.5 years, 6 States
Holden, 1970a	N. Europe, mean, 4 countries
Zitko, 1971	
Jensen *et al.*, 1969	mean, 3 locations
Harvey *et al.*, (pers. comm.)	N. Atlantic, mean 4 locations
,, ,, ,, ,,	N. Atlantic, mean 2 locations
Nimmo *et al.*, 1971	maximum concentration in a heavily contaminated bay
,, ,, ,,	range in a heavily contaminated bay
Koeman *et al.*, 1968	representative estimate, 8 locations
Bugg *et al.*, 1968	median for 2.5 years, 6 States
Hannon *et al.*, 1970	

The PCB's are not important pest control products, but are included in Table 4.4 because of their similarity to organochlorine pesticides as environmental contaminants (see Chapter 1). Unlike the pesticides, PCB's are probably not represented in Table 4.4 in proportion to their importance as environmental contaminants. The existence of PCB residues in natural communities is a comparatively recent discovery; thus, published residue determinations are not abundant, particularly for aquatic invertebrates. Nevertheless, the relatively high residue levels and diverse locations represented in Table 4.4 provide some measure of the probable prevalence of PCB residues in natural systems. The surprisingly high concentrations reported by Harvey *et al.* (personal communication), for oceanic invertebrates, suggest a particularly widespread distribution pattern. Our analyses of zooplankton and euphausiid samples from the Gulf of St. Lawrence (Kerr, Vass, and Addison, unpublished data) indicate that PCB residues were about equal in magnitude to ΣDDT concentrations, a relative level that is met or exceeded by values in the published literature (Harvey *et al.,* personal communication; Holden, 1970a; Jensen *et al.,* 1969; Zitko, 1971).

Taken together, the available information indicates that several organo-chlorine compounds are significant, widespread contaminants of aquatic invertebrates, and that a number of additional pesticides are of lesser, but nevertheless real importance. It remains to consider the various processes that govern the accumulation of residues by aquatic invertebrates.

4.2   THE PRINCIPAL SOURCES OF RESIDUES

Variations in pesticide levels in aquatic ecosystems depend in part on factors governing the *availability* of contaminants to the systems. The same factors appear to dictate our success in conducting experiments that represent processes occurring in nature. Various intermediary mechanisms may intervene between the release of a pesticide or similar contaminant at a given location and its eventual presentation to the organisms of aquatic ecosystems, particularly in habitats remote from the original point of dissemination. Transport within the environment, as well as rates of biological uptake, depend much on the form of residue presentation. In combination, it is these factors which define an 'availability'. Despite decades of experience and ample opportunity to profit from the clear vision of hindsight, we still do not have a quantitative grasp of the salient processes that determine residue availability in aquatic ecosystems.

In general terms, the characteristics of potential ecological contaminants are well enough understood. The contaminant must be relatively resistant to degradation and used in appreciable quantities; it must display affinity for biological systems, the high lipid/water partition coefficient of DDT being a good example; and it must exhibit one or more of the physico-chemical properties that lead to appreciable mobility in natural environments (Wurster, 1969; Macek, 1970). Given these characteristics, the potential for wide occurrence of residues seems assured, a contention amply supported by observations reported in this volume and elsewhere. But broad generalities such as these, do not in themselves provide a sufficient basis for quantitative understanding of the availability of pesticides in aquatic organisms. The generalities must be rephrased in more specific terms.

4.2.1   Initial translocation to water

The principal initial sources of pesticides to the environment have been classified by Westlake and Gunther (1966). Water bodies associated with urban regions undoubtedly receive substantial contaminant burdens from industrial and domestic outfalls (Butler, 1970; Duke *et al.,* 1970; Fay, 1970; Holden, 1970b; Koeman *et al.,* 1968), although conventional agricultural and forestry practices, together with control programmes for human disease vectors, most certainly contribute the major portion of observed aquatic contaminants (Butler, 1970; Goldberg *et al.,* 1971; Westlake and Gunther, 1966). But the present situation is not static. Current trends show important changes in the relative use of specific pesticides. DDT, for example, a major contaminant in many aquatic ecosystems, is increasingly an export commodity in terms of U.S. production, a shift that is likely to be reflected in future environmental distribution. This, and similar changes in use patterns of significant aquatic contaminants, are summarised in the 'Mrak Report' (Mrak, 1969). Unfortunately, the available statistical data on pesticide manufacture and use patterns are insufficient to support more than general consideration of initial contaminant sources. To our knowledge, no nation in the western world requires that detailed records be kept of the sales and use patterns for significant environmental contaminants, which ensures that the scientific community must do without such obviously important data.

Intentional application of pesticides directly to aquatic ecosystems is a residue source that has certainly been important (Harrington and Bidlingmayer, 1958; Hunt and Bischoff, 1960) but rather localised in scale. It is surface runoff and aerial transport that provide the major methods for translocating pesticides to aquatic ecosystems (Peterle, 1970). Spray programmes for a variety of purposes have been followed by direct pesticide runoff to aquatic systems, a process primarily associated in the case of DDT and related compounds with surface erosion of soil particles bearing adsorbed pesticide residues (Bailey and Hannum, 1967; Manigold and Schulze, 1969). It is in the nature of the hydrological cycle that the destination of contaminants introduced to water is ultimately the world's oceans, although this does not occur without some attrition en route. Losses chiefly occur either through feedback to the atmosphere (Acrees, *et al.,* 1963; Bowman *et al.,* 1964) (chapter 10) or sequestration in sediments, a process that assists degradation (Graetz *et al.,* 1970; Yule and Tomlin, 1971), particularly under anaerobic conditions (Hill and McCarthy, 1967).

4.2.2   Translocation to the sea

Nevertheless, it is increasingly clear that runoff does not contribute the major portion of contaminants to marine systems. Risebrough *et al.* (1968) described substantial atmospheric transport of organochlorines from the European-African land masses to Barbados, and Goldberg *et al.* (1971) have estimated that the annual atmospheric input of DDT to the world's oceans corresponds approximately to one quarter of the world's annual production of that compound. The estimate prepared by Goldberg *et al.* does not provide for potential recycling across the air-sea interface and is likely to be excessive for that reason, but their conclusion that aerial fallout greatly exceeds river discharge as the principal source of pesticides to the open ocean, nevertheless, seems valid. Apart from the substantial portions of pesticides that enter the atmosphere directly, by virtue of never reaching the intended target (Mrak, 1969) DDT and related compounds are known to enter the atmosphere by various other methods, including vaporization from plants and soils (Nash and Beall, 1970); thus organochlorines are available to natural systems remote from sites of pesticide application through precipitation scavenging and fallout (Peterle, 1969; Tarrant and Tatton, 1968). Aerial fallout occurs over land and sea alike, but the relevant distinction made here is that airborne contaminants constitute the *principal* residue input to oceanic environments. A particularly clear demonstration of the scale of aerial distribution of organochlorines has been documented by Cohen and Pinkerton (1966); but as Munn and Bolin (1971) have pointed out, our understanding of the physical processes involved is inadequate, in general, to support thoroughgoing quantitative analysis. Two particularly intractable aspects of the problem, concern the degradation patterns of compounds during aerial transport and the surface depletion characteristics of pesticides in the turbulent boundary layer at the air-sea interface.

4.2.3   Events in the aquatic surface layer

There is increasing evidence that events such as transference of pesticides, occurring in the region of the water surface layer are extremely complex, and crucial to our understanding of the dynamics of pesticide translocation to aquatic organisms. Airborne distribution of organochlorine contaminants appears to be predominantly in the form of residues adsorbed to particulate matter (Cohen and Pinkerton, 1966; Risebrough *et al.*, 1968), although the latter authors point out that PCB's were not associated with airborne particles in their analyses, implying that PCB transport may occur primarily in the vapour phase. However, they are transported, airborne contaminants must enter the sea via the surface layer.

Nicholson (1967) noted that concentration of residues in marine organisms occurred despite the enormous dilution potential of the oceans, implying the existence of an efficient process for concentrating pesticide fallout at the ocean surface. Even in the absence of obvious surface slicks, Garret (1967) identified lipid components in marine water surface films, that included free fatty acids, fatty acid esters, fatty alcohols and hydrocarbons. On the basis of observations that natural concentrations of hydrophobic compounds are generally high in the surface microlayer, Williams (1967) suggested that the sea surface could serve as an efficient intermediary sink and transfer zone for airborne organic and inorganic contaminants. Seba and Corcoran (1969) have demonstrated concentrations of organochlorine pesticides in surface slicks sampled near shore, confirming that the putative concentration mechanism is effective for pesticides. The surface microenvironment is known to support a biologically active community (Parker and Barsom, 1970), which must serve to some degree as a means of transferring pesticide residue to the biota of the underlying water column.

Pesticide transfer across the air-sea interface is undoubtedly cyclic, however, with gross transfer exceeding net uptake by the aquatic system. Blanchard and Syzdek (1970), among others, have demonstrated mechanisms capable of returning contaminants from water to the atmosphere. Williams (1967) has suggested that such mechanisms could support a sufficiently high rate of flux across the surface interface to account for otherwise intractable aspects of oceanic nitrogen budgets; pesticides could conceivably be subject to recycling on a similar scale. At present, there is no sound basis for evaluation of exchange rates of pesticides and similar contaminants, between natural waters and the atmosphere, but the existence of mechanisms for vertical transport in the water suggests a net loss to the exchange process by virtue of which the aquatic system would serve as a sink for airborne contaminants.

It should be noted that particular attention has been placed here, and in subsequent sections, upon movement of the persistent organochlorine compounds, because it is these agents which are of major residue significance on a large geographic scale. Products that are soluble in water, and all compounds directly entering the water in their original solvent or other formulation, may of course exhibit different dispersal patterns, at least in detail (Cope, 1966; Macek, 1970).

4.2.4 Translocation beneath the surface

Although pesticide concentrations have been shown to occur in the surface film, the processes by which the residues are transferred to the biota are not at all well understood. Palumbo (1961) concluded, from his review of the literature, that the radioisotope fallout that remains in particulate form occurs in marine organisms in the highest concentrations. This apparently results from direct adsorption of particles to the external surfaces of the plankton, and ingestion of suspended matter bearing adsorbed contaminants. If pesticide fallout is comparable, its availability to aquatic invertebrates will be correlated with the processes governing the genesis and distribution of particulate matter in the near surface zone. For reasons reviewed by Riley (1970), a portion of the particulate matter found near the surface is known to be formed from materials composing the surface slick; we might therefore suppose that micro-particulates in water carry pesticides, a view confirmed by Pfister et al. (1969). Although it has yet to be established that organic particles bearing pesticides are formed at the surface, an appreciable but unknown portion of the particulate matter in the water column is presumably grazed directly by the plankton, indicating a direct means, apart from the neuston, for input of pesticide residues to the pelagic community. Sutcliffe *et al.* (1971) have provided evidence of increased quantities of particulate matter in the convergence areas between Langmuir cells, a specific concentration and transport mechanism that can be extended to convective downdrafts in general. As is pointed out by Riley (1970), particles less than 5 μ in diameter are unlikely to be transported to appreciable depths by sinking or convective downdrafts; but aggregation to produce larger particles, or recycling through aggregation processes at the sea surface, can be suggested as auxiliary mechanisms of indeterminate importance for providing pesticides in concentrated form to the biota of the mixed layer. Parenthetically, it should also be noted that surface macrophytes (*Sargassum* sp.) have been shown to be effective collectors of radioactive fallout (Angino *et al.,* 1965; Shelby, 1963), suggesting a potentially important concentration mechanism for certain oceanic regions.

Translocation of organic particulates beneath the mixed zone is largely a mystery at the present time, although current information favours a dynamic equilibrium between formation and utilization of organic particles with depth (Riley, 1970), suggesting that organic particulate matter in itself affords an unlikely primary mechanism for introduction of pesticides to deeper levels, except for the presumably small quantities of contaminant that could be transported by rapidly-sinking larger particulates. Fowler and Small (1967) have suggested that cast exuvia of *Euphausia* sp., an abundant vertical migrant, may constitute a potential mechanism for vertical transport of ^{65}Zn in the sea; a mechanism that might apply equally well to vertical transport of pesticides. Polikarpov (1966) reviewed earlier literature on the role of organisms in transferring radioisotopes to the deeper layers of the sea. Indeed, we have little information as to whether pesticides do in fact occur offshore in the invertebrates of mesopelagic or benthic regions, beyond some preliminary information such as that obtained by Harvey *et al.* (personal communication) for a mesopelagic decapod collected at two locations (Table 4.2). Harvey and colleagues have also suggested that these and similar vertical migrants may constitute a significant source of pesticide to the depths beneath the mixed zone.

4.2.5 Available forms of residue

All told, there is surprisingly little evidence with which to assess the relative contribution of dissolved and particulate pesticide residues to aquatic organisms. Apart from particulate matter formed from the surface film, both inorganic and organic particulate matter is also formed *in situ* in the water column, and it is reasonable to suppose that this material too provides substrate for contaminants cycled within pelagic ecosystems. Conceivably, organic particulate matter from either source could contain incorporated residues, available to aquatic organisms only by particle ingestion, in addition to adsorbed contaminants that would be available through solvent partitioning on surface contact with organisms. Such evidence as exists favours the view that water-borne pesticides will be chiefly available as residues bound to particulate matter, a suggestion supported by the marked ability of aquatic sediments to concentrate residues.

Bailey and Hannum (1967) reported a marked inverse correlation between grain size and high pesticide residue levels in sediments from rivers and estuaries, but the relationship they observed was confounded somewhat by an autocorrelation with organic content. Conversely, Nimmo *et al.* (1971) observed no evident correlation between sediment grain sizes and their PCB burden, but noted that PCB residues were leached more rapidly from coarse sediments exposed to running water, indicating that an inverse correlation of grain size with residue level might have become established in time. Odum *et al.* (1969) found high DDT residue levels but a relatively weak correlation with sediment particle size in a marsh ecosystem, a difference that is apparently attributable to the higher organic content of the marsh detritus.

Various investigators have shown that pesticide residues tend to be associated with particulate matter suspended in the water column. Meeks (1968) found that DDT was about evenly distributed between filtrate and filter when surface water samples were passed through an 0.45 μ filter in the weeks following treatment of a freshwater marsh. A similar analysis of Detroit River water containing 40 ppm$\times 10^6$ of *p,p'*-DDT showed that 7 ppm$\times 10^6$ remained in the filtrate following passage through an 0.5 μ membrane filter (Hartung and Klingler, 1970). Cox (1971a) inferred from indirect evidence that the major portion of pesticide present in raw seawater may be adsorbed to particulate matter less than 1-2 μ in diameter, but his inference was not convincing in the absence of direct determinations. Pfister *et al.* (1969) provided clear evidence of a number of organochlorine pesticides associated with particles greater than 0.15 μ, obtained from open water in Lake Erie, but their analysis did not identify the proportion of pesticide recovered in particulate form, relative to the total quantity present in the raw water.

Observed distributions of particulate matter in seawater (Bader, 1970; Sheldon *et al.*, in press) indicate that the largest potential surface area for adsorption is provided overwhelmingly by the smaller particles, suggesting on *a priori* grounds that the greater portion of pesticide residues may be initially available to aquatic invertebrates in the form of microparticles less than about 1 μ in diameter. The scanty available data support this conclusion for both large lakes and marine systems, although the suggested threshold particle size might need to be revised upward for strong current systems or similar extenuating factors. It is probable, that freshwater and inshore marine environments would provide a substantially lower proportion of organic particulates, in some instances, than would be the case in marine pelagic zones, a difference that would influence the ease of availability of adsorbed residues. Indeed, the availability of residues in lakes seems to be mainly determined by factors affecting the abundance and distribution of suspended particulate matter. Thus, Terriere *et al.* (1966) noted that camphechlor remained toxic for much longer in the less productive of their study lakes, and that turnover of the water mass resuspended particulate matter bearing adsorbed camphechlor.

Necessarily, inference based on the foregoing cannot be taken as substantiated fact. But taken together, the available information supports the view, that in general, the major source of water-borne pesticide, available to the pelagic biota of aquatic environments, is in the form of residues bound to particulate matter suspended in the water column. The hypothesis is not new, having been suggested at least as long ago as 1963 (Hickey *et al.*, 1966), but a clear need for unequivocal confirmation still exists.

Critical examination of the hypothesis would serve several ends. Pfister *et al.* (1969) have already pointed out that the charcoal extraction method for pesticide analysis of water probably leads to errors because of its failure to detect microparticulate residues; the error may be particularly misleading in the biological sense. More seriously, until we have a better understanding of the ways in which residues become biologically available, it is impossible to assign initial input parameters to the various trophic levels with any confidence, much less undertake analyses offering greater insight into the dynamics of pesticide transfer in aquatic ecosystems. Finally, it appears that existing laboratory analyses of the kinetics of pesticide uptake by various aquatic invertebrates, can only be applied to natural systems with great caution, until we can be certain that the availability of pesticides in natural systems is adequately represented by current experimental methods.

Taken together, the available information on the mechanisms governing pesticide translocation between the initial point of release to the environment, and the ultimate availability of residues to the invertebrates of aquatic communities does not provide grounds for complacency. Suitable mathematical analyses and models of environmental systems for residue transfer are particularly needed, but the descriptive picture is itself weak in various places, offering little hope of rigorous quantitative analyses of the salient transfer processes on a significant geographic scale, in the near future.

4.3   FACTORS INFLUENCING THE BURDEN OF RESIDUE

The residue level in any organism is an observed consequence of a balance of uptake and clearance processes. In reality, however, the complexity of natural ecological transfer systems, with diverse mechanisms operating on various temporal and spatial scales, has steadfastly defied description of the energy and nutrient exchange network sufficient for production forecasts. At least equal difficulties obscure the dynamics of residue formation in natural systems. The potential for complexity is particularly evident in the heterogenous assemblage comprising the aquatic invertebrates. For this reason, it is initially advantageous to consider the process of pesticide accumulation in its simplest terms; to provide an analytical frame for subsequent examination of the various factors that accompany ecological reality in aquatic environments, and as a means of emphasising important weaknesses in our current understanding of the relevant processes in aquatic invertebrates.

4.3.1 Biological uptake of residues

The principal methods of acquisition of pesticide residues by aquatic invertebrates are direct uptake of water-borne residues, and indirect uptake by ingestion of contaminated food. For many invertebrates, particularly the smaller forms, the rate of exposure to the surrounding medium is a measure of the rate of contact with both water-borne residues and contaminated food. The quantitative exposure of invertebrate species to the aquatic medium can be highly variable and exceedingly difficult to estimate. Many invertebrates are primarily exposed to their medium in a passive sense, in that the rate of exposure to contaminants depends upon exogenous micro-turbulence or diffusion processes (Södergren, 1968), which in themselves do not appear to have been sufficiently well defined with respect to invertebrates, to permit general analytic description. But residue intake in such circumstances must in part be a function of surface area, which, depending upon variable morphology, varies approximately as the 2/3 power of body weight. The precise relationship of surface area and body weight would, of course, need to be established for any given organism, using appropriate techniques such as that described by Zeidman (1948).

Endogenous factors, either respiratory or feeding processes for example, could in principle account for the rate of contact of many active organisms with aquatic contaminants. Respiratory exposure to water is correlated with metabolic rate, which is usually observed to vary as the 3/4 to 4/5 power of the body weight, for aquatic invertebrates (Conover, 1968; Sushchenya, 1970). Feeding processes are themselves only indirectly related to metabolic rate, although comparable power functions of body weight can sometimes usefully describe feeding relationships (Sushchenya, 1970).

Accordingly, there is reason to believe that residue uptake, by both passive and active invertebrates, could often be proportional to body weight, raised to some exponent greater than 2/3, but less than unity. Because of the similarity of the predicted exponents, it is difficult, on the basis of existing data, to distinguish uptake processes that are surface-proportional from processes that are directly or indirectly metabolism-proportional, without very careful observation and analysis. Such analyses have yet to be made of pesticide uptake by aquatic invertebrates.

By far the major portion of pesticide uptake investigations using aquatic invertebrates have dealt with whole-body residues observed following exposure to contaminated water. In part, this emphasis may reflect a relative consensus on the importance of direct uptake from water in nature, but if so, the rationale has not usually been made explicit. It seems possible that the convenience of relatively simple experimental design may also have played a role in determining the choice of contaminated water as the form of pesticide administration to organisms. Whichever is true, diverse methods have been used for presenting the comparatively insoluble pesticides to the test organisms, often in unnaturally high concentrations.

Most often, the contaminants have been dispersed in the test media after solution in organic solvents such as acetone or various oils. Occasionally, the contaminant has been solubilised with a surfactant (Wildish and Zitko, 1971), shaken with water (Richards and Cutkomp, 1946), or adsorbed to a substrate which in turn is exposed to water (Rose and McIntire, 1970). The resulting form in which the pesticide is initially presented to the test organisms is then primarily a function of the type of carrier, the amount of agitation used to initiate and sustain the dispersion, the water solubility and final concentration of pesticide, and the amount of particulate matter present or generated as potential substrate. Taken together, methodological variables such as these suggest that different experimental regimes may have provided pesticides in ultimate size units ranging from the molecular to the nearly visible. In a practical sense, it may be difficult to distinguish between absorption through the integument and uptake by ingestion, in circumstances where appreciable particulate substrate is available.

Such criticisms are easily made, but careful control of the mode of pesticide treatment is not so easy. Even the exceptional precautions described by Södergren (1968), which included careful filtration tests, could well be compromised for some purposes, by subsequent aeration and other processes known to generate particulate matter (Riley, 1970). The intention here is not simply to criticise, but rather to emphasise the many uncertainties that necessarily attend extrapolation to natural systems on the basis of existing experimental techniques. Perhaps the exact form in which a pesticide reaches the test organism is of small consequence, but it is more reasonable to suppose that presentation of contaminants in such a wide variety of potential forms will certainly sometimes appreciably affect the experimental results, effects that could be important when making comparisons among experiments or in assessing the relevance of any specific experiment to natural systems.

4.3.2   Residues acquired directly from water

Direct uptake of residues from the surrounding water necessitates surface contact. Despite the wide variety of external body surfaces collectively offered by the aquatic invertebrates as a group, experiments have generally indicated uptake kinetics that are at least consistent with surface adsorption as the initial stage of the uptake process (Hamelink *et al.,* 1971; Richards and Cutkomp, 1946; Wilkes and Weiss, 1971). Surface adsorption presumably requires an intermediate partitioning between particle and integument for particle-borne contaminants, indicating that biological uptake of the dissolved and particulate fractions of contaminants, present in natural environments, may proceed at different rates. By this reasoning, analytical determinations of residue levels in water do not necessarily indicate the amount of residues available for uptake. Some portion of the detected residues could be strongly adsorbed or internally bound to particulate matter, and biologically unavailable for this reason. Thus, Haven (1969) found that concentration of the herbicide diquat, by clams and oysters, was largely in the form of residues adsorbed to inorganic particulate matter accumulated at the integument, an uptake process that was incomplete and potentially reversible in this instance. Nevertheless, contaminants with favourable partition coefficients that adsorb less strongly to inorganics than does diquat could doubtless be acquired from particulate matter contacted by the integument. Consequently, residue uptake in natural systems can be said to depend in part upon an intermediary partitioning process between contaminated particle and organism, before absorption can proceed.

The precautions taken by Södergren (1968), in measuring DDT uptake by both living and dead algal cells, provide assurance that the pesticide was initially in solution and that his measurements are not confounded with adsorbed residues. His observations are consistent with an uptake mechanism involving surface adsorption, followed immediately by very rapid absorption across the cell wall. Absorption is known to be a relatively slower process for other organisms, however, because they have integuments that are less permeable to pesticides than the cell wall of *Chlorella.* O'Brien (1967) noted that the rate of DDT transfer into and through human and cockroach integuments, measured in both cases as 'half-lives' for disappearance from the integument surface of approximately 24 hours, are surprisingly similar, despite obvious differences in integument composition. He interpreted the evidence as support for the view that pesticide uptake through the integument is primarily mediated by solvent partitioning between the surface and underlying body substances, and is not necessarily determined by the composition of the integument itself. Apparently, for this reason, pesticide uptake tends to occur over the entire body surface of smaller organisms, but is confined to particular sites of organisms with more substantial integuments, such as mosquitofish (Ferguson *et al.,* 1966). Wildish and Zitko (1971) examined the effect of extirpating the branchiae of *Gammarus* on the resulting rate of PCB uptake, but could not demonstrate any clear effect, a similar result to that obtained in studies of the effects of comparable mutilations of respiratory organs, on rate of oxygen uptake of various small aquatic invertebrates. But permeability to contaminants may be a highly variable characteristic of invertebrates, even within taxonomic groups. Thus,

sediment, and filter feeders by uptake of particulate matter suspended in the water column. Thus, Odum *et al.* (1969) demonstrated that fiddler crabs could accumulate DDT effectively from the highly organic sediments of a marsh ecosystem, and Butler (1966a, b, c, 1969a, b) has reported extensively on the ability of various lamellibranchs, particularly oysters, to accumulate pesticides during the process of feeding, although the distinction between respiratory and dietary uptake is in this case problematical.

The available evidence provides only partial support for the view that pesticide residues carried by particulate matter are available by ingestion to zooplankton grazers, filter feeders, phagotrophs, and other detritivores. The relative importance of these and other types of feeding process as methods of residue uptake from particulate matter remains to be determined.

Once incorporated into living organisms, pesticides are transferred to other organisms by prey-predator interactions, but few investigators have sought explicit confirmation of the process for aquatic invertebrates. Consequently, there is little comparative information available on uptake rates or efficiencies, or even of the importance of food-chain transfer as a source of residue. On the basis of his experiments, Cox (1971b) estimated that ingestion of contaminated prey was a sufficient mechanism to account for the DDT concentrations he observed in natural populations of *Euphausia pacifica,* but explicit comparison of ingestion and alternative uptake routes under natural conditions is required to resolve the question.

Transfer of pesticides through food-chains is likely to be a generally efficient process. In view of the lipophilic characteristics of the prevalent organochlorine contaminants, Cox (1971b) has suggested that fatty components of animal diets contribute a major portion of fat-soluble pesticide in a form that is suitable for direct incorporation into the depot fats of the predator. Using two slightly different techniques, Cox determined DDT uptake efficiencies from studies in which *Artemia* labelled with C^{14}-DDT were fed to *Euphausia pacifica.* His efficiency estimates of 76 and 62% were determined over two and three day-time periods respectively, and therefore underestimate the true value ·to some unknown extent if appreciable DDT clearance occurred during the experiments. Lasker (1966) estimated 84% total carbon assimilation from the diet in feeding experiments using the same species, but it is not possible to determine whether the disparities are due to underestimation by Cox, or due to real differences between the assimilation efficiencies for DDT and for dietary carbon.

4.3.4 Clearance of residues

Metabolic clearance of a variety of substances often approximates to exponential disappearance. Indeed, the concept of 'half-life' that has been borrowed from the radioisotope literature, and widely applied to pesticide clearance, implies processes that can be described with a single exponential term, although the description is not always acceptable.

Numerous observations of pesticide elimination have been made that approximate to the simple exponential form. Thus, Södergren (1968) found that DDT concentration in *Chlorella* cultures decreased semi-logarithmically with time, an observation that appears comparable to lindane clearance data for both *Chlorella* and *Chlamydomonas* (Sweeney, 1969). Cox (1971b) observed a roughly similar disappearance pattern for DDT in *Euphausia pacifica*, and so have many other workers in representatives of various taxonomic groups. Frequently, however, the observed decline of residue levels with time cannot be satisfactorily described with a single exponential term, and two or sometimes more, exponential terms are required, implying that the organism is inadequately represented as a single homogeneous pool, or compartment. Robinson *et al.* (1969) have provided particularly clear examples of the use of both one- and two-compartment models for describing HEOD clearance from various tissues of the laboratory rat. Ordinarily, the initial rapid phase of clearance is ascribed, in the case of two-compartment systems, to an initial rapid loss from a 'non-assimilated' residue pool, followed by a longer phase of elimination from an 'assimilated' pool. In ecological terms, residue elimination from the assimilated pool is most often of greater interest. Clearly, the half-life concept must be applied judiciously to pesticide clearance.

A prevalent, but almost certainly erroneous, assumption in the pesticide literature has been to treat residue clearance functions implicitly as constant relationships for a given combination of species and pesticide. That is, the implication is that an observed half-life remains constant despite changes in organism size, activity level, feeding rate, and similar variables. Indeed, the effects on pesticide clearance, of such parameters as body size seem to have attracted scant attention, despite the obvious expectation that clearance, being chiefly a metabolic process, should correlate with metabolic rate.

Although data exist which indicate that increased body size has the predictable effect of slowing residue clearance in a marine crustacean (Cox, 1971b), the problem seems not to have been explicitly investigated for any aquatic invertebrate. In the absence of appropriate information on pesticide clearance, it is therefore useful to consult the radioecology literature, where the equivalent phenomenon was recognised and investigated at least a decade ago. Odum (1961) used ^{65}Zn to label three species of invertebrates, including a marine isopod (*Idothea baltica*), and found that the resulting clearance pattern contained a 'fast' and a 'slow' phase, implying a two-compartment system. He observed that the half-life for the slow phase decreased in direct proportion to increasing activity level or temperature, providing clear evidence in support of the expected inverse correlation of half-life with metabolic rate. Mishima and Odum (1966) conducted similar investigations using the snail *Littorina irrorata*, and observed that ^{65}Zn clearance followed a comparable pattern. As with the earlier work, the observed half-life of the slow phase varied inversely with temperature or activity, again implying the expected cor-

where the numerical coefficients correspond to units of ml/organism/day when dry body weight is given in mg. The quantity of food ingested is also a function of its density. Although arbitrary, a numerical prey density of 25,000 cells/ml is likely to represent a comparatively favourable prey density in nature. If *Chlamydomonas* cells, which average $.248 \times 10^{-9}$ grams each (Richman, 1958), provide a suitable prey, it follows that 6.2×10^{-6} grams of prey are ingested per ml of water filtered. Solving equation 4.4 for f, and multiplying by 6.2×10^{-6} gives r_{10} for these feeding conditions, and residue uptake with the diet is then described by:

$$a_0 R_{10} = 12 \times 10^{-6} a_r \ (\ln W_1 + 6.255). \qquad (4.5)$$

A coefficient of residue assimilation, c_r, has not been appropriately estimated for *Daphnia,* but assuming a value of unity will not affect the use to be made of the model.

4.3.7   Residue absorbed through the integument

In itself, absorption is a first order function of the concentration of residue at the integument surface (O'Brien, 1967), while surface adsorption in turn must surely be a function of the rate of exposure to the surrounding medium. That is, at low external residue concentrations, absorption must vary with the relative flow rate of the surrounding water. To avoid having to formulate factors accounting for exogenous water flow, it is convenient to select an active organism, one that creates its own water currents, for illustrative use in the model. Equation 4.4 describes the filtration rate of *Daphnia* at 20°C. By taking 4 as a description of exposure to water, the rate of uptake of water-borne contaminants is given by:

$$a_0 R_{10} = 1.934 a_w \ (\ln W_1 + 6.255), \qquad (4.6)$$

where a_w represents the concentration of water-borne residues, apart from contaminated food. Equation 4.6 suffices as an uptake term for use in the model.

4.3.8   Residue clearance

Explicit formulation of the effect of metabolic rate on pesticide clearance is also necessary for use in the model. The simplest hypothesis is that k_{01}, the rate constant from equation 4.1, varies in direct proportion to metabolic rate per unit of body weight. That is, using the familiar power function of body weight to describe metabolic rate gives:

$$k_{01} = ma W^{\gamma - 1}, \qquad (4.7)$$

where m is constant.

Some assurance of the required constancy of m can be derived from determinations of DDT elimination by euphausiids (Cox, 1971b). Rough estimates of k_{01} were made from graphs and substituted in equation 4.7, together with calculations of metabolism that were made with the general body weight-metabolism equation for crustacea given by Sushchenya (1970). The estimates of m were 20, 19 and 22 for organisms weighing 2, 3 and 10 mg respectively, and considering the crude method of derivation there is some indication of constancy. Unfortunately, the 2-hour period of DDT uptake used in these experiments was so brief that it seems unlikely that appreciable portions of the residue entered the assimilated pool, resulting in an atypically brief half-life. The only available invertebrate information from which to derive an estimate of m for the assimilated pool, appears to be the ^{65}Zn data discussed earlier. Calculations of m based upon data provided by Mishima and Odum (1966) yield .0022, .0024 and .0023 for daily clearance at 15, 25 and 30°C respectively. Other half-life data provided in the paper, and in Odum (1961), were not accompanied by metabolic data and could not be used for direct estimates of m, but further assurance of constancy was provided by calculations using the temperature factors for metabolism set out by Winberg (1956). Assuming that the ^{65}Zn clearance pattern is representative of at least some pesticides, a value of m equal to .002 is adequate for use in the model. The body weight-metabolism relationship estimated by Richman (1958) for *Daphnia* has been printed incorrectly in his paper, but least-squares calculation from his data gives:

$$T = .1104W_1^{-.11},$$

where metabolism, T, is expressed as ml O_2/mg dry body weight. Accordingly, assuming that the residue clearance pattern of *Daphnia* corresponds to the slow phase for *Littorina*, equation 4.7 can be rephrased explicity for *Daphnia* at 20°C as:

$$k_{01} = .0002W_1^{-.11}. \tag{4.8}$$

4.3.9 Evaluation of the model

For constant W_1, equations 4.5 and 4.6 describe residue uptake from the food and from water respectively, in terms corresponding to the a_0R_{10} of equation 4.2. Either 4.5 or 4.6, or both in combination, may therefore be used as an input function to the model. Equation 4.8 provides a rate constant for residue clearance, and is also in terms appropriate to equation 4.2. Accordingly, substitution of equations 4.5, 4.6 and 4.8 in equation 4.2 provides a statement of the model for an hypothetical invertebrate combining pesticide assimilation properties attributable to *Daphnia*, and the principal clearance pattern of *Littorina*. Making the substitution and rearranging gives:

$$a_1 = (.06a_r + 9670a_w)(\ln W_1 + 6.255)W_1^{-.89} - [(.06a_r + 9670a_w)$$

$$(\ln W_1 + 6.255)W_1^{-.89} - a_1(0)]\, e^{-.0002tW_1^{-.11}}. \tag{4.9}$$

residue is therefore necessarily less efficient in nature, in part because of an intermediary partitioning process between organism and particulate substrate. Conversely, uptake by ingestion assumes relatively greater importance in nature, particularly among filter-feeding invertebrates. In terms of the model, equations 4.5 and 4.6 are therefore somewhat artificial. Both are based on filtration rate, but the ingestion equation 4.5 assumes that only prey organisms are ingested, and equation 4.6 assumes implicitly that only soluble contaminants are contacted. While generally true in laboratory environments, neither assumption is likely to be valid for natural systems that are chronically contaminated. Accordingly, for description of natural systems, the numerical coefficient of equation 4.5 should be considerably increased to account for ingestion of non-food or relatively refractory food items; and an assimilation coefficient should be added to 4.6, decreasing the uptake of water-borne residues to account for the substantial portion that is relatively unavailable because of particulate substrate. Apart from the *Daphnia* model, such considerations would be generally true of natural systems, at least to the extent that the availability of water-borne pesticides will be low, relative to nominal residue determinations for water.

If water-borne residues in nature are relatively less available for uptake compared to those in laboratory environments, then food-chain transfer of residues becomes more important. This is a different conclusion to that reached regarding the interpretation of ecological parameters for radio-isotope concentration in aquatic systems. On the basis of his detailed review of the relevant kinetics, Polikarpov (1966) concluded that food-chain concentration of radioisotopes, by marine organisms, is minor compared to direct absorption from water. In comparison with radio-isotopes, the common pesticide contaminants exhibit extremely low water solubility and marked affinity for particulate substrate, which serves to reduce their availability for direct uptake from natural waters. Food-chain transfer could therefore become an important source of pesticide residue, particularly to larger invertebrates.

A second consideration is the relative duration of exposure for large and small organisms. In general terms, generation time of organisms increases steeply as a power function of body size (Bonner, 1965). The smaller invertebrates, composing the lower trophic levels of aquatic communities, therefore manifest generally higher population turnover rates and shorter lifetime exposures to contaminants. Taken together, short life spans and comparatively low availability of water-borne residues for direct uptake, coupled with the modest concentration factors incurred with each food-chain transfer, would provide a sufficient explanation of higher residue concentrations in the larger invertebrates of natural communities, if they do in fact occur.

4.4 HAZARDS CAUSED BY RESIDUES

Residues of certain pesticides are widely distributed in the biosphere. It remains to evaluate the hazards that these, and potential future contaminants, pose to the invertebrate fauna of aquatic communities. Many pesticide hazards are conveniently identifiable as effects on individual species, but the hazards of general significance are chiefly measurable as changes in integral ecological systems. For this reason, we distinguish two general categories of hazard: microsystem effects discernible by observation of individual species, and macrosystem effects expressed at the community level.

4.4.1 Microsystem effects—individual species
Death is the most obvious biological consequence of exposure to toxic compounds. Understandably, a substantial portion of the published literature is given over to evaluation of the lethal doses of diverse pest control products to various species of organism. The published record, in turn, reflects but a small portion of the tolerance evaluations that have been made, particularly by government agencies and the private sector. Data of this kind are needed to meet specific requirements of pesticide development and registration. Much of the available data does not receive wide circulation. Nevertheless, the general scientific literature is sufficient to permit a general appraisal of lethal hazards among the aquatic invertebrates.

The smaller crustacea are killed by relatively brief laboratory exposure to DDT and similar compounds, often at aqueous concentrations in the order of $0.1\text{-}10 \ ppm \times 10^3$ (Anderson, 1945; Grosch, 1967; Naqvi and Ferguson, 1970; Nimmo et al., 1970; Sanders and Cope, 1966). Larger crustacea tend to be more tolerant, although Lowe (1965) reports that blue crabs are killed in a few days by $0.5 \ ppm \times 10^3$ of DDT. Insects too are, of course, sensitive to concentrations of various insecticides in the low 10^{-3} ppm range (Jensen and Gauffin, 1964), but various experiments that tested both insects and crustaceans with similar methods suggest that, with many exceptions, insects may be somewhat more resistant to insecticides than crustaceans (Macek and Sanders, 1970; Richards and Cutkomp, 1946). On average, both taxa are less sensitive to herbicides (Sanders, 1970; Wilson and Bond, 1969), with no obvious distinction between them.

With the exception of coelenterates, most other phyla of aquatic invertebrates appear rather more resistant to pesticides than are arthropods, although resistance is relative, and the majority of aquatic invertebrates are susceptible in varying degrees (Richards and Cutkomp, 1946). The relative tolerance of tubificids to various pesticides (Whitten and Goodnight, 1966) merits particular note, as these worms are sometimes useful biological indicators of polluted waters.

In common with many other stress factors, sensitivity to pesticides differs with developmental stage of the test organism, and with temperature and other environmental identities. Early stages of development are usually less tolerant to pesticide residues (Davis and Hidu, 1969; Davis, 1961). A remarkable instance of invertebrate sensitivity was that development of an adult calanoid copepod from the nauplear stages was completely blocked when hatched from egg-bearing females exposed to 10 ppm×10^6 DDT in seawater (Menzel et al., unpublished). Significant mortality occurred at 5 ppm×10^6 (SCEP, 1970). The effects of factors such as temperature on pesticide tolerance have not been thoroughly investigated for aquatic invertebrates. Data provided by Sanders and Cope (1966) suggest a negative temperature coefficient for the expression of toxic effects, as do observations made by Richards and Cutkomp (1946). The latter authors also report a positive temperature coefficient at higher DDT exposure levels, which led them to suggest that the rate of adsorption of DDT may have limited uptake at the lower experimental concentrations.

It has been known for some time that pesticides can act as a selective factor in the development of resistant strains. Numerous examples have been documented among aquatic insects. For example, increased DDT resistance by blackflies, *Simulium damnosum,* was observed by Walsh (1970). Similarly, Naqvi and Ferguson (1970) describe several populations of freshwater shrimp, *Palaemonetes kadiakenesis,* that developed increased resistance to a variety of pesticides, compared to a population from a relatively uncontaminated environment. Cory et al. (1970) have recently suggested a correlation between observable chromosome changes in wild *Drosophila pseudoobscura* and exposure to environmental DDT residues. Although other factors can also be important, development of resistance to contaminants obviously occurs most rapidly for organisms with a high reproductive potential and short generation time. Among the aquatic invertebrates, capacity for genetic response to pesticides is therefore quite variable, in the sense that an ecological time scale to some invertebrates is an evolutionary time scale to others. On the basis of the differing potential of different species for development of resistant strains, there is considerable latitude for a modified species composition in aquatic invertebrate communities exposed to chronic or repeated pesticide residues.

Apart from immediate or delayed expression of acute toxicity, exposure to pesticides in natural communities may exert various indirect effects on the capacity of an organism to survive. Many authors have noted behavioural effects on various major taxa, generally including irritability, loss of co-ordination, and finally immobility, that precede death or occur with sublethal pesticide exposure (Anderson, 1945; Grosch, 1967; Odum et al., 1969; Sanders, 1970; Wilson and Bond, 1969). Although we are unaware of evidence that behavioural malfunction induced by sublethal pesticide exposure can, in fact, play a role in invertebrate prey-predator relations by, for example, increasing vulnerability to predators, the potential seems real, and deserving of explicit investigation. By similar reasoning, differential sensitivity to pesticides by competing species can assist the more resistant species by removing competitors. Thus, Waugh and Ansell (1956) noted improved oyster production as a consequence of the relatively greater susceptibility of barnacles than oysters to DDT. Although DDT initially inhibited growth of oyster spat, the absence of competing barnacles was the more important factor under the circumstances.

The rate of production and species composition of aquatic invertebrate communities is greatly influenced by the level of production and species composition of the primary producers. Studies with a number of phytoplankton species have determined that common contaminants, including mercurials and organochlorine pesticides, inhibit primary production in laboratory culture (Harris *et al.*, 1970; Menzel *et al.*, 1970; Ukeles, 1962; Wurster, 1968). The effective concentrations required to demonstrate inhibition of most species were in excess of the probable residue levels that could be attained in natural waters by many of the contaminants, but, in each of the studies cited, there were instances of inhibition that seem possible in natural systems, a conclusion that appears equally valid in nature for the stable 'terminal residues' of organochlorine photodecomposition (Batterton *et al.*, 1971). But in making such judgments, we must remember that we have very little information on the potential availability of water-borne residues to organisms in natural systems. Potential availability may depend only partly upon the water solubility of pesticides.

The preceding effects of pesticides, and other effects, are due to changes induced by contaminants in the endogenous physiological processes of the susceptible organisms. As well as neurotoxic effects, various pesticides, alone or in combination, are known to affect diverse biochemical functions of organisms, and although these aspects have not been well investigated for aquatic invertebrates, many similarities with other taxa can be expected. Of particular interest, are observations of whole body responses to pesticides, that have indicated interference with energy metabolism. Butler (1966b, 1969a) noted a reduced pumping rate and increased shell movements of bivalves in response to water-borne pesticide residues. Costa (1970) recorded an increased heart rate of *Cardina prista,* a freshwater shrimp, in response to several organochlorine insecticides. Engel *et al.* (1970) found that lindane and DDT interfered with glucose metabolism of the Quahog, *Mercenaria mercenaria,* and Okubo and Okubo (1965) reported that imposition of a metabolic load induced decreased resistance to pentachlorophenate by the clam, *Venerupis japonica.* Pentachlorophenol is among the better examples of contaminants that interfere with energy metabolism, because it is known to uncouple oxidative phosphorylation, and has been shown to impair energy metabolism of cichlid fishes (Krueger *et al.,* 1968). Taken together, there is evidence that pesticides may impose energy losses in varying ways; thus, growth and other functions of aquatic invertebrates can be expected to suffer significantly from exposure to some pesticides at least. Grosch (1967) reports that DDT affected the reproductive pattern of *Artemia salina* through its debilitating effect on the adult female, and impaired growth of laboratory cultures has been reported to occur after exposure to a variety of pesticides (Butler, 1963; Davis, 1961; Davis and Hidu, 1969; Södergren, 1968).

The available evidence clearly indicates that pesticides can potentially affect aquatic invertebrates in a variety of ways and in varying degrees. As well as direct mortality, the species composition can be indirectly influenced by pesticides, through diverse effects on predator-prey and competitive interactions. Specifically, it appears that the additional energy losses associated with sublethal pesticide exposure may affect the production efficiency of a population, and therefore its potential for survival. It remains to evaluate the consequences of these microsystem effects on the distinctive properties of aquatic communities.

4.4.2 Macrosystem effects—communities

Communities and ecosystems are hierarchical systems with emergent properties, stability, diversity, and the like, that in part reflect the integral effect of interactions among system components. In theory, evaluation of community tolerance to pesticides therefore requires direct observation of changes in the emergent properties, or assessment of the important changes in interactions among species. In practice, explicit studies of the ecology of pesticide effects are hampered by the present inability of ecological theory to provide appropriate criteria for observing the effects of perturbations of complex natural systems. For this reason, most ecological studies of pesticide effects have consisted of necessarily incomplete observations of effects on groups of species and their interactions. Nevertheless, pesticides constitute a sufficiently powerful cause of gross ecological disturbance, that important community effects are often readily discernible (Moore, 1967).

On the basis of laboratory observations of pesticide effects on individual species, changes in species abundance and production are predictable in communities exposed to pesticide stress. Many such observations have been reported for aquatic invertebrate communities. Various authors have concluded that production by selected community components is impaired by sublethal and lethal exposure to pesticides (Butler, 1969b; Woodwell *et al.*, 1967), and numerous studies have shown profound effects on the species composition of stressed communities (Harrington and Bidllingmayer, 1958; Hastings *et al.*, 1961; Hopkins *et al.*, 1966; Ide, 1967). The examples provided are admittedly dramatic, but the sampling problems imposed by the complexity and scale of natural communities make unequivocal detection of lesser effects difficult, although, by implication, these are surely the more common phenomena.

Differential changes in species abundance and productivity have obvious implications for the structure and efficiency of aquatic food webs. Nicholson *et al.* (1962) observed that bluegills, *Lepomis macrochirus,* changed from feeding on aquatic insects to feeding on plankton crustacea when standing crops of their preferred prey were depleted by parathion. Ide (1967) reported that numbers of stream-dwelling arthropods were severely decreased in a watershed sprayed with DDT. Small species, notably chironomids, recovered quickly to become abundant by the end of the spray year, but larger species took several years to recover, with caddisflies being among the last to establish pretreatment abundance. Keenleyside (1967) reported results on changes in the dietary habits of Atlantic salmon, resident in the same stream, with parr being the slowest to recover normal food habits, because of preference for the larger insect species. A common theme in these and similar observations is the tendency of the available prey species to be smaller after exposure to pesticides.

Change in size of organisms after pesticide exposure, that result in communities composed of smaller species, is a relatively general ecological response to exogenous stress (Woodwell, 1970). Among the many probable causes for changed size composition of aquatic communities exposed to pesticides is the generally higher reproductive potential of smaller species, stress-induced changes in reproductive pattern resulting in resistant life stages in small organisms (Grosch, 1967), tendency to higher gross growth efficiency for smaller predators (Paloheimo and Dickie, 1966; Kerr, 1971a), and the more rapid development of resistant strains when generation time is short. Young (1968) provided some instructive calculations on the potential effects of pesticides on the population structure of long and short-lived organisms. Large vertebrate predators can be especially vulnerable to elimination by pesticides because of the potential for concentration of residues through the food chain, as shown for example by Butler *et al.* (1970). Shrimp are relatively susceptible to DDT and die before accumulating appreciable residue levels, whereas Menhaden do not. Butler *et al.* found that Seatrout, *Cynoscion nebulosus,* reproduced normally on a shrimp diet, but a population feeding on Menhaden failed to reproduce normally because of high DDT residues that they acquired from their diet. As well as predator mortality, pesticide stress could also affect the production efficiency of predators through changes in the size composition of the prey. Kerr and Martin (1970) and Kerr (1971b) have shown that prey size can be as important as prey abundance in determining predator growth efficiency in nature, implying a further cause for modified size composition of stressed communities.

Changes in the size composition of communities can also affect the diversity and abundance of the smaller organisms, particularly if higher predators are eliminated. Herbivore populations are controlled for the most part by predators (Hairston *et al.,* 1960). Where species eruptions occur following pesticide treatment of ecological systems, it is generally herbivorous species which become overabundant. The phenomenon has long been recognised as a consequence of predator mortality in crop treatments (Newsom, 1967), but more subtle factors may also be involved with aquatic invertebrates, and presumably other community types. Thus, Paine (1966) observed that herbivores in a littoral community decreased in diversity following removal of predatory starfish, indicating that predation had previously checked herbivore abundance sufficiently to allow co-existence of competing species. The view that such indirect responses to stress may be a common phenomenon with ecological systems is supported by the theoretical system model provided by Verhoff and Smith (1971).

In affecting species abundance, diversity, and productivity of aquatic communities, pesticides also affect community stability, although the phenomenon has yet to be examined explicitly and rigorously. Ide (1967) and others concluded, from their observations of DDT stress on stream communities, that restoration of pre-treatment conditions occurred after about five years. These particular communities are therefore stable in the face of periodic DDT stress spaced at relatively long time intervals. Ide notes that repetition of the stress at more frequent intervals would result in effects comparable to the original treatment. In the sense that recovery would not then be possible, the community is therefore unstable if stress is frequent or chronic. Unfortunately, we are unaware of any satisfactory method of assessing ecological stability that does not require a considerable measure of such hindsight.

The factors contributing to ecological stability have recently received considerable attention. Stability is undoubtedly correlated, in some degree, with species diversity, although there is scant evidence of a causal relationship linking the two. A common correlate of both is ecosystem heterogeneity. Habitat diversity has been known for some time as a requisite factor for sustained culture of competing species. Smith (in press), among others, has examined the effects of spatial heterogeneity with a simple mathematical model and found it a sufficiently powerful stabilizing factor to account for the observed stability of natural systems. Sources of stabilizing heterogeneity, other than spatial effects, also seem possible, but have not been demonstrated. Thus, it appears that spatial heterogeneity contributes to ecological stability. At least, a useful method of appraising stability under pesticide stress, might therefore combine measures of system heterogeneity with duration and intensity of stress. It may be that sufficient data for an empirical evaluation of these factors now exist in the literature. Certainly, there is clear need for some general measure of the tolerance of communities to pesticide stress.

4.4.3 Appraisal of hazards

Clearly, the available information on the effects of pesticides on aquatic communities is weakest with respect to the more important questions. Immediate toxic effects of various compounds on particular species can be determined with reasonable assurance, as can numerous categories of sublethal effects. But the general hazards are more difficult to evaluate, chiefly because suitable criteria for observing ecosystem responses to stress are, for the most part, lacking. The dearth of genuinely ecological methods of hazard evaluation seems attributable to current weaknesses in ecological theory. Unlike physical theory, ecological theory is remarkably deficient in ways of dealing with macrosystem properties, quite aside from methods of combining microsystem and macrosystem properties, in the sense that quantum mechanics and thermodynamics are combined in statistical mechanics (Rosen, 1969). Nevertheless, the important general questions must be answered in terms of emergent ecological properties, or some combination of these and species responses.

Hopefully, practical yardsticks for hazard evaluation may be sufficient, and these are more readily available than a robust general theory. In addition to established methods for appraisal of hazards, several less well-developed, but promising, methods of appraisal can be suggested. First, important effects of pesticides at the community level can be discerned, although not understood, through application of any of a number of cor-relational methods. For example, various systems of indicator species have been used as measures of pesticide impact on ecosystems. As the processes underlying the biological effects of pesticides become better defined, indicator systems will doubtless improve, although these may never be both sensitive and satisfactorily general. Secondly, it seems possible that general ecological indicators, the tendency of stressed communities to change in size structure, for example, or the potential for development of an empirical correlation between ecological heterogeneity and tolerance to stress, could provide useful indices of hazard, but their interpretation may prove to be difficult. Thirdly, increased understanding of the ecologically significant transfer and accumulation mechanisms will undoubtedly lead to increasingly comprehensive quantitative models of the behaviour of important environmental contaminants (Woodwell et al., 1971). Perhaps the greatest contribution of such systems would be to provide clear defini-tion of sampling requirements, making sampling programmes easier to design, and environmental monitoring a more useful exercise than is often the case. Truly predictive models would also obviate much of the present need for hindsight in problem identification, and could assist correction of existing hazards by identifying critical points in the transfer process. Fourthly, development of model ecosystems, for laboratory evaluation of the ecological properties of toxic agents, is a field that promises to provide useful, rapid methods of screening compounds for environmental hazards (Kapoor et al., 1970; Metcalf et al., 1971). As noted earlier, current laboratory models have serious shortcomings as mimics of natural communities, but these appear to be surmountable difficulties.

Acknowledgments

Drs. L. M. Dickie, D. C. Gordon, and W. H. Sutcliffe, Jr. criticised an earlier version of the manuscript. We also thank H. Wilson and J. Dale for valuable assistance, and M. Trapnell, Scientific Information and Library Services, Bedford Institute.

REFERENCES

F. Acree, Jr., M. Beroza and M. C. Bowman, *J. Agr. Food Chem.*, **11**, 278 (1963)

B. G. Anderson, *Science N. Y.*, **102**, 539 (1945)

E. E. Angino, J. E. Simek and J. A. Davis, *Publ. Inst. Mar. Sci. Univ. Tex.*, **10**, 173 (1965)

P. J. Aucamp, J. H. Henry and G. H. Stander, *Mar. Pollut. Bull.*, **2**, 190 (1971)

G. J. Bacher, *Fish. Wildlife. Pap. Vict.*, **1**, 1 (1971)

H. Bader, *J. Geophys. Res.*, **75**, 2822 (1970)

T. E. Bailey and J. R. Hannum, *Proc. Sanit. Div. Amer. Soc. Civ. Eng.*, **93** 27 (1967)

J. C. Batterton, G. M. Boush and F. Matsumura, *Bull. Environ. Contam. Toxicol.*, **6** 589 (1971)

D. C. Blanchard and L. Syzdek, *Science N. Y.*, **170**, 626 (1970)

J. T. Bonner, Princeton University Press, Princeton, New Jersey (1965)

M. C. Bowman, F. Acree, Jr., C. S. Lofgren and M. Beroza, *Science N. Y.*, **146**, 1480 (1964)

W. R. Bridges, B. J. Kallman and A. K. Andrews, *Trans. Amer. Fish. Soc.*, **92**, 421 (1963)

N. J. Brown and A. W. A. Brown, *J. Wildlife Manage.*, **34**, 929 (1970)

J. C. Bugg, Jr., J. E. Higgins and E. A. Robertson, Jr., *Pestic. Monit. J.*, **1**, 9 (1967)

R. Burnett, *Science N. Y.*, **174**, 606 (1971)

P. A. Butler, *U.S. Fish Wildlife Serv. Circ.*, 167 (1963)

P. A. Butler, *N. Amer. Wildlife Natur. Resour. Conf. Trans.*, **31**, 184 (1966a)

P. A. Butler, *J. Appl. Ecol.*, **3** (Suppl.), 253 (1966b)

P. A. Butler, *Amer. Fish. Soc. Spec. Publ.*, **3**, 110 (1966c)

P. A. Butler and J. D. Newsom (Editors), Proc. marsh and estuary management symposium, Louisiana State Univ., July 19-20, 1967. Thos. J. Moran's Sons, Inc., Baton Rouge, Louisiana (1968)

P. A. Butler, *Bioscience,* **19**, 889 (1969a)

P. A. Butler. M. W. Miller and G. G. Berg (Editors), Chemical Fallout, p. 205. Charles C. Thomas, Springfield, Illinois (1969b)

P. A. Butler and J. W. Gillett (Editor), The biological impact of pesticides in the environment. Environmental Health Sciences Center, Oregon State University, Corvallis, Oregon (1970)

P. A. Butler, R. Childress and A. J. Wilson, Jr., F.A.O. technical conference on marine pollution and its effects on living resources and fishing. Rome, Italy (December 1970)

V. L. Casper, *Pestic. Monit. J.*, **1**, 13 (1967)

J. M. Cohen. C. Pinkerton and R. F. Gould (Editors), Organic pesticides in the environment. *Advances in Chemistry Series* 60, American Chemical Society, Washington, D.C. (1966)

H. Cole, D. Barry, D. E. H. Frear and A. Bradford, *Bull. Environ. Contam. Toxicol.*, **2**, 127 (1967)

R. J. Conover, *Amer. Zool.*, **8**, 107 (1968)

O. B. Cope, *J. Appl. Ecol.*, **3** (Suppl.), 33 (1966)

L. Cory, P. Fjeld and W. Serat, *Nature* (Lond.), **229**, 128 (1971)

H. H. Costa, *Hydrobiologia*, **35**, 469 (1970)

J. L. Cox, *Science N.Y.*, **170**, 71 (1970)

J. L. Cox, *Fish. Bull.*, **69**, 443 (1971a)

J. L. Cox, *Fish. Bull.*, **69**, 627 (1971b)

R. A. Croker and A. J. Wilson, *Trans. Amer. Fish. Soc.*, **94**, 152 (1965)

D. G. Crosby and R. K. Tucker, *Environ. Sci. Technol.*, **5**, 714 (1971)

H. C. Davis, *Commer. Fish. Rev.*, **23**, 8-23 (1961)

H. C. Davies and H. Hidu, *U.S. Fish Wildlife Serv. Fish Bull.*, **67**, 393 (1969)

J. B. Dimond, R. E. Kadunce, A. S. Getchell and J. A. Blease, *Ecology*, **49**, 759 (1968)

T. W. Duke, J. I. Lowe and A. J. Wilson, Jr., *Bull. Environ. Contam. Toxicol.*, **5**, 171 (1970)

L. L. Eberhardt, R. L. Meeks and T. J. Peterle, Atomic Energy Commission Research and Development Report BNWL-1297. UC-48 (1970)

C. A. Edwards, *Critical Reviews in Environmental Control*, **1**, 7 (1970)

R. H. Engel, M. J. Neat and R. E. Hillman, Rome (December 1970)

R. C. Fay, *Mar. Pollut. Bull.*, **1**, 183 (1970)

D. E. Ferguson, J. L. Ludke and G. G. Murphy, *Trans. Amer. Fish. Soc.*, **95**, 335 (1966)

S. W. Fowler and L. F. Small, *Int. J. Oceanol. Limnol.*, **1**, 237 (1967)

F. J. H. Fredeen and J. R. Duffy, *Pestic. Monit. J.*, **3**, 219 (1970)

V. H. Freed. J. W. Gillett (Editor), The Biological Impact of Pesticides in the Environment. Environmental Health Sciences Center, Oregon State University, Corvallis, Oregon, 210 p. (1970)

J. H. Gakstatter and C. M. Weiss, *Trans. Amer. Fish. Soc.*, **96**, 301 (1967)

W. D. Garrett, *Deep-Sea Res.*, **14**, 221 (1967)

P. J. Godsil and W. C. Johnson, *Pestic. Monit. J.*, **1**, 21 (1968)

E. D. Goldberg, P. Butler, P. Meier, D. Menzel, G. Paulik, R. Risebrough and L. F. Stickel, National Academy of Sciences, Washington, D.C. (1971)

D. A. Graetz, G. Chesters, T. C. Daniel, L. W. Newland and G. B. Lee, *J. Water Pollut. Contr. Fed.*, **42**, R76 (1970)

D. S. Grosch, *Science N.Y.*, **155**, 592 (1967)

A. R. Grzenda, D. F. Paris and W. J. Taylor, *Trans. Amer. Fish. Soc.*, **99**, 385 (1970)

A. R. Grzenda, W. J. Taylor and D. F. Paris, *Trans. Amer. Fish. Soc.*, **100**, 215 (1971)

N. G. Hairston, F. E. Smith and L. B. Slobodkin, *Amer. Natur.*, **94**, 421 (1960)

J. L. Hamelink, R. C. Waybrant and R. C. Ball, *Trans. Amer. Fish. Soc.*, **100**, 207 (1971)

M. R. Hannon, Y. A. Greichus, R. L. Applegate and A. C. Fox, *Trans. Amer. Fish. Soc.*, **99**, 496 (1970)

R. W. Harrington, Jr. and W. L. Bidlingmayer, *J. Wildlife Manage.*, **22**, 76 (1958)

R. C. Harriss, D. B. White and R. B. Macfarlane, *Science N.Y.*, **170**, 736 (1970)

R. Hartung and G. W. Klingler, *Environ. Sci. Technol.*, **4**, 407 (1970)
E. Hastings, W. H. Kittams and J. H. Pepper, *Ann. Ent. Soc. Amer.*, **54**, 436 (1961)
D. S. Haven, *Va. J. Sci.*, **20**, 51 (1969)
J. J. Hickey, J. A. Keith and F. B. Coon, *J. Appl. Ecol.*, **3** (Suppl.), 141 (1966)
D. W. Hill and P. L. McCarty, *J. Water Pollut. Contr. Fed.*, **39**, 1259 (1967)
A. V. Holden, *Pestic. Monit. J.*, **4**, 117 (1970a)
A. V. Holden, *Nature* (Lond.), **228**, 1220 (1970b)
C. L. Hopkins, H. V. Brewerton and H. J. W. McGrath, *N.Z. J. Sci.*, **9**, 236 (1966)
E. G. Hunt and A. I. Bischoff, *Calif. Fish Game*, **46**, 91 (1960)
F. P. Ide, *J. Fish. Res. Bd. Canada*, **24**, 769 (1967)
R. H. Janicki and W. B. Kinter, *Science N.Y.*, **173**, 1146 (1971)
L. D. Jensen and A. R. Gaufin, *Trans. Amer. Fish. Soc.*, **93**, 357 (1964)
S. Jensen, A. G. Johnels, M. Olsson and G. Otterlind, *Nature* (Lond.), **224**, 247 (1969)
I. P. Kapoor, R. L. Metcalf, R. F. Nystrom and G. K. Sangha, *J. Agr. Food Chem.*, **18**, 1145 (1970)
J. A. Kawatski and J. C. Schmulbach, *Trans. Amer. Fish. Soc.*, **100**, 565 (1971)
M. H. A. Keenleyside, *J. Fish. Res. Bd. Canada*, **24**, 807 (1967)
S. R. Kerr, *J. Fish. Res. Bd. Canada*, **28**, 809 (1971a)
S. R. Kerr, *J. Fish. Res. Bd. Canada*, **28**, 815 (1971b)
S. R. Kerr. N. V. Martin and J. H. Steele (Editors), Marine food chains, p. 365. Oliver and Boyd, Edinburgh (1970)
J. H. Koeman, J. Veen, E. Brouwer, L. Huisman-de Brouwer and J. L. Koolen, *Helgolanderwiss. Meersunters.*, **17**, 375 (1968)
H. M. Krueger, J. B. Saddler, G. A. Chapman, I. J. Tinsley and R. R. Lowry, *Amer. Zool.*, **8**, 119 (1968)
R. Lasker, *J. Fish. Res. Bd. Canada*, **23**, 1291 (1966)
J. I. Lowe, *Ecology*, **46**, 899 (1965)
K. J. Macel. J. W. Gillett (Editor), The Biological Impact of Pesticides in the Environment. Environmental Health Sciences Center, Oregon State University, Corvallis, Oregon (1970)
J. K. Macek, C. R. Rodgers, D. L. Stalling and S. Korn, *Trans. Amer. Fish. Soc.*, **99**, 689 (1970)
K. J. Macek and H. O. Sanders, *Trans. Amer. Fish. Soc.*, **99**, 89 (1970)
D. B. Manigold and J. A. Schulze, *Pestic. Monit. J.*, **3**, 124 (1969)
R. L. Meeks, *J. Wildlife Manage.*, **32**, 376 (1968)
D. W. Menzel, J. Anderson and A. Randtke, *Science N.Y.*, **167**, 1724 (1970)
R. L. Metcalf, G. K. Sangha and I. P. Kapoor, *Environ. Sci. Technol.*, **5**, 709 (1971)
J. Michima and E. P. Odum, *Limnol. Oceanogr.*, **8**, 39 (1966)
J. C. Modin, *Pestic. Monit. J.*, **3**, 1 (1969)
N. W. Moore, *Adv. Ecol. Res.*, **4**, 75 (1967)
R. J. Moubry, J. M. Helm and G. R. Myrdal, *Pestic. Monit. J.*, **1**, 27 (1968)
E. M. Mrak, Report of the Secretary's commission on pesticides and their relationship to environmental health. Parts I and II. U.S. Dept. of Health, Education and Welfare, U.S. Govt. Printing Office, Washington, D.C. (1969)

R. E. Munn and B. Bolin, *Atmos. Environ.*, **5**, 363 (1971)
S. M. Naqvi and D. E. Ferguson, *Trans. Amer. Fish. Soc.*, **99**, 696 (1970)
R. G. Nash and M. L. Beall, Jr., *Science N.Y.*, **168**, 1109 (1970)
L. D. Newsom, *Ann. Rev. Entomol.*, **12**, 257 (1967)
H. P. Nicholson, *Science N.Y.*, **158**, 871 (1967)
H. P. Nicholson, H. J. Webb, G. L. Lauer, R. E. O'Brien, A. R. Grzenda and D. W. Shanklin, *Trans. Amer. Fish. Soc.*, **91**, 213 (1962)
D. R. Nimmo, A. J. Wilson, Jr. and R. R. Blackman, *Bull. Environ. Contam. Toxicol.*, **5**, 333 (1970)
D. R. Nimmo, P. D. Wilson, R. R. Blackman and A. J. Wilson, Jr., *Nature* (Lond.), **231**, 50 (1971)
R. D. O'Brien, Insecticides: action and metabolism. Academic Press Inc., New York (1967)
E. P. Odum, *Biol. Bull. (Woods Hole)*, **121**, 371 (1961)
W. E. Odum, G. M. Woodwell and C. F. Wurster, *Science N.Y.*, **164**, 576 (1969)
K. Okubo and T. Okubo, *Bull. Tokai Reg. Fish. Res. Lab.*, **44**, 31 (1965)
R. T. Paine, *Amer. Natur.*, **100**, 65 (1966)
J. E. Paloheimo and L. M. Dickie, *J. Fish. Res. Bd. Canada*, **23**, 1209 (1966)
R. F. Palumbo, Recent Advances in Botany. Vol. II, p. 1367. University of Toronto Press (1961)
B. Parker and G. Barsom, *Bio Science*, **20**, 87 (1970)
T. J. Peterle, *Nature* (Lond.), **224**, 620 (1969)
T. J. Peterle. J. W. Gillett (Editor), The Biological Impact of Pesticides in the Environment, p. 11. Environmental Health Sciences Center, Oregon State University, Corvallis, Oregon (1970)
R. M. Pfister, P. R. Dugan and J. I. Frea, *Science N.Y.*, **166**, 878 (1969)
G. G. Polikarpov, Radioecology of Aquatic Organisms. Reinhold Book Division, New York (1966)
C. L. Prosser and F. A. Brown, Comparative Animal Physiology, 2nd ed. Saunders, Philadelphia (1961)
A. G. Richards and L. K. Cutkomp, *Biol. Bull. (Woods Hole)*, **90**, 97 (1946)
S. Richman, *Ecol. Monogr.*, **28**, 273 (1958)
G. A. Riley, *Advan. Mar. Biol.*, 8, 1 (1970)
R. W. Risebrough, R. J. Huggett, J. J. Griffin and E. D. Goldberg, *Science N.Y.*, **159**, 1233 (1968)
R. W. Risebrough, D. B. Menzel, D. J. Martin, Jr. and H. S. Olcott, *Nature* (Lond.), **216**, 589 (1967)
J. Robinson, A. Richardson, A. N. Crabtree, J. C. Coulson and G. R. Potts, *Nature* (Lond.), **214**, 1307 (1967)
J. Robinson, M. Roberts, M. Baldwin and A. I. T. Walker, *Food Cosmet. Toxicol.*, **7**, 317 (1969)
F. L. Rose and C. D. McIntire, *Hydrobiologia*, **35**, 481 (1970)
R. Rosen, L. L. White, A. G. Wilson and D. Wilson (Editors), Hierarchical Structures, p. 179. American Elsevier Publishing Company, Inc., New York (1969)
H. O. Saunders, *J. Water Pollut. Contr. Fed.*, **42**, 1544 (1970)
H. O. Sanders and O. B. Cope, *Trans. Amer. Fish. Soc.*, **95**, 165 (1966)
S.C.E.P. Man's Impact on the Global Environment. Report of the Study of Critical Environmental Problems. The M.I.T. Press, Cambridge, Massachusetts (1970)

D. B. Seba and E. F. Corcoran, *Pestic. Monit. J.*, **3**, 190 (1969)

C. A. Shelby, *Publ. Inst. Mar. Sci. Univ. Tex.*, **9**, 33 (1963)

M. G. Sheldon, R. B. Findley, Jr., M. H. Mohn and G. H. Ise, *U.S. Fish Wildlife Serv. Circ.*, 167, p. 63 (1963)

R. W. Sheldon, A. Prakash and W. H. Sutcliffe, Jr., *Limnol. Oceanogr.* (in press)

L. B. Slobodkin. J. H. Steele (Editor), Marine Food Chains, p. 537. Oliver and Boyd, Edinburgh (1970)

F. E. Smith *Proc. Connecticut Academy of Arts and Sciences* (in press)

A. Södergren, *Oikos*, **19**, 126

J. B. Sprague and J. R. Duffy, *J. Fish. Res. Bd. Canada*, **28**, 59 (1971)

V. F. Stout, *Bull. Environ. Contam. Toxicol.*, **3**, 240 (1968)

L. M. Sushchenya. J. H. Steele (Editor), Marine Food Chains, p. 127. Oliver and Boyd, Edinburgh (1970)

W. H. Sutcliffe, Jr., R. W. Sheldon, A. Prakash and D. C. Gordon, Jr., *Deep-Sea Res.*, **18**, 639 (1971)

R. A. Sweeney, *Proc. Conf. Great Lakes Res.*, **12**, 98 (1969)

K. R. Tarrant and J. O'G. Tatton, *Nature* (Lond.), **219**, 725 (1968)

J. O'G. Tatton and J. H. A. Ruzicka, *Nature* (Lond.), **215**, 346 (1967)

L. C. Terriere, U. Kiigemagi, A. R. Gerlach and R. L. Borovicka, *J. Agr. Food Chem.*, **14**, 66 (1966)

R. Ukeles, *Appl. Microbiol.*, **10**, 532 (1962)

F. H. Verhoff and F. E. Smith, *J. Theor. Biol.*, **33**, 131 (1971)

J. F. Walsh, *Bull. World Health Organ.*, **43**, 316 (1970)

G. D. Waugh and A. Ansell, *Ann. Appl. Biol.*, **44**, 619 (1956)

W. E. Westlake and F. A. Gunther. R. F. Gould (Editor), Organic Pesticides in the Environment, p. 110. *Advances in Chemistry Series* 60, American Chemical Society, Washington, D.C. (1966)

B. K. Whitten and C. J. Goodnight, *J. Water Pollut. Contr. Fed.*, **42**, 1544 (1966)

D. J. Wildish and V. Zitko, *Mar. Biol.*, **9**, 213 (1971)

F. G. Wilkes and C. M. Weiss, *Trans. Amer. Fish. Soc.*, **100**, 222 (1971)

P. M. Williams, *Deep-Sea Res.*, **14**, 791 (1967)

D. C. Wilson and C. E. Bond, *Trans. Amer. Fish. Soc.*, **98**, 438 (1969)

G. G. Winberg, Nauch. Tr. Belorusskovo Gosudarstvennovo Universiteta imeni V. I. Lenina, Minsk. 253 pp. (Transl. from Russian by Fish. Res. Board Can. Transl. Ser. No. 194, 1960) (1956)

G. M. Woodwell, *Sci. Amer.*, **216**, 24 (1967)

G. M. Woodwell, *Science N.Y.*, **168**, 429 (1970)

G. M. Woodwell, P. P. Craig and H. A. Johnson, *Science N.Y.*, **174**, 1101 (1971)

G. M. Woodwell, C. F. Wurster, Jr. and P. A. Isaacson, *Science N.Y.*, **156**, 821 (1967)

C. F. Wurster, Jr., *Science N.Y.*, **159**, 1474 (1968)

C. F. Wurster, Jr., *Biol. Conserv.*, **2**, 123 (1969)

H. Young, *Ecology*, **49**, 991 (1968)

W. N. Yule and A. D. Tomlin, *Bull. Environ. Contam. Toxicol.*, **5**, 479 (1971)

I. Zeidman, *Science N.Y.*, **108**, 214 (1948)

V. Zitko, *Bull. Environ. Contam. Toxicol.*, **6**, 464 (1971)

Chapter 5

Pesticide Residues in Fish

D. W. Johnson

Department of Biology,
Idaho State University,
Pocatello, Idaho

The chemical nature of our environment has been altered for the past 20 years by the addition of several hundred thousand synthetic chemical compounds. Some of these were used as pesticides although their general lack of specificity suggests that the term biocide may be more appropriate (Carson, 1962) The environmental persistence, accumulation, and effects (singular, additive, or synergistic) of these chemicals in the biosphere are poorly understood. As a beginning toward understanding the biological significance of this chemical alteration, we must determine if, how, and in what amounts these contaminating compounds or their residues are being incorporated into ecosystems. Only after these questions are answered, can we ask legitimate questions regarding the effects of these chemicals on living systems—individuals, populations, and communities. All of these contaminants or residues reach the aquatic environment. The persistence and low water solubility of many pesticides contribute to their concentration in fish tissues. Residue analyses may in themselves provide us with indications of the action and effects of pesticides on fish and their ecosystems.

5.1   RESIDUES

Residue surveillance on major waters of the U.S.A. is provided by a National Pesticide Monitoring Programme. The residues determined are meant to provide an alert to potential fishery problems by detecting increasing trends or levels previously shown to be potentially harmful (Murray, 1971). This programme is restricted to *major waters*; as a result of station location and dilution its warning value is minimized (Johnson, 1968; Johnson and Lew, 1970). This may be illustrated by examining camphechlor residue values reported as a product of this programme. In Mississippi and Arizona, camphechlor has been heavily used on cotton lands, where it has been a principal substitute for DDT since 1966. In 1967 and 1968, only three of 27 fish species analysed in the Monitoring Programme contained camphechlor residues (0.01 to 0.02 ppm); in 1969, no camphechlor residues were reported (Henderson et al., 1969 and 1971). On the other hand, at other stations in Arizona, in waters influenced by run-off from agricultural land and waste water, camphechlor residues up to 25.0 ppm were reported for five fish species (Johnson and Lew, 1970). The reference values provided by the Monitoring Programme are, however, of interest. Fish collected in autumn, 1969, contained DDT residues ranging from 0.03 to 57.8 ppm, dieldrin from 0.01 to 1.59 ppm, and BHC from 0.01 to 4.37 ppm; these levels surpassed those of 1967 and 1968. DDT residues in fishes from Imperial Reservoir, Colorado River, were up by a factor of four in Arizona's first year of a moratorium against the use of DDT.

5.1.1   Condition
Much of the intraspecific variation in accumulation and elimination of pesticides, and much of the importance of residues to the survival of fish

populations may be a function of fish condition based on lipid content. An apparent relationship between fat content and residue concentration has been observed in *Salmo salar*. Those in poor condition had DDT residues of 3.8 ppm and those in good condition 8.3 ppm (Anderson and Everhart, 1966). A good correlation between lipid content and DDT residues has also been shown for most San Francisco Bay fish (Earnest and Benville, 1971). Dieldrin residues were greatest in the visceral fat, and least in the muscle of aldrin-exposed *Carassius auratus* (Gakstatter, 1968). Whereas the highest residue concentrations are most often found in visceral fat, the amount of residue in muscle tissue may be more closely related to mortality (Schoettger, 1970). Although Grzenda *et al.* (1970) found mesenteric adipose, testicular and nerve tissues of *Carassius auratus*, exposed to ^{14}C-DDT for 192 days, with 23.0, 13.5 and 12.5 times the muscle DDT residue level respectively, they reported no correlation between tissue lipid content and residue distribution. Experiments with *C. auratus* and dieldrin yielded similar results and conclusions. In this more recent investigation, the residue content of the ovary increased more than six times as the lipid content increased with maturity (Grzenda *et al.*, 1971). Their conclusion, of an absence of positive correlation between residue and lipid content, is difficult to understand on the basis of their data. Endosulfan (Thiodan®) also appears to accumulate in the fat of *C. auratus*, the highest residues being in fat, brain, and liver (Schoettger, 1970). *Poecilia latipinna* exposed to dieldrin (in water) for 238 days, had brain residues exceeded only by those in the gills, which were slightly higher than blood levels, and these in turn were greater than in the liver. Muscle levels were lowest, being only about 1/4 to 1/3 the level of brain residues (Lane and Livingston, 1970). It appears that residue concentrations may indeed be a function of the fat coefficient of a fish. The lipid content of resistant' *Gambusia affinis* was greater (1.8 times) than that of susceptible fish (Fabacher and Chambers, 1971). This difference did not exist between 'removed-resistant' (transplanted) and susceptible fish, casting some doubt on the nature of this resistance. Lipid content, together with susceptibility, varied within populations as well as seasonally. The 'resistance' of these fish seems to be a function of their nutritional state. With the lipid content of salmonids differing greatly, the organochlorine residues in a fish with a 14% lipid content could be increased by a factor of 40 with a drop to 0.35% lipid content (Holden, 1966). Changes such as these might be associated with migratory activities or the elimination of an important food item from a fish's diet, i.e. as might occur with some environmental change (contamination-pollution).

5.1.2 Trophic level

Biological magnification of residues is dependent on the balance established between the processes of accumulation and elimination as reviewed by Hamelink *et al.* (1971). While interspecific variation in residue levels is often explained in terms of biological magnification, the factor most likely responsible in many of those cases is size rather then trophic level. Whole-body DDT residues do increase with size of fish (Buhler *et al.*, 1969; Buhler and Shanks, 1970). An equilibrium between fish residues and water residues should produce similar residues in fish independent of size; however, if loss of residue from fish to water is insignificant relative to uptake of residues from the gut contents, then higher residue concentrations should be found in larger fish (Cox, 1970). Included in the variables that limit the utilization of such generalizations, are fish condition, reproductive state, and activity patterns. Although mercury accumulates within a few days in *Esox lucius*, its elimination is slow, with a half-life of 70 days, and the highest concentrations in larger older fish (Bligh, 1970). In *Salvelinus namaycush*, the total mercury residues and the proportion of methyl mercury also increased with age (Bache *et al.*, 1971). An increase in ovarian (0.87, 6.10 and 6.00 ppm) and liver (0.65, 1.55 and 1.99 ppm) DDT residues with age, has been noted in *Salmo salar* (Anderson and Everhart, 1966). Age and fat content were interdependent in their influence on DDT residues in *S. salar*. Residue levels decreased with age in fish with high fat contents, and remained constant with age in those with low fat contents (Anderson and Fenderson, 1970). Fish in higher trophic levels were found by Hannon (1969) to have the greatest percentages of the pesticide metabolites, DDD and DDE, heptachlor epoxide, and dieldrin. Whereas no correlation between residues in fish and their sex or with season was found, a relationship with age and fat content was established.

5.1.3   Size and season
In many instances, the residue and size correlation may be more important than trophic level, in the accumulation of residues. In such primary consumers as *Cyprinus carpio* and *Mugil cephalus* this appears true (Johnson and Lew, 1970), perhaps as a result of their high lipid content. Butler (1966) tells of DDT residues in 1963 year-class fish that were twice those from the 1964 year-class, as well as a seasonal difference. Summer levels were four times those found in winter. Smith (1969) reported *Pseudopleuronectes americanus* to have maximum DDT residue in summer (0.31 ppm) and minimum in winter (0.01 ppm), whereas DDE residues were reversed, that is high in winter (1.10 ppm) and lowest in summer (0.3 ppm). An apparent seasonal correlation between fish residues and pesticide usage is common (Kelso *et al.*, 1970).

5.1.4   Metabolites
Many pesticides are metabolised to detectable metabolites. Differences appear to exist in this metabolism of residues between tissues, species, length of exposure, and possibly most important, between habitats. The rate of DDT metabolism in *Salmo clarki* appears highest in the liver and gut, where DDD levels are highest (Allison *et al.*, 1963). DDE is the principal DDT-derived residue in eggs of fish (Morris and Johnson, 1969).

Bridges *et al.* (1963) found DDT-derived brain residues were limited almost entirely to DDE; however, Allison *et al.* (1963) reported high levels of DDT in brain tissues. Bridges *et al.* (1961) suggested that fat might be the only tissue with significant amounts of unaltered DDT. They also reported a 2:1 ratio of DDE to DDT in *Micropterus salmoides* as compared to a 1:1 ratio in *Lepomis macrochirus* and *Ictalurus nebulosus.* Sprague and Duffy (1971a) found that unaltered DDT was common in the viscera of marine fish. Stucky (1970) found in *Ictalurus punctatus* fat, residues up to 258.7 ppm, with DDT the principal residue.

Fish from lower trophic levels are said to have predominantly DDT residues, as opposed to an accumulation of DDE in those from higher trophic levels (Woodwell *et al.,* 1967). Data in Table 5.1 indicate a predominance of metabolites (DDE, DDD) over the parent compound (DDT) in Ictaluridae and Centrarchidae, whereas the reverse appears to occur in the Catostomidae. Fish subjected to chronic exposure have little or no unmetabolised DDT residues (DeWitt *et al.,* 1960). This may be substantiated by data in Table 5.1 where DDE is the dominant residue twice as often as DDD or DDT when DDT and metabolites are less than 1 ppm; if total residues exceed 2 ppm, the parent compound, DDT, is less often the dominant residue. It appears, however, that generalisations of this kind are extremely difficult to support. The single most important factor determining the ratio of DDT and its metabolites might well be a result of geographic region and unrelated to the biology of the exposed fish. DDE is the dominant metabolite in fish in most areas. In the Columbia River Basin, DDT never exceeded DDE, while in the Mississippi River Basin DDT, DDE, and DDD were equally abundant (Table 5.1). In marine fish, DDE is the dominant metabolite, comprising 61-73% of the total DDT residues; this is reduced to 24-35% in estuarine fish (Murphy, 1970; Sprague and Duffy, 1971). However, this generalisation was contradicted by the results of Stout (1968) who found approximately equal amounts of DDT metabolites in marine fish of the Pacific Northwest. Hannon *et al.* (1970) reported a DDT to DDT metabolites ratio of 2:3, aldrin to dieldrin of 1:12, and heptachlor to heptachlor epoxide of 1:6, in fish from a South Dakota lake. Aldrin is rapidly converted to dieldrin and accumulated by animals in this form (Johnson, 1968). However, data of Henderson *et al.* (1969) indicated a more complex metabolic pattern, with aldrin residues commonly exceeding those of dieldrin. Aldrin was more abundant then dieldrin in *Cyprinus carpio, Ictalurus punctatus,* and *Micropterus salmoides* in the Colorado River Basin, while in California streams, dieldrin was the dominant residue in these species. Although local pesticide practices may be a most important factor, when these species were collected from the same station on the Mississippi River, aldrin was most abundant in *Cyprinus carpio* while dieldrin was dominant in the others. Explanations of such occurrences are outside our present understanding of the controlling factors.

TABLE 5.1. Some recent whole fish residues of DDT (maximums in ppm) by species and area. (1)

Area	<0.10	0.11-0.50
Alaska (2)	Oncorhynchus kisutch b O. nerka Salmo gairdneri a, b Salvelinus namaycush	Coregonus clupeaformis Esox lucius b
Arizona (3)		
Atlantic Streams (2)	Erimyzon sucetta a	Amia calva b Esox niger b Micropterus dolomieui
California Streams (2)	Lepomis cyanellus b	Ptychocheilus oregonen Ictalurus melas
Canada Atlantic (4)	Microgadus proximus	Osmerus mordax
Colorado R. Basin (2)	Lepomis cyancllus b Xyrauchen texanus L. macrochirus b	Micropterus salmoides
Columbia R. Basin (2)	Salmo gairdneri b Ictalurus melas b	Oncorhynchus kisutch b Prosopium williamsoni Ictalurus nebulosus c Pomoxis annularis b
Florida (5)		Leiostomus xanthurus Micropogan undulatus
Great Lakes (6)		Notropis hudsonius Noturus flavus Pungitius pungitius
Gulf Coast (2)		Amia calva a, c Ictiobus cyprinellus b, Pylodictis olivaris b
Hudson Bay Drainage (2)		Cyprinus carpio b Lota lota
Interior Basin (2)	Gila bicolor b Catostomus macrocheilus b Archoplites interruptus b	Cyprinus carpio b Catostomus tahoensis Ictalurus nebulosus b Ictalurus punctatus a Morone chrysops c
Mississippi R. Basin (2) (7)	Salmo gairdneri b, c Cyprinus carpio Catostomus commersoni b, c C. macrocheilus Perca flavescens b, c Aplodinotus grunniens a, b	Esox lucius a Carpiodes sp. a, b Catostomus columbian Stizotedion canadense S. vitreum vitreum a
Pacific Streams (2)	Catostomus rimiculus b Lepomis gibbosus b, c	Ictalurus nebulosus b Micropterus salmoides
Pennsylvania (8)		
St. Lawrence (9)		
San Francisco Bay (10)	Citharichthys stigmaeus b, c	Cymatogaster aggregata Micropterus minimus c Phanerodon furcatus b Parophrys vetulus c Platichthys stellatus c Leptocottus armatus b
South Dakota (11)	Notropis hudsonius Pimephales promelas Catostomus commersoni Ictalurus melas Lepomis macrochirus Pomoxis nigromaculatus Perca flavescens Stizostedion vitreum vitreum	Esox lucius Cyprinus carpio Ictiobus cyprinellus Ictalurus punctatus Morone chrysops Pomoxis annularis

0.51-0.99	1.00-1.99	2.00-2.99
opium cylindraceum a	*Thymallus arcticus* a	
stomus catostomus a, b		
	Dorosoma petenense	
stomus commersoni a, c	*Lepomis cyanellus* b	*Cyprinus carpio* b, c
omis auritus b		*Catostomus columbianus* b
il cephalus a, c		*Minytrema melanops* a, c
	Morone saxatilis b	*Ictalurus nebulosus*
	Pomoxis nigromaculatus c	
nber scombrus a		
stomus latipinnis		*Cyprinus carpio* b
urus punctatus b	*Acrocheilus alutaceus* b	*Catostomus macrocheilus* b
opterus dolomieui b	*Pomoxis nigromaculatus* b	
	Orthopristis chrysoptera c	
	Lagodon rhomboides a	
	Bairdiella chrysura a	
opium cylindraceum b	*Petromyzon marinus*	*Osmerus mordax*
ropis athermoides	*Oncorhynchus kisutch*	*Morone chrysops*
opsis omiscomaycus	*Catostomus commersoni* a	*Cottus cognatus*
omis auritus b	*Aplodinotus grunniens* a	
dinotus grunniens b		*Dorosoma petenense* b
		Hiodon alosoides b
	Ictalurus catus b, c	
	Micropterus salmoides b	
on alosoides b	*Hiodon tergisus* b	*Moxostoma* sp. c
stomus catostomus c	*Ictalurus nebulosus* c	*Micropterus salmoides* c
trema melanops c	*Pomoxis annularis*	
mis macrochirus b		
a flavescens b	*Salmo clarki* b	*Salmo gairdneri* b
	Catostomus columbianus a	*Cyprinus carpio* b
	C. macrocheilus a	
	Micropterus dolomieui c	
loplites rupestris c	*Cyprinus carpio* c	
	Ictalurus nebulosus c	

TABLE 5.1 (continued)

Area	3.00-4.99	5.00-9.99
Alaska (2)		
Arizona (3)		
Atlantic Streams (2)	*Lepomis gibbosus* c *Pomoxis annularis* c *Perca flavescens* b	*Ictalurus nebulosus* c *I. punctatus* a, b
California Streams (2)	*I. punctatus*	*Cyprinus carpio* b *Micropterus salmoides*
Canada Atlantic (4)		*Carpiodes* sp.
Colorado R. Basin (2)	*I. punctatus*	
Columbia R. Basin (2)	*Ptychocheilus oregonensis* b *Cyprinus carpio* b	
Florida (5)		
Great Lakes (6)	*Alosa pseudoharengus* *Dorosoma cepedianum*	*Coregonus artedii* *C. clupeaformis* b *C. hoyi* b *Salvelinus namaycush* b *Ictalurus punctatus* b, c *Ambloplites rupestris* b, c *Perca flavescens* a *Stizostedion vitreum vitreum*
Gulf Coast (2)	*Minytrema melanops* a, b	*Carpiodes* sp. b, c *Ictalurus punctatus* b
Hudson Bay Drainage (2)	*Catostomus commersoni* a *Stizostedion canadense*	
Interior Basin (2)		
Mississippi R. Basin (2) (7)	*Pylodictis olivaris* a	*Ictiobus bubalus* a, c *I. cyprinellus* c *Ictalurus punctatus* c *Lepomis cyanellus* a, c (7)
Pacific Streams (2)		
Pennsylvania (8)		
St. Lawrence (9)	*Esox lucius*	*Perca flavescens* c
San Francisco Bay (10)		
South Dakota (11)		

stomus clarki
ssius auratus c   *Morone americana* b
opterus salmoides b

(1) Predominant metabolite: a = DDT; b = DDE; c = DDD;
     no notation = equal amounts
(2) Henderson *et al.* (1969)
(3) Johnson and Lew (1970)
(4) Sprague and Duffy (1971)
(5) Hansen and Wilson (1970)
ne americana b   (6) Reinert (1970)
(7) Finley *et al.* (1970)
(8) Cole *et al.* (1967)
(9) Fredeen and Duffy (1970)
(10) Earnest and Benville (1971)
(11) Hannon *et al.* (1970)

rus furcatus b
opterus salmoides b
cephalus a
inus carpio c

Gambusia affinis a, c (7)

linus fontinalis b
stomus commersoni

TABLE 5.2. Recent Dieldrin residues (maximums in ppm) in fish, by species and area.

Area	ND	<0.05	0.06-0.09
Alaska Streams (1)	Catostomus catostomus Prosopium cylindraceum Esox lucius Oncorhynchus kisutch Salmo gairdneri Salvelinus namaycush Oncorhynchus nerka	Coregonus clupeaformis Thymallus arcticus	
Atlantic Coast Streams (1)		Esox niger Notemigonus crysoleucas Micropterus dolomieui Lepomis auritus	
California Streams (1)	Ictalurus nebulosus P. oregonensis L. cyanellus	Ictalurus catus Morone saxatilis	
Columbia River System (1)	Prosopium williamsoni Ictalurus melas I. punctatus Oncorhynchus kisutch	Salmo gairdneri Arocheilus alutaceus Pomoxis annularis Ictalurus nebulosus	Cyprinus carpio Catostomus macroche Ptychocheilus oregon Micropterus dolomiei
Colorado River System (1)		Cyprinus carpio Xyrauchen texanus Catostomus latipinnis Ictalurus punctatus	Micropterus salmoide
Great Lakes Drainage (1)		Catostomus commersoni Moxostoma sp. Ambloplites rupestris Cyprinus carpio Prosopium cylindraceum Salvelinus namaycush	Perca flavescens Aplodinotus grunnier Coregonus clupeaforr
Great Lakes Drainage (2)	Cyprinus carpio Percopsis omiscomaycus Ictalurus nebulosus Notropis antherinoides Notropis hudsonius Noturus flavus	Catostomus commersoni Ambloplites rupestris Petromyzon marinus Aplodinotus grunniens Oncorhynchus kisutch Prosopium cylindraceum Pungitius pungitius	Perca flavescens Dorosoma cepedianu Ictalurus punctatus Carassium auratus
Gulf Coast Streams (1)		Aplodinotus grunniens Pylodictis olivaris	
Hudson Bay Drainage (1)	Lepomis cyanellus L. macrochirus	Lota lota	Cyprinus carpio Catostomus commers
Interior Basins (1)	Siphateles bicolor C. macrocheilus Archoplites interruptus	Cyprinus carpio Catostomus tahoensis Ictalurus nebulosus Morone chrysops	Micropterus salmoide
Mississippi River System (1)	Ictalurus melas Perca flavescens Salmo gairdneri Stizostedion canadense	Ictalurus nebulosus Moxostoma sp. Lepomis macrochirus Minytrema melanops Catostomus commersoni C. catostomus C. columbianus Hiodon alosoides Esox lucius Aplodinotus grunniens	
Pacific Coast Streams (1)	Catostomus rimiculus Catostomus columbianus Lepomis gibbosus Micropterus salmoides M. dolomieui	Salmo gairdneri Salmo clarki	Ictalurus nebulosus Perca flavescens
Pennsylvania (3)			

0.10-0.29	0.30-0.49	0.50-1.49	>1.50

mis gibbosus   *Catostomus commersoni*   *Morone americana*   *Perca flavescens*
opterus salmoides   *Ictalurus nebulosus*   *Cyprinus carpio*   *Ictalurus catus*
urus punctatus   *Carassium auratus*   *Mugil cephalus*
mis macrochirus
trema melanops
ostoma sp.
* calva
soma cepedianum
inus carpio   *Ictalurus punctatus*
opterus salmoides
oxis nigromaculatus

                        (1) Henderson, Johnson and Inglis (1969)
                        (2) Reinert (1970)
urus melas   (3) Cole, Barry and Frear (1967)

stedion vitreum vitreum   *Morone americana*
gonus hoyi   *Ictalurus punctatus*

* pseudoharengus   *Coregonus clupeaformis*
rus mordax
gonus hoyi
vi
edii
linus namaycush
us cognatus
ne chrysops
nericana
stedion vitreum v.
trema melanops   *Ictalurus punctatus*
nus carpio   *Micropterus salmoides*
* cephalus   *Carpiodes* sp.
us cyprinellus
calva
soma cepedianum
rus furcatus
on alosoides   *Ictalurus punctatus*
stedion canadense

odes sp.   *Ictalurus punctatus*   *Cyprinus carpio*
us bubalus   *Micropterus salmoides*   *Ictiobus cyprinellus*
on tergisus   *Pomoxis annularis*
stedion vitreum v.   *Ictalurus melas*
nadense   *Pylodictis olivaris*

                Catostomus macrocheilus *Cyprinus carpio*

inus fontinalis   *Catostomus commersoni*

5.1.5 Herbicides

Endothal derivative herbicide residues, in contrast to those of DDT, apparently do not accumulate in fish. Although these residues were toxic to *Notropis umbratilis,* tissue residues could not be identified (Walker, 1963). Sodium arsenite, used to control submerged plants, produced residues of up to 0.47 ppm in fish muscle and 0.78 ppm in viscera within 11 days after exposure (Ullmann *et al.,* 1961). Although *Lepomis macrochirus* do not accumulate 2,4-D (Cope *et al.,* 1970), they have been shown to produce and excrete it after exposure to 4—(2,4—DB), a herbicide used for aquatic weed control (Gutemann and Lisk, 1965).

5.1.6 Interspecific variability

In addition to variation in residue accumulation with lipid content and age, relative residue levels of an area can, in general, be predicted in different taxonomic groups. There appears to be an inverse correlation between susceptibility of a taxonomic group and pesticide concentrations in their tissues. Ictaluridae and Cyprinidae seem among the least susceptible, while Salmonidea appear one of the most susceptible to pesticides (Macek and McAllister, 1970). Specific variability in residue accumulation by hatchery trout was in the order *Salmo gairdneri* >*S. trutta*>*Salvelinus fontinalis* (Cole *et al.,* 1967). The autumn 1969 National Pesticide Monitoring Programme reported consistently high residues in *Ictalurus punctatus, Micropterus salmoides,* and *Morone americana,* whereas *Lepomis macrochirus, Ictalurus melas,* and *I. nebulosus* usually had some of the lowest (Henderson *et al.,* 1971). Catostomidae, *Ictalurus melas, I. nebulosus,* and *Lepomis* sp. had consistently low DDT residues; residues in Salmonidea appeared in the high range nearly as often as in the low, while *Cyprinus carpio, Ictalurus punctatus,* and *Micropterus salmoides* most commonly contained the highest residue levels (Table 5.1). Although *Morone americanus* and *Mugil cephalus* are more restricted geographically, their pesticide residues are consistently among the highest reported. It is only among these last five named species that levels of aldrin, endrin or heptachlor have commonly exceeded 0.05 ppm, although seldom exceeded 1.5 ppm (Henderson *et al.,* 1969). By contrast, many species in the U.S. are characterised by dieldrin residues in this range (Table 5.2). This is also true of heptachlor epoxide, lindane, and chlordane (Henderson *et al.,* 1969). On the basis of DDT residues in fat, reported by Hannon *et al.* (1970), these taxonomic generalisations do not hold (Table 5.3). This data is in better agreement with the trophic level explanation, although, on a whole body basis, *Cyprinus carpio* and *Ictalurus punctatus* were highest in average total residues.

5.1.7 Tissues

Residues in fish tissues differ greatly from organ to organ. The affinity of DDT for fat is well illustrated by the high residues typical of fish oil, adipose, and brain tissue. Liver tissue also commonly carries high DDT residues, presumably a function of its role in residue elimination (Table 5.3).

5.2 SOURCES OF RESIDUES

5.2.1 Atmosphere
Pesticides are often introduced to the environment via a spray directed toward pest vegetation (weeds), vegetation to be protected from pests, or vegetation harbouring pests. Much of this spray remains in the atmosphere (see Chapter 10), sometimes in excess of 50% of that sprayed by aeroplane in forests of the Northeastern U.S.A. (Woodwell *et al.,* 1971). The vapour pressure of DDT is high enough for some to enter the atmosphere from soil and plant residues; this, together with a half-life of approximately 20 years, provides a rather constant source for the contamination of the aquatic habitat (see Chapter 4).

5.2.2 Sediment
Most of the residues in aquatic systems are probably washed from cropland in run-off or in irrigation water. Stout (1968) found greater residues in fish at the mouth of the Columbia River than in those from ocean waters distant from agricultural land. Odum (1970) has described the great sorptive capacity of estuarine sediments and their ability to act as a pollutant sink, thus providing a continuing source of pesticide residues. These may reach fish by direct ingestion of sediment or by the maintenance of solubility equilibrium with the water. Substantial dieldrin and aldrin residues have been found in agricultural run-off one month after aldrin had been applied. These residues could be traced through stream run-off back to the site of application (Morris *et al.,* 1969). From 0.10 ppm$\times 10^3$ to 0.007 ppm$\times 10^3$ of endrin has been analysed between growing seasons in irrigation waste water returned to Tule Lake National Wildlife Refuge and this became concentrated up to 198 ppm$\times 10^3$ during growing seasons in *Siphatales bicolor*. Between growing seasons this fell to 4 ppm$\times 10^3$ (Godsil and Johnson, 1968).

5.2.3 Effluent
Some pesticide residues, found as aquatic contaminants, originate from industrial sources. Dieldrin and DDT have been widely used in woollen mills and by dry cleaners for moth-proofing. Organic mercury compounds have been used extensively in paper pulp mills to control slime, and also in herbicides and fungicides. All forms of mercury introduced to water are apparently converted to methyl mercury, the residue found in fish. Conversion of mercury to methyl mercury has been demonstrated in *Esox lucius* mucus within four hours of exposure, with a concentration factor from water of approximately 3,000X (Stroud, 1970).

Pesticides have been found concentrated in a surface microlayer, 'hydroepidermis', in both fresh water and sea water, with much higher surface concentrations than at a few centimetres depth (Parker and Barsom, 1970). While most areas of the marine environment have undetectable levels (<1 ppm $\times 10^{-6}$), residues have been found concentrated in surface slicks (13 ppm$\times 10^3$) and the associated fish, including Clupeidae, Engraulidae and their predators (Seba and Corcoran, 1969).

5.2.4   Solubility

Organochlorine residues accumulate in the aquatic biota as a result of their low solubility in water and high solubility in fats. Phytoplankton generally do not find environmental residue levels acutely toxic and they can concentrate residues for instance as much as 270 times the water concentration (Vance and Drummond, 1969). Vascular aquatic plants also concentrate residues; *Najos* sp. and *Potamogeton* sp. treated with 0.5 ppm diquat and paraquat concentrate as much as 80 times the amount in water (Coats *et al.*, 1964). *Cottus perplexus* most readily accumulated dieldrin from water (0.5 ppm$\times 10^3$) with no more than 16% of total body residue burden from food (7 ppm), although they reached equilibrium when exposed to either contaminated food or water independently (Chadwick and Brocksen, 1969). Adsorbed DDT represented about 1% of that absorbed by *Notemigonus crysoleucas* with high residue levels at the site of adsorption (Reed, 1969).

Fish residues originate from both food and water. Murphy (1971), using *Gambusia affinis,* found that small fish could remove four times more ^{14}C-labelled DDT from water than larger fish, demonstrating circumstantially the importance that the branchial route can assume. *Salvelinus fontinalis,* on the other hand, accumulated 10 times more available ^{14}C-labelled DDT from food than water (Macek and Korn, 1970). Commercial fish food from four different sources, fed to hatchery trout, yielded residues of heptachlor, heptachlor epoxide, dieldrin, DDT, DDD and DDE in the fish (Cole *et al.,* 1967). In a natural situation, such residues could be introduced through contaminated dead insects. In field collections from a population of *G. affinis,* endrin residues of 6.8 to 11.9 ppm were found in August, while levels were down to 0.73 to 0.88 ppm in February. Laboratory tests provided no evidence to indicate pesticide metabolism or alteration (Ferguson *et al.,* 1966). This data supports the hypothesis that fish residue levels are largely controlled through a solubility equilibrium with water residues, which are highest at the time of year when pesticides are most heavily used, and when fish lipid content is highest. Ferguson *et al.* (1966) found that the difference in residue loss from live, compared with dead, fish would not support excretion as an elimination route; further evidence for the solubility-regulated hypothesis.

5.3   UPTAKE AND PERSISTENCE

Pesticide solubility, interaction, chemical structure, fish condition, variations in metabolic patterns, and habitat have all been implicated in the regulation of pesticide uptake and persistence. *Salmo gairdneri* has accumulated 20 to 25% of available dietary DDT, whereas comparable figures for dieldrin were 9 to 11%. Dieldrin and DDT residue levels in *S. gairdneri* have been shown to be dose-dependent, with a linear increase in tissue concentration, followed by an equilibrium state that is maintained with continued exposure. This equilibrium state, in skeletal muscle as well as liver and brain tissue, was reached in 28 days for both dieldrin and DDT residues. Tissues edible by human beings accounted for 70% of total residues. While whole-body dieldrin residues were similar after 14 and 28 days, accumulation then resumed until 140 days and then remained stable until 168 days. These residue plateaus were not observed for whole-body DDT residues, nor for either dieldrin or DDT in the pyloric caecae (Macek *et al.*, 1970). This difference in uptake may be related to the respective solubilities of DDT and dieldrin, although the lower retention of dieldrin was suggested to be due to a greater ability of the trout to metabolise and/or excrete dieldrin than DDT.

5.3.1   Interactions

The action of additional (other/new) pesticides can and does increase DDT residues. When dogs chronically exposed to DDT were treated with aldrin, they reacted by accumulating increased DDT, DDE, and DDD residues in their blood and fat (Deichmann *et al.*, 1971). An interaction between dieldrin and DDT affected accumulation in pyloric caecae of *Salmo gairdneri*, so that DDT residues increased and dieldrin residues decreased over the amounts taken up from the same doses of single insecticides (Macek *et al.*, 1970). When DDT was incorporated in the diet, DDT residues accumulated, but these increased by 157% and DDE residues by 328% when dieldrin was added. Methoxychlor accumulation also increased in the presence of either dieldrin or DDT. On the other hand, dieldrin accumulation decreased in the presence of either DDT or methoxychlor. I quote: '. . . ultimate effects of these interactions are yet to be recognised in the natural environment' (Mayer *et al.*, 1970). Practical implication of the persistence of pesticides is the increased susceptibility to future doses (Gakstatter and Weiss, 1967). Combinations of pesticides (e.g. DDT and methyl parathion) have been shown to be more toxic to 'resistant' than to 'susceptible' *Gambusia affinis* (Ferguson and Bingham, 1966). These 'resistant' fish inhabit agricultural drainage ditches described as 'extremely fertile' and are characterised by increased susceptibility to pesticides during winter and early spring (Finley, 1970). These factors, plus greater lipid content in these 'resistant' *G. affinis* (Fabacher and Chambers, 1971), are explicable on the basis of the fat solubility of organochlorines and the ability of fish with high lipid content to accumulate massive concentrations of pesticide residues (Ferguson, 1967).

5.3.2 Turnover

A relatively short DDT residue half-life of three to five weeks has been established in *Carassius auratus*, utilizing ^{14}C-labelled DDT (Grezenda *et al.*, 1970). This could account for the lower susceptibility to pesticides of this species (Macek and McAllister, 1970). Within 32 days of terminating exposure of *C. auratus* to dieldrin, the residue levels in the fish decreased by 49 to 82% (Gakstatter, 1968). *Oncorhynchus tshawytscha* and *O. kisutch* lost, respectively, 45 to 68% and 19 to 35% of their DDT residues within 35 days after termination of exposure to DDT (Buhler *et al.*, 1969). DDT residues in *Salmo salar* increased for three or four months following spraying and then gradually declined until they were undetectable after a few months. DDT metabolites, however, remained in the fish for several years (Sprague and Duffy, 1971b). *S. salar* parr can degrade absorbed DDT to DDE and DDD within hours, while adsorbed DDT remains unaltered. While DDE is accumulated, DDD is rapidly excreted (Greer and Paim, 1968). Macek *et al.* (1970), using ^{14}C-labelled DDT and ^{14}C-labelled dieldrin, predicted 'half-lives' of 160 days for DDT and 40 days for dieldrin in *S. gairdneri*. A month after DDT contamination of a stream, *Salvelinus fontinalis* and *Catostomus commersoni* living in it had increased residues of from 0.88 to 20.4 ppm and from 6.1 to 17.3 ppm respectively. Three months later, these values were 0.77 ppm and 0.61 ppm (Cole *et al.*, 1967). This study may indicate either rapid residue elimination, or an elimination of contaminated fish with recruitment of fish from uncontaminated areas. *Carassius auratus*, exposed to ^{14}C-labelled aldrin, contained dieldrin residues within eight hours (Gakstatter, 1968). Dieldrin residues in *Cottus perplexus* decreased from 2.5 to 1 ppm, 60 days after termination of exposure (Chadwick and Brocksen, 1969).

Less persistence of organophosphates is indicated by *Cyprinus carpio* exposed to malathion. An equilibrium was reached after four days of exposure and the residue half-life is 12 hours (Bender, 1969a). Hepatic phosphatases, in *Lepomis macrochirus* and *Ictalurus punctatus*, are capable of degrading some organophosphates. The presence of manganic ions increases the hydrolysing potential by approximately three times (Hogan and Knowles, 1968). Macek (1970) recently reviewed the persistence and magnification of pesticide residues in fish.

5.3.3 Season and habitat

Seasonal residue fluctuations in fish could be a product of variation in adrenal sex steroid level correlated with reproductive activity. Corticosteroids might also be involved in the reduced accumulation of DDT and DDE residues in *Gambusia affinis* maintained in 15‰ seawater (Murphy, 1970). Endocrine-pesticide interaction is an area considered in need of further research.

5.4   REDUCTION AND ELIMINATION

5.4.1   Effect of turbidity

Turbidity and high organic content of water have long been known to reduce the availability of organochlorine residues to solution (Hoffman, 1959). Terriere *et al.* (1966) found that toxaphene remained toxic to fish for six years (acute toxicity in *S. gairdneri* within two weeks) in an oligotrophic lake. Toxaphene amounts, soon after application to this lake, were 40×10^{-3} ppm, and after six years it was 0.84 ppm$\times 10^{-3}$. In the water of a eutrophic lake, used for comparison, toxaphene residues were <0.2 ppm$\times 10^3$ three years after application with an initial level of 88 ppm$\times 10^3$. Although tissue residues in *S. gairdneri* exceeded 13 ppm, six to nine months after treatment of the eutrophic lake, there were no apparent adverse effects to trout. Dieldrin muscle residues in *Ictalurus punctatus,* from some Iowa streams which received drainage from aldrin-treated cropland, exceeded the Food and Drug Administration guideline (0.3 ppm) by as much as a factor of five. These excessive levels were not found in other streams of the area (Morris and Johnson, 1971). These investigators discussed three factors they considered contributed to dangerous residue levels in our freshwater environment. These were: unnecessary use of pesticides, poor soil conservation providing a silt carrier, and dams which increase the amount of habitat over a silted bottom.

5.4.2   Degradation

DDT has been shown to be converted to DDD by fungi and bacteria from the gut of *Engaulis mordax,* with fungi thought to be responsible for further degradation to a water-soluble product (Malone, 1970). Thirty-three days after introduction into a model ecosystem, 54% of DDT residues in *Gambusia affinis* were DDE, with a concentration factor of 30,000 to 50,000; in contrast, methoxychlor was rapidly metabolised and stored at low levels, only about one-hundredth of the DDT residues (Metcalf *et al.,* 1971). A similar DDT concentration factor was seen, after five weeks, in *Lagodon rhomboides* and *Micropogan undulatus,* with most residues being acquired in the first two weeks of exposure. After eight weeks in uncontaminated water, DDT residues were reduced up to 87%, but DDD and DDE residues were unchanged during the experiment (Hansen and Wilson, 1970).

5.4.3   Interactions

Pesticide residues also interact to affect the elimination of one another. Cole *et al.* (1967) hypothesised that DDT treatment may have been responsible for the absence of dieldrin in experimental *Salvelinus fontinalis.* Mayer *et al.* (1970) confirmed a similar effect on *Salmo gairdneri,* although it was not as pronounced as with rats. While Macek *et*

fewer ova at high doses. Residues in these treated fish were similar to those found in wild populations (Macek, 1968a). Egg residues and survival data agreed with predictions by Burdick *et al.* (1964) who indicated that 5 ppm DDT residue would produce approximately 15% mortality. At 5 ppm Azinphos-methyl (Guthion®), diazinon, endrin, dieldrin, DDT, or chlordane affected *Cyprinus carpio* egg mortality by between 50 and 100%. The incubation time of eggs exposed to <1 ppm of chlordane was reduced by about one third (Malone and Blaylock, 1970). Chronic exposure of *Poecilia reticulata* caused maternal mortality at birth (Cairns *et al.*, 1967). *Lepomis macrochirus* exposed to 2,4-D were delayed two weeks in spawning; there was, however, no apparent effect on number of fry produced (Cope *et al.*, 1970). Prolonged heptachlor exposure had no effect on reproduction in the same species. The transfer of residues from generation to generation via egg lipids may result in increased susceptibility in later generations. This phenomenon has been demonstrated with three generations of *Cyprinodon variegatus* (Holland, 1970).

5.5.3 'Resistance'
The heavy use of pesticides in cotton-producing areas results, relatively commonly, in fish kills (Ferguson, 1965). As a by-product of this heavy exposure, 'resistant' populations, characterised by unusually high pesticide residue levels, have been described. Endrin levels in *Gambusia affinis* of 214 ppm and muscle of *Lepomis cyprinellus* with 26 ppm have been reported (Ferguson, 1967). *Gambusia affinis* populations showed 'resistance' to only the toxaphene-endrin related organochlorines on the basis of 48-hour bioassays of 28 insecticides from five major groups (Culley and Ferguson, 1969). The investigators of these 'resistant' populations have stressed the potential ecological hazard represented by these concentrations of residues (Ferguson *et al.*, 1967). DDD residues of 40 ppm in fat from *Cyprinus carpio* and of 2,500 ppm in *Ictalurus nebulosus* were seen in the classic Clear Lake, California, incident (Hunt and Bischoff, 1960). Did these residues indicate 'resistance' or surviving fish of increased susceptibility to subsequent pesticide contamination? Has the behaviour, growth, fecundity or some other unmeasured property of these 'resistant' populations been altered? If fish containing these unusually high residues are indeed affected, do they carry a sublethal or slowly lethal dose? In a more complex environment, that is, one with predators and other stresses, i.e., osmotic, reproductive, migratory, would these fish be 'resistant'? In terms of residue levels in 'resistant' fish, 'people kills', referred to by Egler (1964), are not inconceivable. Pesticide poisoning of fish-eating birds is well-documented and discussed elsewhere in this volume (chapter 7).

5.5.4 Reduced productivity
Growth of fish exposed to heptachlor was inversely proportional to

exposure level (Andrews *et al.*, 1966). Mirex has been also shown to inhibit growth of *L. macrochirus* (VanValin *et al.*, 1968). Dieldrin exposure has reduced both growth and fecundity of *Poecilia latipinna* (Lane and Livingston, 1970). *Poecilia reticulata*, exposed to dieldrin for a long time, gained more weight than control fish (Cairns *et al.*, 1967). Exposure to 2,4-D caused an apparent dose-dependent increase in growth of *L. macrochirus* which was at least partly attributable to decreased population density resulting from acute toxicity (Cope *et al.*, 1970). DDT and dieldrin increased lipogenesis in *Salmo gairdneri*, although total growth was unaffected (Macek *et al.*, 1970). DDT, however, did increase weight gain in *Salvelinus fontinalis* (Macek, 1968b). The long-term effect may have been demonstrated when these fish were placed on a reduced diet (10%) and mortality was 88.7% compared with 1.2% for control fish. The increased susceptibility to DDT of smaller slow-growing fish (*Oncorhynchus kisutch*) (Buhler and Shanks, 1970) points to the possibility of fish populations being eliminated by much lower levels of exposure than are currently believed to be toxic. This increased sensitivity could follow decreased population productivity resulting from other factors and effects and environmental degradation. These might include decreased invertebrate populations caused by pesticide levels which do not currently produce acute toxicity in fish.

5.5.5 Interactions
In this era of thermal pollution, it is not comforting to find that while oxygen consumption is doubled in *Gambusia affinis* at 20°, compared with 5°C, accumulation of DDT triples (Murphy and Murphy, 1971). This factor will doubtlessly most acutely affect amphidromous and diadromous fish that are currently suffering from increasingly stressful conditions in estuarine and impounded waters. While fish, generally, are capable of avoidance responses with regard to thermal pollution [recent reviews include Chittenden and Westman (1970) and Coutant (1970)], temperature selection by *Salmo salar* is affected by DDT exposure (Ogilvie and Anderson, 1965). Chronic exposure to dieldrin also increases oxygen consumption, while decreasing cruising speed of *Lepomis gibbosus* (Cairns and Scheier, 1964). Is this significant for migrating fish confronted by heated effluent from thermal power generators, reduced oxygen in slack-water reservoirs, and fish ladders with accompanying delay at each sequential dam? Exposure to 'sublethal' doses of DDT has also been shown to inhibit the development of conditioned response to light (Anderson and Prins, 1970). This behavioural effect could seriously affect feeding patterns, vulnerability to predation, migratory behaviour, and reproductive success.

The complete chemistry of most of the pesticides being used, their metabolites, and biochemical effects, are not yet known. Even so, the effects of a few mixtures have been investigated. A mixture of DDT and methyl parathion appeared to act synergistically in increasing their acute toxicity to *Gambusia affinis* (Ferguson and Bingham, 1966). While linear alkyl benzene sulfonate (LAS) did not affect the toxicity of DDT or endrin, it did act synergistically with parathion in producing acute toxity to exposed *Pimephales promelas* (Solon *et al.*, 1969). Polychlorinated biphenyl (PCB) residues as environmental contaminants may exceed by 2-3 times the organochlorine pesticide residues in some fish (Zitko, 1971). Although not included in the U.S. monitoring programme, these PCB residues are common in aquatic systems (Holden, 1970; Schmidt *et al.*, 1971). *Lagodon rhomboides* and *Leiostomus xanthurus*, chronically exposed to low levels, have increased susceptibility to disease (Hansen *et al.*, 1971). Interaction with pesticide residues to give an additive effect would seem not unlikely. The biological interaction of insecticides and plasticisers has been demonstrated in experiments with flies, the presence of PCB increasing dieldrin and DDT toxicity (Lichtenstein *et al.*, 1969). Pesticide metabolites may also contribute to increased toxicity. Malathion undergoes rapid hydrolysis to diethyl fumarate and DMPTA. Dimethyl fumerate, the basic metabolite, is more toxic to *P. promelas* than the parent malathion and in combination they exert a synergistic toxicity (Bender, 1969b). Bartha (1969) established that the breakdown products of two herbicides could combine in soil to form unexpected and unpredictable hybrid products, thereby emphasising the complexity of the problem if there is massive environmental contamination with a great variety of synthetic pesticides. Bender (1969b) found that after continuous exposure to malathion and its metabolites, the tolerance level of *P. promelas* decreased. These are examples of unpredicted synergistic actions and other environmental effects that may become increasingly common.

5.6   RESEARCH NEEDS

5.6.1   Effect of residues

Abelson (1971) used mercury as an example of the ability of fish to concentrate environmental contaminants because from environmental levels of 0.5 ppm$\times 10^3$, tuna obtained residues of 0.03 to 2.0 ppm$\times 10^3$ He called for an increase in monitoring efforts to provide earlier warning of impending ecological tragedies. Although monitoring systems can provide helpful reference values, they cannot be used wisely as an alarm system, unless we know the biological effects of the monitored residue levels. We have not yet completed the research required to define the lower limits of biologically significant and potentially dangerous levels of contamination. Sprugel (1971) asked, 'Shouldn't a concerted effort be made to find out exactly if, how, and why the pesticide molecule produced these effects?' Sufficient allocation of research support could supply an affirmative reply to this question.

Before the definitive answers required, relative to significance of our contamination of the earth with pesticide residues, can be supplied, there is need for much more information in the areas of fish physiology, bio-chemistry, histophysiology and histopathology, and aquatic ecology. Criteria must be developed for the establishment of pesticide threshold levels that initiate population declines and sublethal effects. We must establish the effects of residue concentrations under conditions of either intermittent or continuous exposure as well as clarifying the fate of pesticides after application, the influencing factors and the ratio between the amount applied and the amount available to fish.

5.6.2 Fate of residue

Modelling can be used to assess the movement of pesticides through ecosystems and for prediction of population changes produced by this environmental contamination. A model developed by Harrison *et al.* (1970) indicated that, probably, we will not know for many years the effects of present environmental pesticide loads; Metcalf *et al.* (1971) reported that a 30-day study, utilising radio-isotope labelled compounds and a model laboratory system, yielded results similar to those from a 20-year study of Lake Michigan. Such a laboratory ecosystem may have predictive capability for the determination of at least some of the environ-mental effects of new pesticides before their introduction, rather than 20 years later. This approach appears to hold great promise.

5.6.3 Enzyme relationships

The investigation of the response of several enzyme systems to pesticide contamination has contributed and will contribute to our understanding of possible sublethal effects of fish residues. Weiss (1958, 1959, 1961) and his colleagues have quite comprehensively described the inhibitory effect of organo-phosphorus compounds on brain cholinesterase. Recently, the effect of pesticides on sodium and potassium-activated adenosine triphosphatase has been investigated. The importance of this system, in the maintenance of hydromineral balance in fish, has been demonstrated (Epstein *et al.*, 1967; Kamiya and Utida, 1968; Zaugg and McLain, 1969). Treatment of intestine from marine *Anguilla anguilla* with DDT *in vitro*, inhibited intestinal Na^+ and K^+ ATPase, as well as water movement from the intestine (Janicki and Kinter, 1971). *In vitro* enzyme inhibition by dicofol, endosulfan, and DDT, has also been shown for *Salmo gairdneri* brain, gill, and kidney tissue (Davis and Wedenmeyer, 1971). Adenosine triphosphatase sensitivity to DDT had been reported earlier in rat brain synapses (Matsumura and Patil, 1969). Carbonic anhydrase inhibition has not been reported in fish; carbonic anhydrase has a major role in branchial excretion of sodium in sea water and its uptake in fresh water (Hodler *et al.*, 1955; Maetz and Garcia Romeu, 1964). Dvorchik *et al.* (1971) have described an *in vitro* effect of pesticides on human erythrocytes. Another *in vitro* study, using *Lepomis macrochirus* liver mitochondria, suggested that the primary effect of DDT residue is the inhibition of electron flow from succinic acid to the cytochrome chain (Hiltibran, 1971). Solubility

difficulties, as well as dilution and distortion of the drug-enzyme relationship, have been described as problems in studies of *in vitro* treatments
(Dvorchik *et al.*, 1971). An investigation of the effects of DDT, dieldrin,
and endrin on intermediary metabolism in fish is under way (Grant and
Mehrle, 1970; Mehrle *et al.*, 1971).

5.6.4 Endocrine Relationships

There are many indications that fish endocrinology might provide some
very important answers regarding sublethal effects of pesticides.
Wassermann *et al.* (1970) described an interaction between rat
adrenocorticoids and the accumulation of DDT residues and reported
that adrenalectomised rats stored more DDT residues. In rats, organochlorine residues increase when organophosphate pesticides inhibit
the metabolism of steroids (Kupfer, 1969). Oxidative liver enzymes are
stimulated by both steroids and insecticides, and contribute to the breakdown of both (Risebrough, 1969). Might not then DDT residues disrupt
the seasonal timing of gonad development and/or the hydromineral
acclimation of diadromous fish, resulting either in spawning failures and/or
estuarine die-offs? Deichmann and Radomski (1968) found that hyper-
tensive human subjects were characterised by increased storage of organochlorine insecticides. Chronic hyperactivity of the interrenal tissue, as a
result of organochlorine contamination in a diadromous fish, could leave
the fish without the ability to increase corticosteroid secretion when its
osmotic gradient fluctuates with migratory movement (Johnson, 1973).
The structural similarity of DDT and synthetic estrogens, as well as
the effects of DDT on estrogenic activity of rats, is established (Bitman
and Cecil, 1970). A persistent estrus syndrome has been induced in rats by
DDT (Heinricks *et al.*, 1971). This supports conjecture that the reproductive cycles of populations may be disrupted by pesticide residues, to
the detriment of many fisheries. Oshima and Gorbman (1969) found that
estradiol influenced the olfactory responses of *Carassius auratus.*
Therefore, might not DDT affect olfactory-guided anadromous fish?
Chronic exposure to both DDT and dieldrin significantly increases lipogenesis, and their effect is additive (Buhler *et al.*, 1969; Macek *et al.*,
1970). Natural fluctuations in fat content are common in fish (Perkins and
Dahlberg, 1971) and investigation as to their hormonal control (Meier,
1970), and interaction with pesticides, seems a fertile field for additional
research. Cortisol and prolactin are important hormones in the regulation
of teleost hydromineral balance, and they may prove to have interactions
with pesticides. The endocrine control of Na^+, K^+ ATPase probably
involves both prolactin and cortisol (Johnson, 1973). Several investigators
have believed that alteration of ion balance was an effect of pesticide
residues (Walker, 1963; Eisler and Edmunds, 1966; Macek, 1968a). Many
of the above findings provide strong support for increased investigations of
the response of gill, gut and renal function to pesticide residues. To provide understanding required to protect fish populations from possible

extermination, the interaction of many factors remains to be investigated and more thoroughly considered.

5.7   OUTLOOK

With increases in the number of synthetic toxicants introduced into the environment and the complexity of possible interactions, detrimental effects on fish populations and the frequency of fish kills are apparently increasing. In the 1960's, over 144 million fish were reported to have been killed in 4,200 incidents. In the 1969 annual voluntary census, 41 million fish were reported killed in 45 states as a result of pollution. Kills were principally between May and September, coinciding with peak crop production. Industrial operations and agriculture ranked number one and two as contributory causes. During 1970, 231 kills, totalling 50 million fish, were reported affecting 860 miles of stream and 10,700 surface acres (Mackenthun, 1971). An estimated 80% of Lake Michigan and 42% of the Great Lakes fish catch in 1970, had DDT residues exceeding the Food and Drug Administration's legal tolerance for food (Lueschow, 1970).

The concept of proving a fish kill, by finding a threshold toxicant level in a critical tissue, assumes that the fish population killed will have a higher residue concentration than a similar population exposed but not killed. This residue technique is credited with 80 to 90% success in separating tissue from exposed live fish from tissue of fish killed by the toxicant in question (Stephan, 1971). The use of pesticide residue levels in the diagnosis of fish kills must, however, consider both interspecific and population variability in susceptibility. *Ictalurus melas,* with body residues of 8-15 ppm toxaphene, have displayed symptoms of poisoning (Kallman *et al.,* 1962). Levels in this range have been reported for field fish collections of *I. punctatus,* with whole fish residues as high as 25.0 ppm in *Catostomus clarki* (Johnson and Lew, 1970). The endrin concentration in the blood of laboratory-exposed *Dorosoma cepedianum* indicated a critical level of 0.10 ppm, above which less than 5% of the fish survived. Residues in blood from dead fish averaged 0.24 ppm, while those from live exposed fish were 0.06 ppm (Brungs and Mount, 1967). A critical endrin blood level of 0.28 ppm has been defined for *Ictalurus punctatus* (Mount *et al.,* 1966). DDT residues of 0.16 and 0.19 ppm have been reported from field collections of this species (Table 5.2). Dieldrin has a critical value of 6.0×10^{-3} ppm in blood of *Lepomis cyanellus.* Fish that survived exposure, but suffered severe symptoms, had higher residues than those characterised by lesser symptoms (Hogan and Roelofs, 1971). The data of Ludke *et al.* (1968) shows an apparent direct correlation between endrin residues in the blood of *Notemigonus crysoleucas,* and water residue concentration and length of exposure. Stickel (1969) also stated the need for identifying diagnostic tissues, and correlating tolerance limits for pesticides separately and in combination with other environmental contaminants. We need to

G. L. Greer and V. Paim, *J. Fish. Res. Bd. Canada,* **25**, 2321 (1968)
A. R. Grzenda, D. F. Parris and W. J. Taylor, *Trans. Am. Fish. Soc.,* **99**, 385 (1970)
A. R. Grzenda, W. J. Taylor and D. F. Paris, *Trans. Am. Fish. Soc.,* **100**, 215 (1971)
W. H. Gutenmann and D. J. Lisk, *N. Y. Fish Game J.,* **12**, 108 (1965)
J. L. Hamelink, R. C. Waybrant and R. C. Ball, *Trans. Am. Fish. Soc.,* **100**, 207 (1971)
M. R. Hannon, M.S. Thesis, South Dakota State Univ. (1969)
M. R. Hannon, Y. A. Greichus, R. L. Applegate and A. C. Fox, *Trans. Am. Fish. Soc.,* **99**, 496 (1970)
D. J. Hansen and A. J. Wilson, Jr., *Pestic. Monit. J.,* **4**, 51 (1970)
D. J. Hansen, P. R. Parrish, J. I. Lowe, A. J. Wilson, Jr. and P. D. Wilson, *Bull. Environ. Contam. Toxicol.,* **6**, 113 (1971)
H. L. Harrison, O. L. Loucks, J. W. Mitchell, D. F. Parkhurst, C. R. Tracy and D. G. Watts, *Science N.Y.,* **170**, 503 (1970)
W. L. Heinrichs, R. J. Geller, J. L. Bakke and N. L. Lawrence, *Science N.Y.,* **173**, 642 (1971)
C. Henderson, W. L. Johnson and A. Inglis, *Pestic. Monit. J.,* **3**, 145 (1969)
C. Henderson, A. Inglis and W. L. Johnson, *Pestic. Monit. J.,* **5**, 1 (1971)
R. C. Hiltibran, *Trans. Ill. State Acad. Sci.,* **64**, 46 (1971)
J. Hodler, H. Heinemann, A. Fisher and H. Smith, *Am. J. Physiol.,* **183**, 155 (1955)
C. H. Hoffman, *Trans. 2nd Seminar Biol. Prob. Water Pollution,* SEC Tech. Rpt. W60-3, 51 (1959)
J. W. Hogan and C. O. Knowles, *J. Fish. Res. Bd. Canada,* **25**, 1571 (1968)
R. L. Hogan and E. W. Roelofs, *J. Fish. Res. Bd. Canada,* **28**, 610 (1971)
A. V. Holden, *J. Appl. Ecol.,* **3**, (Suppl.), 45 (1966)
A. V. Holden, *Nature* (Lond.), **228**, 1220 (1970)
H. T. Holland, *Bull. Environ. Contam. Toxicol.,* **5**, 362 (1970)
C. L. Hopkins, S. R. B. Solly and A. R. Ritchie, *N.Z. J. Mar. Freshwater Res.,* **3**, 220 (1969)
E. G. Hunt and A. I. Bischoff, *Calif. Fish Game,* **46**, 91 (1960)
R. H. Janicki and W. B. Kinter, *Science N.Y.,* **173**, 1146 (1971)
D. W. Johnson, *Trans. Am. Fish. Soc.,* **97**, 398 (1968)
D. W. Johnson, *Amer. Zool.* (in press).
D. W. Johnson and S. Lew, *Pestic. Monit. J.,* **4**, 57 (1970)
B. J. Kallman, O. B. Cope and R. J. Navarre, *Trans. Am. Fish. Soc.,* **91**, 14 (1962)
M. Kamiya and S. Utida, *Comp. Biochem. Physiol.,* **26**, 675 (1968)
J. R. M. Kelso, H. R. MacCrimmon and D. J. Ecobichon, *Trans. Am. Fish. Soc.,* **99**, 423 (1970)
J. O. Keith, *J. Appl. Ecol.,* **3**, (Suppl.), 71 (1966)
D. Kupfer, *Ann. N.Y. Acad. Sci.,* **160**, 244 (1969)
C. E. Lane and R. J. Livingston, *Trans. Am. Fish. Soc.,* **99**, 489 (1970)
E. P. Lichtenstein, K. R. Schulz, T. W. Fuhremann and T. T. Liang, *J. Econ. Entomol.,* **62**, 761 (1969)
J. L. Ludke, D. E. Ferguson and W. D. Burke, *Trans. Am. Fish. Soc.,* **97**, 260 (1968)
L. Lueschow, AAAS Program, 125 (1970)

K. J. Macek, *J. Fish. Res. Bd. Canada*, **25**, 1787 (1968a)

K. J. Macek, *J. Fish. Res. Bd. Canada*, **25**, 2443 (1968b)

K. J. Macek, Environmental Health Sciences Series No. 1. Oregon State University (1970)

K. J. Macek and S. Korn, *J. Fish. Res. Bd. Canada*, **27**, 1496 (1970)

K. J. Macek and W. A. McAllister, *Trans. Am. Fish. Soc.*, **99**, 20 (1970)

K. J. Macek, C. R. Rodgers, D. L. Stalling and S. Korn, *Trans. Am. Fish. Soc.*, **99**, 689 (1970)

K. M. Mackenthun, E.P.A./F.W.Q.A., Fish Kill Investigation Seminar, Ohio State Univ. (1971)

J. Maetz and F. Garcia Romeau, *J. Gen. Physiol.*, **47**, 1209 (1964)

C. R. Malone, *Nature* (Lond.), **227**, 848 (1970)

C. R. Malone and B. G. Blaylock, *J. Wildl. Mgmt.*, **34**, 460 (1970)

F. Matsumura and K. C. Patil, *Science N. Y.*, **166**, 121 (1969)

F. L. Mayer, Jr., J. C. Street and J. M. Neuhold, *Bull. Environ. Contam. Toxicol.*, **5**, 300 (1970)

P. M. Mehrle, D. L. Stalling and R. A. Bloomfield, *Comp. Biochem. Physiol.*, **38**, 373 (1971)

A. H. Meier, *Proc. Soc. Exp. Biol. Med.*, **133**, 1113 (1970)

R. L. Metcalf, G. K. Sangha and I. P. Kapoor, *Environ. Sci. Technol.*, **5**, 709 (1971)

K. P. Milne, J. N. Ball and I. Chester Jones, *J. Endocrinol.*, **49**, 177 (1971)

R. L. Morris, L. G. Johnson and W. Patton, State Hygienic Lab., Iowa State Conservation Commission, Rpt. 70-10 (1969)

R. L. Morris and L. G. Johnson, State Hygienic Lab., Iowa State Conservation Commission, Rpt. 70-12 (1969)

R. L. Morris and L. G. Johnson, *Pestic. Monit. J.*, **5**, 12 (1971)

D. I. Mount, L. W. Vigor and M. L. Schafer, *Science N.Y.*, **152**, 1388 (1966)

M. Meuller, *Science N. Y.*, **164**, 936 (1969)

P. G. Murphy, *Bull. Environ. Contam. Toxicol.*, **5**, 404 (1970)

P. G. Murphy, *Bull. Environ. Contam. Toxicol.*, **6**, 20 (1971)

P. G. Murphy and J. V. Murphy, *Bull. Environ. Contam. Toxicol.*, **6**, 581 (1971)

W. S. Murray, *Pestic. Monit. J.*, **5**, 36 (1971)

W. E. Odum, *Trans. Am. Fish. Soc.*, **99**, 836 (1970)

D. M. Ogilvie and J. M. Anderson, *J. Fish. Res. Bd. Canada*, **22**, 503 (1965)

K. Oshima and A. Gorbman, *Gen. Comp. Endocrinol.*, **13**, 92 (1969)

B. Parker and G. Barsom, *Bio. Sci.*, **20**, 87 (1970)

R. J. Perkins and M. D. Dahlberg, *Ecology*, **52**, 359 (1971)

T. J. Peterle, *J. Appl. Ecol.*, **3** (Suppl.), 181 (1966)

G. E. Pickford, R. W. Griffith, J. Torretti, E. Hendlez and F. H. Epstein, *Nature*, **228**, 378 (1970)

J. K. Reed, Water Resources Res. Instit., Tech. Compl. Rpt. 18 pp. (1969)

H. A. Reigier, *Trans. Am. Fish. Soc.*, **100**, 804 (1971)

R. E. Reinert, *Pestic. Monit. J.*, **3**, 233 (1970)

R. W. Risebrough. M. W. Miller and G. C. Berg (Editors), C. C. Thomas Publ., p. 5. Springfield, Ill. (1969)

C. A. Rodgers, *Weed Sci.*, **19**, 50 (1970)

such compounds as lead arsenate, copper sulphate, sodium arsenite, sodium cyanide and phenolic mixtures. In addition to these, naturally-occurring organic compounds derived from plants, such as pyrethrum, derris and nicotine, were widely used as insecticides. Derris, the active constituent of which is rotenone, has also been known as a fish poison since ancient times, being used by natives in Asia and South America, in the form of a paste or extraction of plants of the genus *Derris* (in Asia) or *Lonchocarpus* (S. America).

With the development of the first two of a new generation of synthetic organic pesticides, DDT and lindane (γ-BHC), in the 1940's, a new era of pest control began, but the very property of these chemicals, and the multitude of successors produced over the following thirty years, with high toxicity to life processes, has resulted in environmental problems never anticipated by the discoverers of the insecticidal properties of DDT and lindane. Very few pesticides can be regarded as specific for a selected pest, and in consequence, many other forms of life frequently fall victim to their action./Fish are particularly sensitive to a wide variety of pesticide chemicals, and toxic conditions may arise, not only from the spillage or deliberate discharge of these chemicals into rivers and lakes, but also from approved agricultural practices if their use is excessive. Many applications outside agriculture, such as sylviculture, horticulture or public health, can also lead to a detrimental influence on fish populations.

The nature of the effects on fish can be very variable. Apart from causing death, either directly, or due to starvation, by destruction of food organisms, many pesticides have been shown to affect growth rate, reproduction and behaviour, with evidence of tissue damage. While under experimental conditions, affected fish may survive, but in their natural environment, such influences can hardly be other than detrimental, rendering the fish more vulnerable to predators, less able to compete with other fish and less able to withstand the normal stresses such as seasonal temperature variation, reproduction, or temporary starvation. The young growth stages of fish are particularly susceptible to some pesticides, and as survival at this point in the life cycle is minimal for most species under natural conditions, a further decrease in the survival rate may be disastrous for the future of the stock.

With the recognition that all pesticides are potentially lethal to fish even at relatively low concentrations, by comparison with those commonly used in spray applications, it is now normal practice to test all new chemicals for their toxicity to fish. A variety of test procedures is in use, there being little or no standardisation of test conditions, but the usual screening method assesses the aqueous concentration of the chemical in question (LC_{50}) which will kill 50% of the test fish in 24, 48 or 96 hours. In contrast to tests on mammals or birds, the dose ingested or otherwise absorbed is not estimated, but only the environmental aqueous concentration to which the fish are exposed. Yet the amount absorbed is clearly in some way related to the effect produced, and estimates of tissue concentrations may be more valuable for the assessment of situations in the natural environment. Some pesticides, such as the organochlorines, with their inherent chemical stability and lipophilic properties, present a further problem in the assessment of toxicity. They are accumulated in tissues to an increasing degree, as the exposure of the fish to aqueous concentrations is extended (see chapter 5), and concentrations which are without apparent effect in the normal short-term test, can be lethal if the test period is prolonged for several weeks or months. Such test conditions, however, present additional problems for the research worker, with the difficulty of maintaining constant concentrations, providing adequate food without introducing an alternative source of contamination, and maintaining otherwise acceptable conditions for fish survival.

This chapter seeks to present the main features of our present knowledge of the effects of pesticides on fish, both that derived from experimental work and that from field experience in many parts of the world. Most of the available information, however, has been published in North America, where not only is the use of pesticides perhaps most widespread, but where facilities for the study of ecological effects have been developed to an extent which has enabled such effects to be identified. It must be admitted that the large-scale use of pesticides in crop protection, and in safeguarding human health, in Africa, India, Asia and South America, has not been followed by an adequate study of environmental effects, and little is known of the influence which such practices have had on fisheries.

6.2 PESTICIDE CONCENTRATIONS IN NATURAL WATERS

While high concentrations of pesticides frequently result from the accidental spillage of pesticide formulations into lakes and rivers, the consequent fish-kill is often no greater than would be expected from a similar incident involving other chemicals. The persistent chemicals may present a long-term hazard by virtue of their stability, but many pesticides are hydrolysed or rapidly flushed from the river system, and create only a temporary problem.

The contamination of waters from the accepted agricultural or other use of pesticide chemicals, on the other hand, may lead to a long-term situation more difficult to identify, in the absence of a mass mortality, but potentially more detrimental to fish populations. The concentration of a chemical in run-off water, or in a lake sprayed from the air, may be calculated, for comparison with concentrations used in long-term toxicity testing, or in experiments to establish the sublethal effects of pesticides on fish. Experiments of this type are only of academic value if the effect-producing concentrations are of a significantly lower magnitude than those which may occur in practice. Thus, a spray application of 1 lb/acre (1.12 kg/ha) of DDT, falling on water one foot deep, such as in a pond or drainage channel, would, if mixed, give an aqueous concentration of DDT of 0.37 mg/l, about twenty times the 24-hour LC_{50} for brown trout. Holden (1964) calculated the concentrations, in water one foot deep, of various pesticides sprayed at rates recommended for crop use on adjacent land, and compared them with the 24-hour LC_{50} values for brown trout. Several pesticides, particularly the organochlorines, such as aldrin, dieldrin and DDT, would reach concentrations liable to be very toxic to fish. The herbicide pentachlorophenol, and the fungicide copper oxychloride are also potentially lethal in such circumstances.

Most documented examples of the effects of aerial spraying on aquatic life have related to DDT, but few of the reports have included measurements of the DDT concentrations reached in the rivers and streams. The use of DDT at 1 lb/acre (1.12 kg/ha) on forests in Montana (Cope and Park, 1957) caused a significant reduction in the aquatic insect population of streams, but no effect was observed on trout. A concentration of 0.10 ppm DDT immediately after spraying and 0.33 ppm 30 minutes after spraying, were recorded in the water, but after 27 hours, no DDT was detectable. Bridges and Andrews (1961), studying the effect of the same spray rate in other Montana streams, measured a maximum of 0.01 ppm DDT after 30 minutes, and again no immediate effects were noted on the resident fish. Cole et al. (1967) found 0.024 ppm of DDT in stream water one hour after a spray application of 0.5 lb/ac (0.56 kg/ha), but only 0.00012 ppm after two weeks. Although brief periods of high concentrations of DDT probably occur in such operations, it seems likely that the average concentration will be well below 0.01 ppm over the first 24 hours (see chapter 11). Yule and Tomlin (1971), studying the effect of a DDT application of 0.25 lb/ac (0.28 kg/ha) on forests, found a post-spray maximum of only 0.0017 ppm in stream waters, decreasing to below 0.001 ppm after a few hours. At such a spray rate, fish-kills are often minimal or non-existent, although there are several recorded instances of fish mortality occurring after several months. Fish and aquatic insects in the Yellowstone River died over a 90-mile stretch, four months after a 1 lb/ac (1.12 kg/ha) spray application of DDT (Graham, 1960). In 1954, DDT sprayed at 0.51 lb/ac (0.56 kg/ha) in New Brunswick, resulted in the virtual elimination of salmon fry (Keenleyside, 1959). Older fish died over periods of several months, seemingly as the result both of direct poisoning and the ingestion of poisoned insects (Elson and Kerswill, 1966). It seems likely that fish-kills following such spraying operations are due more to the accumulation of pesticide from contaminated food than to direct assimilation from water.

The effects of pesticides so far described, are those resulting from the application of the chemicals to land, particularly in large-scale forest spraying. Pesticides have also been used in some instances to control aquatic insects, in circumstances where fish have either been unimportant, or where fish-kills were unintended. The application of DDD to Clear Lake, California (Hunt and Bischoff, 1960), which was a classic example of the phenomenon of biological magnification, or progressive accumulation of an organochlorine pesticide through an eco-system, was intended to control a gnat, *Chaoborus astictopus*. The treatment began in 1949, and was repeated in 1954 and 1957. Western grebes, *Aechmophorus occidentalis*, were dying by September 1954, and more died subsequently for several years. In this example, there was no evidence of fish mortality, although various species of fish were found to have accumulated very high concentrations of DDD. Consumption of the fish by the grebes is presumed to have led to the deaths of the birds by a further accumulation of the pesticide in the avian tissues. The concentration of DDD applied to the water was 0.014 ppm in 1949, and 0.020 ppm in 1955 and 1957.

A similar concentration of DDT, 0.020 ppm was used by Bridges *et al.* (1963) in a farm pond, to study the subsequent distribution of DDT in the environment. The effect on invertebrates, if any, was estimated to be only slight (see chapter 4), but rainbow trout began to show distress one day after treatment, swimming erratically near the surface, with difficulty in maintaining equilibrium. Mortality appears to have been limited, and was restricted mainly to the smaller fish, occurring only during the first week. Water analyses indicated that the concentration of DDT fell rapidly from 0.080 ppm after 30 minutes to 0.004 ppm in 8 hours, and none was detectable after the second day. Concentrations in the rainbow trout reached a total of 4.15 ppm (DDT+DDD+DDE) in one month and remained between 2 and 4 ppm for at least 16 months. Similar levels were found in black bullheads, *Ictalurus melas*, but no evidence of mortality was found in this species. Most of the degradation of DDT to DDD and DDE occurred during the first few months, and no significant adverse effects on the fish or subsequent progeny were observed, apart from the few early deaths.

The use of DDT to control mosquitoes has largely been superseded by other chemicals, particularly organophosphorus insecticides. Mulla *et al.* (1966) used parathion at 1.0 and 0.1 lb/ac (1.12 and 0.11 kg/ha) in duck ponds, and at the higher concentration, mosquito fish (*Gambusia affinis*) were almost eliminated, concentrations of parathion up to 27 ppm being found in their tissues. The lower concentration had no apparent effect on the fish, although the maximum residue level in the fish, 22.5 ppm, was not much below that found at the higher dose rate. In field operations where the destruction of fish populations is required, the most commonly-used chemicals are rotenone (as an extract of derris powder) and toxaphene (chlorinated camphene), although in Asia the chlorinated insecticide endrin has also been used. The concentrations of these chemicals are usually 0.01-0.04 ppm of rotenone, 0.01-0.02 ppm of toxaphene, and 0.001-0.003 ppm of endrin. Toxaphene is much more persistent than the other two chemicals, and may enable the more resistant species to be controlled. The concentrations quoted nevertheless give some indication of the levels necessary for producing fish mortalities under field conditions. Factors such as adsorption to sediments or on vegetation, and

TABLE 6.1. Toxicity of pesticides to freshwater fish in static conditions (concentrations in $\mu g/l$).

Pesticide	Species	Median lethal concentration (LC_{50}) 24-hr.	96-hr.	Reference
Insecticides				
DDT	Bluegill	–	16	Katz (1961)
	,,	–	8	Macek and McAllister (1970)
	,,	7.4	4.5	U.S.D.I. (1964)
	Black bullhead	65	27	,, ,,
	Channel catfish	4.2	2.9	,, ,,
	Rainbow trout	4.2	2.1	U.S.D.I. (1971)
	,, ,,	–	7	Macek and McAllister (1970)
TDE (DDD)	Bluegill	56	42	U.S.D.I. (1965)
	Rainbow trout	30	–	U.S.D.I. (1963)
Dicofol	Rainbow trout	110	–	,, ,,
Perthane®	Rainbow trout	9	5	U.S.D.I. (1965)
Methoxychlor	Bluegill	67	53	Macek et al. (1969)
	Rainbow trout	20	–	U.S.D.I. (1963)
	,, ,,	45	42	Macek et al. (1969)
Aldrin	Bluegill	10	–	U.S.D.I. (1963)
	,,	9.6	5.2	U.S.D.I. (1965)
	,,	16	5.8	Macek (1969)
	Rainbow trout	36	31	U.S.D.I. (1965)
	,, ,,	42	18	Katz (1961)
	,, ,,	8.1	3.3	Macek (1969)
Dieldrin	Bluegill	5.5	2.8	U.S.D.I. (1965)
	,,	24	14	Macek et al. (1969)
	Black bullhead	11	10	U.S.D.I. (1964)
	Rainbow trout	16	9.9	Katz (1961)
	,, ,,	19	13	U.S.D.I. (1965)
	,, ,,	3.1	1.1	Macek et al. (1969)
Endrin	Bluegill	0.4	–	U.S.D.I. (1963)
	,,	0.35	0.25	U.S.D.I. (1965)
	,,	1.5	0.41	Macek et al. (1969)
	Black bullhead	1.3	1.1	U.S.D.I. (1964)
	Channel catfish	0.45	0.29	,, ,,
	Rainbow trout	1.8	0.86	U.S.D.I. (1965)
	,, ,,	5.3	1.4	Macek et al. (1969)
BHC (lindane)	Bluegill	61	–	U.S.D.I. (1964)
	,,	100	51	Macek et al. (1969)
	Rainbow trout	49	–	U.S.D.I. (1964)
	,, ,,	30	22	U.S.D.I. (1965)
Ovex®	Rainbow trout	860	620	U.S.D.I. (1965)
Chlordecone	Rainbow trout	66	20	U.S.D.I. (1965)
Heptachlor	Bluegill	83	13	U.S.D.I. (1971)
	Black bullhead	76	34	U.S.D.I. (1964)
	Rainbow trout	15	8	U.S.D.I. (1965)
Chlordane	Bluegill	58	40	U.S.D.I. (1965)
	,,	170	77	Macek et al. (1969)
	Rainbow trout	20	10	U.S.D.I. (1971)
Toxaphene	Bluegill	7.2	2.6	U.S.D.I. (1964)
	,,	6.8	2.6	Macek et al. (1969)
	Black bullhead	7.7	5.8	U.S.D.I. (1964)
	Rainbow trout	–	11	Macek and McAllister (1970)
Rotenone	Bluegill	24	22	U.S.D.I. (1964)
	Rainbow trout	32	26	,, ,,
Pyrethrum	Bluegill	78	70	U.S.D.I. (1964)
	Rainbow trout	56	54	,, ,,
Malathion	Bluegill	45-120	–	U.S.D.I. (1963)
	,,	140	55	Macek et al. (1969)

TABLE 6.1. (continued)

Pesticide	Species	Median lethal concentration (LC_{50}) 24-hr.	96-hr.	Reference
	Rainbow trout	100	–	U.S.D.I. (1963)
	,, ,,	160	94	U.S.D.I. (1971)
Methyl parathion	Bluegill	–	5,720	Macek and McAllister (1970)
	Rainbow trout	–	2,750	Macek and McAllister (1970)
Diazinon	Bluegill	59	22	U.S.D.I. (1965)
	Rainbow trout	380	90	,, ,,
Dichlorvos	Bluegill	1,000	480	U.S.D.I. (1965)
	Rainbow trout	500	–	U.S.D.I. (1963)
Dimethoate	Bluegill	28,000	6,000	U.S.D.I. (1965)
	Rainbow trout	20,000	8,500	U.S.D.I. (1964)
Phosdrin	Bluegill	41	23	U.S.D.I. (1965)
	Rainbow trout	34	12	,, ,,
Azinphos-methyl	Bluegill	–	22	Macek and McAllister (1970)
	Rainbow trout	–	14	Macek and McAllister (1970)
Carbaryl	Rainbow trout	3,500	–	U.S.D.I. (1963)
	,, ,,	3,940	1,090	U.S.D.I. (1971)
Zectran®	Bluegill		11,200	Macek and McAllister (1970)
	Rainbow trout		10,200	Macek and McAllister (1970)
Herbicides				
2,4-D (PGBEE)	Rainbow trout	1,200	1,100	U.S.D.I. (1964)
2,4-DB (acid)	Rainbow trout	13,500	5,400	U.S.D.I. (1965)
2,4,5-TP	Bluegill	19,000	9,600	U.S.D.I. (1965)
	Rainbow trout	23,000	14,800	,, ,,
2,4,5-TP (PGBEE)	Bluegill	500	450	U.S.D.I. (1964)
	Rainbow trout	1,500	1,300	,, ,,
Dalapon-Na	Bluegill	115,000	105,000	U.S.D.I. (1965)
Diuron	Bluegill	12,000	4,000	U.S.D.I. (1965)
	,,	17,000	7,600	Macek et al. (1969)
Chlorfenac	Bluegill	61,000	41,000	U.S.D.I. (1965)
Chlorfenac (Na salt)	Bluegill	26,000	14,000	U.S.D.I. (1965)
Simazine	Bluegill	130,000	118,000	U.S.D.I. (1965)
	Rainbow trout	100,000	25,000	U.S.D.I. (1971)
Trifluralin	Bluegill	100	68	U.S.D.I. (1965)
	,,	360	120	Macek et al. (1969)
	Rainbow trout	210	86	U.S.D.I. (1965)
	,, ,,	239	152	Macek et al. (1969)
Sodium arsenite	Bluegill	58,000	30,000	U.S.D.I. (1965)
	Rainbow trout	100,000	26,000	,, ,,
Copper sulphate	Bluegill	2,800	2,800	U.S.D.I. (1965)
Copper chloride	Bluegill	1,100	980	U.S.D.I. (1965)

TABLE 6.2. Toxicity of pesticides to marine fish in continuous-flow tests (concentrations in $\mu g/l$).

Pesticide	Species*	Median lethal concentration (LC_{50})	
		24-hr.	48-hr.
Insecticides			
DDT	S	5	2
	M	0.8	0.4
	K	5.5	5.5
Methoxychlor	S	30	30
	M	55	55
Aldrin	S	8.2	5.5
	M	3.1	2.8
Dieldrin	S	5.5	5.5
	M	7.8	7.1
Endrin	S	4.4	0.6
	M	2.6	2.6
	K	0.3	0.3
BHC (Lindane)	S	30	30
	M	30	30
	K	300	240
Chlordecone	M	500	55
	K	300	84
Heptachlor	S	55	25
	M	4.8	3
Chlordane	M	43	5.5
Toxaphene	S	2.2	1
	M	5.5	5.5
Mirex	S	>2,000	>2,000
Thiodan	S	0.9	0.6
	M	5	0.6
Malathion	S	550	550
	M	950	570
Methyl parathion	C	>1,000	>1,000
Diazinon	M	250	250
Dichlorvos	S	550	550
Dimethoate	K	>1,000	>1,000
Phosdrin	C	830	830
Phorate	K	3.2	0.4
Azinphos-methyl	S	55	50
	M	5.5	5.5
Ethion	C	420	69
Fenthion	S	1,720	1,220
Demeton	S	550	550
Zectran®	C	>1,000	>1,000
Herbicides			
2,4-D (PGBEE)	K†	5,000	4,500
2,4,5-T (PGBEE)	S	320	320
Dalapon-Na	K	>1,000	>1,000
Diuron	M†	10,800	6,300
Monuron	M†	20,000	16,300

From U.S.D.I. reports 1962, 1963, 1964, 1965.
  * S=juvenile spot (*Leiostomus xanthuras*); K=longnose killifish (*Fundulus similis*); M=juvenile white mullet (*Mugil curema*); C=sheepshead minnow (*Cyprinodon variegatus*).
  † Static test.

TABLE 6.3. Toxicity of different pesticide formulations.

Pesticide	% composition	LC$_{50}$ (μg/l) 24-hr.
p,p'-DDT	100	13
o,p'-DDT	100	30
Commercial DDT	75 *p,p'*-DDT / 20 *o,p'*-DDT	1.4
Shell 'Arkotine' DDT	18 DDT	200
Murphy 'De De Tane 25'	25 DDT / 3 emulsifier / 48 naphtha / 24 water	140
Murphy 'De De Tane Paste'	50 DDT / 24 suspending agent / 26 water	10,700
2,4-D (BEE)	100	1,000
2,4,5-T (BEE)	100	1,000
Econal 13086	50 2,4-D / 25 2,4,5-T	230

From Alabaster, 1969b.
Test Species: *Rasbora heteromorpha*

TABLE 6.4. Toxicity of various phenoxy herbicides (LC$_{50}$ values expressed as mg/l acid equivalent).

Form	24-hr.
2,4-D, alkanolamine	450-900
dimethylamine	166-542
di-N-N-dimethylcocoamine	1.5
acid+emulsifiers	8.0
iso-octyl ester	8.8-66.3
propylene glycol butyl ether ester	2.1
butoxyethanol ester	2.1
Butyl ester	1.3
isopropyl ester	0.9
ethyl ester	1.4
2,4,5-T, dimethylamine	144
iso-octyl ester	10.4-31
propylene glycol butyl ether ester	17
butoxyethanol ester	1.4
Fenoprop, potassium salt	83
iso-octyl ester	1.4-15.5
propylene glycol butyl ether ester	19.9
butoxyethanol ester	1.2

From Hughes and Davis, 1963.

manufacturers. Hughes and Davis (1962) compared the toxicities of other types of herbicide, amitrole, dalapon, diquat and endothal being among the least toxic to bluegills. A number of other herbicides were also tested in 24- and 48-hour static tests, using chinook (*Oncorhynchus tshawytscha*), coho salmon (*O. kisutch*) and large-mouth bass (*Micropterus salmoides*), by Bond *et al.* (1960). Some, such as dalapon and amino-triazole, gave LC_{50} values of several hundred mg/l, although 2,3-dichloro-naphthoquinone, sometimes used as an algicide, was toxic to large-mouth bass at less than 0.1 mg/l.

The toxicity of some combinations of insecticides to mosquito fish was studied by Ferguson and Bingham (1966a) using DDT, endrin, toxaphene and methyl parathion. No super-additive effects were found, but endrin and toxaphene were relatively more toxic to strains generally susceptible to organochlorines, and DDT and methyl parathion relatively more toxic to resistant strains.

Toxicity data for marine fish have been obtained by the U.S. Bureau of Commercial Fisheries and published in the annual reports of the Fish and Wildlife Service, the system in this instance using continuous flow testing. The species of marine fish most commonly used was spot (*Leiostomus xanthurus*), and typical data are shown in Table 6.2. Comparison of values in this table with those for the same chemical in Table 6.1 suggests that there is no inherently greater or lower sensitivity to pesticides in the marine environment, but different species must be used for the tests. Butler (1965) reported on the toxicity of several herbicides to marine fish and invertebrates. He considered herbicides to be the least toxic to fish, of the various groups of synthetic pesticides tested, but a few had 48-hour LC_{50} values of less than 1´ppm. Katz (1961) tested thirteen insecticides for toxicity to three-spined sticklebacks (*Gasterosteus aculeatus*) at two levels of salinity (five and twenty-five parts per thousand), over periods of 24 to 96 hours, but found little difference between LC_{50} values at the two levels, except for azinphos methyl, which was more than twice as toxic in the more saline water. Eisler (1970), using static bioassays, determined the 24, 48 and 96-hour LC_{50} values of twelve insecticides to seven species of estuarine fish, and found that endrin was consistently the most toxic, and methyl parathion the least toxic, of the compounds tested. The 96-hour LC_{50} values for endrin were in the range 0.05-3.1 μg/l, and for methyl parathion 5,700-7,580 μg/l, the organophosphorus compounds being in general less toxic than the organochlorines.

6.3.2   Effect of flow
A comparison of the LC_{50} values obtained by static and continuous flow systems was described by Burke and Ferguson (1969), using mosquito fish, either susceptible to, or resistant to, certain types of pesticides. One organophosphorus insecticide, parathion, and three organochlorines, toxaphene, DDT and endrin, were tested. The mortality of susceptible fish increased with pesticide concentration, and was greater and more rapid in

flowing water. Resistant fish followed the same pattern, except with parathion, which was more toxic in static conditions. The mortality of susceptible fish was, as expected, greater than that of resistant fish at a given concentration. The apparently lower toxicity in static tests is assumed to be due to reduction of the initial concentration by adsorption of the chemical on the test vessel surfaces, and its removal and metabolism by the fish. Holden (1962) found that ^{14}C-labelled DDT was removed by both tank surfaces and fish, rapidly reducing the original aqueous concentration under static conditions. The increased toxicity of parathion in the static tests of Burke and Ferguson was attributed by them to the production of a more toxic metabolite, which would be flushed out in a flowing system. Weiss (1965) described the conversion by fish of parathion to the more toxic para-oxon.

A comparative study of static and dynamic systems was also described by Lincer *et al.* (1970), using endrin and DDT on fathead minnows (*Pimephales promelas*). With endrin, the 48-hour LC_{50} value under dynamic conditions was 74% of that under static conditions. After 96 hours, the value was 51%. The situation with DDT was reversed, however, as the 48-hour LC_{50} for the dynamic system was more than 5.4 times greater than that for the static system. It was postulated that low oxygen and high ammonia concentrations in the static tanks may have caused increased respiration rate in the fish, resulting in a greater assimilation of DDT via the gills; also synergism between DDT and ammonia could not be discounted. The difference between DDT and endrin, in this respect, could not be explained. In a separate experiment, the ammonia concentration increased from about 9 μg/l to 47 μg/l in the static tanks, but neither concentration could be regarded as toxic in itself. Dissolved oxygen, however, fell from 7.5 mg/l to 1.5 mg/l in the static tanks and to 6 mg/l in the dynamic tanks, the low level in the former probably contributing to the toxic conditions.

This experiment underlines the importance of dynamic tests, at least for obtaining a useful measurement of the LC_{50} parameter. The fact that steady state conditions seldom occur naturally, or that it would be common to find decreasing concentrations in practice, seems to the author to be somewhat irrelevant. Test conditions to simulate such natural (and very variable) conditions would be impossible to establish in standard form in routine examinations. The LC_{50} concentrations determined by any bioassay technique are assumed to have been constant for the duration of the test, as would theoretically be true for dynamic tests. However, it was observed by Lincer *et al.* that there was an initial slight decrease in the actual concentration in the dynamic tests with DDT, during the first 24 hours, before becoming stabilised. A similar initial decrease has been observed in tests at the Salmon and Freshwater Fisheries Laboratory of the Ministry of Agriculture, Fisheries and Food in London (R. Lloyd, personal communication).

LC_{50} as determined in toxicity tests, whether static or dynamic. The presence of other potentially toxic conditions, such as high ammonia or low dissolved oxygen concentrations, has already been mentioned, but the temperature, hardness of the water used, the size or stage of development of the fish, and sometimes even the ratio of the mass or volume of fish to that of the water in the tank, are also influential. The physiological condition of the fish, particularly in respect of their fat content when organochlorines are tested, can also be important, fish in poor condition being more susceptible to poisoning. The pH of the water, possibly in association with hardness, is also known to affect LC_{50} values. The influence of the various parameters is discussed in the following sections.

6.3.5 Effect of temperature

The effect of temperature on the toxicity, to fish, of a large number of chemicals has been well documented. Among the studies made on pesticides, Iyatomi et al. (1958) found that endrin was 84 times more toxic to carp (*Cyprinus carpeo*) at 27-28°C than at 7-8°C. Bridges (1965) examined the toxicity of the insecticides heptachlor and chlordecone to redear sunfish (*Lepomis microlophus*). Heptachlor was tested at a series of times from 6 to 96 hours, at five temperatures from 45-85°F, using static tests. Chlordecone was tested for all periods at each of the five temperatures. The increase in toxicity of chlordecone from 45-85°F was about five-fold for periods from 24-96 hours, but non-linear, being proportionately greater from 45-55°F than from 75-85°F. At shorter times, the difference in toxicity between 45°F and 85°F was greater than at 24 hours. For heptachlor, the 24-hour LC_{50} was 0.092 mg/l at 45°F and 0.022 mg/l at 85°F, the relationship again being non-linear. However, at 75°F, the toxicity in 96 hours was 3.7 times greater than in six hours, whereas with chlordecone, the ratio was about thirty times greater.

The toxicity of a number of pesticides at different temperatures was studied also by Macek et al. (1969) using bluegills and rainbow trout, again in static tests. Generally, the toxicity increased with temperature, but there were several exceptions. With methoxychlor, the susceptibility of both species of fish decreased with increasing temperatures. Bluegills were unaffected by temperature in tests with lindane and azinphos methyl. After 96 hours, the effect of temperature was less than in 24-hours tests, except with trifluralin on rainbow trout (Table 6.5). Cope (1965a) found that with bluegills the effect of temperature on the toxicity of trifluralin was considerable, although less in 96-hour tests than in 24-hour tests. At 7.2°C (45°F), the 24-hour and 96-hour LC_{50} values were 1.300 mg/l, and 0.280 mg/l, but at 29.4°C (85°F) the corresponding values were 0.010 mg/l and 0.0084 mg/l. Such a wide range in LC_{50} values causes considerable difficulty in predicting the effect under field conditions.

Respiration, and thus pesticide intake via the gills, will be more rapid at higher temperatures, the oxygen demand being greater but oxygen solubility less. However, as Macek and his colleagues indicated, in the

TABLE 6.5. LC_{50} values (μg/l) at three temperatures for rainbow trout in static tests.

	Compound	1.6°C	7.2°C	12.7°C
24 hours	Endrin	15	5.3	2.8
	Dieldrin	13	3.1	3.1
	Endosulfan	13	6.1	3.2
	Aldrin	24	8.1	6.8
	Heptachlor	17	12	13
	Methoxychlor	55	45	74
	Dursban ®	550	110	53
	Trifluralin	328	239	98
96 hours	Endrin	2.5	1.4	1.1
	Dieldrin	2.4	1.1	1.4
	Endosulfan	2.6	1.7	1.5
	Aldrin	3.2	3.2	2.2
	Heptachlor	7.7	7.0	7.3
	Methoxychlor	30	42	62
	Dursban ®	51	15	7.1
	Trifluralin	210	152	42

From Macek *et al.*, 1969.

static tests the pesticide available is limited, and after 96 hours the effect of temperature could be expected to be less than after 24 hours, a trend which was observed also by Bridges (*loc. cit.*). Increased metabolism and oxygen uptake would, however, result in lower dissolved oxygen concentrations and the accumulation of higher concentrations of waste products, thus increasing susceptibility. This complication in determining the true LC_{50} value for the chemical in question, unaffected by other changes in the chemical conditions, lends further support to the continuous-flow testing procedure.

Toxicity has, in some instances, been found to increase with decreasing temperature. Cope (1964) gave data for the 48-hour LC_{50} of DDT for bluegills, showing that the value at 45°F was 0.0024 mg/l, and at 85°F 0.0064 mg/l. In 96 hours, the values were 0.0016 and 0.0056 mg/l respectively. A similar, but less pronounced, relationship was found for DDT with rainbow trout (Cope, 1965b), but endrin, lindane, dieldrin and aldrin, tested against bluegills, all showed an increase in toxicity with temperature. Although the reduction of DDT toxicity with rising temperature might have been due to rapid degradation to TDE by micro-organisms in static tanks, Schoettger (1970) apparently found the same type of negative activity coefficient in continuous-flow tests. With rainbow trout, DDT was twice as toxic at 4.4°C (40°F) as at 12.8°C (55°F). Cope (1968) also reported that methoxychlor was more toxic to bluegills at low temperatures than at high temperatures.

Chlorinated insecticides are, in general, more stable than organo-phosphorus compounds, and show a greater difference in toxicity between 24-hour and 96-hour tests.

Bridges (1965) found a much greater increase in the toxicity of chlordecone to redear sunfish with time than with the toxicity of hepta-chlor, the ratio of the 24-hour to 96-hour LC_{50} values being 5.5 for chlordecone and 2.0 for heptachlor at 75°F. Cope (1965b) reported the results of static tests of rainbow trout and bluegills, using a large number of formulated insecticides. Many showed little change in toxicity between 24 and 96 hours, but in a few instances, e.g. diazinon with rainbow trout, Dylox with both species, and dimethoate and diuron with bluegills, the increase is at least three-fold. Trichlorophon showed a twenty-fold increase for rainbows and bluegills between 24 and 96 hours.

6.3.8   Size of test fish

Few studies have been made of the influence of size of fish on the susceptibility to pesticides, although in static tests this factor may be significant, in that larger fish will remove proportionately more of the chemical than smaller fish. Most investigations suggest that susceptibility decreases with increasing size, an effect which would be expected in static bioassays if the weight/volume relationship was not held constant. Pickering et al. (1962) found that larger fish were less susceptible in static bioassays, and Mount (1962) reported 96-hour LC_{50} values of 0.27 $\mu g/l$ endrin, for using 30 mm long fish and 0.47 $\mu g/l$ using 60 mm long fish (both guppies and bluntnose minnows, Pimephales notatus). However, Cope (1969) reported that, in contrast to earlier studies, DDT showed no size dependence in static bioassays over the size range 0.5 to 40.0 g, although no data were given. This problem is likely to be resolved more effectively with suitably-maintained concentrations of toxicant, in continuous flow bioassays.

The density of fish in the test tanks is perhaps a factor related to that of size of the individual fish, but it has been found that increasing the numbers of fish per tank usually increases the estimated LC_{50} values. Prevost et al. (1948) found that an increase in volume of the test solution, for the same number of fish, produced an increase in mortality in tests with rotenone. Katz and Chadwick (1961) found that the 96-hour LC_{50} for endrin, when five bluegills were tested in 15 litres of water, was 0.60 $\mu g/l$ but when 20 fish were tested in 15 litres the 96-hour LC_{50} was 1.1 $\mu g/l$. Shiff et al. (1967) also found, in static bioassays with harlequin fish, that an increase in the number of fish per tank resulted in reduced mortality, when the molluscicide N-tritylmorpholine was tested. These results all suggest that the fish were removing a significant propor-tion of the substance under test.

6.3.9   Threshold of toxicity

Some investigators have determined the threshold toxicity value LC_0, the highest concentration at which no long-term lethal effect is expected. The value is calculated from a graphical plot of the proportion of fish dying at various time intervals. For example, Konar (1969) determined 168-hour LC_0 values for two organophosphorus compounds, dichlorvos and phosphamidon, using several species of fish in India. The LC_0 values were generally one-third to two-thirds of the LC_{50} values obtained in the same test. Meyer (1965) showed a somewhat similar relationship for azinphos methyl, with five species of fish, while the LC_{100} concentration in 48 hours was not more than twice the LC_{50} value. MacPhee and Ruelle (1969) calculated the LC_0 values for salmonids, for comparison with LC_{100} values for squawfish, in developing a chemical selectively lethal for the latter, while virtually safe for the salmonids.

6.4   LETHAL EFFECTS OF LONG-TERM EXPOSURE

One difficulty arising from tests involving long exposure in static conditions, is that if the fish are fed, a proportion of the chemical under test may be adsorbed to the food and ingested, rather than being assimilated via the gills. In the absence of feeding, tests are usually not extended beyond 96 hours. A further complication occurs as a result of the removal by the fish of a significant quantity of the pesticide under test. Most workers ensure that the ratio of fish biomass to water volume is as small as possible, a maximum of 1 gram of fish to 1 litre of water often being quoted. Yet it is doubtful if this ratio is small enough, because fish may concentrate organochlorines at least a thousand-fold, from water to tissues, in a few days. Thus, 1 gram of fish exposed to 1 microgram of DDT in 1 litre of water could, in theory, remove all the DDT and contain 1 ppm in its tissues, although this would probably take longer than 96 hours. Nevertheless, significant loss of DDT from the water would occur during a period of 96 hours, from this factor alone (aeration and adsorption to surfaces being capable of removing a further fraction) in such static tests. Gakstatter and Weiss (1967), using radio-actively labelled DDT and dieldrin, found that about 60% of the insecticides were removed within five hours by 60 bluegills (345 g) in 100 litres of water, but the decline in the aqueous concentration ceased when the fish were removed.

Mount (1962) exposed bluntnose minnows and guppies to endrin for 291 days in a continuous flow system, but less than 50% of either species survived 0.5 $\mu g/l$ for more than 30 days. At 0.25 and 0.1 $\mu g/l$ there was little mortality, and Mount did not consider endrin to be cumulative in its effect on fish subjected to chronic exposure. In acute toxicity tests with bluntnose minnows under continuous flow, Mount found that the mean value for the 96-hours LC_{50} of endrin was 0.34 $\mu g/l$ ppm, suggesting that the fish in the acute toxicity tests were less tolerant than those used in the chronic toxicity tests. He suggested that the fish used in the chronic toxicity tests were perhaps in better physical

ppm. Hatchery fish were more sensitive than wild fish, but as all the concentrations tested were extremely toxic, it is perhaps not surprising that the eventual mortality was not dependent on concentration in the range 0.5-10 ppm.

The eggs of fish are usually very resistant to toxic substances, but some research has been carried out, in the United States, on their susceptibility to pesticides. Hiltibran (1967) found that none of 25 herbicides used for aquatic weed control appeared to be toxic to the fertilized eggs of green sunfish at concentrations used in the field. Lennon (1967) reported that thiodan is toxic to newly-fertilized rainbow trout eggs after 29 days at 0.2 mg/l. (This compound is, however, among the most toxic of all substances to fish.) Antimycin at 4 μg/l was lethal to carp eggs when applied in a lake, but the period of exposure was not stated. The pesticide Juglone® was reported by Lennon (1970) to be lethal to recently-spawned rainbow trout eggs at 0.10 mg/l in five days, and at 0.05 mg/l in 25 days.

Malone and Blaylock (1970) exposed fertilized eggs of carp to DDT, chlordane, dieldrin, endrin, diazinon and azinphos methyl at concentrations between 0.001 and 10.0 mg/l. Below 1 mg/l, the viability of embryos was not affected, but at 5 mg/l, all six insecticides caused 50-100% mortality. Chlordane, below 1 mg/l, actually stimulated embryo development and reduced incubation time.

6.6   CHRONIC TOXICITY OF PESTICIDES

The evidence presented so far, has related solely to the lethal effects of pesticides and the factors which may affect the measurement of the susceptibility of fish. Long-term chronic studies, leading to death or to other symptoms of toxicity, whether behaviour or physiological, are more difficult to perform, but in recent years a number of such investigations have been reported. An example of the delayed lethal action, which may arise if tests are prolonged, was when Alabaster (1969a) found that harlequins, exposed for only thirty minutes to a solution of a diquat formulation, which would have been lethal in about eight hours, died a week later in clean water. If the initial exposure period was increased, mortality in clean water occurred earlier.

This delayed action, following a short period of exposure to a concentration of toxicant which would have been lethal over a somewhat longer period, probably occurs in streams where the initial exposure is brief but intense, following a spillage or overspray. The ingestion of further quantities of pesticide, from invertebrates killed by the chemical and eaten by the fish, increases the hazard. Elson and Kerswill (1966) found that the mortality of Atlantic salmon, in the Miramichi River in New Brunswick, was abnormally high in the early winter following the spraying of forests with DDT in the summer. The fish may have absorbed quantities of DDT both from the water and the insects, but the lethal action was delayed until the temperature fell in the winter.

6.6.1 Effects on growth and population structure

These two aspects are discussed together, because one effect of a population reduction could be the increased growth of survivors. Such a result has been observed in field operations, as after the spraying of forests in Maine with DDT at 1 lb/ac (1.12 kg/ha) (Warner and Fenderson, 1962). Numbers of brook trout fry in the streams of the area were seriously reduced, but in the following year, growth was above-average. By the second year after spraying, however, many streams had reverted to normal populations.

Most observations on the effect on growth have been made in experimental studies. Crandall and Goodnight (1962) found that sodium pentachlorophenate depressed the growth of guppies at sublethal levels (0.5 ppm) after 60 days. Allison *et al.* (1964) found that many cut-throat trout (*Salmo clarki lewisi*), exposed to 0.3 or 1.0 ppm DDT solutions for 30 minutes at 28-day intervals, were killed, but survivors were larger than control fish or fish exposed to lower concentrations. Fish which survived the highest concentrations in their feed were also larger than normal. These workers believed that the fish were naturally stronger, as no increase in growth rate was observed, and the increase in mean size was therefore due to the mortality of the smaller, weaker fish.

Andrews *et al.* (1966), who exposed bluegills to heptachlor in ponds at concentrations from 0.0125 to 0.050 mg/l, found that the survivors from the higher concentrations (0.025 mg/l and above) grew fastest. On the other hand, Macek (1968) fed brook trout with DDT in pelleted food for 156 days, but found no significant effects on growth except at the highest dosage (2 mg/kg per week), the males being significantly longer than control fish. Grant and Mehrle (1970), in feeding goldfish with food pellets containing endrin, found that only the highest dose (430 μg/kg of body weight per day) produced a significant decrease in growth rate after 157 days. Fish receiving this dose began to leave food after two weeks, and 14 of the 30 fish in the test died during the full period of 157 days.

Walker (1963) found that endothal, by killing weeds and making more food available to fish, caused an increase in fish growth. This increased growth, following the destruction of vegetation, was also found by Cope *et al.* (1969) in fish pond experiments with the herbicide dichlobenil, but the effect seems to have been the resultant effect of both serious fish mortality, reducing competition for food, and increased availability of food. A similar situation was reported by Cope *et al.* (1970), in experiments with 2,4-D; bluegills that survived the highest treatment level in ponds (10 mg/l) showed the greatest increase in growth. Gilderhus (1966) observed a decrease in the growth rate of bluegills in pools treated with high concentrations of sodium arsenite, and the invertebrate fauna was also reduced. This effect on the fish may have been partly due to the directly toxic action of the arsenic accumulated and retained by the fish, whereas with the dichlobenil experiments, referred to above, accumulation of chemical in the fish was only temporary. In experiments with methoxychlor on bluegills, Kennedy *et al.* (1970) could detect no

and convulsions followed. These increased until equilibrium was lost, but the fish often 'barrel-rolled' or spiralled at intervals before respiration eventually ceased.

These effects are typical of poisoning by many pesticides, although there may also be paralysis and inability of the fish to move off the bottom of the test tank. Toxic chemicals, other than pesticides, may also produce some of the symptoms described, particularly if interference with respiration results, as with cyanide. Irritants such as phenol or some suspended solids may also increase ventilation rate of the gills.

The effects of sublethal concentrations of pesticides on behaviour, although also resulting from their influence on the central nervous system, are in most instances quite different from those described above for lethal concentrations. Warner et al. (1966) described the results of experimental work in a specially-designed apparatus to study the effects of exposure of goldfish to concentrations of toxaphene of 1.8 $\mu g/l$ and 0.44 $\mu g/l$. The fish had been trained beforehand to respond to a series of stimuli, and their ability to continue to respond in the presence of the toxicant over periods up to 264 hours was studied. A continuous-flow system was employed to ensure constant concentrations of toxicant. The lower concentration, 1/25th of the 96-hour LC_{50}, produced definite evidence of behavioural pathology in this period, but even at 264 hours the fish exposed to the higher concentration were able to perform some complex tasks. An organophosphorus insecticide, TEPP, produced evidence of behavioural pathology at toxicant levels less than 1/200th of the LC_{50}. The experiments are valuable in demonstrating that effects on behaviour were different at different doses, and could occur at concentrations which might otherwise be regarded as too low to have any biological consequences. Although the programmed stimuli, to which the fish were subjected in these experiments, were necessarily artificial, there is a clear implication that pesticides may affect learning ability, or response to natural stimuli, in such a way that the survival of fish in their normal environment might be affected.

Ogilvie and Anderson (1965) showed that 24-hour exposures of Atlantic salmon (S. salar) underyearlings to 0.005 to 0.050 mg/l of DDT, concentrations regarded by these workers as sublethal (although the highest value must have been near-lethal), resulted in a change in the selected (or preferred) temperature of the fish. Two groups of fish were acclimated to 8°C and 17°C, and the effect of DDT was more marked on the latter group. The warm-acclimated fish appeared to be extremely sensitive to cold water after exposure to 0.10 mg/l or more. At the lowest concentrations tested, the fish appeared to prefer a lower temperature than that selected by control fish, but at higher DDT concentrations the preferred temperature increased to values several degrees *above* that of the controls.

The inability to withstand sudden decreases in temperature was related by Anderson (1971) to the observations of a mass mortality of young

Atlantic salmon in the Miramichi River, in the fall (autumn), following aerial spraying of the catchment areas with DDT in the previous spring during forestry operations (Elson and Kerswill, 1966). The deaths occurred when the stream temperatures fell suddenly to $5°C$ or less. Surviving fish from the sprayed streams died in the laboratory when transferred to cold water, but those from unsprayed areas were not affected.

Anderson (1968) found evidence that one receptor in the nervous system of fish, the lateral line, is strongly influenced by sublethal DDT concentrations, and Anderson and Peterson (1969) showed that the cold-blocking temperature of the propeller tail reflex, of which the spinal chord is the control mechanism, is similarly influenced. Brook trout can be trained to avoid light or dark sides of a tank, but these workers also found that sublethal DDT concentrations resulted in an inability to learn this type of avoidance. Thus, both central and peripheral nervous systems are affected by sublethal exposure to DDT.

The ability of fish to avoid pesticide concentrations was studied by Hansen (1969). He presented untrained sheepshead minnows (*Cyprinodon variegatus*), in a specially-designed apparatus, to DDT, endrin, malathion, Dursban,[®] carbaryl and 2,4-D (BEE) at concentrations both above and below the 24-hour LC_{50} value. The fish did not avoid malathion and carbaryl, at the concentrations tested, but avoided the other four chemicals, some at concentrations well below the LC_{50} value. They were not able to distinguish between concentrations of a pesticide, however.

6.6.4 Resistant strains

Following the discussion of the recorded effects on fish behaviour, it is appropriate to consider the evidence for the development of their resistance to certain groups of pesticides. Although the phenomenon of resistance among insects is well known, there have been few instances of an increase in tolerance in vertebrates as the result of the selective breeding of the appropriate characteristic. The frequency with which successive generations are produced, is clearly a major factor in the development of a resistant strain, some species of insects being particularly well-adapted in this respect. Among fish, it might be expected that a species which matures rapidly, and occurs in high density populations, would be the most likely to produce pesticide-tolerant strains in an environment where the exposure to a particular pesticide was continuous, or at least very frequent.

The first evidence for such resistance was reported by Vinson *et al.* (1963), who found a DDT-resistant population of mosquito fish near heavily-sprayed cotton fields in Mississippi. Subsequently, several other species of fish were found to be resistant to DDT and some other insecticides. Ferguson *et al.* (1964) found that golden shiners, bluegills and green sunfish, from the Mississippi Delta, had developed resistance to toxaphene, aldrin, dieldrin and endrin, but not to DDT. Ferguson *et al.* (1965) failed to find evidence for resistance to endrin in black bullheads in the

Mississippi River, and concluded that either the species could not develop resistance or that the necessary exposure of previous populations to lethal concentrations had been lacking. Yellow bullheads (*Ictalurus natalis*), from a ditch in the cotton fields, did develop resistance to endrin, however (Ferguson and Bingham, 1966b), by a factor of ×60, on the basis of 36-hour LC_{50} values in comparison with normal populations.

A number of factors are involved in the development of resistant populations. The original populations must be exposed frequently or continuously, to lethal concentrations of the pesticide in question, but a few survivors bearing the resistance characteristic must always remain to propagate the species. It has been observed by Ferguson and his co-workers that the top piscivores, such as large-mouth bass, are absent in heavily-contaminated environments, perhaps as the result of consuming smaller species which had carried a high body-burden of insecticides. This, in turn, ensures an even higher population density, giving more favourable conditions for reproducing the strain. Finley *et al.* (1970) examined possible selective mechanisms, and found that direct exposure from the water, rather than through food chains, was a major selective factor.

The direct correlation between a high lipid content and endrin tolerance in mosquito fish, and increased susceptibility in winter, led Fabacher and Chambers (1971) to evidence that organochlorine residues may induce lipid deposition in susceptible fish. Other resistance factors may include a decrease in uptake of pesticides, development of 'resistant' nervous tissue and a tolerance to stress.

Cross-resistance, in which exposure to one pesticide results in resistance to a number of pesticides, was demonstrated by Boyd and Ferguson (1964). Resistant fish are found to concentrate the particular insecticide to a very high degree, the resistance not being due to either an ability to reject the chemical or to metabolize it (Ferguson *et al.*, 1966). Resistant fish, accumulating high concentrations on immersion in appropriate solutions, can subsequently release appreciable amounts when placed in clean water, in quantities sufficient to kill susceptible fish. Moreover, other fish, fed on the resistant fish containing large amounts of residues, are likely to die as a result of ingestion of the prey. The oxygen consumption of resistant and susceptible mosquito fish was found to be the same in the absence of endrin as in the presence of endrin, until the susceptible fish began to show evidence of poisoning, at which time the oxygen consumption increased rapidly until the onset of death. Ludke *et al.* (1968) found that endrin-resistant golden shiners contained concentrations of endrin in the blood, 64 times higher than those in endrin-killed susceptible shiners, and pointed out that caution is needed in diagnosing death from endrin poisoning by blood analysis alone.

Culley and Ferguson (1969) compared resistance to 28 insecticides in susceptible and resistant populations of mosquito fish. The resistant fish were so only towards one group of insecticides, comprising toxaphene, Strobane,® lindane, chlordane, heptachlor, dieldrin, aldrin and endrin.

There was no significant resistance to DDT, Perthane,[®] kelthane or methoxychlor (all related to DDT) or to any organophosphorus compounds tested. The resistance factor (ratio of LC_{50} values between the two types of fish ranged from 42 - 568 in the toxaphene-endrin group, but only from 0.8 - 5.1 among the other 20 chemicals. The failure to find evidence of a resistance to DDT in these populations of mosquito fish is contrary to the findings originally recorded by Vinson *et al.* (1963), which were referred to earlier in this section.

Ferguson *et al.* (1967) warned of the potential danger to consumers of high concentrations of pesticides in resistant fish. Fish, eating resistant mosquito fish died after accumulating high levels of DDT and its metabolites. Other fish, from the same area, which were likely to be consumed by the local human population, were found in experiments to be capable of accumulating 11-26 mg/kg of endrin, levels which would render them unfit for human consumption.

The evidence for such resistance, and the consequences which result, have so far been recorded only for the warm-water species described, although, in other areas of the world, with species of similar characteristics (dense populations, high reproduction rate) and high pesticide usage it is likely that the same phenomenon occurs. No reports of similar effects among species in the temperate zone have been found, and conditions in the marine environment would not be conducive to the phenomenon.

6.6.5 Biochemical changes

A number of changes in the biochemistry of fish as the result of exposure to pesticides have been noted, but probably the most familiar effect is of organophosphorus chemicals on the activity of the enzyme acetylcholinesterase, which is involved in nerve-impulse conduction. Weiss and Gakstatter (1964) used the inhibition of acetylcholinesterase activity in bioassay determinations of organophosphorus pesticides in water. They found bluegills more sensitive than other fish, and significant inhibition was produced by a 15-day exposure at 0.001 mg/l of several such pesticides, but the more toxic compounds azinphos methyl and parathion produced detectable inhibition in 30 days at 0.0001 mg/l. Weiss (1961) found a 40-70% inhibition to be lethal and the same author (1964) proposed the determination of fish-brain acetylcholinesterase activity as a means of detecting pollution by organophosphorus chemicals.

Distressed Atlantic menhaden (*Brevoortia tyrannus*), in a polluted river, showed 47% inhibitions of AChE, while apparently normal fish had 16.5% as compared with controls (Williams and Sova, 1967). Holland *et al.* (1967) used the technique, in an estuarine environment, with spot and sheepshead minnows, and considered that a ±10% variation in activity in natural populations was normal. Values of 73-88% of normal were considered to be the result of organophosphorus waste discharges. However, Gibson *et al.* (1969), using bluegills, studied the variation in specific activity among different regions of the brain, and concluded that the technique was too inaccurate for use in monitoring. Fish becoming moribund in 0.75 mg/l of parathion showed only 25% inhibition, while those moribund in 0.020 mg/l showed 57% inhibition, and even 90% inhibition, and failed to develop pronounced symptoms of organophosphorus poisoning, the fish recovering completely in fresh water.

Several studies have been made of the effects of organochlorine insecticides on blood and tissue chemistry of fish. Eisler and Edmunds (1966) exposed a marine species, the northern puffer (*Sphaeroides maculatus*), to concentrations of endrin from 0.05-10 μg/l in static tanks. 10 μg/l was lethal to all fish in 24 hours, but at 1 μg/l or less, no mortality occurred in 96 hours. There were no consistent changes in the levels of serum chloride, gamma-globulin or uric acid, but the concentrations of sodium, potassium, calcium and cholesterol in serum were consistently higher after exposure. In their liver, sodium, potassium, calcium, magnesium and zinc were consistently lower, the transfer from liver to serum indicating impaired liver function. Eisler (1967), using the same species, examined the effects of methoxychlor and methyl parathion. Exposure to 30 μg/l of methoxychlor, for up to 45 days, did not produce any significant change in the whole blood or serum content of sodium, potassium, calcium, magnesium, zinc and iron, although haematocrit values were lowered. Exposure to 20.2 mg/l of methyl parathion produced lethargy, and the zinc content was lowered in the serum and liver. Serum esterases were inhibited, and a number of other changes in serum proteins and electrophoretic patterns were observed. Ishihara et al. (1967) found that pentachlorophenol, dimethylthio-phosphate, endrin and phenyl mercuric acetate all caused changes in the component ratio of serum protein and lipoprotein in the blood of carp.

Relationships between the serum enzymes of fish and pesticides were also found by Hing (1967). Lane and Scura (1970) found that the activity of glutamic oxaloacetic transaminase, in the sailfin molly, increased significantly on exposure to dieldrin at a low concentration (0.003 mg/l), but fish suffering an early death in 0.012 mg/l (a concentration 100% lethal in 72 hours) showed a lower activity. There appears to have been no significant correlation between enzyme activity and the proportion of deaths over the range 0.003-0.012 mg/l. Grant and Mehrle (1970) found that serum sodium, in the diet of goldfish exposed to endrin for three to four months, was slightly raised, but considered that, as osmo-regulation is very precise, the main electrolytes are normally well-regulated and such variations as were found in the serum constituents were poor indications of chronic intoxication. They believed, however, that endrin affects specific physiological functions which could be detrimental, such as ineffective feeding behaviour.

Thus, it would seem that although changes may occur in electrolyte concentrations in both serum and liver, and in serum enzymes, the correlation is not distinct, and moreover, the changes occur only at exposure levels relatively near those which are lethal. This suggests that the parameters so far examined, would not be capable of indicating that effects were being caused at exposure levels well below the lethal dose. The observation that DDT and related compounds can induce a decrease in eggshell calcium in birds has led to much speculation about the mechanisms which might be responsible (Bitman *et al.*, 1969). No similar studies, on the disturbance of electrolyte balance due to enzyme induction or other processes in fish, that can be related to such a significant ecological effect as is eggshell thinning in birds, appear to have been reported.

6.6.6 Physiological effects

As well as the effects on growth and reproduction already described, other physiological effects have been recorded. One early symptom of acute pesticide toxicity (although not specific to pesticides) is respiratory distress, and a number of investigations of the influence on oxygen consumption have been reported. Entry of pesticides into a fish is largely through the gills (Holden (1962), Premdas and Anderson (1963), Ferguson and Goodyear (1967)). With the onset of symptoms of poisoning, the rate of oxygen consumption increases. Ferguson *et al.* (1966) found this to be so for mosquito fish exposed to endrin, but noted that just before death occurred there was a marked decrease in oxygen consumption. The increased respiration rate probably increases the rate of intake of any pesticides in the water. Huner *et al.* (1967) found that bluegills exposed to 0.1 µg/l of endrin (sublethal) increased their oxygen consumption, but 1.0 µg/l (lethal) decreased it. Exercise had no effect on oxygen consumption, but affected mucus production and hastened death. Lee (1969) exposed goldfish to methyl parathion and dieldrin, and found that while the former tended to depress the respiration rate, the latter only caused fluctuations in respiration. Cairns and Scheier (1964) examined the chronic effects of dieldrin on the pumpkinseed sunfish (*Lepomis gibbosus*) and found that an exposure to 0.0017 mg/l (one-ninth of the 24-hour LC_{50}), for twelve weeks, affected the cruising speed and oxygen consumption of the fish. As these tests were of the static type, and the fish were fed, it is likely that the aqueous concentration was even lower than intended, but some oral intake probably occurred.

Investigating the massive fish-kill due to endrin in the Mississippi and Atchafalaya Rivers in 1963, Mount and Putnicki (1966) found that all the dying fish suffered convulsions, loss of equilibrium, had haemorrhagic areas on the body, and were swimming on the surface. Only channel catfish (*Ictalurus punctatus*) had distension of the abdomen, which was caused by gas and liquid contained in the alimentary tract. In all dying fish, of several other species, this tract was empty, but the excessive deposits of visceral fat, and length/weight ratios, indicated that the physical condition of the fish was excellent. Hematocrit values were about 50% lower in affected than in the control fish, and counts of both red and white cells were low, suggesting that disease was not the cause of death. Confirmation that endrin was the most likely cause of death was obtained by exposing channel catfish to endrin solutions and comparing the concentrations of endrin in the blood with those found in the dead Mississippi fish (Mount et al., 1966).

6.6.7 Pathology

Allison et al. (1964) examined the chronic effects of DDT on cut-throat trout, but could not detect any pathology due to the pesticide. Hematocrit measurements likewise showed no difference between exposed and control fish, and Gilderhus (1966) found the same with the exposure of bluegills to sodium arsenite. Other workers, however, have noted a significant reduction in hematocrit values. Eisler (1967) found that both methoxychlor and methyl parathion produced this effect on northern puffers. Conversely, Andrews et al. (1966) observed an increase in hematocrit values of bluegills exposed to 0.05 mg/l of heptachlor for four hours, but they returned to normal after 28 days. Degenerative liver lesions were caused both by aqueous exposure to, and by feeding on, heptachlor, but only at concentrations close to the 24-hour LC_{50} (0.035 mg/l).

King (1962) described, in detail, various histopathological symptoms of guppies and brown trout fry, exposed to sublethal concentrations of DDT, mainly in the liver, intestine and kidneys, and particularly the appearance of cell vacuolation. Christie and Battle (1962) observed oedematous changes in the gills of trout and lampreys (Petromyzon marinus) exposed to 3-trifluromethyl-4-nitrophenol (sodium salt). Mount and Putnicki (1966), investigating a fish-kill due to endrin, found tissue abnormalities in the fish kidneys, where large vacuolated cells were seen in the glomeruli. Hematocrit values were 50% of normal. Cope (1966) reported that the herbicide, norbromide, at levels sublethal to bluegills, caused degenerative lesions in liver and testes, while dichlobenil caused fusion of gill lamellae. He also noted lesions in gill, liver and kidney tissues, caused by a number of other pesticides. Other observations of chronic effects on bluegills, include the development of haemorrhagic globules on gill lamellae by diuron (McCraren et al., 1969), and a number of changes caused by 2,4-D (PGBEE), including depletion of liver glycogen, globular deposits in blood vessels, and vascular stasis of the central nervous system (Cope et al., 1970). Kennedy et al. (1970) reported similar effects caused by chronic exposure to methoxychlor. Pathological lesions, in the brain, spinal chord, liver, kidneys and stomach of spot, were produced in the survivors of a three weeks exposure to 0.075 μg/l of endrin, but seven months at 0.05 μg/l had no obvious effect (Lowe, 1966). The 24-hour LC_{50} for this species was 0.45 μg/l, so that a long-term exposure to one-sixth of this value apparently resulted in the pathology noted above. This concentration was, however, lethal to 65% of the population tested. One further pathological effect of pesticides, the production of hepatomas, seems only to have been observed after a high dietary dosage of DDT (75 mg/kg) or rainbow trout (Halver et al., 1962).

The various physiological and histopathological effects described, although having been found in living fish, nevertheless, seem to have occurred at exposure levels which were lethal to other fish, and which could have been near the lethal dose for the fish examined. Several investigators reported the presence of parasitic or bacterial infections in the fish, making it difficult to establish whether the tissue damage found was due directly to the pesticide, or to the infections.

6.7 DISCUSSION

The wide variety of the effects of the many different types of pesticide on fish is evident from the previous sections. Yet it is obvious that the exact causes of death, or of other symptoms, are still unknown. The early stages of poisoning are usually manifested by considerable activity on the part of the fish, but subsequently, activity decreases, and the fish becomes inert, with only an occasional spasm of activity, until death ensues. The most likely explanation is that pesticides enter the central nervous system after a period in which the peripheral nervous system has been affected, but during which the blood-brain barrier has effectively prevented a significant penetration of the brain. Nevertheless, the paralysis of even the peripheral nerves could incapacitate or handicap the fish in such a manner that its survival in the natural environment was jeopardized. Even behavioural changes, rendering the fish incapable of the normal pattern of response to external stimuli, could be sufficient to make it vulnerable to physical influences, or to predation.

There are still considerable gaps in our knowledge, on the one hand, of toxicity as measured by short-term acute exposure tests, which may be suitable for the comparison of the toxic properties of different chemicals, and, on the other hand of long-term, low-dosage chronic exposure. The latter may differ in ultimate effect from the former, because the metabolic and other processes of the fish may take different courses at the lower concentrations involved. The order in which the various effects occur also differ considerably. With acute exposure, for example, death may ensue before the behavioural abnormalities noted in sublethal exposure could develop.

It is still difficult to establish, with any degree of confidence, the magnitude of the difference in concentration levels between those producing death in, say, 24 or 96 hours (the acutely-toxic concentrations) and those from which chronic symptoms of toxicity have been recorded during exposures of at least several months. It seems likely that there is often less than one order of magnitude between the two levels, and thus, continuous exposure to a concentration of one-sixth of the 24-hour LC_{50} might be considered safe. With some pesticides, particularly those which can be excreted or metabolised at a significant rate, this is probably true, but with others, such as the more stable of the organochlorines, a long-term accumulation in tissues could result in a lethal dose being gradually acquired, taking effect particularly in a subsequent period of stress. If the toxicity of such chemicals is estimated only by the short-term exposure of juvenile or adult fish, a concentration which might be considered safe for all stages of growth, including fry, and to be without any behavioural or physiological effects, could be about only one-hundredth of the 48-hour LC_{50} value. Continuous exposure at low concentrations might, in some circumstances, be more dangerous than a brief period at a high concentration, because survivors from the latter might be able to metabolize or excrete the toxicant over a subsequent period, if no further exposure occurred.

It must be emphasised, however, that fish are able to concentrate some pesticides, and particularly, organochlorines, rapidly from very low concentrations, by factors of 10^3 or 10^4 (See chapter 5), making them potentially vulnerable to such concentrations if the compounds are not easily degraded. More information is required to determine the implications of this accumulation, in respect not only of fish, but of predators such as birds or seals. The problem may arise only with the persistent organochlorines, but nothing is known of the metabolites or degradation products of many other pesticides, some of which may also be stable and persistent. All pesticides should be tested for their toxicity to several species of aquatic fauna, and at various stages of growth, in order to assess their potential danger in eco-systems generally, rather than to fish alone. The destruction of an important food species could be as important to fish as a directly toxic action.

The concentrations of some pesticides which can exert a toxic action on fish are extremely small, the 96-hour LC_{50} of endrin, for example, being less than 1 μg/l for several species. On the other hand, many weedkillers and a number of organophosphorus insecticides have a relative low toxicity. The environmental dangers may arise when the recommended spray rate for crops is sufficient to produce a concentration toxic to fish in shallow waters, or when repeated use leads to an accumulation in streams, ditches or ponds after continued soil erosion. Resistant strains of mosquito fish, studied by Ferguson and his colleagues, have developed in such areas.

It is incorrect to suggest that all pesticides are dangerous to fish, as many are unstable in water, or have a low toxicity relative to other chemicals. Concentrations toxic to fish can never arise with the normal agricultural usage of many such chemicals, although spillages of concentrates can, as with other chemicals, cause fish-kills. Yet certain stable, highly toxic, lipophilic pesticides are among the most toxic chemical known, and it is not suprising that chemicals such as DDT, dieldrin, endrin and toxaphene have been subject to widespread studies by fishery biologists. Despite this, our knowledge of the mechanism of toxicity, the extent of metabolism to which these pesticides are subject, and the nature of the metabolites themselves is still meagre. The determination of LC_{50} values is somewhat variable, data produced under static conditions often being inaccurate and misleading, and continuous flow data for long-term periods being available only for relatively few pesticides.

The various studies of pesticides effects, resulting from aqueous or oral exposure, which have been described in this chapter, have not always included a measurement of the associated pesticide concentrations in the tissues of the experimental fish. These concentrations, rather than those in the water or food, must be directly related to the biochemical reactions resulting in the observed symptoms, and a more detailed knowledge of the levels in tissues at which effects occur would assist considerably in assessing the biological significance of concentrations in field specimens. Observations in natural habitats can rarely include aberrant behaviour, or the adequate assessment of such parameters as growth rate or population dynamics, but chemical analyses of fish tissue are constantly in demand. Yet, apart from their relevance to limits of tolerance in human food, where such limits are in existence, analytical data cannot in most instances, as yet, indicate whether a potential biological hazard to the fish, or to other ecological dependents, may exist.

REFERENCES

F. S. H. Abram, *Lab. Pract.*, **9**, 797 (1960)

J. S. Alabaster, *Proc. 5th Br. Insect. Fungic. Conf.*, 370 (1969a)

J. S. Alabaster, *Pest Technol.*, March-April, 29-35 (1969)

J. S. Alabaster and F. S. H. Abram, *2nd Int. Conf. Wat. Poll. Res. Tokio*, 1964 (1965)

D. Allison, B. J. Kallman, O. B. Cope and C. Van Valin, U.S. Bur. Sport Fish. Wildl. Res. Rep., No. 64 (1964)

J. M. Anderson, *J. Fish. Res. Bd. Can.*, **25**, 2677 (1968)

J. M. Anderson, *Proc. R. Soc.*, **177**, 307 (1971)

J. M. Anderson and M. R. Peterson, *Science, N.Y.*, **164**, 440 (1969)

A. K. Andrews, C. C. Van Valin and B. E. Stebbings, *Trans. Am. Fish. Soc.*, **95**, 297 (1966)

A.P.H.A. American Public Health Association, Standard Methods for the Examination of Water and Wastewater. 11th Edn., New York (1960)

A.P.H.A. American Public Health Association, Standard Methods for the Examination of Water and Wastewater. 12th Edn., New York (1965)

V. C. Applegate, J. H. Howell and A. E. Hall, Jr., Spec. scient. Rep. U.S. Fish Wildl. Serv., No. 207, 157 pp. (1957)

C. E. Bond, R. H. Lewis and J. L. Fryer, Trans. *2nd Seminar on Biol. Problems in Wat. Poll.* R. A. Taft Sanit. Eng. Center, 96 (1960)

C. E. Boyd, *Progve Fish Cult.*, **26**, 138 (1964)

C. E. Boyd and D. E. Ferguson, *J. econ. Ent.*, **57**, 430 (1964)

W. R. Bridges, Trans. *3rd Seminar on Biol. Problems in Wat. Poll.* U.S. Publ. Health. Serv., Publ. PHS-999-WP.25, 247 (1965)

W. R. Bridges and A. K. Andrews, Spec. scient. Rep. U.S. Fish Wildl. Serv., No. 391 (1961)

W. R. Bridges, B. J. Kallman and A. K. Andrews, *Trans. Am. Fish. Soc.*, **92**, 421 (1963)

D. R. Buhler, M. E. Rasmusson and W. E. Shanks, *Toxic. appl. Pharmac.*, **14**, 535 (1969)

D. R. Buhler and W. E. Shanks, *J. Fish. Res. Bd. Can.*, **27**, 347 (1970)

G. E. Burdick, H. J. Dean, E. J. Harris, J. Skea and D. Colby, *N.Y. Fish and Game J.*, **12**, 127 (1965)

G. E. Burdick, E. J. Harris, H. J. Dean, T. M. Walker, J. Skea and D. Colby, *Trans. Am. Fish. Soc.*, **93**, 127 (1964)

W. D. Burke and D. E. Ferguson, *Trans. Am. Fish. Soc.*, **97**, 498 (1968)

W. D. Burke and D. E. Ferguson, *Mosquito News*, **29**, 96 (1969)

P. A. Butler, *Proc. 5th Weed. Control Conf.*, **18**, 576 (1965)

J. Cairns, J. R. and A. Scheier, *Notul. Nat.*, No. 370 (1964)

R. M. Christie and H. I. Battle, *Can. J. Zool.*, **41**, 51 (1962)

H. Cole, D. Barry, D. E. H. Frear and A. Bradford, *Bull. Environ. Contam. Toxicol.*, **2**, 127 (1967)

O. B. Cope, Circ. Fish Wildl. Serv., Wash., No. 199, 29 (1964)

O. B. Cope, *Proc. 18th Mtg. Southern Weed Conf.*, Texas, 439 (1965)

O. B. Cope, Circ. Fish Wildl. Serv., Wash. No. 226, 51 (1965)

O. B. Cope, *J. appl. Ecol.*, **3** (Suppl.), 33 (1966)

O. B. Cope, U.S. Bur. Sport Fish. Wildl., Resource Publ., No. 64, 125 (1968)

O. B. Cope, U.S. Bur. Sport Fish. Wildl., Resource Publ., No. 77, 92 (1969)

O. B. Cope, J. P. McCraren and L. Eller, *Weed Sci.*, **17** 158 (1969)

O. B. Cope and B. C. Park, Prog. Rep. U.S. Fish Wildl. Serv., 56 pp. (1957)

O. B. Cope, E. M. Wood and G. H. Wallen, *Trans. Am. Fish. Soc.*, **99**, 1 (1970)

C. A. Crandall and C. J. Goodnight, *Limnol. Oceanogr.*, **7**, 233 (1962)

J. P. Cuerrier, J. A. Keith and E. Stone, *Naturaliste Can.*, **94**, 315 (1967)

D. D. Culley Jr., and D. E. Ferguson, *J. Fish. Res. Bd. Can.*, **26**, 2395 (1969)

J. C. Dacre and D. Scott, *N.Z. J. Mar. Freshwat. Res.*, **5**, 58 (1971)

P. Doudoroff, B. G. Anderson, G. E. Burdick, P. S. Galtsoff, W. B. Hart, R. Patrick, E. R. Strong, E. W. Surber and W. M. Van Horn, *Sewage Ind. Wastes*, **23**, 130 (1951)

R. Eisler, U.S. Bur. Sport Fish. Wildl. Tech. Paper No. 17, 15 pp. (1967)

R. Eisler, U.S. Bur. Sport Fish. Wildl. Tech. Paper No. 46 (1970)

R. Eisler and P. H. Edmunds, *Trans. Am. Fish. Soc.*, **95**, 153 (1966)

P. F. Elson and C. J. Kerswill, *3rd Int. Conf. Wat. Poll. Res.*, Munich, 1966 (1967)

D. L. Fabacher and H. Chambers, *Bull. Environ. Contam. Toxicol.*, **6**, 372 (1971)

D. E. Ferguson and C. R. Bingham, *Bull. Environ. Contam. Toxicol.*, **1**, 97 (1966a)

D. E. Ferguson and C. R. Bingham, *Trans. Am. Fish. Soc.*, **95**, 325 (1966b)

D. E. Ferguson, W. D. Cotton, D. T. Gardner and D. D. Culley, *J. Miss. Acad. Sci.*, **11**, 239 (1965)

D. E. Ferguson, D. D. Culley, W. D. Cotton and R. P. Dodds, *Bio Science*, **14**, 43 (1964)

D. E. Ferguson and C. P. Goodyear, *Copeia* (2), 467 (1967)

D. E. Ferguson, J. L. Ludke, M. T. Finley and G. G. Murphy, *J. Miss. Acad. Sci.*, **13**, 138 (1967)

D. E. Ferguson, J. L. Ludke and G. G. Murphy, *Trans. Am. Fish. Soc.*, **95**, 335 (1966)

M. T. Finley, D. E. Ferguson and J. L. Ludke, *Pest. Monit. J.*, **3**, 212 (1970)

J. H. Gakstatter and C. M. Weiss, *Trans. Am. Fish. Soc.*, **96**, 301 (1967)

J. R. Gibson, J. L. Ludke and D. E. Ferguson, *Bull. Environ. Contam. Toxicol.*, **4**, 17 (1969)

P. A. Gilderhus, *Trans. Am. Fish. Soc.*, **95**, 289 (1966)

R. J. Graham, Trans. *2nd Seminar on Biol. Problems in Water Poll.* R. A. Taft. Sanit. Eng. Center, 62 Cincinnati (1960)

B. F. Grant and P. M. Mehrle, *J. Fish. Res. Bd. Can.*, **27**, 2225 (1970)

J. E. Halver, C. L. Johnson and L. M. Ashley, *Fed. Proc.*, **21**, 390 (1962)

D. J. Hansen, *Trans. Am. Fish. Soc.*, **98**, 426 (1969)

C. Henderson and Q. H. Pickering, *Trans. Am. Fish. Soc.*, **87**, 39 (1958)

C. Henderson, Q. H. Pickering and C. M. Tarzwell, *Trans. Am. Fish. Soc.*, **88**, 23 (1959)

C. Henderson, Q. H. Pickering and C. M. Tarzwell, *Trans. 2nd Seminar on Biol. Problems in Wat. Poll.* R. A. Taft Sanit. Eng. Center, 76 (1960)

C. Henderson and C. M. Tarzwell, *Sewage Ind. Wastes*, **29**, 1002 (1957)

F. Herr, E. Greslin and C. Chappel, *Trans. Am. Fish. Soc.*, **96**, 320 (1967)

R. C. Hiltibran, *Trans. Am. Fish. Soc.*, **96**, 414 (1967)

C. L. Hing, *Diss. Abstr.*, **27**B, 1966 (1967)

A. V. Holden, *Ann. appl. Biol.*, **50**, 467 (1962)

A. V. Holden, *J. Proc. Inst. Sew. Purif.*, (4), 361 (1964)

H. T. Holland, D. L. Coppage and P. A. Butler, *Bull. Environ. Contam. Toxicol.*, **2**, 156 (1967)

D. R. Hubble and B. Reiff, *Bull. Environ. Contam. Toxicol.*, **2**, 57 (1967)

J. S. Hughes and J. T. Davis, *Proc. La Acad. Sci.*, **25**, 86 (1962)

J. S. Hughes and J. T. Davis, *Weeds*, **11**, 50 (1963)

J. V. Hunter, B. F. Dowden and H. J. Bennett, *Proc. La Acad. Sci.*, **30**, 80 (1967)

E. G. Hunt and A. I. Bischoff, *Calif. Fish Game*, **46**, 91 (1960)

T. Ishihara, M. Yasuda and O. Tamura, *Bull. Fac. Fish. Nagasaki Univ. (Japan)*, **24**, 65 (1967)

K. Iyatomi, T. Tamura, Y. Itazawa, I. Hanyu and S. Sugiura, *Progve Fish Cult.*, **20**, 155 (1958)

D. W. Johnson, *Trans. Am. Fish. Soc.*, **97**, 398 (1968)

M. Katz, *Trans. Am. Fish. Soc.*, **90**, 264 (1961)

M. Katz and G. C. Chadwick, *Trans. Am. Fish. Soc.*, **90**, 394 (1961)

M. H. A. Keenleyside, *Can. Fish Cult.*, No. 24, 17 (1959)

H. D. Kennedy, L. L. Eller and D. F. Walsh, U.S. Bur. Sport Fish. Wildl. Tech. Paper No. 53, 18 pp. (1970)

S. F. King, Spec. scient. Rep. U.S. Fish Wildl. Serv., No. 399 (1962)

S. K. Konar, *Trans. Am. Fish. Soc.*, **98**, 430 (1969)

C. E. Lane and R. J. Livingston, *Trans. Am. Fish. Soc.*, **99**, 489 (1970)

C. E. Lane and E. D. Scura, *J. Fish. Res. Bd. Can.*, **27**, 1869 (1970)

E. L. Lee, *Diss. Abs.*, **30**B, 1727 (1969)

Chapter 7

Pesticide Residues in Birds and Mammals

L. F. Stickel

Patuxent Wildlife Research Center,
Laurel, Maryland

7.1 INTRODUCTION

Chemicals have long been used as tools to prevent insects from making undue inroads into man's food supply and to break the vector chain for diseases carried by insects to man and domestic animals. Widespread applications to fields, forests, and marshlands has introduced the materials directly into the environment over large areas. Waste disposal or leakage from manufacturing and formulating plants, and from subsequent users, has introduced the materials directly into waterways. Chemicals have moved from the sites of application by air, water, and other routes.

The result is a widespread distribution of man-made chemicals and their breakdown products in all living organisms, wild and domestic, including man.

The purpose of this chapter is to describe the occurrence and distribution of organochlorine pesticides in birds and mammals, and to summarize research on the effects of these chemicals on organisms and populations.

Birds and mammals will be discussed separately under the headings of Occurrence and Distribution of Pesticide Residues, Kinetics of Pesticides, and Effects of Pesticides on Physiology and Populations.

7.2   OCCURRENCE AND DISTRIBUTION OF ORGANOCHLORINE PESTICIDES IN WILD BIRDS

Organochlorine pesticides accumulate in the fat of birds from a relatively small exposure. DDE, a primary environmental breakdown product of DDT, is universally distributed, so that exposure is essentially continuous, and few, if any, wild birds are free from this compound. The parent compound, DDT, is present less frequently, and so is DDD, another metabolite that is also used as an insecticide. DDE is more prevalent because it is more resistant to degradation in animals or the environment. Dieldrin occurs in wild birds almost as often as DDE, but in smaller amounts. Most dieldrin residues arise from use of aldrin, which converts rapidly to dieldrin, although dieldrin also is used directly. Heptachlor epoxide occurs occasionally, arising either from heptachlor or from the heptachlor component of chlordane. Other fractions of chlordane (alpha-chlordane, gamma-chlordane) also occur occasionally. Gamma-BHC, a persistent fraction of commercial BHC and the main component of lindane, occurs occasionally in small amounts. Endrin and camphechlor occur rarely. Probably not all of the fat-soluble metabolites of these chemicals have been identified, and analytical standards are not available even for all of the identified forms. For this reason, and related analytical problems, the 'discovery' of new contaminants in bird tissues will predictably continue, although major unidentified peaks are now rare on gas-liquid chromatograms of residues in bird tissues.

Even the chemicals that are not common, however, build up rapidly in birds living in areas where these chemicals are applied for insect control. Birds respond to pesticide use very quickly.

This is most strikingly shown when applications are high enough to kill birds outright. DDT, applied for control of Dutch elm disease, for example, killed large numbers of many kinds of birds in many communities in the United States (Hickey and Hunt, 1960; Wallace et al., 1961, 1964; Wurster et al., 1965). Dieldrin, applied to control pest insects in Illinois, caused extremely heavy mortality among birds and even among mammals (Scott et al., 1959). Dieldrin used to control white-fringed beetles also caused considerable mortality among a wide variety of birds (Dustman and Stickel, 1966).

Heptachlor, used throughout the southeast U.S.A. in the 1950's to control the imported fire ant, caused direct mortality of a variety of birds (Rosene, 1965). It also resulted in the widespread occurrence of sublethal quantities of heptachlor epoxide in tissues of many species of birds (Stickel et al., 1965b; Rosene et al., 1962; DeWitt et al., 1960). Woodcock wintering in the South accumulated heptachlor epoxide from worms in treated fields and carried the residues north with them in the spring (Wright, 1965). Heptachlor epoxide was transmitted to the eggs, so that young birds on the Canadian breeding grounds contained heptachlor epoxide residues at hatching. After discontinuation of the programme, heptachlor epoxide residues gradually disappeared from the environment and residues were reduced in magnitude and frequency (McLane et al., 1971).

In the late 1960's, mirex, an unusually persistent insecticide, was applied in several southern areas for fire-ant control. Its accumulative capacity was remarkable in view of the low rate of application, which was only 5 grams per acre. It was found in the tissues of all of the bird species and 78% of the individuals that were sampled, despite movements of birds in and out of the treated area (U.S. B.S.F.W., 1971). Residues in muscle exceeded 1 ppm in 15 of the birds sampled; the maximum, 17 ppm, occurred in a kingfisher. Residues in herons and egrets declined little over a 2-year period following the treatment.

DDT, applied once at the rate of 1 pound per acre, persisted in the upper soil litter of Maine forests with little change throughout a 9-year period (Dimond et al., 1970). DDT and its metabolites, in robins living in the forest, also continued at significantly higher levels than in surrounding areas, indicating a continuous availability of residues through the food chain. Robins, sampled the year after treatment, contained an average of 13.53 ppm in their bodies, an average of 4.50 ppm the year after treatment, and 3.55 ppm 9 years after treatment. Residues in robins from untreated areas averaged 0.47 ppm.

DDT applied to a forested area in Montana at the rate of 0.5 lb/acre (0.56 kg/ha) resulted in an average of 46 ppm (1.5-280 ppm) of DDT and its metabolites in the fat of 28 blue grouse (Mussehl and Finley, 1967). Concentrations of up to 80 ppm appeared in fat within 7 days of spraying. Residues were still high in the second season, averaging 22 ppm (1-110 ppm) in 35 samples. In the third season, only two birds were sampled, but they contained residues of 15 and 21 ppm.

Organochlorine pesticides, and related compounds, also may reach the environment in wastes from factories. Chemicals from such sources have on occasion escaped into the environment on a large scale, permeating the ecosystem over large areas. For example, dieldrin and 'Telodrin®' spread from a plant near the mouth of the River Rhine to contaminate much of the Netherlands coast and the Wadden Sea (Koeman, 1971). Large residues accumulated in several kinds of birds and mortality was extensive. When the leakage was stopped, residues declined and the pronounced bird mortality from this cause ceased.

Organochlorine pesticides and related materials escaped into the environment in substantial amounts, in Hardeman County, Tennessee, as a result of waste disposal by a pesticide manufacturer (Mount and Putnicki, 1966; Rima et al., 1967). Environmental contamination from DDT-manufacturing plants has occurred on both the east and west coasts of the United States (Stickel and Heath, 1965; Burnett, 1971).

Quantities of pesticides in wild birds are primarily correlated with exposure, although birds differ in their ability to absorb and metabolize chemicals.

Birds that eat other birds or fish usually have higher residues than those that eat seeds and vegetation. Compilations that compare residues in bird species of different food habits have been published by Cramp et al., 1964; Moore and Walker, 1964; Ratcliffe, 1965; Cramp and Olney, 1967; Keith and Gruchy, 1970; Vermeer and Reynolds, 1970; and Prestt et al., 1970.

In Great Britain, eggs of herons and grebes, which are freshwater fish-eaters, contained residues in concentrations that were tens of times those in the eggs of seabirds (Moore and Tatton, 1965). Grey herons contained higher residues than any other species, including other fisheaters as well as terrestrial birds of prey (Prestt et al., 1970). In the United States, bald eagles, which eat both fish and birds, contained median amounts of 5-16 ppm of DDE, and maxima of 91-263 ppm, in five sampling years from 1964-1968 (Reichel et al., 1969; Mulhern et al., 1970). By contrast, golden eagles, which feed primarily on mammals, contained a median DDE residue of 0.49 ppm and a maximum of 2.9 ppm. At the extreme, pheasants in South Dakota contained only trace amounts of DDE, even in their fat (Linder and Dahlgren, 1970).

Alaskan peregrine falcons, which feed primarily on birds, contained far higher residues than the small birds in the same area (Cade et al., 1968). Scaup, which feed more heavily upon animal material than mallards, accumulated residues that were two to four times as great when both were placed on a DDT-treated marsh for the same periods of time (Dindal, 1967; Dindal and Peterle, 1968). Black ducks, which also eat more animal material than mallards, contained higher DDE residues in a nationwide monitoring survey that included both species (Heath, 1969).

Several publications, in recent years, have drawn upon the literature to summarize the occurrence and concentrations of pesticidal residues in eggs and tissues of wild species (for example, Stickel, 1968; Edwards, 1970; Keith and Gruchy, 1970).

Residues in eggs reflect residues in tissues of birds at a specific time in the annual cycle. They also measure the exposure of the embryo. They thereby provide a useful means to compare species and geographic areas. A compilation from the literature is assembled in Table 7.1, to illustrate and extend the subjects discussed earlier in this section.

7.3   KINETICS OF PESTICIDES IN BIRDS

Kinetics, in its sense of movement and change, is a suitable term to describe the actions and metabolism of chemicals entering the animal body, making changes and being changed therein, accumulating in tissues, and ultimately leaving again. The pattern of events is somewhat different when exposure is at sublethal levels than when it is enough to cause death, and these two conditions will be discussed separately.

7.3.1   Sublethal accumulation
7.3.1.1   Uptake, loss, and equilibrium
As early as 1947, experimental studies showed that hens fed diets containing as little as 2.3 ppm of DDT, accumulated DDT in their tissues, transferred residues to their eggs, and higher dosages produced higher residues (Bryson et al., 1950). Earlier studies had shown that DDT accumulated in eggs in proportion to the DDT dosage received by the hens (Rubin et al., 1947); turkeys also accumulated DDT in their tissues in amounts that increased with increased dosage (Marsden and Bird, 1947). Other studies confirmed these results (Draper et al., 1952).

Accumulation of DDT residues in tissues of wild birds in the field was first shown when deaths of birds followed DDT applications and was followed by analyses of their tissues. The principle that DDT accumulated in bird tissues was thus established and accepted, within a few years of the general introduction of DDT for insecticidal use. In the intervening years, much research was done to clarify the basic principles involved in the kinetics of pesticidal chemicals in the animal body, and to understand the practical implications.

A classic study of the accumulation and loss of pesticidal chemicals in birds was made by Cummings et al. (1966, 1967) who employed small amounts of a mixture of chemicals. Sixty laying hens were fed 0.05, 0.15, or 0.45 ppm of DDT, lindane, heptachlor epoxide, dieldrin and endrin, in combination, for 16 weeks and then untreated food for 4 weeks, to show the rates of accumulation and loss of residues.

Residues in their eggs were above the control level by 3 days after pesticide feeding began. Residues first increased sharply, then gradually, until they reached a plateau. Residues declined continuously during the 4-week period after untreated food was given. DDE also increased as DDT was metabolised, but the DDE residues did not reach a plateau and continued to increase even during the feeding on untreated food, showing that stored DDT was being continuously converted to DDE.

DDT accumulation in the abdominal fat of hens paralleled that in the eggs. Residues in abdominal fat levelled off at about 10 times those in eggs. Since eggs contained an average of 10-12% fat, the concentration in the abdominal fat and in the egg fat was about the same. DDE residues began to decline in the fat during the withdrawal period, indicating that this compound was also gradually being lost, although time had not permitted the loss to show in eggs. DDD also appeared in the fat, as did o,p'-DDT. Although o,p'-DDT constitued 30% of the technical grade DDT that was fed to the hens, and p,p'-DDT constituted 70%, o,p'-DDT had much less of a propensity for storage, and maximum levels in the tissues were only about 10% of the total of the two compounds. During the withdrawal period, o,p'-DDT disappeared from the tissues more rapidly than did p,p'-DDT. These two isomers were not measured separately in the eggs, which were analysed early in the study, and so the rates and extent of transfer to eggs were not determined.

DDT and its metabolites were stored in tissue in greater amounts at the higher doses, but storage was not strictly proportional to the feeding levels, probably as a result of the complexities introduced by the conversions to metabolites. All the other chemicals showed a traditional dose-storage relationship. Dieldrin and heptachlor epoxide had the greatest propensity for storage, endrin next, and lindane least. The five chemicals can be rated for accumulation in the tissues of hens in the order of heptachlor epoxide≥dieldrin>endrin>DDT>lindane.

Other studies of poultry, in which chemicals were fed separately, have tended to array the storage potential of the different compounds in much the same sequence as reported by Cummings. For example, dieldrin, heptachlor, DDT, and lindane, fed separately to different groups of hens, accumulated at levels corresponding to the order in which they are listed (Stadelman et al., 1965). DDE, dieldrin, and heptachlor epoxide persisted in samples of tissues even after 26 weeks of untreated diet.

Studies with mammals have shown that chemicals fed in combination may be stored at different rates than they would be if they were fed separately (Street, 1964; Deichmann et al., 1969, 1971). A few comparisons can be made between Cummings' results and those of others. In Cummings' studies, for example, concentrations of dieldrin and heptachlor epoxide in fat were about 10 times the dietary concentration, and concentrations in whole eggs were about equal to the dietary concentration. Dietary concentrations were expressed on a dry-weight basis, and egg residues on a wet-weight basis, so that the accumulation factor was probably about three. Lindane, in contrast, accumulated to about only one tenth the degree of dieldrin or heptachlor epoxide.

When lindane was fed to chickens for 60 days at 10 ppm in the diet, 21 ppm of lindane accumulated in the fat of the birds (Ware and Naber, 1961). A dietary dosage of 1 ppm produced 1.2 ppm in the fat. However, results of another study, in which DDT and lindane were fed both together and separately, suggested that lindane may have enhanced the storage of DDT (Ritchey et al., 1967).

TABLE 7.1 (continued)

Name, place, and time[2]	Number	DDE		Dieldrin		Source
		Mean	Range[3]	Mean	Range[3]	
Order Falconiformes (continued)						
Everglade kite (*Rostrhamus sociabilis*)						
Florida, 1966	1	0.33	–	0.05	–	Lamont and Reichel, 1970
Cooper's hawk (*Accipiter cooperii*)						
California, 1968	1	22.7	–	–	–	Risebrough et al., 1968
Sparrow hawk (*Accipiter nisus*)						
SE England, [1963-1969]	8	29.53	13.2-70	4.59	1.4-8.9	Ratcliffe, 1970
N&W, Great Britain, [1963-1969]	96	11.20	0.5-50	1.84	0.1-10	”
Buzzard (*Buteo buteo*)						
Great Britain [1963-1969]	30	0.78	tr-3.1	0.72	0-4.4	Holt and Sakshaug, 1968
Norway, 1966-1967	1	0.07	–	0.06	–	” ”
Rough-legged hawk (*Buteo lagopus*)						
Alaska, 1967	3	0.92	–	–	–	Lincer et al., 1970
Alaska, 1967-1969	25	22.5	2.2-109	–	–	Cade et al., 1971
Norway, 1966-1967	3	0.17	0.1-0.2	<0.05	–	Holt and Sakshaug, 1968
Golden eagle (*Aquila chrysaëtos*)						
California, 1968	1	2.0	–	–	–	Risebrough et al., 1968
Norway, 1966-1967	6	0.24	0.07-0.6	<0.05	–	Holt and Sakshaug, 1968
W. Scotland, 1963-1965	47	0.78	tr-3.0	0.87	0.05-6.9	Ratcliffe, 1970
W. Scotland, 1967-1968	13	0.48	tr-0.50	0.34	tr-1.20	”
E. Scotland, 1964-1968	8	0.11	tr-0.2	0.06	tr-0.10	” ”
Bald eagle (*Haliaeetus leucocephalus*)						
United States, 1962-1963	5	–	1-28	–	0.5-1.0	Stickel et al., 1966a
United States, 1962-1963	4	–	4.6-13	–		” ”
Wisconsin, 1968	10	4.76	1.81-15.27	0.37	0.07-0.65	Krantz et al., 1970 and Wiemeyer et al., 1972
Maine, 1968	5	21.76	13.20-27.55	1.41	0.31-3.12	”
Maine, 1969	3	14.95	11.86-20.55	0.22	0.15-0.29	”
Florida, 1968	6	10.72	4.29-27.93	0.21	0.11-0.28	”
Florida, 1969	2	18.37	18.21-18.52	1.11	0.68-1.54	”
Minnesota, 1969-1970	4	9.57	5.37-21.62	0.99	0.16-2.29	”
Michigan, 1969	1	39.46		1.02		”

Osprey (*Pandion haliaetus*)

Location, Year						Reference
New York, 1966	1	8.8	—	—	—	Woodwell *et al.*, 1967
Maine, 1963	3	1.7	—	—	—	Ames, 1966
Rhode Island, 1962	1	7.4	—	—	—	„
Connecticut, 1962	6	6.7	—	—	—	„
Connecticut, 1963	15	4.7	—	—	—	„
New Jersey, 1963	2	5.1	—	—	—	„
Maryland, 1963	25	2.3	—	—	—	„
Merlin (*Falco columbarius*)						
Great Britain, [1963-1969]	12	10.47	3.23-19.8	2.77	0.39-7.0	Ratcliffe, 1970
Peregrine falcon (*Falco peregrinus*)						
Alaska, 1966	2	12.5	11.4-13.5	0.7	0.4-0.8	Cade *et al.*, 1968
Alaska, 1967	3	27.4	—	—	—	Lincer *et al.*, 1970
Alaska, Colville, 1967-1969	19	889*	159-3130*	—	—	Cade *et al.*, 1971
Alaska, Yukon and Interior, 1968-1969	14	673*	210-1343*	—	—	„
Alaska, Amchitka, 1969-1970	11	167*	97-420*	—	—	„
Canada, Ungava, 1967	10	12.7	—	—	—	Berger *et al.*, 1970
Canada, W., 1968	5	17.8	4.4-31.2	0.8	nd-1.7	Enderson and Berger, 1968
Canada, W., 1968	2	27.9	nd-204.9	0.5	nd-1.8	„
Great Britain, E. and Central Highlands, [1963-1969]	16	3.25	tr-8.6	0.29	tr-1.6	Ratcliffe, 1970
Great Britain, other regions, [1963-1969]	42	13.67	0.2-33.0	0.57	0-2.6	„
Prairie falcon (*Falco mexicanus*)						
Colorado-Wyoming, 1967-1968	34	27.5	—	1.9	—	Enderson and Berger, 1970
Gyrfalcon (*Falco rusticolus*)						
Alaska, 1968-1969	13	3.88*	nd-20.3*	—	—	Cade *et al.*, 1971
American kestrel (*Falco sparverius*)						
California, 1968	6	0.19	—	—	—	Risebrough *et al.*, 1968
Hobby (*Falco subbuteo*)						
Great Britain, [1963-1969]	1	4.4	—	0.2	—	Ratcliffe, 1970
Kestrel (*Falco tinnunculus*)						
Great Britain, [1963-1969]	29	3.64	0.1-24.0	0.23	tr-1.8	„

Order Galliformes

Bobwhite quail (*Colinus virginianus*)

Location, Year						Reference
Alabama, 1967	67	3.45	—	4.78	—	U.S.D.A., 1969

(continued

TABLE 7.1 (continued)

Name, place, and time[2]	Number	DDE		Dieldrin		Source
		Mean	Range[3]	Mean	Range[3]	
Order Galliformes (continued)						
Pheasant (*Phasianus colchicus*)						
Great Britain, before 1964	7	0.03	–	0.05	–	Moore and Walker, 1964
Iowa, 1969	8	0.014	0.008-0.026	0.266	0.009-1.4	Johnson *et al.*, 1970
South Dakota, 1967	67	0.019	0.01-0.06 (44)	0.099	0.01-1.58 (54)	Linder and Dahlgren, 1970
Purple gallinule (*Porphyrula martinica*)						
Louisiana, 1966	56	0.43	0.06-5.81	6.51	0.49-15.35	Causey *et al.*, 1968
Common gallinule (*Gallinula chloropus*)						
Louisiana, 1966	14	0.31	0.12-1.05	9.37	1.13-22.12	''
Louisiana, 1966	14	0.10	0.05-0.20 (11)	0.28	–	''
Coot (*Fulica americana*)						
Canada (Prairie Provinces), 1968-1969	1	0.41	–	0.004	–	Vermeer and Reynolds, 1970
Iowa, 1969	6	0.145	0.039-0.450	0.060	0.037-0.074	Johnson *et al.*, 1970
Order Charadriiformes						
Golden plover (*Pluvialis apricaria*)						
Great Britain, [1963-1969]	2	1.53	0.43-2.62	0.10	0.09-0.11	Ratcliffe, 1970
American avocet (*Recurvirostra americana*)						
Canada (Prairie Provinces), 1968-1969	2	–	3.16-3.32	–	0.171-0.252	Vermeer and Reynolds, 1970
Herring gull (*Larus argentatus*)						
Canada (Prairie Provinces), 1968-1969	7	–	2.67-17.2	0.34	0.012-0.514	''
Great Britain, 1965	5	<0.02	–	1.0	–	Robinson *et al.*, 1967c
Michigan, 1965	112	106.5	–	–	–	Ludwig and Tomoff, 1966
Wisconsin, 1964	9	202	122-281	–	–	Keith, 1966
California gull (*Larus californicus*)						
Canada (Prairie Provinces), 1968-1969	15	–	4.07-21.6	–	nd - 1.52	Vermeer and Reynolds, 1970
Ring-billed gull (*Larus delawarensis*)						
Canada (Prairie Provinces), 1968-1969	17	–	1.26-12.3	–	0.260-1.27	''
Lesser black-backed gull (*Larus fuscus*)						
Great Britain, 1965	6	0.43	–	0.63	–	Robinson *et al.*, 1967c

Great Britain, [1963-1969]	53	0.49	tr-2.2	0.09	0-0.6	"
Great Britain, 1965 (North Shields)	26	0.10	–	0.25	–	Robinson et al., 1967c
Great Britain, 1965 (Farne Is.)	2	0.11	0.08-0.14	0.22	0.15-0.28	"
Great Britain, 1965 (Farne Is.)	6	0.15	–	0.31	–	"
Great Britain (E. Coast), 1967	7	0.2	–	0.1	–	Prestt et al., 1970
Great Britain (E. Coast), 1967	6	0.7	–	0.2	–	"
Roseate tern (Sterna dougallii)						
Great Britain, 1965	5	0.15	–	0.55	–	Robinson et al., 1967c
Ireland, 1965	6	0.27	0.13-0.45	0.06	0.03-0.12	Koeman, 1971
Common tern (Sterna hirundo)						
Alberta, 1969	68	7.57	0.64-104.0	–	–	Switzer et al., 1971
Canada (Prairie Provinces), 1968-1969	7	–	2.04-25.2	–	0.017-0.396	Vermeer and Reynolds, 1970
Germany, 1965	14	0.54	0.25-1.5	0.70	0.31-3.5	Koeman, 1971
Germany, 1966	6	0.44	0.09-0.86	0.29	0.08-0.55	"
Great Britain, 1965 (Coquet Is.)	5	<0.02	–	0.04	–	Robinson et al., 1967c
New York, 1966	5	3.56	–	–	–	Woodwell et al., 1967
Sandwich tern (Sterna sandvicensis)						
Great Britain, 1965	8	0.17	–	0.75	–	Robinson et al., 1967c
Great Britain, 1965	6	0.30	–	0.64	–	"
Great Britain, 1966	6	0.55	0.26-1.2	0.17	0.08-0.41	Koeman, 1971
Ireland, 1965	5	0.61	0.43-0.97	0.12	0.10-0.15	"
Netherlands, 1965	8	0.61	0.29-1.2	0.8	0.36-2.4	"
Netherlands, 1966	25	0.61	0.26-1.1	0.23	0.09-0.79	"
Netherlands, 1967	10	0.43	0.25-0.85	0.19	0.05-0.42	"
Netherlands, 1969	10	0.75	0.35-1.4	0.093	0.03-0.19	"
Netherlands, 1969	5	1.1	0.32-2.8	0.068	0.01-0.19	"
Netherlands, 1970	10	0.4	0.21-0.88	0.065	0.034-0.089	"
Razorbill (Alca torda)						
Great Britain, E. Coast, 1967	5	1.0	–	0.5	–	Prestt et al., 1970
Great Britain, W. Coast, 1967	6	1.1	–	0.5	–	"
Great Britain, 1965	3	0.35	–	0.64	–	Robinson et al., 1967c
Great Britain, [1963-1969]	49	2.15	0.9-5.4	0.84	0.2-3.0	Ratcliffe, 1970

(continued

TABLE 7.1 (continued)

Name, place, and time[2]	Number	DDE Mean	DDE Range[3]	Dieldrin Mean	Dieldrin Range[3]	Source
Order Charadriiformes (continued)						
Guillemot (*Uria aalge*)						
Great Britain, [1963-1969]	66	2.12	0.5 - 6.5	0.28	n.d - 1.4	Ratcliffe *et al.*, 1967
Great Britain, 1965	3	0.11	–	1.30	–	Robinson *et al.*, 1967c
Great Britain, 1965	5	0.60	–	1.88	–	”
Great Britain, E. Coast, 1967	8	0.7	–	0.1	–	Prestt *et al.*, 1970 ”
Great Britain, W. Coast, 1967	7	2.0	–	0.3	–	”
Puffin (*Fratercula arctica*)						
Great Britain, 1965	1	0.30	–	0.17	–	Robinson *et al.*, 1967c
Great Britain, 1965	5	0.33	–	0.40	–	”
Order Columbiformes						
Mourning dove (*Zenaidura macroura*)						
Alabama, 1967	–	15.00	–	0.10	–	U.S.D.A., 1969
Ground dove (*Columbigallina passerina*)						
Alabama, 1967	–	1.15	–	1.60	–	U.S.D.A., 1969
Order Strigiformes						
Barn owl (*Tyto alba*)						
California; 1968	5	3.2	–	–	–	Risebrough *et al.*, 1968
Tawny owl (*Strix aluco*)						
Norway, 1966-1967	1	0.7	–	<0.05	–	Holt and Sakshaug, 1968
Order Passeriformes						
Blue jay (*Cyanocitta cristata*)						
Alabama, 1966	–	15.15	–	0.08	–	U.S.D.A., 1969
Alabama, 1967	–	6.41	–	0.64	–	U.S.D.A., 1969
Raven (*Corvus corax*)						
Great Britain, [1963-1969]	10	0.95	0.08-2.63	0.78	0.2-1.46	Ratcliffe, 1970
Carrion crow (*Corvus corone*)						
Great Britain, [1963-1969]	26	0.34	0.07-0.88	0.14	tr - 0.59	

Great tit (*Parus major*)	3	0.4	0.4-0.5	<0.05			
Norway, 1966-1967	3	—	—	—	—	—	,, ,,
Mockingbird (*Mimus polyglottos*)							
Mississippi, 1966	—	17.9	—	0.13	—	—	U.S.D.A., 1969
Alabama, 1967	—	22.11	—	0.42	—	—	,, ,,
Brown thrasher (*Toxostoma rufum*)							
Mississippi, 1966	—	29.43	—	0.10	—	—	,, ,,
Alabama, 1966	—	59.45	—	1.38	—	—	,, ,,
Alabama, 1967	—	15.89	—	0.10	—	—	,, ,,
House sparrow (*Passer domesticus*)							
Mississippi, 1966	—	5.24	—	0.09	—	—	,, ,,
Blackbirds (unidentified)							
Mississippi, 1966	—	29.70	—	1.14	—	—	,, ,,
Red-winged blackbird (*Agelaius phoeniceus*)							
Mississippi, 1966	—	10.91	—	0.14	—	—	,, ,,
North Dakota, 1966	—	1.90	—	–	—	—	,, ,,
Alabama, 1966	—	20.81	—	2.52	—	—	,, ,,
Alabama, 1967	—	54.28	—	5.94	—	—	,, ,,
Illinois, 1964 (control area)	12	nd	—	1.7	—	—	Graber *et al.*, 1965
Illinois, 1964 (treated area)	5, 6, 4	nd	—	5.7, 6.3, 0.2	—	—	,, ,,
Orchard oriole (*Icterus spurius*)							
Alabama, 1967	—	1.16	—	0.19	—	—	U.S.D.A., 1969
Purple grackle (*Quiscalus quiscula*)							
Mississippi, 1966	—	23.52	—	0.08	—	—	,, ,,
Alabama, 1966	—	4.85	—	0.90	—	—	,, ,,
Alabama, 1967	—	23.92	—	3.20	—	—	,, ,,
Cardinal (*Richmondena cardinalis*)							
Alabama, 1967	—	7.87	—	5.62	—	—	,, ,,

[1] Only DDE and dieldrin residues are included in this table although other pesticides and also polychlorinated biphenyls are included in the original papers. DDT and DDD were omitted because they are ordinarily present in only small amounts in wild animals and so do not serve well for the general comparisons of species and areas. Also, their measurements were distorted by the presence of polychlorinated biphenyls (PCB's) in analyses reported in some publications from about 1963-1964 (when gas chromatography became general) until about 1968 (by which time separations from PCB's were generally made). Methodology in individual papers of this period must be scrutinised carefully to determine whether this interference occurred.

[2] Scientific names follow the A.O.U. Checklist (American Ornithologists' Union, 1957) and Peterson *et al.* (1965).

[3] Numbers in parentheses show the number of samples in which the chemical was detected, if this is different from the total number examined; nd = not detected; tr = trace; — = information not given in publication.

residues. Chickens stored more DDT residues in their fat and liver after oral dosing, than after intraperitoneal or intramuscular injection, the reverse of the storage pattern for rats, sheep, rabbits, or guinea pigs (McCully *et al.*, 1968). Residues produced in fat by oral or intramuscular dosing, were considerably higher in chickens than in any of the other species. Residues produced by intraperitoneal dosing were slightly lower than in the rat and higher than all others.

Losses of chemicals from the body tissues of birds when dosages were stopped have been shown in many of the studies described above, and the rates of loss have paralleled the rates of accumulation.

Acceleration of loss rates, by reducing the food supply and so speeding the mobilisation of fat reserves, has been attempted as a practical measure to permit earlier return of accidentally contaminated poultry and products to market. After 15 weeks of dietary feeding of DDT to hens, Noakes and Benfield (1965) subjected the birds to 3 weeks of partial starvation. Residues of DDE continued to increase during that period although those of DDT and DDD decreased. In other studies, lindane, dieldrin, and DDT were fed in combination, and DDT was fed alone, and the hens were subsequently forced to moult to induce weight loss and mobilisation of chemicals (Smith *et al.*, 1970).

The mobilisation of residues into blood, which is the first step in residue depletion, has been shown for DDT in cockerels (Ecobichon and Saschenbrecker, 1969), for DDT in pigeons (Findlay and de Freitas, 1971), and for dieldrin in chickens (Davison *et al.*, 1971). The loss of DDT residues from hens and eggs was increased by partial starvation and the forced moult that resulted (Wesley *et al.*, 1966). Loss of residues was considerably greater when the partially starved hens were fed a high protein diet, than when they were fed a low protein diet. Control birds, on normal adequate diets, lost residues much more slowly than the experimental groups, despite continuing to lay eggs, although the experimental birds did not lay for periods of 4-6 weeks. Seven different management methods for depleting residues were studied in subsequent experiments, including injection of androgen, varying protein levels, varying lengths of starvation periods, adding fat to the diet, and adding vitamins to the diet (Wesley *et al.*, 1969). Androgen increased retention of residues, starvation decreased it, and changed protein levels showed no significant overall effect. The addition of fat to the diet increased residue storage, but addition of vitamins had no statistically significant effects.

In younger chicks, the effects of starvation on DDT loss were complicated by its effects on growth (Donaldson *et al.*, 1968). Residue concentrations in blood and tissues increased with starvation. However, the overall body burden was decreased by starvation, and concentrations were reduced when full feeding was resumed.

Loss of DDT residues also was speeded among cowbirds that lost weight (Van Velzen *et al.*, 1972). In a 2-week period, birds on full rations lost 26% of the total body burden of DDT+DDD+DDE, whereas birds subjected to an initial 4-day period of reduced rations (43% of normal) lost 36% of their total residues, although DDE increased.

7.3.1.2 *Metabolism*

Chemical changes of organochlorine insecticides within the animal body are inherent and critical parts of the kinetic process. These changes may drastically alter the toxicity, solubility, storage potential, and other characteristics of the parent compound. An evaluation of the effects of a chemical cannot be complete without an evaluation of the effects of its metabolites. Despite a vast amount of research and an enormous literature, the complete series of breakdown stages is not fully known for even one of the organochlorine insecticides. Furthermore, the rates and the pathways of degradation are probably altered by dosage levels, by the degree of toxication of the organism, by differences between species, and by other factors. A review of this subject is outside the scope of the present chapter. Only a few critical papers and findings will be mentioned; some of the better known metabolites are discussed in other appropriate sections of this chapter and in chapter 13. A good review and documentation, including chemical structural formulae, is given by Menzie (1969).

The metabolism of DDT, DDD, DDE, and DDMU in pigeons, including rates of elimination of the compounds from various tissues, has been reported in two careful and thorough studies by Bailey *et al.* (1969a, 1969b). DDT was eliminated from all tissues at the same rate (half-life of 28 days) and gave DDE and DDD by separate pathways. DDD had a half-life of 24 days and transformed mainly to DDMU. The half-life of DDE was 250 days and no metabolites were found. *In vitro* studies with pigeon liver have also shown conversion of DDT to DDE and DDD; DDD predominated under anaerobic conditions (Bunyan *et al.*, 1966).

The metabolism of DDT in chicks was studied both by feeding the parent chemical and two of its primary metabolites, DDE and DDD, to them and by injecting these chemicals into fertile eggs prior to incubation (Abou-Donia and Menzel, 1968b). Eleven metabolites were detected from p,p'-DDT, eight from p,p'-DDD, and one from p,p'-DDE, many being common.

Ecobichon and Saschenbrecker (1968) concluded from a pharmacodynamic study of DDT in cockerels, that there were two major routes of metabolism of DDT, the initial change being either to DDE or to DDD. Studies of grackles (Walley *et al.*, 1966) and chickens (McCulley, 1968) resulted in the same conclusion. Oral treatment resulted in a greater overall conversion of p,p'-DDT to p,p'-DDD than did intraperitoneal or intramuscular injection, and modified the metabolite ratios.

The o,p'-DDT in technical DDT appeared to be broken down into p,p'-DDT and then to DDE in living tissues of pigeons. In the anaerobic conditions after death, the o,p'-DDT was metabolised to DDD (French and Jefferies, 1969). Post-mortem degradation of p,p'-DDT to p,p'-DDD also

Birds that die in captivity during treatment with high dosages and birds that are sacrificed will usually have high residues in all their tissues. Comparable residues in carcass, liver, or other tissues of wild birds, of course, indicate hazardous exposure and the likelihood of loss of part of the population. The residues in the liver, in particular, are closely correlated with a regular or recent dose.

Birds exposed substantially, but sublethally, may survive for some time, continuously depleting their body residues, but then, under some stress, mobilize sufficient chemical to kill them. This will be discussed in the next section.

7.3.2.2 *Lethal mobilisation*

Organochlorine pesticides are stored in body fat and this may provide a mechanism for preventing immediate toxic effects. For example, coturnix quail that had lost weight, and so contained less fat, were considerably more susceptible to DDT poisoning than were other birds of the same group that were at full weight (Gish and Chura, 1970). Wild robins are most susceptible to DDT poisoning in spring, when they are undergoing weight loss (Hunt, 1969). The body condition of woodcock was a critical factor in their susceptibility to pesticide poisoning (Stickel *et al.*, 1965a).

If time permits, a chemical will be mobilised gradually from the body and eliminated at the normal rates governed by the basic physiological processes. Any events that speed this mobilisation may kill apparently healthy animals long after exposure to the chemical, or at dosages that would not normally kill them.

Chemical poisoning was caused by loss of weight after discontinuation of dosage of house sparrows with DDT (Bernard, 1963), of turkeys fed aldrin (Anderson *et al.*, 1951), and of woodcock fed heptachlor-dosed earthworms (Stickel *et al.*, 1965b). Stickel *et al.* (1966b) reported DDT-induced mortality of cowbirds after dosage had ceased, but it was not known whether weight loss was involved. In another study of cowbirds, the stress of disturbance was sufficient to cause death. Only two of 50 birds died during 8 weeks' dietary dosage of 40 ppm of DDT, after which a clean diet was restored. Additional mortality, and tremors, occurred 1, 8, 11, 15 (three birds), and 28 days post-treatment; six of the seven deaths followed the disturbance caused by workers entering the cage, or the one adjacent, for the purpose of catching birds (Stickel, 1965). When coturnix quail on a dosage of 25 ppm of DDT for 6 months without any mortality were induced to cease breeding and go into moult by reducing the day-length, 5 of 12 males and 3 of 11 females died, but no mortality occurred among control birds (Stickel and Rhodes, 1970). In a similar study of coturnix quail fed with 10 ppm of dieldrin, two of three males and one of three females died after moulting and no deaths occurred among undosed birds (Stickel *et al.*, 1969).

The concentration of organochlorine pesticides in the blood of female eider ducks increased by a factor of about 20 in the course of the incubation period (Koeman, 1971). This was related to the fact that the female eiders feed little during this period. Mortality of adults was evidently the result of mobilization of Telodrin® and dieldrin to the point that critical levels were reached.

Less food greatly increased the mortality of pheasants that received DDT in their diet (Hunt, 1966). In a group of chickens completely deprived of food, so that all birds were starved until they died, after 12 weeks of dietary dosage of dieldrin, birds fed 20 ppm died sooner than those fed 5 ppm or less, and those fed 20 or 10 ppm showed signs of poisoning during the fasting period (Davison et al., 1971). Seifert et al. (1968) also found that chickens, fed diets containing 20, 40, or 80 ppm of dieldrin, died sooner than control birds during prolonged fasting. Ecobichon and Saschenbrecker (1969) reported mortality of cockerels fed smaller rations after DDT dosage, and related the tissue residues to mortality.

Under sublethal conditions of continuous exposure, residues in the different tissues are, ordinarily, directly correlated with each other, as shown by Dindal and Peterle (1968) for DDT and its metabolites, and by Robinson et al. (1967b), for dieldrin, as well as in other studies. Lethal mobilisation disrupts these normal relationships. For example, residues of dieldrin in brain and liver of song thrushes that would have been 7.8 and 3.1 ppm, rose to 17.9 and 16.9 ppm respectively, during lethal mobilisation (Jefferies and Davis, 1968).

Lethal mobilization of DDT and the kinetic processes involved have been shown in an experiment with cowbirds (Van Velzen et al., 1972). When less food was given 2 days after dosage ceased, and again 4 months later, both treatments resulted in mortality of dosed birds, yet no mortality occurred among control birds subjected to the same food restriction. Brains of DDT-dosed birds, that were sacrificed immediately before food restriction, had concentrations of DDT and DDD that were far below the lethal range (DDT, 1.3 ppm, DDD 3.2 ppm). Seven of 20 birds died during 4 days' food restriction (43% of normal intake). Residues of DDT in brains of birds that died were 23 ppm and those of DDD were 30 ppm, showing the very rapid and pronounced mobilisation of chemical and its deposition in brains. The total body burden of DDT+DDD+DDE was reduced 16% in birds that died, by comparison with those in the pre-weight loss sample. The birds that died also lost 21% of their weight and 81% of their fat, showing that loss of residue fell proportionately far short of loss of fat.

7.4 PESTICIDES, PHYSIOLOGY, AND POPULATIONS OF BIRDS

7.4.1 The thin eggshell problem
7.4.1.1 *Ecological studies*
In 1967, a classic paper, 'Decrease in Eggshell Weight in Certain Birds of Prey', by Derek Ratcliffe, revolutionized thought and introduced a new era

of research into the problem of the effects of organochlorine chemicals on wild birds. In this paper, Ratcliffe reported a synchronous, rapid, and widespread decline in weight and thickness of shells of eggs laid by British peregrine falcons and sparrowhawks. The decline occurred in the mid-forties, and suggested the possibility of a causal chain between this phenomenon, the frequency of egg breakage, subsequent status of breeding population, and exposure to persistent organic pesticides.

The full details of the phenomenon and a scholarly consideration of other possible factors were set forth in a subsequent paper (Ratcliffe, 1970). There, he also described the observations and events that led to his study of eggshell thickness.

'Before the late 1940's it was a rare event for broken eggs to be found in the nests of peregrine (*Falco peregrinus*) and sparrowhawk (*Accipiter nisus*), but from 1951 onwards, both species have shown a high frequency of clutch depletions in which eggs were either found broken, or disappeared in circumstances where human interference was not indicated as the cause. In most instances it appeared that the parent birds destroyed their own eggs, either by èating or by breaking and/or ejecting them' (Ratcliffe, 1970).

These observations, along with increased egg breakage by the Scottish golden eagle, in association with reduced reproductive success, suggested a possible relationship between these phenomena and organochlorine pesticide contamination of birds and their eggs (Lockie and Ratcliffe, 1964).

In Great Britain, shell thickness (as expressed by a defined thickness index) decreased significantly among nine of 17 species; decreases were 5-19% (Ratcliffe, 1970). The peregrine showed a 19% change; the sparrowhawk, 17%; the merlin, 13%; the shag, 12%; and the golden eagle in West Scotland, 10%. Declines of 5% occurred in the shell thickness of eggs of the kestrel, hobby, carrion crow, and rook. No thinning was found in eggshells of the common buzzard, raven, guillemot, razorbill, kittiwake, black-headed gull, golden plover, or greenshank.

Eggshell-thinning was not confined to birds in Great Britain, for in 1968, Hickey and Anderson reported eggshell-thinning of 18-26% in regional populations of three species of raptorial birds that had declined catastrophically in the United States. The period of decline coincided with that in Great Britain. Shell thickness of eggs laid by herring gulls in five colonies from Maine to Minnesota declined consistently as the DDE content increased. A graph of shell thickness plotted against the logarithms of the residue concentrations showed a linear relationship.

Measurements of the shell thickness of United States birds were extended to include more than 23,000 eggs of 25 species (Anderson and Hickey, 1970). Some degree of shell thinning was found among 22 species representing seven orders of birds. Nine of the species sustained shell thickness and shell weight decreases of 20% or more, at least for brief periods: the peregrine falcon in at least three regions, the marsh hawk and brown pelican in two regions, and the prairie falcon, Cooper's hawk, double-crested cormorant, black-crowned night heron, bald eagle, and osprey in at least one region each. They reached 15-19% in Ontario common loons, 14-16% in some white pelicans, 7-9% in some great blue herons, 8-12% in California goshawks, 9-13% in some sharp-shinned hawks, and 10% in Great Lakes herring gulls. Changes generally under 10% were observed in red-tailed hawks, red-shouldered hawks, great horned owls, and common crows.

These eggshell changes did not occur in broad-winged hawks and rough-legged hawks. Whooping cranes similarly showed no change in shell thickness in old and modern samples (Anderson and Hickey, 1970; Anderson and Kreitzer, 1971). No shell thickness changes were apparent in mourning doves (Kreitzer, 1971).

Eggs, from 34 clutches of prairie falcons collected in Colorado and Wyoming in 1967-1968, had shells 14% thinner than eggs of 32 clutches collected in the same general region between 1909 and 1940 (Enderson and Berger, 1970). DDE in the recently collected eggs averaged 27.5 ppm and dieldrin averaged 1.9 ppm.

In the Upper Yukon Valley, in the taiga area of Alaska, 14 peregrine falcon eggs collected after 1946 had shells 17% thinner than 20 eggs collected earlier (Cade and Fyfe, 1970; Cade et al., 1971).

Twenty-three peregrine eggs collected along the Colville River in the Alaskan tundra area in 1967-1969, had shells 22% thinner than 18 eggs collected before 1947 (Cade et al., 1971). Eleven eggs from the Aleutians had shells 7.5% thinner than 30 eggs collected before 1947. Tundra, peregrine eggs contained an average of 889 ppm of DDE (lipid base); taiga, peregrine eggs contained 673 ppm and Aleutian peregrine eggs contained 167 ppm. Tundra and taiga peregrines had fledged progressively fewer young each year since 1966.

In contrast, the shell thickness of eggs of Alaskan rough-legged hawks had decreased 3.3%, and DDE residues averaged 22 ppm. The shell thickness of eggs of gyrfalcons had not decreased and DDE residues in them averaged 4 ppm. There has been no apparent decrease in numbers of gyrfalcons and rough-legged hawks in northern Alaska since the first studies in the 1950's. Peregrine eggs with higher DDE residues had thinner shells.

Eggs of ungava peregrines, collected in 1967 and 1970, were about 20% thinner than museum eggs collected in the eastern Arctic from 1900 to 1940 (Berger et al., 1970). Although the recent samples included mainly cracked or broken eggs and some with well-developed dead embryos, it was apparent that this bias was insufficient to discount the degree of thinning that was observed. Reproductive success was poor in the 15 eyries that were studied. Ten eggs analysed for DDE contained 12.7±8.7 ppm of this compound (other chemicals not reported).

In southern Saskatchewan and Alberta in 1966-1968, the shell thickness of prairie falcon eggs was less in areas where eggs contained higher concentrations of organochlorine residues; reproductive success also was poorer in these areas (Fyfe *et al.,* 1969). Organochlorine pesticide residues averaged 8.45 and 8.13 ppm in the two areas of poorest reproductive success, 5.16 and 2.09 ppm in the two areas of greatest success. Residues were primarily DDE, but also included dieldrin, heptachlor epoxide, DDD, and DDT. Eggs laid by great blue herons in Alberta had thinner shells and higher DDE residues in two samples from Chip Lake than in 38 samples from other areas (Vermeer and Reynolds, 1970).

Grey herons nesting in the Troy heronry, Lincolnshire, England, in 1966-1968, laid eggs with shells 20% thinner than in 1931-1935 (Prestt, 1969). In addition, 35-40% of the birds destroyed their own eggs. Persistent re-nesting resulted in production of young by 70-78% of the pairs despite egg destruction. The eggs contained an average of about 6 ppm of DDE, 3 ppm of dieldrin, and 6 ppm of PCB's. Individual eggs contained considerably higher residues.

Livers of herons from this and other areas contained residues of both DDE and dieldrin that were considerably higher than recorded in other species. Overall populations appeared not to be affected, paralleling the situation of the herring gull in the United States (Keith, 1966). Eggs laid by herring gulls in colonies in north-western Lake Michigan in 1964, had exceptional embryonic mortality. In 115 nests, 30-35.7% of the embryos died. An unusual feature of many dead eggs and one live egg was that large portions, commonly one-third, of the shell had been fractured into small sections and chipped off. Despite obvious reproductive problems, the herring gull populations in the area were expanding. It is evident that both herring gulls and grey herons, by several population mechanisms, are able to compensate for reproductive impairment although building up high residues of organochlorines (Prestt, 1969; Keith, 1966).

Double-crested cormorants in the Prairie States and Provinces in 1965 laid eggs with shells 8.3% thinner than in the pre-1940 period (Anderson *et al.,* 1969). Shell thickness decreased as residues of DDE and PCB's increased. In 29 first nests, correlations with DDE residues were closer, but in 6 second-nestings, correlations with PCB's were closer. DDE residues averaged 11 ppm in the first nests, 7 ppm in re-nests. PCB residues were estimated at 9 ppm in first nests and 5 ppm in re-nests. The correlation between DDE and PCB was statistically significant, suggesting similar contamination patterns for the two compounds; both would then naturally correlate with shell changes, even if only one had an effect. The colony at Lake DuBay, Wisconsin, where eggshell thickness was 25% lower than in the pre-1940 period, and where DDE residues averaged 45 ppm and PCB's 28 ppm, had failed completely by 1967. The status of three Manitoba colonies, that showed pronounced shell thinning and high residues, was not known. Four apparently stable colonies had lowest residues and least shell-thinning.

White pelicans, from the same areas, laid eggs with shells 4.5% thinner than in the pre-1940 period. DDE residues in these eggs averaged 1.7 ppm and PCB's averaged an estimated 0.5 ppm. Shell thickness decreased as

DDE increased, but there was no apparent correlation with PCB's. The population level in these pelicans apparently was steady. Eggs of the common murre from the Farallon Islands, California, collected in 1968 and 1970, were 13% thinner than eggs collected in 1913. Lipid extracts of the recent eggs contained DDE residues that averaged 297 ppm and PCB's 168 ppm (Gress et al., 1971).

The brown pelicans of Anacapa Island, California, in 1969, laid eggs with shells so thin that the eggs collapsed. In a colony of 300 nests visited in April, only 12 nests contained intact eggs (Keith et al., 1970). In a second colony observed one month later, adults were apparently incubating normally, but only 19 of 339 nests contained intact eggs. Of the remaining 320 nests, about two-thirds were empty and one-third had a collapsed, dehydrated egg in or near the nest. Shells of nine intact eggs measured 0.38 mm, 34% thinner than the pre-1940 norm. Shells of an additional 10 eggs collected in April also averaged 0.38 mm thick. They contained 50.6-135 ppm of DDE (mean 75.6 ppm) (Lamont et al., 1970). Average concentrations of other chemicals were p,p'-DDD, 1.4 ppm; p,p'-DDT, 2.0; dieldrin, 0.12; o,p'-DDT, 0.14; o,p'-DDD, 0.12. PCB's averaged 5 ppm, ranging from <0.07 to 12 ppm.

Brown pelicans of the Eastern United States, also laid eggs with abnormally thin shells (Blus, 1970). In the South Carolina colonies, where populations had declined from more than 5,000 pairs in the early 1960's to little more than 1,200 pairs in 1969, shells were 17% thinner than the pre-1947 norm. Brown pelicans nesting on Florida's Atlantic coast laid eggs with shells 9% thinner than before 1947 and those on the Gulf Coast 6% thinner. Populations of brown pelicans in Florida were not in obvious difficulty, although no population estimates were available for the earlier years.

Shell thinning of brown pelican eggs was closely correlated with DDE content (Blus et al., 1971). The correlation was closer for DDE than for any of the other chemicals considered, including DDD, DDT, dieldrin, PCB's, and mercury. Residues of all but mercury were correlated with those of DDE, so that the concentrations of these other chemicals also increased as shell thickness decreased. The correlation coefficient for DDE, however, was higher than for the other chemicals. A more severe test, using multiple regression statistics, showed that DDE was the only chemical that accounted for a significant percentage of the variability in shell thickness, and indicated DDE as the critical chemical.

The shell thickness of brown pelican eggs decreased as the DDE concentration increased. The relationship was linear when the logarithms of the DDE concentrations were plotted against the shell thickness (Blus et al., 1972). The percentage change in shell thickness was greater per unit of DDE when the concentration of DDE was lower, showing that lower concentrations were proportionately more effective. For example, the calculated percentage of thinning, per ppm of DDE, was 4.2 at 1 ppm, 3.0 at 5 ppm, 1.9 at 10 ppm, and 0.4 at 100 ppm. The susceptibility of the brown pelican to shell-thinning appeared to be similar to that of the prairie falcon, and greater than that of the double-crested cormorant or herring gull.

The common tern at Chip Lake, Alberta, showed no correlative relationship between shell thickness and DDE content of individual eggs (Switzer *et al.*, 1971). Shell thickness varied from 0.147-0.220 mm (mean, 0.185 mm) and DDE content ranged from 0.64-104.0 ppm (mean 7.57; only two eggs significantly above 22 ppm). PCB's were present in roughly a 1:9 ratio to DDE. Terns in this particular colony were reproducing at a rate inadequate for maintenance of the colony. Egg disappearance accounted for 58% of the clutch failures; egg fracture, 16%; nest abandonment, 18%; and miscellaneous factors, 8%. Historical data concerning shell thickness or population status apparently were not available.

Shell thickness (thickness index) of eggs laid by the Sandwich tern and the eider duck in the Netherlands, was 5% lower in 1964-1970 than in the period before 1950 (Koeman, 1971). There were no indications that the eggshell changes coincided with reduction in reproductive potential.

Bald eagle eggshells collected in 1968-1970 were significantly thinner than pre-1946 norms (Krantz *et al.*, 1970; Wiemeyer *et al.*, 1972). The percentage change was 10-11% in shells of seven eggs from Kodiak, Alaska, 14 eggs from the Great Lakes States, six from Florida, and four from Maine. Three eggs from the Admiralty Island area, Alaska, had shells not significantly different in thickness from the earlier norm. The eggs contained DDE (and smaller amounts of DDT or DDD), dieldrin, heptachlor epoxide, and PCB's, in amounts that differed greatly in the different areas. Smallest quantities were in eggs from Alaska. Clearcut relationships between degree of shell-thinning and residue content were not apparent.

7.4.1.2 *Experimental studies*

The discovery of shell-thinning among natural populations of wild species, and the hypothesis that this thinning was related to the environmental occurrence of organochlorine pesticides (Ratcliffe, 1967; Hickey and Anderson, 1968), stimulated experimental studies to determine whether a cause-and-effect relationship existed.

The first clearcut experimental demonstration, that DDE caused thin eggshells, was provided in studies of mallard ducks (Heath *et al.*, 1969). Dietary dosages were given a few weeks before the first breeding season and continued through the second season, as follows: DDE at 10 and 40 ppm, DDD at 10 and 40 ppm, DDT at 2.5, 10, and 40 ppm (reduced to 25 ppm in the second season) (dry weight). DDE at 40 ppm reduced shell thickness by 13.2% in the first season, 13.5% in the second; produced 19% and 25% cracked eggs (as compared with 5% among controls); and resulted in 55% and 34% of the number of control ducklings. At 10 ppm of DDE, shells were not significantly thinner than controls in the first season, but were 10.8% thinner in the second. Ten percent of the eggs were cracked in the first year, 24% in the second; 66% of the number of ducklings as in the controls were produced the first year, 52% the second. Since only sound eggs were measured, the recorded thickness changes are minimal.

DDD, by contrast, produced no significant shell-thinning or egg cracking and only a marginally significant reduction in duckling production. DDT at 2.5 and 10 ppm had no significant effects on any of the measurements in either year. The 40 ppm dosage (reduced to 25 ppm) resulted in 13.2% shell-thinning in the second year, an increase in egg crackage and a reduction in duckling production. Duckling survival was impaired in both years.

Black ducks proved to be even more susceptible to DDE-induced shell-thinning than were mallards (Longcore et al., 1971b). Diets containing DDE at 10 and 30 ppm (dry weight, the wet weight would be about one-third of these amounts) decreased shell thickness by 17.6% and 23.5% in the standard measurements made at the equator. Thicknesses of the cap and apex were reduced even more, and 25% of the cracked eggshells from dosed hens actually were collapsed. Numbers of cracked eggs reached 10% among ducks fed 10 ppm of DDE, and 21% among those fed 30 ppm.

The number of eggs laid per hen did not differ between dosed birds and controls. However, the survival of ducklings to 3 weeks of age was 30% (10 ppm DDE) and 55% (30 ppm DDE) poorer than among controls, even though the young birds received only untreated food. Residues of DDE in eggs averaged 46 ppm and 144 ppm in the two groups. By contrast with DDE, dieldrin at 10 ppm in the diet of mallard ducks, produced only about 4.5% shell-thinning in comparison with controls, and a lesser degree of thinning at 1.6 or 4 ppm (Lehner and Egbert, 1969). American kestrels, fed diets containing p,p'-DDT and dieldrin in combination, laid eggs with shells thinner than normal and fledged fewer young (Porter and Wiemeyer, 1969). There were two dietary dosages, the lower containing 5 ppm of DDT and 1 ppm of dieldrin (dry weight), and the higher containing 15 ppm of DDT and 3 ppm of dieldrin. Wet-weight equivalents were about one-third of these. Shells of eggs laid by first-generation birds were 8% and 10% thinner than controls; shells of eggs laid by their offspring (continued on the diet of the parents) were 15% and 17% thinner than controls. In another study, American kestrels fed diets containing DDE at 10 ppm (dry weight) laid eggs with shells 9.7% thinner than those laid prior to dosage (Wiemeyer and Porter, 1970). Controls laid eggs 2.1% thinner the second year than the first, a difference that was not statistically significant.

Screech owls, fed a dietary dosage of 10 ppm (dry weight) of DDE, laid eggs with shells 13% thinner than controls and 12% thinner than they themselves had laid in the season prior to dosage (McLane and Hall, 1972).

By contrast to the results with ducks and birds of prey, experiments with gallinaceous birds suggest that they may be considerably less susceptible to shell-thinning by chemicals of the DDT group. Coturnix quail fed 2.5, 10, or 25 ppm of p,p'-DDT in a normal diet containing 3.5% calcium for 26 weeks, laid eggs with shells 6.0%, 6.4%, and 7.3% thinner than

Ring doves were injected intraperitoneally with 150 mg/kg of p,p'-DDE or 30 mg/kg of dieldrin, within a day of the laying of their first egg (Peakall, 1970b). This caused eggshell weights to be reduced 23% by DDE but to be unaffected by dieldrin. Carbonic anhydrase activity was reduced 59% in the oviducts of birds receiving DDE but was unaffected in those treated with dieldrin. The oviducts of birds given DDE contained an average of 77 ppm of DDE and the eggs contained 17 ppm. Oviducts of dieldrin-dosed birds contained 5 ppm of dieldrin and the eggs 6 ppm.

Coturnix quail, fed 100 ppm of p,p'-DDT or p,p'-DDE in their diet for 3 months, and killed prior to deposition of their first eggs, showed 16% and 19% decreases respectively, in carbonic anhydrase activity in their shell glands, 22% and 44% in their blood (Bitman et al., 1970). Eggshell calcium was reduced by 8% in both groups. The eggs of birds fed DDT contained 196 ppm of DDT and 48 ppm of DDE. Those of birds fed DDE contained 196 ppm of DDE.

An effect of p,p'-DDT on the avian thyroid has been postulated as a mechanism for production of thin eggshells (Jefferies, 1969; Jefferies and French, 1969, 1971). Experiments with homing pigeons showed that p,p'-DDT caused an increase in the thyroid weight and a decrease in colloid content of the follicles (Jefferies and French, 1969). The birds were fed capsule doses of p,p'-DDT every other day for 42 days, at rates of 18, 36, and 72 mg/kg in each dose. The highest dose approached lethality, because tremors were observed.

Bengalese finches, fed p,p'-DDT at various doses, laid lighter eggs than control birds (Jefferies, 1969). Eggs that weighed less had proportionately heavier shells. This was true for both dosed and undosed groups, considered separately. The heaviness of the shells of the eggs laid by the dosed birds, however, was greater than mathematically predicted on the basis of the egg size alone. This increase in thickness suggested the possibility of an induced hyperthyroidism among the DDT-dosed birds, and was supported by the fact that heart weight increased and weight gains decreased, both suggestive of hyperthyroidism.

A further exploration of the effects of p,p'-DDT dosage on the thyroid of homing pigeons, showed an increase in thyroid weight and reduction of colloid content of the follicles that was similar at dosage levels from 3-54 mg/kg/day (Jefferies and French, 1971). However, examination of body temperatures, oxygen consumption, and vitamin A storage showed that pigeons fed low dose rates were in a hyperthyroidal condition, whereas those fed high dose rates were in a hypothyroidal condition. As hypothyroidism is known to result in thin-shelled eggs, the hypothyroidism was suggested as an explanation for the occurrence of thin eggshells among wild birds. A further examination of effects of DDT dosage, on beating rate, amplitude, and weight of the heart of homing pigeons and Bengalese finches, supported the likelihood of both hypo- and hyperthyroidism among pigeons, and hyperthyroidism only among Bengalese finches (Jefferies et al., 1971).

Premature egg extrusion was not induced in coturnix quail fed with either DDT or DDE (Cecil *et al.,* 1971). All birds were very regular in the lengths of time between successive ovipositions.

Various physiological effects on birds, including enzyme induction, have been shown experimentally. Some of these may be directly or indirectly related to eggshell-thinning, but others may be independent.

Pigeons fed with oestradiol and DDE in their diets had medullary bone formation reduced 48% as compared with birds given only oestradiol, suggesting a breakdown of oestradiol due to increased enzyme induction (Oestreicher *et al.,* 1971). This finding was consistent with a reduced level of circulating oestradiol that was found in ring doves given p,p'-DDT, as well as a decreased amount of calcium in their tibiae and femurs, and increased hepatic enzyme activity (Peakall, 1970b).

In an experiment with zebra finches, uptake of calcium from the gut was not affected by DDT, indicating that the absorption mechanism was not altered (Peakall, 1969). Ring doves fed 10 ppm of *p,p'*-DDT showed no differences in breakdown of vitamin D although metabolism of oestradiol was greatly increased.

Steroid metabolism was increased in pigeons by feeding them DDT (10 ppm) or dieldrin (2 ppm) for a week, indicating stimulation of enzyme activity (Peakall, 1967). The chromatographic patterns showed that the metabolites formed were different, and suggested that different enzyme systems were involved. The increase in metabolism caused by dieldrin was slightly greater than that by DDT, and the effect of the two together was essentially additive.

Similarly, metabolism of oestradiol was increased in pigeons by intramuscular injection of DDE, DDT, and PCB's (Risebrough *et al.,* 1968). Gas chromatographic profiles, obtained after enzyme induction using DDE and DDT, were essentially the same, but the profile produced by PCB's (Aroclor 1262) was different, indicating that a different metabolite was formed. PCB's, injected into pigeons at 20 mg/kg, increased metabolism much more than DDT or DDE at 40 mg/kg. The effect of DDT was somewhat greater than that of DDE.

Wild European kestrels were shown to contain liver enzymes capable of metabolising certain drugs (Wit, 1969). Significant induction of hepatic microsomal ethylmorphine N-demethylase activity occurred among American kestrels fed DDT combined with dieldrin in their diet (Gillett *et al.,* 1970).

Dieldrin, fed at 5 ppm to week-old coturnix quail chicks without killing them, increased aldrin epoxidase activity of hepatic microsomes, but did not affect cytochrome P-450 or NADPH-neotebrazolium reductase levels (Gillett and Arscott, 1969; Gillett *et al.,* 1970). Epoxidase activity declined with age, whereas P-450 levels increased in maturing chicks. DDT, fed to mature quail at 100 ppm for several months, caused depression of their hepatic microsomal activities and also in those of pheasants from a DDT-treated area. DDT and DDE, fed at 100 ppm to coturnix quail for 21 days, increased both their aniline oxidase activity and cytochrome P-450; DDE produced the greater increases (Bunyan *et al.,* 1970b).

A contrast in enzyme-stimulating activity of DDT and DDE was found in coturnix quail fed for 28 days on diets containing either of these compounds at 5, 20, 50, or 100 ppm (Bunyan *et al.,* 1970a). DDT increased hepatic glucose-6-phosphate dehydrogenase (G-6-P) and decreased 6-phosphogluconate dehydrogenase (6-P-G) levels, whereas DDE decreased G-6-P and had little effect on 6-P-G. Tests also were made with other analogues, and led to the postulate that the effects were due to interference in protein metabolism primarily by the unsaturated analogues and metabolites.

Experiments with domestic chickens showed that pretreatment with a combination of DDT and DDE increased phenobarbital-induced sedation time, suggesting an inhibition of microsomal enzymes (Stephen *et al.,* 1971). Dieldrin decreased sedation time in young chicks but did not alter it in adult birds. Limited tests with young birds suggested that *o,p'*-DDT and *p,p'*-DDE increased their sleeping time, and that lindane, chlordane, and toxaphene decreased their sleeping time. Hens fed 20 ppm of DDT and 20 ppm of DDE combined, for 39 days, laid eggs that showed no significant change in shell calcium (expressed as percentage of egg weight), either with or without the addition of vitamin D.

In another study with adult chickens, liver aniline hydroxylase activity was reduced by diets containing 100 or 200 ppm of *p,p'*-DDT for 12 weeks (Sell *et al.,* 1971). Neither N-demethylase activity nor cytochrome P-450 was affected. Dieldrin fed at 20 ppm had no effect on aniline hydroxylase activity but increased demethylase activity and cytochrome P-450. However, the ability of chick microsomal enzymes to dealkylate some carbamate insecticides was found to be increased in chicks hatched from eggs injected with DDT or its metabolites (Abou-Donia and Menzel, 1968a). DDT produced the greatest increase in enzyme action (56%), DDD next (34%), and DDE least (19%).

A significant aspect of pesticide-enzyme interactions is the potential of one compound to affect the storage of another. However, Street *et al.* (1966) found no dieldrin-DDT storage interactions in chickens, despite its occurrence in rats, swine, and sheep.

Estrogenic activity of *o,p'*-DDT has been demonstrated in both chickens and coturnix quail (Bitman *et al.,* 1968). For instance, *o,p'*-DDT produced the same effects as estradiol in the oviducts of both, with 100% increase in oviduct weight and 150-175% increase in glycogen content. Little, if any, estrogenic activity was shown by *p,p'*-DDT. Chlordecone, fed at a high dosage (300 ppm) in the diet, had an estrogenic effect on immature female quail, but not on laying hens (McFarland and Lacy, 1969), and the testes of the males were affected also.

Mallard drakes fed DDT at 2.5, 25, or 250 ppm in their diet, showed no alteration in pre-established peck-order positions (Nauman, 1969). There were no indications of changes in adrenal weight, percentage of cortex in the adrenal, or plasma corticosterone levels. However, thyroid weight increased consistently as DDT dosage level increased. Diets containing 2.5 and 25 ppm of DDT caused a decrease in the percentage of ^{131}I uptake by the thyroid gland, whereas those containing 250 ppm caused a significant increase.

Injections of duck hepatitis virus (DHV) into mallard drakes, decreased their hepatic microsomal mixed-function oxidase activity (tested by N-demethylation of ethylmorphine), and DDT increased it (Ragland et al., 1971). Injection of DHV, prior to exposure to DDT, did not prevent this enzyme stimulation and may have enhanced it. Previous studies had shown that ducks exposed to DHV, 48 hours prior to DDT, were less susceptible to DDT intoxication than mallards treated with DDT alone (Friend, 1971).

7.4.3   Other reproductive effects

7.4.3.1   Laboratory studies

Feeding Hungarian partridge with 3 ppm of dieldrin in their diet, for 90 days during the breeding season, decreased egg production and lowered hatchability, due to 'dead in shell' embryos (Neill et al., 1969). Parathion (8 ppm) in food lowered hatchability markedly. Fertility, chick survival, and growth were not altered in either group.

Bengalese finches, fed DDT in amounts approximating 8, 32, and 274 ppm, or DDE in amounts approximating 4, 38, and 191 ppm in the total diet, had reduced fertility, hatchability, and fledging success (Jefferies, 1971). There were longer periods prior to ovulation, longer incubation and rearing periods, lighter eggs and smaller newly hatched young. The phenomenon of delayed ovulation had been shown earlier in the same species (Jefferies, 1967).

Only very large doses of DDT affect the reproduction of chickens. Rubin et al. (1947) found that 310 ppm of DDT in the diet of hens lowered egg production; 620 ppm and 1,250 ppm lowered both egg production and hatchability; 2,500 ppm killed some adults. Residues in the eggs were 360 and 320 ppm at the two higher doses.

Hens fed 20 ppm of DDT in their diet for 10 weeks showed no effects on egg production, hatchability, or survival of the chicks, and those fed 200 ppm showed only slightly reduced chick survival (Weihe, 1967). A dietary dosage of 1,000 ppm resulted in toxic symptoms in the hens, reduced egg production and hatchability, and tremor and mortality of 80-95% of the chicks in the first week after treatment. Hens fed dieldrin at dietary dosages of up to 5 ppm showed no effects on production or hatchability of eggs and there was no chick mortality, even with egg residues of 4-5 ppm (Graves et al., 1969).

Hens, fed chlordecone at 75 and 150 ppm in their diet for 16 weeks, laid fewer eggs than control birds, and hatchability of the eggs laid by those birds on the higher dosage was reduced (Naber and Ware, 1965). Hens fed at the higher dosage lost weight. Chicks exhibited a nervous syndrome characterized by quivering extremities and inability to walk or stand, and their survival was decreased. In the fifth week, residues in egg yolk were 163 ppm in eggs laid by hens fed 75 ppm, and 336 ppm in those laid by hens fed 150 ppm, and did not increase significantly thereafter. After 3 weeks of untreated food, residues were 26 and 70 ppm respectively, in the two groups.

Hens fed 300 and 600 ppm of mirex produced normal numbers of eggs, but their hatchability was reduced and survival was less in the higher dosage group. Hens on the higher dosage lost weight. Residues levelled off at least by the fifth week, when concentrations in the yolk were 620 ppm for the lower dosage group and 1,905 ppm for the higher group. After 3 weeks of untreated food, residues were 193 and 500 ppm in the two groups.

Pheasants are also resistant to many of the organochlorine pesticides. Reproductive effects were produced by 100 and 400 ppm of DDT, 25 and 50 ppm of dieldrin, and 100 and 300 ppm of camphechlor in their diet (Genelly and Rudd, 1956). Similar results were obtained by Azevedo et al. (1965) in studies with DDT. Adverse effects were pronounced only at near-lethal doses. Pheasant hens dosed with dieldrin once each week in capsules at rates of 2, 4, or 6 mg per week for 12 weeks, showed no reproductive effects beyond producing fewer eggs at the highest dosage, which also resulted in lowered food consumption (Atkins and Linder, 1967). Residues in eggs ranged from 8-52 ppm after 11 weeks.

However, in a second generation of pheasants whose parents had received 4 or 6 mg of dieldrin per week for 13 weeks, fertility and hatchability of eggs were significantly less than control birds, even for those that received no dieldrin except through the egg (Baxter et al., 1969). In visual cliff tests, chicks that hatched from eggs laid by hens receiving 8 mg doses of dieldrin, tended to jump more frequently off the deep side, or to make no choice, than did control chicks. Pheasants that were fed BHC with their diets at an average daily rate of 87 ppm, reached peak egg production later than control birds and produced somewhat fewer eggs, although hatchability of their eggs was not altered (Ash and Taylor, 1964).

Pheasant eggs, collected from areas heavily treated with insecticides in California, produced more crippled chicks than did eggs from untreated areas, and mortality of young birds was also higher in the samples from the treated areas (Hunt and Keith, 1962). Residues of DDT and its metabolites in egg yolks were in the order of hundreds of ppm, whereas those of dieldrin were 0-1.3 ppm. Residues in the fat of hens were correspondingly high, up to thousands of ppm of DDT and metabolites, and 0.1-25 ppm of dieldrin.

Coturnix quail, 5 weeks old, that were fed 100 or 200 ppm of DDT in the diet for 60 days, did not differ from control birds in mortality, egg production, fertility, or hatchability (Smith et al., 1969). A dosage of 400 ppm resulted in 50% mortality within 30 days, but none during the remaining 30 days, although fertility and hatchability were reduced.

Effects of DDT on spermatogenesis in chickens (Albert, 1962) and eagles (Locke *et al.*, 1966) have been produced only at doses that were in the lethal range for the adults. High dosages of injected DDT also reduced secondary sex characters and a change in the testes size of cockerels (Burlington and Lindeman, 1950).

7.4.3.2   Field observations

The eggs of purple gallinules that nested in Louisiana rice fields in 1966, contained dieldrin residues averaging 6.5 ppm (0.5-15.4 ppm), and those of common gallinules contained 9.4 ppm (1.1-22.1) (Causey *et al.*, 1968). Only one of fourteen egg samples from an untreated area had dieldrin (0.3 ppm). DDE residues averaged 0.4 ppm (0.06-5.81) in the eggs of purple gallinules, 0.3 ppm (0.12-1.05) in the eggs of common gallinules from the treated areas, and 0.10 ppm (0.05-0.20) in eggs from the control area. Neither clutch size nor hatchability differed in field-observed nests of the common gallinule in either treated or control areas.

A correlative relationship, between breeding success of woodcock and DDT-treatment of forested areas in New Brunswick, was reported by Wright (1965). Breeding success varied inversely with the amount of DDT used and with the total area sprayed. If Nova Scotia is considered as a control area, it appeared that the birds in the spray zone had a lower breeding success than those outside it.

Field observations in Wisconsin showed that whenever elms were sprayed at greater than 10 lb/a (11.2 kg/ha) of DDT, wild robin populations had over 85% spring mortality, and virtualy no reproduction occurred, mainly because of losses of adults (Hunt and Sacho, 1969). Immediately after methoxychlor was first used instead, 17 of 22 nests on a 61-acre study area were successful, with 16 producing at least 44 fledglings.

Studies over a number of years at Clear Lake, California, provided strong inferential evidence that pesticides, mainly DDD, have been responsible for reproductive failure in Western grebes (Herman *et al.*, 1969).

Shag on the Farne Islands (Northumberland, England), in 1964-1967 laid eggs containing 0.5-3.0 ppm of dieldrin and 1-9 ppm of DDE, with no obvious relationship between their residue content and eggs per clutch or chicks per brood (Potts, 1968). There may however, have been some relationship between clutch viability and dieldrin content, since six of eight clutches, in which eggs contained 2-3 ppm of dieldrin, had no chicks surviving to the tenth day, whereas at lower concentrations, only one of three clutches failed. There was no apparent relationship between DDE content and clutch failure.

Poor reproduction of golden eagles in west Scotland was correlated with the amounts of organochlorine residues ingested with their food (Lockie and Ratcliffe, 1964). Varying amounts of dieldrin (mean 1.34 ppm), DDE (mean 0.99 ppm), gamma BHC (mean 0.08 ppm) and heptachlor epoxide (mean 0.006 ppm) were present in the eggs. Comparison of

TABLE 7.2.   Organochlorine residues ppm in aquatic mammals.[1]

Name, place, and time	Sample		DDE		DDD	
	Type	No.	Mean	Range	Mean	Range
Mustelidae						
Sea Otter						
(*Enhydra lutris*)						
California, 1970	fat	7	10.3	0.39-34	0.54	0.007-1.
Otariidae						
Northern Fur Seal						
(*Callorhinus ursinus*)						
Alaska and Wash., 1968-1969	liver (wet)	23	0.68	0.04-5.1	0.14	0.05-0.47
Alaska, 1969	blubber (pups)	5	14.5	0.35-45	0.67	0.07-1.9
Alaska, 1969	milk	5	2.6	0.04-5.1	0.09	0.02-0.1
Phocidae						
Harbour Seal						
(*Phoca vitulina*)						
California, 1970	blubber	2		15-142		0.76-7.1
Canada, 1967[2]	fat	6	5.9	1.2-17.3	0.78	0.35-2.1
Canada, 1967[2]	fat	2	1.2	0.67-1.8	0.12	0.08-0.1
Netherlands, 1963	fat	3	8.9	5.4-14.0	1.8	0.7-3.6
Scotland, E., 1965-1966[2]	fat	18	5.5	1.0-11.1	1.2	0.27-3.0
Scotland, N.W., 1965-1966[2]	fat (ad)	6	3.4	1.2-7.0	0.41	0.17-0.8
Scotland, N.W., 1966[2]	fat (pups)	9	2.5	1.0-4.0	0.32	0.13-0.5
Sweden, 1968[3]	blubber (lipid)	2				
Ringed Seal						
(*Pusa hispida*)						
Sweden, 1968	blubber (lipid)	2				
Grey Seal						
(*Halichoerus grypus*)						
England, S.W., 1969	blubber	3	6.7	3.2-8.7	0.69	0.27-1.1
England, S.W., 1969	blubber	3	4.7	3.3-6.0	0.66	0.46-0.8
England, S.W., 1969	blubber	2	9.5	8.6-10.3	1.16	1.15-1.1
Sweden, 1968	liver (lipid)	1				
Sweden, 1968	blubber (lipid)	3				
Sweden, 1968	muscle (lipid)	2				
Sweden, 1968	milk (lipid)	1				
Harp Seal						
(*Pagophilus groenlandicus*)						
Canada, 1968	milk (lipid)	1		0.47		0.11
Crabeater Seal						
(*Lobodon carcinophagus*)						
Antarctica, 1964	fat	1		0.017		0.007
Weddell Seal						
(*Leptonychotes weddelli*)						
Antarctica, 1965, 1967	fat	20	0.02	0.005-0.045		–

DDT		Dieldrin		PCB		Source
Range	Mean	Range	Mean	Range		
0.015-0.87	–	–	–	–		Shaw, 1971
0.02-0.38 (18)	0.04	0.02-0.09 (3)	–	–		Anas and Wilson, 1970a
0.22-1.4	0.06	0.04-0.09 (4)	tr	tr		Anas and Wilson, 1970b
0.03-0.20	0.02	0.01-0.03 (4)	tr	tr		Anas and Wilson, 1970b
2.6-0.11	–	–	–	–		Shaw, 1971
2.1-15.6	0.07	0.03-0.10	ca 257	of DDD and 107 of DDT		Holden and Marsden, 1967
0.08-0.63	0.03	0.02-0.04	ca 257	of DDD and 107 of DDT		Holden and Marsden, 1967
3.5-9.8	1.3	0.3-2.3	–	–		Koeman and van Genderen, 1965, 1966
1.5-23.3	0.79	0.15-2.1	ca 257	of DDD and 307 of DDT		Holden and Marsden, 1967
1.7-7.7	0.20	0.08-0.39	ca 257	of DDD and 307 of DDT		Holden and Marsden, 1967
1.1-3.7	0.18	0.06-0.44	ca 257	of DDD and 307 of DDT		Holden and Marsden, 1967
110-150	–	–	30	16-43		Jensen et al., 1969
110-130	–	–	13	9.7-16		Jensen et al., 1969
1.7-5.0	0.25	0.08-0.44	160	118-187		Bonner, 1970
2.6-6.5	0.38	0.20-0.58	34	25-44		Bonner, 1970
7.2-7.6	0.57	0.54-0.59	48	46-50		Bonner, 1970
	–	–	44			Jensen et al., 1969
97-310	–	–	30	16-56		Jensen et al., 1969
41-43	–	–	6.5	6-7		Jensen et al., 1969
	–	–		4.5		Jensen et al., 1969
8		–		–		Cook and Baker, 1969
5		–		–		Sladen et al., 1966
4 0.020-0.060		–		–		Brewerton, 1969

Table 7.2 (continued)

Name, place, and time	Sample		DDE		DDD	
	Type	No.	Mean	Range	Mean	Range
Eschrichtiidae						
Grey Whale						
(*Eschrichtius gibbosus*)						
California, 1968-1969	blubber	23	0.14	0.04-0.36 (6)	0.07	0.029-0.1
Delphinidae						
Harbor Porpoise						
(*Phocoena phocoena*)						
Canada, 1969, 1970	blubber	36	81.9	15.0-181	36.5	7.2-10
Florida, 1967	blubber	2				
Scotland, Ork., 1967	fat	1		1.1		0.58
Scotland, E., 1965, 1967	fat	3	12.8	9.6-15.3	8.9	5.2-14
Physeteridae						
Sperm Whale						
(*Physeter catodon*)						
California, 1968	fat	6	3.6	0.74-6.0	0.52	0.22-0.

[1] Where PCB's are not mentioned, it is uncertain whether they were separated fully from and DDD in the analyses. In those cases, comparisons of residues in the DDT group shou based primarily on DDE. Only fat residues are tabulated, unless other tissues were the only analysed. In the Range columns, numbers in parentheses show the number of samples in whic chemical was detected, if this is different from the total number examined. A dash (−) column indicates that the chemical was not mentioned. Scientific and common names follow and Scheffer (1968).

[2] Species not separated. Includes also the gray seal and the harp seal.

[3] Species not separated. Includes also the gray seal.

[4] Includes DDE and DDD.

[5] Also 12.4 (0.1-26.7) ppm of *o,p'*-DDT.

[6] Also endrin.

sampled (kangaroo rat, *Dipodomys merriami*; and two species of pocket mice, *Perognathus penicillatus* and *P. intermedius*). In August, in the area nearest cottonfields, average residues were: 5.2 ppm of DDE, 5.8 ppm of DDD, 5.7 ppm of DDT, 5.1 ppm of parathion, 4.6 ppm of methyl parathion, and 0.5 ppm of benzene hexachloride. In June, in the same area, residues were one-half to one-fourth of these. In the least contaminated area in June, average residues of any one chemical did not exceed 0.1 ppm, and all but BHC increased to 0.2-0.3 ppm in August.

Residues in muscle tissue of rodents collected in the summer of 1968, at eight sites within the Big Bend National Park, and in one area three miles west of the park, were conspicuously higher in places nearer cotton-growing areas and at campsites (Applegate, 1970). Residue levels in soils and plants were highest in the same areas. Average residues of DDT, DDD, and DDE combined were 6.6 ppm in eight pocket mice (*Perognathus*

DDT		Dieldrin		PCB		Source
n	Range	Mean	Range	Mean	Range	
6	0.022-0.13 (6)	0.06	0.044-0.075 (4)	–	–	Wolman and Wilson, 1970
s	14.1-175	7.0	0.1-13.1	1 to 3	times DDE	Gaskin et al., 1971
	>200[4]	–	0.05-2.0[6]	–	–	Wilson, 1967
	2.2		0.59	ca 257	of DDD and 307 of DDT	Holden and Marsden, 1967
	13.1-25.7	9.9	4.9-18.0	ca 257	of DDD and 307 of DDT	Holden and Marsden, 1967
	0.86-2.6	0.018	0.016-0.019 (2)	–	–	Wolman and Wilson, 1970

penicillatus) from the three areas with highest residues, and 0.3 ppm in 11 mice from four other areas. Residues in the cotton rat (Sigmodon hispidus) and two species of ground squirrel (Citellus variegatus and C. mexicanus) were generally similar to those in the pocket mice from the same areas.

Heptachlor, aldrin, or dieldrin applied at high rates in certain control programmmes may kill mammals as well as birds. High residues of heptachlor epoxide and dieldrin were present in the tissues of mammals found dead, in areas of the southeast U.S.A. treated with dieldrin or heptachlor in the late 1950's for fire ant control (DeWitt et al., 1960). Species in this compilation included the cotton rat, whitefooted mouse (Peromyscus sp.), cottontail rabbit (Sylvilagus floridanus), raccoon (Procyon lotor), armadillo (Dasypus novemcinctus), and red fox (Vulpes fulva).

In a treatment programme employing dieldrin against Japanese beetles (Popillia japonica) in Illinois, ground squirrels (Citellus tridecemlineatus and C. franklinii), muskrats (Ondatra zibethica), and cottontail rabbits were virtually eliminated (Scott et al., 1959). Short-tailed shrews (Blarina brevicauda), fox squirrels (Sciurus niger), woodchucks (Marmota monax) and meadow voles (Microtus ochrogaster) appeared to have suffered heavy losses. Other mammals found dead, and believed to have been killed by dieldrin treatment, included opossums (Didelphis marsupialis), moles (Scalopus aquaticus), and domestic cats (Felis catus). Although some white-footed mice (Peromyscus leucopus) were killed, they did not appear to be particularly vulnerable to dieldrin poisoning and populations began to recover rapidly.

A study in Maryland also suggested that white-footed mice were relatively resistant to DDT (Stickel, 1951). Five annual treatments of a forested area at 2 lb/acre (2.25 kg/ha) of DDT resulted in no population reduction (Stickel, 1951). In another study, muskrats seemed to be relatively resistant to DDT used in marsh treatments (Wragg, 1954).

Dieldrin applications against white-fringed beetles in Tennessee, killed both birds and mammals, including cotton rats and cottontail rabbits (Stickel et al., 1969). Other examples of mortality of wild mammals due to aldrin, dieldrin, or endrin treatments have been reported (Dustman and Stickel, 1966).

Meadow voles (*Microtus pennsylvanicus*), in a Saskatchewan grassland area, declined in numbers significantly, after endrin was applied at ½ lb/acre (0.56 kg/ha) to an experimental plot (Morris, 1970). The population recovered rapidly by invasion of new individuals, who survived well. By contrast, numbers of deer mice (*Peromyscus maniculatus*), in the same area, declined abruptly after the application, and the populations did not recover during the next two years of the study.

Increased mortality of British wild foxes (*Vulpes vulpes*), in the spring of 1959 and 1960, was apparently due to insecticide poisoning caused by the foxes eating birds that had died from eating treated grain (Blackmore, 1963). Livers of 10 foxes found dead, contained dieldrin or heptachlor epoxide or both; 8 of 10 contained 2.6-13.2 (av. 9.1) ppm of dieldrin and 9 of 10 contained 9.4-90.6 (av. 27.0) ppm of heptachlor epoxide.

Wild pine mice (*Pitymys pinetorum*), from Virginia orchards, with a history of 11 years' treatment with endrin, exhibited a 12-fold greater tolerance to the pesticide than did mice from Maryland orchards with no history of endrin treatment (Webb and Horsfall, 1967).

However, cotton rats from Mississippi areas heavily treated with DDT, and those taken from untreated areas, were equally susceptible to DDT poisoning (Ferguson et al., 1964).

Often, herbivorous small mammals accumulate only very low residues even in areas of heavy pesticide use (U.S.D.A., 1969). This was shown particularly clearly by analyses of a number of species collected in agricultural areas in Mississippi, Arkansas, and Alabama. In terms of average carcass residues in an area in any given year, rabbits (*Sylvilagus* sp.), rice rats (*Oryzomys palustris*), and muskrats usually contained less than the detectable limit of total DDT or dieldrin. Fox squirrels and chipmunks (*Tamias striatus*) contained <0.1 ppm of total DDT and no detectable dieldrin. Whitefooted mice (*Peromyscus* sp.) and wood rats (*Neotoma floridana*) contained <0.5 ppm of total DDT or dieldrin. Cotton rats contained <0.5 ppm with one exception of 1.3 ppm of total DDT. Harvest mice (*Reithrodontomys* sp.) and house mice (*Mus musculus*), both field dwellers, contained higher residues. In samples from three areas in 2 years, house mice contained 0.05-3.94 ppm of total DDT, 0.01-0.13 ppm of dieldrin, ND-0.14 ppm of endrin, and ND-0.02 ppm of heptachlor epoxide. Harvest mice contained 0.87 and 1.27 ppm of total DDT in 2 years, 0.05 and 0.24 ppm of dieldrin, 0.003 and 0.06 ppm of heptachlor epoxide, and 0.01 ppm of chlordane. Opossums (*Didelphis marsupialis*), which are omnivorous, contained 2.2 and 8.76 ppm of total DDT in 2 years, and no detectable dieldrin.

Residues in the fat of big game animals have been reported in a number of studies. Residues are often low, even by comparison with the tolerance limits set by the U.S. Food and Drug Administration for residues in the fat of beef sold for human consumption. Residues are conspicuously higher, however, in animals from some other areas and, in particular, in areas where DDT has been applied for forest insect control. Table 7.3 summarises these data.

7.5.2   Experimental studies of organochlorine pesticides in relation to wild mammals

Metabolism, kinetics, and physiological effects of organochlorine pesticides on laboratory and domestic mammals are the subject of a very large body of literature that is far beyond the scope of the present paper.

Relatively few laboratory studies have been conducted with wild species of mammals. In terms of lethality, big brown bats (*Eptesicus fuscus*) proved to be remarkably sensitive to DDT, far more so than reports for any other mammals (Luckens and Davis, 1964). The sensitivity did not carry over to dieldrin and endrin, however, the brown bats being similar to laboratory rats in sensitivity to these compounds (Luckens and Davis, 1965).

Mortality of adult deer mice (*Peromyscus maniculatus osgoodi*) increased with increasing dietary doses of endrin (Morris, 1968). Adult animals that survived a 6-month feeding period included: controls, 85%; 1 ppm, 86%; 2.4 ppm, 64%; 4.4 ppm, 14%; and 7.3 ppm, 14%. The reproductive performance of the survivors did not appear to differ from that of controls except for greater mortality of young from parents given the higher levels of endrin.

As in birds, residues of organochlorine pesticides in brains of mammals seem to provide the best diagnostic criteria of toxic effects. Hazardous levels of DDT and of dieldrin, at least, are of the same magnitude in both groups (Dale *et al.*, 1962; Stickel *et al.*, 1966b, 1969).

In a 3-year study of the effects of dieldrin on the physiology and reproduction of white-tailed deer (*Odocoileus virginianus*), fawns from dieldrin-fed does were smaller at birth and had greater post-partum mortality than fawns from does fed untreated food (Murphy and Korschgen, 1970). Treatments were 5 and 25 ppm of dieldrin in the diet. Fawns from does fed dieldrin had significantly poorer weight gains in 2 out of 3 years. The fertility of the male progeny was not affected. Dieldrin-treated females that were immature when the study was begun, grew more slowly than controls. The pituitary glands were smaller and thyroids were larger in dieldrin-fed deer than in control animals. Livers were englarged (as indicated by liver/body weight ratio) in deer fed 25 ppm of dieldrin. Hematologic values and serum protein concentrations were not significantly related to dosage. Whole milk, from does fed 25 ppm of dieldrin in their diet, contained 17 ppm of dieldrin. Fawns also received dieldrin by placental transfer. Concentrations of residues in tissues were correlated with residues in the diet, as would be expected. An equilibrium between ingestion and storage or excretion occurred prior to 200 days, and continued until 1,000 days.

SUMMARY

Residues of organochlorine pesticides and their breakdown products are present in the tissues of essentially all wild birds throughout the world. These chemicals accumulate in fat from a relatively small environmental exposure. DDE and dieldrin are most prevalent. Others, such as heptachlor epoxide, chlordane, endrin, and benzene hexachloride also occur, the quantities and kinds generally reflecting local or regional use. Accumulation may be sufficient to kill animals following applications for pest control. This has occurred in several large-scale programmes in the United States. Mortality has also resulted from unintentional leakage of chemical from commercial establishments.

Residues may persist in the environment for many years, exposing successive generations of animals. In general, birds that eat other birds, or fish, have higher residues than those that eat seeds and vegetation.

The kinetic processes of absorption, metabolism, storage, and output differ according to both kind of chemical and species of animal. When exposure is low and continuous, a balance between intake and excretion may be achieved. Residues reach a balance at an approximate animal body equilibrium or plateau; the storage is generally proportional to dose. Experiments with chickens show that dieldrin and heptachlor epoxide have the greatest propensity for storage, endrin next, then DDT, then lindane. The storage of DDT was complicated by its metabolism to DDE and DDD, but other studies show that DDE has a much greater propensity for storage than either DDD or DDT. Methoxychlor has little cumulative capacity in birds.

Residues in eggs reflect and parallel those in the parent bird during accumulation, equilibrium, and decline when dosage is discontinued.

Residues with the greatest propensity for storage are also lost most slowly. Rate of loss of residues can be modified by dietary components and is speeded by weight loss of the animal.

Under sublethal conditions of continuous exposure to an organochlorine pesticide, the concentrations of residues in the different tissues are ordinarily directly correlated with each other. When the dosage is at lethal levels, or when stored residues are mobilised to lethal levels, the balanced relationship is disrupted. The concentrations of residues in the brain provide the most rigorous criteria for diagnosis of death due to these chemicals, and levels are generally similar across a wide range of species of birds and mammals. Residues in liver are closely correlated with recent dose, either from direct intake or from mobilisation from storage, and so reflect hazardous exposure. Residues in the whole carcass show the storage reserve, and so indicate the potential for adverse effects from lethal mobilisation or from the continuous slow mobilisation that occurs during the normal processes of metabolism and excretion.

A synchronous, rapid, and widespread decline in weight and thickness of shells of eggs laid by many species of wild birds occurred in the late 1940's and has persisted. Birds of prey were primarily affected; exceptions apparently are the result of lesser exposure because of different food habits. Many species of fish-eating birds are also affected. Others, however, appear to be more resistant and to accumulate much higher residues before shell-thinning occurs. Seed-eating birds do not appear to have been generally affected; their exposure is ordinarily lower, but physiological factors also seem to be involved. A relationship between shell-thinning and population decline has been established for many species. In exceptional cases, such as the herring gull, persistent re-nesting and other population reactions have overcome adverse effects at the population level.

The discovery of shell-thinning among natural populations, and the hypothesis that this thinning was related to the occurrence of organo-chlorince pesticides, stimulated experimental studies to determine whether a cause-and-effect relationship existed.

In these experiments, DDE caused substantial shell-thinning in three major bird groups: the Order Anseriformes (mallard ducks, black ducks), the Order Falconiformes (American kestrels), and the Order Strigiformes (screech owls). In the studies of ducks, adverse effects on reproduction were associated with shell-thinning; the studies with hawks and owls were not carried beyond egg laying. DDT and dieldrin, fed in combination, resulted in shell-thinning and poor hatching success of captive American kestrels.

In studies with mallards, DDD produced no significant shell-thinning and DDT produced thinning only after a long period of dosage, by which time the involvement of DDE was possible. Dieldrin produced only a small percentage thinning of mallard eggshells and no thinning of pheasant egg-shells. Only modest degrees of shell-thinning have been produced by DDT or its metabolites in coturnix quail or domestic chickens, even when the dosage was high and the dietary calcium low. DDT fed to Bengalese finches resulted in delayed ovulation, but shells of the eggs were thicker, not thinner, than those produced by controls.

Statistical studies of field-collected eggs, in which the residue contents were compared with amount of shell-thinning, have supported the experimental studies in implicating DDE as the primary factor in shell-thinning.

M. J. Bryson, C. I. Draper, J. R. Harris, C. Biddulph, D. A. Greenwood, L. E. Harris, W. Binns, M. L. Miner and L. L. Madsen, *Advan. Chem. Series*, **1**, 232 (1950)

P. J. Bunyan, J. Davidson and M. J. Shorthill, *Chem. Biol. Interactions*, **2**, 175 (1970a)

P. J. Bunyan, J. M. J. Page and A. Taylor, *Nature* (Lond.), **210**, 1048 (1966)

P. J. Bunyan, A. Taylor and M. G. Townsend, *Biochem. J.*, **118**, 51P (1970b)

H. Burlington and V. F. Lindeman, *Proc. Soc. Exp. Biol. Med.*, **74**, 48 (1950)

R. Burnett, *Science N. Y.*, **174**, 606 (1971)

T. J. Cade and R. Fyfe, *Can. Field-Naturalist*, **84**, 231 (1970)

T. J. Cade, J. L. Lincer, C. M. White, D. G. Roseneau and L. G. Swartz, *Science N. Y.*, **172**, 955 (1971)

T. J. Cade, C. M. White and J. R. Haugh, *Condor*, **70**, 170 (1968)

M. K. Causey, F. L. Bonner and J. B. Graves, *Bull. Env. Contam. Toxicol.*, **3**, 274 (1968)

H. C. Cecil, J. Bitman and S. J. Harris, *Poultry Sci.*, **50**, 657 (1971)

A. H. Conney, R. M. Welch, R. Kuntzman and J. J. Burns, *Clin. Pharmacol. Ther.*, **8**, 2 (1967)

H. W. Cook and B. E. Baker, *Can. J. Zool.*, **47**, 1129 (1969)

D. Cotton and J. Herring, *24th Ann. Mtg. SE Ass. Game and Fish Commissioners*, p. 14 (1970)

S. Cramp, P. J. Conder and J. S. Ash, The risks to bird life from chlorinated hydrocarbon pesticides, September 1962—July 1963, p. 24. *Royal Soc. Prot. Birds* (1964)

S. Cramp and P. J. S. Olney, The sixth report of the Joint Committee of the British Trust for Ornithology and the Royal Society for the Protection of Birds on Toxic Chemicals in collaboration with the Game Research Association, July 1964—December 1966, p. 26. *Royal Soc. Prot. Birds* (1967)

D. D. Culley, Jr. and H. G. Applegate, *Pest. Monit. J.*, **1**, 21 (1967a)

D. D. Culley, Jr. and H. G. Applegate, *Tex. J. Sci.*, **19**, 301 (1967b)

J. G. Cummings, M. Eidelman, V. Turner, D. Reed and K. T. Zee, *J. Ass. Offic. Anal. Chem.*, **50**, 418 (1967)

J. G. Cummings, K. T. Zee, V. Turner and F. Quinn, *J. Ass. Offic. Anal. Chem.*, **49**, 354 (1966)

R. B. Dahlgren and R. L. Linder, *J. Wildl. Mgmt.*, **34**, 226 (1970)

W. E. Dale, T. B. Gaines and W. J. Hayes, Jr., *Toxicol. Appl. Pharmacol.*, **4**, 89 (1962)

K. L. Davison, J. L. Sell and R. J. Rose, *Bull. Env. Contam. Toxicol.*, **5**, 493 (1971)

W. B. Deichmann, M. Keplinger, I. Dressler and F. Sala, *Toxicol. Appl. Pharmacol.*, **14**, 205 (1969)

W. B. Deichmann, W. E. MacDonald and D. A. Cubit, *Science N. Y.*, **172**, 275 (1971)

J. B. DeWitt, C. M. Menzie, V. A. Adomaitis and W. L. Reichel, *Trans. 25th N. Amer. Wildl. Conf.*, 277 (1960)

J. B. Dimond, G. Y. Belyea, R. E. Kadunce, A. S. Getchell and J. A. Blease, *Can. Entomol.*, **102**, 1122 (1970)

J. B. Dimond and J. A. Sherburne, *Nature* (Lond.), **221**, 486 (1969)

D. L. Dindal, Kinetics of ^{36}Cl-DDT in wild waterfowl, p. 214. Ph.D. Thesis, Ohio State U. (1967)

D. L. Dindal and T. J. Peterle, *Bull. Env. Contam. Toxicol.*, **3**, 37 (1968)

W. E. Donaldson, T. J. Sheets and M. D. Jackson, *Poultry Sci.*, **47**, 237 (1968)

C. I. Draper, J. R. Harris, D. A. Greenwood, C. Biddulph, L. F. Harris, F. Mangelson, W. Binns and M. L. Miner, *Poultry Sci.*, **31**, 388 (1952)

E. H. Dustman and L. F. Stickel, Pesticides and their Effects on Soils and Water, pp. 109-121. Amer. Soc. Agron. Spec. Publ. No. 8, *Soil Sci. Soc. Amer.* (1966)

E. H. Dustman, L. F. Stickel, L. J. Blus, W. L. Reichel and S. N. Wiemeyer, *Trans. 36th N. Amer. Wildl. Natural Resources Conf.*, 118 (1971)

D. J. Ecobichon and P. W. Saschenbrecker, *Can. J. Physiol. Pharmacol.*, **46**, 785 (1968)

D. J. Ecobichon and P. W. Saschenbrecker, *Toxicol. Appl. Pharmacol.*, **15**, 420 (1969)

C. A. Edwards, Critical Reviews in Environmental Control, Vol. 1, Issue 1, pp. 7-67. Chem. Rubber Co. (1970)

J. H. Enderson and D. D. Berger, *Condor*, **70**, 149 (1968)

J. H. Enderson and D. D. Berger, *BioSci.*, **20**, 355 (1970)

D. E. Ferguson, R. L. Callahan and W. D. Cotton, *J. Miss. Acad. Sci.*, **11**, 229 (1964)

G. M. Findlay and A. S. W. DeFreitas, *Nature* (Lond.), **229**, 63 (1971)

M. C. French and D. J. Jefferies, *Nature* (Lond.), **219**, 164 (1968)

M. C. French and D. J. Jefferies, *Science N.Y.*, **165**, 914 (1969)

M. Friend, Pesticide-infectious disease interaction studies in the mallard, Ph.D. Thesis, U. Wis., vi+287 pp. (1971)

R. W. Fyfe, J. Campbell, B. Hayson and K. Hodson, *Can. Field-Naturalist*, **83**, 191 (1969)

D. E. Gaskin, M. Holdrinet and R. Frank, *Nature* (Lond.), **233**, 499 (1971)

R. E. Genelly and R. L. Rudd, *Auk*, **73**, 529 (1956)

J. W. Gillett and G. H. Arscott, *Comp. Biochem. Physiol.*, **30**, 589 (1969)

J. W. Gillett, T. M. Chan and L. C. Terriere, *J. Agr. Food Chem.*, **14**, 540 (1966)

J. W. Gillett, R. D. Porter, S. N. Wiemeyer, T. H. Gram, D. H. Schroeder and J. R. Gillette, The Biological Impact of Pesticides in the Environment, pp. 59-64. Oreg. State U. Env. Health Sci. Ser., No. 1 (1970)

C. D. Gish and N. J. Chura, *Toxicol. Appl. Pharmacol.*, **17**, 740 (1970)

R. R. Graber, S. L. Wunderle and W. N. Bruce, *Wilson Bull.*, **77**, 168 (1965)

J. B. Graves, F. L. Bonner, W. F. McKnight, A. B. Watts and E. A. Epps, *Bull. Env. Contam. Toxicol.*, **4**, 375 (1969)

R. J. Greenwood, Y. A. Greichus and E. J. Hugghins, *J. Wildl. Mgmt.*, **31**, 288 (1967)

F. Gress, R. W. Risebrough and F. C. Sibley, *Condor*, **73**, 368 (1971)

F. E. Guthrie and W. E. Donaldson, *Toxicol. Appl. Pharmacol.*, **16**, 475 (1970)

R. G. Heath, *Pest. Monit. J.*, **3**, 115 (1969)

Cromartie, G. E. Bagley and R. M. Prouty, *Pest. Monit. J.*, **4**, 141 (1970)

D. A. Murphy and L. J. Korschgen, *J. Wildl. Mgmt.*, **34**, 887 (1970)

T. W. Mussehl and R. B. Finley, Jr., *J. Wildl. Mgmt.*, **31**, 270 (1967)

E. C. Naber and G. W. Ware, *Poultry Sci.*, **44**, 875 (1965)

L. E. Nauman, Endocrine response to DDT and social stress in the male mallard. Ph.D. Thesis, Ohio State U., ix+132 pp. (1969)

D. D. Neill, J. V. Shutze and H. D. Muller, *Abst. 58th Ann. Meeting Poultry Sci. Ass.*, Ft. Collins, Colo, 74 (1969)

D. N. Noakes and C. A. Benfield, *J. Sci. Food Agr.*, **16**, 693 (1965)

M. I. Oestreicher, D. H. Shuman and C. F. Wurster, *Nature* (Lond.), **229**, 571 (1971)

C. E. Olney, W. E. Donaldson and T. W. Kerr, *J. Econ. Entomol.*, **55**, 477 (1962)

D. B. Peakall, *Nature* (Lond.), **216**, 505 (1967)

D. B. Peakall, *Nature* (Lond.), **224**, 1219 (1969)

D. B. Peakall, *Sci. Amer.*, **222**, 72 (1970a)

D. B. Peakall, *Science N.Y.*, **168**, 592 (1970b)

R. Peterson, G. Mountfort and P. A. D. Hollom, A field guide to the birds of Britain and Europe (rev. edn.,) Collins, London, xxxv+344pp. (1965)

R. E. Pillmore and R. B. Finley, Jr., Trans. *28th N. Amer. Wildl. Natural Resources Conf.*, 409 (1963)

R. D. Porter and S. N. Wiemeyer, *Science N.Y.*, **165**, 199 (1969)

G. R. Potts, *Nature* (Lond.), **217**, 1282 (1968)

I. Prestt, I.U.C.N. *11th Tech. Meeting, New Delhi, India, Pap. and Proc.*, 95 (1969)

I. Prestt and D. J. Jefferies, *Bird Study*, **16**, 168 (1969)

I. Prestt, D. J. Jefferies and N. W. Moore, *Env. Poll.*, **1**, 3 (1970)

W. L. Ragland, M. Friend, D. O. Trainer and N. E. Sladek, *Res. Commun. Chem. Pathol. Pharmacol.*, **2**, 236 (1971)

D. A. Ratcliffe, *Brit. Birds*, **58**, 65 (1965)

D. A. Ratcliffe, *Nature* (Lond.), **215**, 208 (1967)

D. A. Ratcliffe, *J. Appl. Ecol.*, **7**, 67 (1970)

W. L. Reichel and C. E. Addy, *Bull. Env. Contam. Toxicol.*, **3**, 174 (1968)

W. L. Reichel, E. Cromartie, T. G. Lamont, B. M. Mulhern and R. M. Prouty, *Pest. Monit. J.*, **3**, 142 (1969)

D. W. Rice and V. B. Scheffer, A list of the marine mammals of the world, U.S. Bur. Comm. Fish. Spec. Sci. Rep. – Fish. No. 579. iii+16 pp. (1968)

D. R. Rima, E. Brown, D. F. Goerlitz and L. M. Law, Potential contamination of the hydrologic environment from the pesticides waste dump in Hardeman County, Tennessee, iii+41 pp. U.S. Dept. Int. Geol. Surv. (1967)

R. W. Risebrough, P. Reiche, D. B. Peakall, S. G. Herman and M. N. Kirven, *Nature* (Lond.), **220**, 1098 (1968)

S. J. Ritchey, R. W. Young and E. O. Essary, *J. Food Sci.*, **32**, 238 (1967)

J. Robinson, The Biological Impact of Pesticides in the Environment, pp. 54-58. Oreg. State U. Env. Health Sci. Ser. No. 1 (1970)

J. Robinson, A. Richardson and V. K. H. Brown, *Nature* (Lond.), **213**, 734 (1967b)

J. Robinson, A. Richardson, A. N. Crabtree, J. C. Coulson and G. R. Potts, *Nature* (Lond.), **214**, 1307 (1967c)

J. Robinson, V. K. H. Brown, A. Richardson and M. Roberts, *Life Sci.* (pt. I), **6**, 1207 (1967a)

W. Rosene, Jr., *J. Wildl. Mgmt.*, **29**, 554 (1965)

W. Rosene, Jr., P. Stewart and V. Adomaitis, *Proc. 15th Ann. Conf. SE Ass. Game and Fish. Commiss.*, 107 (1962)

M. Rubin, H. R. Bird, N. Green and R. H. Carter, *Poultry Sci.*, **26**, 410 (1947)

T. G. Scott, Y. L. Willis and J. A. Ellis, *J. Wildl. Mgmt.*, **23**, 409 (1959)

J. H. Seifert, K. L. Davidson and J. L. Sell, *N. Dak. Acad. Sci.*, **22**, 36 (1968)

J. L. Sell, K. L. Davison and R. L. Puyear, *J. Agr. Food Chem.*, **19**, 58 (1971)

S. B. Shaw, *Calif. Fish and Game*, **57**, 290 (1971)

J. A. Sherburne and J. B. Dimond, *J. Wildl. Mgmt.*, **33**, 944 (1969)

W. J. L. Sladen, C. M. Menzie and W. L. Reichel, *Nature* (Lond.), **210**, 670 (1966)

S. I. Smith, C. W. Weber and B. L. Reid, *Poultry Sci.*, **48**, 1000 (1969)

S. I. Smith, C. W. Weber and B. L. Reid, *Poultry Sci.*, **49**, 233 (1970)

W. J. Stadelman, B. J. Liska, B. E. Langlois, G. C. Mostert and A. R. Stemp, *Poultry Sci.*, **44**, 435 (1965)

B. J. Stephen, J. D. Garlich and F. E. Guthrie, *Bull. Env. Contam. Toxicol.*, **5**, 569 (1971)

L. F. Stickel, *J. Wildl. Mgmt.*, **15**, 161 (1951)

L. F. Stickel, Organochlorine Pesticides in the Environment. U.S. Bur. of Sport Fish. Wildl., Spec. Sci. Rep. – Wildl. No. 119, iv+32 pp. (1968)

L. F. Stickel, N. J. Chura, P. A. Stewart, C. M. Menzie, R. M. Prouty, and W. L. Reichel, *Trans. 31st N. Amer. Wildl. Natural Resources Conf.*, 190 (1966a)

L. F. Stickel and R. G. Heath, Effects of Pesticides on Fish and Wildlife, pp. 3-30. U.S. Fish and Wildl. Serv. Cir. 226 (1965)

L. F. Stickel and L. I. Rhodes, The Biological Impact of Pesticides in the Environment, pp. 31-35. Oreg. State U. Env. Health Sci. Ser., No. 1 (1970)

L. F. Stickel and W. H. Stickel, *Ind. Med. Surg.*, **38**, 44 (1969)

L. F. Stickel, W. H. Stickel and R. Christensen, *Science N.Y.*, **151**, 1549 (1966b)

W. H. Stickel, Effects of Pesticides on Fish and Wildlife, p. 17. U.S. Fish and Wildl. Serv. Cir. 226 (1965)

W. H. Stickel, W. E. Dodge, W. G. Sheldon, J. B. DeWitt and L. F. Stickel, *J. Wildl. Mgmt.*, **29**, 147 (1965a)

W. H. Stickel, D. H. Hayne and L. F. Stickel, *J. Wildl. Mgmt.*, **21**, 132 (1965b)

W. H. Stickel, L. F. Stickel and F. B. Coon, *Pesticides Symposia*, pp. 287-294. Halos Assoc., Miami (1970)

W. H. Stickel, L. F. Stickel and J. W. Spann. N. W. Miller and G. G. Berg (Editors). Chemical Fallout, pp. 174-204, C. C. Thomas, Springfield, Ill. (1969)

J. C. Street, *Science N.Y.*, **146**, 1580 (1964)

J. C. Street, R. W. Chadwick, M. Wang and R. L. Phillips, *J. Agr. Food Chem.*, **14**, 545 (1966)

B. Switzer, V. Lewin and F. H. Wolfe, *Can. Zool.*, **49**, 69 (1971)

L. C. Terriere, G. H. Arscott and U. Kiigemagi, *J. Agr. Food Chem.*, **7**, 502 (1959)

R. F. Thomas and J. G. Medley, *J. Ass. Offic. Anal. Chem.*, **54**, 681 (1971)

E. M. Thompson, G. J. Mountney and G. W. Ware, *J. Econ. Entomol.*, **60**, 235 (1967)

U.S. Bureau of Sport Fisheries and Wildlife, Mirex residues in birds and raccoons of South Carolina estuaries. U.S. Bur. Sport Fish. Wildl., Div. Wildl. Serv. Atlanta, Georgia, Spec. Report, 35 pp. (1971)

U.S. Department of Agriculture, Monitoring agricultural pesticide residues, 1965-1967. U.S. Dept. Agr., Plant Pest Control Div. A.R.S. 81-32, ii+97 pp. (1969)

A. C. Van Velzen, W. B. Stiles and L. F. Stickel, *J. Wildl. Mgmt.* (1972, in press)

K. Vermeer and L. M. Reynolds, *Can. Field-Naturalist*, **84**, 117 (1970)

K. C. Walker, D. A. George and J. C. Maitlen, U.S. Dept. Agr., A.R.S. 33-105, p. 21 (1965)

G. J. Wallace, A. G. Etter and D. R. Osborne, *Mass. Audubon*, **48**, 116 (1964)

G. J. Wallace, W. P. Nickell and R. F. Bernard, *Cranbrook Inst. Sci. Bull.*, **41**, 44 pp. (1961)

W. W. Walley, D. E. Ferguson and D. D. Culley, *J. Miss. Acad. Sci.* **12**, 281 (1966)

G. W. Ware and E. C. Naber, *J. Econ. Entomol.*, **54**, 675 (1961)

R. E. Webb and F. Horsfall, Jr., *Science N.Y.*, **156**, 1762 (1967)

M. Weihe, *Acta Pharmacol. Toxicol.*, **25** (suppl. 4), 54 (1967)

R. L. Wesley, A. R. Stemp, R. B. Harrington, B. J. Liska, R. L. Adams and W. J. Stadelman, *Poultry Sci.*, **48**, 1269 (1969)

R. L. Wesley, A. R. Stemp, B. J. Liska and W. J. Stadelman, *Poultry Sci.*, **45**, 321 (1966)

D. M. Whitacre and G. W. Ware, *J. Agr. Food Chem.*, **15**, 492 (1967)

S. N. Wiemeyer, B. M. Mulhern, F. J. Ligas, R. J. Hensel, J. E. Mathisen, F. C. Robards and S. Postupalsky, *Pest. Monit. J.* (in press)

S. N. Wiemeyer and R. D. Porter *Nature* (Lond.), **227**, 737 (1970)

A. J. Wilson, Jr., Rep. Bur. Comm. Fish. Biol. Lab., Gulf Breeze, Fla., 5, U.S. Dept. Interior Cir. 260 (1967)

E. L. Wisman, R. W. Young and W. L. Beane, *Poultry Sci.*, **46**, 1606 (1967)

J. G. Wit, *Comp. Biochem. Physiol.*, **30**, 185 (1969)

A. A. Wolman and A. J. Wilson, Jr., *Pest. Monit. J.*, **4**, 8 (1970)

G. M. Woodwell, C. F. Wurster, Jr. and P. A. Isaacson, *Science N.Y.*, **156**, 821 (1967)

L. E. Wragg, *Can. Field-Naturalist*, **68**, 11 (1954)

B. S. Wright, *J. Wildl. Mgmt.*, **29**, 172 (1965)

C. F. Wurster, Jr. and D. B. Wingate *Science N.Y.*, **159**, 979 (1968)

D. H. Wurster, C. F. Wurster, Jr. and W. N. Strickland, *Ecology N.Y.*, **46**, 488 (1965)

APPENDIX

Index to names of birds (following A.O.U. checklist, 1957 and Petersen *et al.*, 1965).

Common Name	Scientific Name
Buzzard, common	*Buteo buteo*
Cormorant, double-crested	*Phalacrocorax auritus*
Cowbird	*Molothrus ater*
Crane, whooping	*Grus americana*
Crow, carrion	*Corvus corone*
Crow, common	*Corvus brachyrhynchos*
Dove, mourning	*Zenaidura macroura*
Dove, ring	*Streptopelia risoria*
Duck, black	*Anas rubripes*
Duck, eider	*Somateria mollissima*
Duck, mallard	*Anas platyrhynchos*
Duck, scaup	*Aythya affinis*
Eagle, bald	*Haliaeetus leucocephalus*
Eagle, golden	*Aquila chrysaetos*
Peregrine	*Falco peregrinus*
Falcon, prairie	*Falco mexicanus*
Finch, Bengalese	*Lonchura striata*
Finch, zebra	*Poephila guttata*
Gallinule, common	*Gallinula chloropus*
Gallinule, purple	*Porphyrula martinica*
Goshawk	*Accipiter gentilis*
Grackle	*Quiscalus quiscula*
Grebe, western	*Aechmophorus occidentalis*
Greenshank	*Tringa nebularia*
Grouse, blue	*Dendragapus obscurus*
Guillemot	*Uria aalgae*
Gull, black-headed	*Larus ridibundus*
Gull, herring	*Larus argentatus*
Gyrfalcon	*Falco rusticolus*
Hawk, broad-winged	*Buteo platypterus*
Hawk, Cooper's	*Accipiter cooperii*
Hawk, marsh	*Circus cyaneus*
Hawk, red-shouldered	*Buteo lineatus*
Hawk, red-tailed	*Buteo jamaicensis*
Hawk, rough-legged	*Buteo lagopus*
Hawk, sharp-shinned	*Accipiter striatus*
Heron, black-crowned night	*Nycticorax nycticorax*
Heron, grey	*Ardea cinerea*
Heron, great blue	*Ardea herodias*
Hobby	*Falco subbuteo*
Kestrel (American)	*Falco sparverius*
Kestrel (European)	*Falco tinnunculus*

Common Name	Scientific Name
Kingfisher	*Megaceryle alcyon*
Kittiwake	*Rissa tridactyla*
Loon, common	*Gavia immer*
Merlin	*Falco columbarius*
Murre, common	*Uria aalgae*
Osprey	*Pandion haliaetus*
Owl, great horned	*Bubo virginianus*
Owl, screech	*Otus asio*
Partridge, Hungarian	*Perdix perdix*
Pelican, brown (eastern)	*Pelecanus accidentalis carolinensis*
Pelican, brown (western)	*Pelecanus occidentalis californicus*
Pelican, white	*Pelecanus erythrorhynchos*
Pheasant	*Phasianus colchicus*
Pigeon	*Columba livia*
Plover, golden	*Pluvialis apricaria*
Quail, coturnix	*Coturnix coturnix japonica*
Raven	*Corvus corax*
Razorbill	*Alca torda*
Robin	*Turdus migratorius*
Rook	*Corvus frugilegus*
Shag	*Phalacrocorax aristotelis*
Sparrow, house	*Passer domesticus*
Sparrow hawk (British)	*Accipiter nisus*
Tern, common	*Sterna hirundo*
Tern, Sandwich	*Sterna sandwicensis*
Thrush, song	*Turdus ericetorum*
Turkey	*Meleagris gallopavo*
Woodcock	*Philohela minor*

Chapter 8

Pesticide Residues in Man

J. E. Davies

*Pesticides of Dade County,
Miami, Florida*

There are three types of human exposure to pesticides. Acute exposure which is usually the result of accidental contamination by excessive amounts of pesticides, chronic exposure which most frequently occurs in pesticide workers by virtue of their occupation, and incidental exposure which is the consequence of the ubiquity of pesticides and their presence in trace amounts in air, water, food and dust. The former exposure results in typical acute symptoms and signs reflective of the toxicological properties of the material; chronic occupational exposure to pesticides results in patterns of illness which are usually less well-defined, and often the consequences of multiplicity of chemical insults. Incidental exposure, the exposure to which the population at large is experiencing is even less well documented, and with the persistent pesticides the only certain effect is the acquisition of these residues in human tissues. This chapter will seek to summarise the characteristics of these human residues and to discuss the significance and interpretation of these levels in health and disease.

8.1 AMOUNT OF PESTICIDES CURRENTLY OCCURRING

Over the last decade, as a result of the growing list of new pesticides and the sophistication of chemical analytical procedures, both the qualitative and quantitative characteristics of human pesticide residues has increased. The greatest body of knowledge has been obtained on the organochlorine pesticides. Adipose tissues and blood have been most widely used to measure these traces, and DDT and its metabolites have received the major share of interest, both because of their widespread use and dissemination and also on account of the ease with which this pesticide and its metabolites have been identified in all human adipose tissues. Table 8.1 summarises some of the more recent general population surveys of DDT-derived materials in human fat in various countries. Serological surveys of DDT and its metabolites and dieldrin have also been reported and most studies indicate good correlation between adipose and serological surveys. Table 8.2 lists some of the serum or whole blood organochlorine prevalence surveys in different parts of the world. Table 8.3 summarises prevalence in adipose tissues of other organochlorine pesticides.

Trace amounts of polychlorinated biphenyls (PCB's), one of the most abundant synthetic pollutants, are being increasingly detected in human adipose residue surveys. Dr H. Enos (private communication) estimated that 30% of residues in adipose tissues from monitoring programmes in the U.S.A. have trace amounts of these contaminants with 26% having an average concentration greater than 1 ppm. Because of the analytical quality control of the national U.S.A. human monitoring programmes (organised by the Community Studies Pesticide Programs in the United States), the possibility of misinterpretation of PCB's as DDT residues has been largely obviated. PCB's have recently been reported in human milk. In whole milk, the content of PCB's and benzene hexachloride was 0.103 and 0.153 ppm respectively, and in fat, the content was 5.7 and 6.3 ppm respectively (Acker and Schulte, 1970).

Traces of pentachlorophenol (PCP), a pesticide widely used as a fungicide in textiles and in the construction and lumber industries, to control mould and termites, have been identified in urine and blood, both from the occupationally exposed and from persons accidentally exposed to treated materials (Bevenue and Beckman, 1967).

Trace metals and various inorganic compounds have been used for pest control since the sixteenth century (Perkow, 1956). Human residues of these materials in the general population are seldom the result of the agricultural use of these chemicals. Metals and their human residue concentrations will not be further discussed here.

8.2   INTERPRETATIVE USES OF HUMAN
PESTICIDE RESIDUES

The human pesticide residue is a biological index of pesticide exposure which may be acute, occupational, or incidental. In acute intoxication, the residue provides diagnostic information; in the occupationally exposed, the residue is a surveillance tool reflective of industrial exposure. In the general population, the residue is a measure of incidental exposure and average levels of the persistent pesticides in fat and blood have been used to express the level of pollution of the population at large by these chemicals. International comparisons have been made and surveys repeated in time have been used as expressions of secular trends. More recently, a different interpretative potential can be seen in the literature. Levels have been studied in disease to determine whether concentrations are greater or less than the healthy population. Thus, some reports have indicated higher concentrations of DDT and its metabolites in a variety of pathological conditions including cancer, hypertension and disease of the liver (Radomski *et al.*, 1968). Casarette *et al.* (1968) found high DDT fat residues associated with the triad of liver disease, cachexia, and carcinoma. O'Leary *et al.* (1970) observed greater blood levels of p,p'-DDE in premature babies than in full-term babies. Dacre (1970) noted high residues of DDT and its metabolites in 14 lung cancer patients. Other workers have found no positive correlation of fat residues of DDT with disease. Thus, Maier-Bode (1960) and Robinson *et al.* (1965) detected no differences in the total DDT-derived materials or dieldrin in a comparison of 50 biopsies and 50 autopsy specimens. Similarly, Hoffman (1967) in a review of 688 autopsies found that there was no significant correlation between the levels of DDE plus DDT and of benzene hexachloride in human fat and the presence or absence of abnormalities in these tissues.

These new and strictly epidemiologic uses of human pesticide residue data emphasise the urgent need for the understanding of the pesticide residue distribution in the healthy normal population before interpretations can be made of the level in disease.

In addition to unusually high levels of pesticides in disease, significantly low levels have also been shown to have some clinical connotations. A report by Davies *et al.* (1969a) indicated that patients taking phenobarbitone and/or diphenylhydantoin for periods longer than three months, had strikingly lower blood levels of DDE than the general population. Healthy control patients, not taking the drugs, had a mean concentration of 9.1×10^{-3} ppm of DDE in blood, compared with levels of 3.5 and 1.9×10^{-3} ppm respectively in outpatients taking phenobarbitone and diphenylhydantoin alone, and 1.7×10^{-3} ppm for those patients taking both drugs. Adipose levels of DDT and its metabolites in severely mentally retarded children indicated a mean of 2.7 ppm for DDT-derived material in those children not taking medication, compared to only 0.17 ppm with one or both drugs. The lowering of persistent pesticide residues by these drugs was attributed to liver microsomal enzyme induction. Jager (1970) utilised significantly low serum DDE residues as a diagnostic test of liver microsomal enzyme induction, and finding p,p'-DDE levels below the detectable level of 0.005 $\mu g/ml$ in 15 out of 29 endrin factory workers, ascribed this finding to the enzyme-inducing potential of endrin. Others have reported that the occurrence of low serum DDE levels was the finding which made it possible to recognise individuals taking phenobarbitone and diphenylhydantoin (Schoor, 1971; Kwalick, 1971).

Besides studying the associations of human residues with disease, others have studied the prevalence of DDT in populations on different diets, to determine the relative importance of different articles of foods. Thus, Hayes *et al.* (1961) found significantly lower DDT pesticide residues in meat abstainers than those in meat eaters.

8.3 THE PRINCIPAL SOURCES OF THE HUMAN PESTICIDE RESIDUE

Information on the various environmental sources of the human pesticide residue have been obtained from three sources of data:

1. Environmental residue data and concentrations in food, air and water.
2. Controlled human exposure studies.
3. Descriptive epidemiologic studies of human pesticide residues.

8.3.1 Environmental residue data

Those pesticides which are detected in humans are to be found in food, air, water, in dust and also in a variety of other materials such as clothing and bedding. Man is thus exposed from several environmental sources. Since other chapters in the book are covering residue data in the various environmental media, this discussion will merely review the relative significance of these various sources.

Insofar as DDT is concerned, there is ample evidence that pesticide traces in food can be a significant determinant of the human DDT residue. Thus, Walker *et al.* (1954) found trace amounts of DDT and DDE in all meals that they analysed. Campbell *et al.* (1965) believed that the total dietary intake accounts for most, if not all, of the DDT and its metabolite DDE stored in the body fat as a result of incidental exposure. Kraybill (1969) believed that food contributes 85% of the DDT body burden.

Epidemiological studies which have supported the significance of dietary sources in food include the relatively low levels of DDT in fat and blood surveys, in population groups ingesting diets relatively free from DDT contamination, such as meat abstainers (Hayes *et al.*, 1961), Eskimoes (Durham *et al.*, 1961) and institutionalised patients (Edmundson *et al.,* 1970). An overall estimate of the relative role of air, water, food and other sources, derived from a number of studies in the U.S.A., is summarised in the 'Report of the Secretary's Commission on Pesticides and Their Relationship to Environmental Health' (Mrak Report, 1969). The combined studies for the annual intake of DDT plus DDE were calculated at 0.03 mg taken in from the air, 0.01 mg from the water, 44.8 mg from food and 5.0 mg from other sources, which includes intake from individual household usage. The report goes on to suggest that the category 'other' may become the predominant component of pesticide intake in many instances.

8.3.2 Controlled human exposure studies

Classical toxicological information on residue build-up following ingestion of DDT has been obtained through human volunteer studies. Much has been learned of pesticide pharmacodynamics in man by such investigations.

It is well recognised that DDT degradation may proceed through one of two alternate pathways – dehydrochlorination to DDE and probable excretion as DDE through the biliary system. Peterson and Robinson (1964) have shown that DDE degradation to DDA does not occur. Alternatively, DDT may degrade by the substitution of hydrogen for one chlorine atom to form DDD, with subsequent breakdown through several intermediary metabolites to DDA (bis (*p*-chlorophenyl) acetic acid). Edmundson *et al.* (1969) have suggested that the latter mechanism only comes into play after excessive DDT exposure and is reflective of a secondary detoxification mechanism following exceptional exposure. Increased DDA excretion had been noted after DDT exposure of pesticide formulators and after other types of occupational exposure (Wolfe *et al.,* 1970; Durham *et al.,* 1965; Roan *et al.,* 1971). Hayes *et al.* (1965) reported two human volunteer studies wherein subjects ingested 3.5 mg and 35 mg of technical and recrystallised DDT daily. There was a 32-fold increase in fat concentrations of DDT after 21.5 months in the latter group. More recently, Morgan and Roan (1971) reported an average 20-fold increase in adipose concentrations of DDT when two volunteers were fed 10 and 20 mg technical DDT daily for six months. A separate study, wherein another subject ingested 5 mg of *p,p'*-DDE daily for ninety-two days, exhibited a 4 to 5-fold increase. of adipose and serum residue levels of DDE. DDE ingestion increased serum DDE levels 30 times as fast per unit dose as did DDT ingestion. They concluded from these studies that levels of DDE reflected DDE rather than DDT ingestion. Their dose of ingested pesticide, however, was far in excess of the usual general population intake, and their failure to produce a significant DDE residue following DDT-ingestion might well have reflected the compensatory DDA mechanism coming into play, as was suggested by Edmundson *et al.* (1969) and noted in other intensive occupational exposures. Epidemiological studies suggest that serum or blood levels of DDT reflect recent exposure, whereas DDE levels are reflective of long-term DDT exposure and its subsequent conversion to DDE (Davies *et al.*, 1969b). Similar increases of adipose residues of dieldrin have been documented by Hunter and Robinson (1967) and Hunter *et al.* (1969). Thirteen volunteers were given oral daily doses of 10, 50 and 211 µg of dieldrin. A four-fold increase in adipose pesticide residues of this pesticide was noted in subjects on the 50 µg schedule, and a tenfold increase in those on the 211 µg schedule after eighteen months.

8.3.3   Descriptive epidemiologic data
In the last three or four years, two events have occurred which have clarified our understanding of human pesticide residues and facilitated the epidemiologic appraisal of environmental sources of human pesticide residues.

The first of these events has been the growth in the number and stratification of adipose residue population surveys. The second has been the development of reproducible analytical methods for organochlorine measurement in whole blood, plasma or serum. As will be seen from Table 8.1 and Table 8.2, because of analytical and logistic problems inherent in the acquisition of fat from autopsy or biopsy specimens, expressions of national prevalences have been perforce, largely unstratified, and unrepresentative of the population at large.

The availability of larger and more stratified samples in the U.S.A. now suggests that while the distribution of dieldrin in the population is essentially homogeneous, this is not true for DDT and its metabolites, and prevalence information on these pesticide residues clearly indicates that there are important and significant demographic and geographical distributions, which suggest differences in environmental determinants of the residue. In order to illustrate these differences, I have combined the adipose residue data for DDT and its metabolites and dieldrin published in the Mrak Report (1969) with data collected from the Florida Community Pesticide Studies project, thereby making data available from 23 states with information on 5,234 specimens. Using these, and other sources of data, the descriptive characteristics of person and place surveys in organochlorine residue epidemiology are reviewed hereinafter.

8.3.3.1   *Variations between people*
Age and race effects – Most surveys have lacked residue data from a sizeable number of younger persons. In Israel, the mean total DDT in fat from 71 children (0.9 yrs.) was 10.2 ppm, compared to 18.1 ppm from 133 persons in older age groups (Wasserman *et al.*, 1967). In Florida, comparable figures were 5.5 ppm and 7.8 ppm for 17 white and 17 black children 5 years and under respectively, compared to 8.4 ppm and 16.7 ppm for 90 white and 35 black older persons (Davies *et al.*, 1968). Table 8.3 and Fig. 8.1 summarise the age distribution of residues from the U.S. Human Monitoring Program and the Florida data (5,234 persons).

TABLE 8.1. Concentrations of DDT and its metabolites in adipose tissue (ppm).

Country	Year	No. of samples	Analytical method	DDT	DDE as DDT	Total as DDT	DDE as DDT (% of Total)	Reference
England	1965-67	248	GLC	0.7	2.2	3.0	74	Abbott et al., 1968
France	1961	10	Colorimetric	1.7	3.5	5.2	67	Hayes et al., 1963
Belgium	–	20	GLC	1.2	2.1	3.3	64	Maes and Heydrickx, 1966
Holland	–	11	GLC	0.3	1.8	2.2	86	de Vlieger et al., 1968
Denmark	1965	18	GLC	0.6	2.7	3.3	82	Weihe, 1966
West Germany	1958-59	60	Colorimetric	1.0	1.3	2.3	57	Maier-Bode, 1960
German Democratic Republic (Russian Zone)	1966-67	100	GLC	3.7	9.4	13.1	71	Engst, 1967
Czechoslovakia	1963-64	229	Colorimetric	5.5	4.1	9.6	43	Halacka et al., 1965
Poland	1965	72	Colorimetric	–	–	13.4	–	Bronisz et al., 1967
Hungary	1960	48	Colorimetric	5.7	6.0	12.4	48	Denes, 1962
Italy	1965	22	GLC & TLC	4.6	10.6	15.4	68	DelVecchio and Leoni, 1967
Spain	1966	41		6.5	9.2	15.7	59	Llinares and Wasserman, 1968
South Africa	1969	114	GLC and EC			6.38	–	Wasserman et al., 1970
Israel	1965-66	133	Colorimetric	8.2	9.9	18.1	54	Wasserman et al., 1967
India	1964	67	Colorimetric	16.0	10.0	26.0	39	Dale et al., 1965
Australia	1965-66	46	Colorimetric	3.6	6.6	10.2	64	Wasserman et al., 1968
New Zealand	1966	52	GLC	1.6	4.2	5.8	72	Brewerton and McGrath, 1967
Canada	1966	47	GLC and ELC	1.0	2.9	4.3	67	Brown, 1967
United States: North-White	1968	2,835	GLC		3.3	4.8	–	Mrak Report, 1969
,, ,, North-Negro	1968	291			5.3	7.9	–	,, ,, 1969
,, ,, South-White	1968	1,476			6.6	9.2	–	,, ,, 1969
,, ,, South-Negro	1968	588			9.8	14.4	–	,, ,, 1969

TABLE 8.2.   Concentrations of organochlorine pesticides in blood (ppm×10³)

Exposure group	Year	No. of samples	Type	p,p'-DDT	p,p'-DDE	p,p'-DDD	Total DDT*	BHC	Dieldrin	Hep. Epox.	Reference
General population England	1964	44	Whole blood	–	–	–	13.0	–	1.4		Robinson and Hunter, 1966
Occupationally England	1964	136	,, ,,	–	–	–	–	–	46.3		,, ,,
General population Males – U.S.	1965	10	,, ,,	6.8	11.4	–	19.3	3.1	1.4	0.8	Dale *et al.*, 1966
General population Males – U.S.	1965	10	Plasma	13.2	25.7	–	41.5	3.4	1.9	1.1	,, ,,
General population Females – U.S.	1965	10	Serum	9.3	15.2	–	26.0	4.2	1.3	0.8	,, ,,
Occupationally DDT manufacturing	1967	35	,,	214.0	224.0	90.3	590.5	4.7	5.9	1.3	Laws *et al.*, 1967
General population Males	1967	10	Plasma	25.3	25.7	8.3	59.3	5.7	2.8		Dale *et al.*, 1967
General population Females	1967	10	,,	8.6	13.5	3.7	25.8	0.8	0.9		,, ,,
General population Idaho	1967-68	1000	Serum	4.7	22.0	0.2	26.9	–	0.5		Watson *et al.*, 1970

Occupationally										
Israel	1969	108	,,	–	–	–	109.5	11.0	1.8	,,
General population										
White (Dade County, Fla.)	1969	293	Serum	5.0	17.5	–	22.5	–	1.6	J. E. Davies, (unpublished data)
General population										
Black (Dade County, Fla.)	1969	209	,,	10.6	32.2	–	32.8	–	1.1	,,
General population										
White – California	1969	18	,,	12.0	35.0	–	51.0	–	–	Poland *et al.*, 1970
Occupational										
California	1969	18	,,	573.0	506.0	97.0	1,359.0	–	–	,, ,,
General population										
Dade County	(1970-71)									
Males by Social Class:										
White Social Class I†		50	,,	5.5	22.2	–	30.2	–	1.2	Davies *et al.* (in press)
,, ,, ,, II		52	,,	6.1	28.9	–	38.4	–	1.3	,,
,, ,, ,, III		31	,,	6.4	30.6	–	41.4	–	1.0	,,
,, ,, ,, IV		104	,,	7.3	31.8	–	42.8	–	1.7	,,
,, ,, ,, V		36	,,	8.1	38.1	–	50.5	–	2.0	,,
Black Social Class I†		13	,,	7.5	34.5	–	45.9	–	1.0	,,
,, ,, ,, II		7	,,	7.8	40.0	–	52.4	–	1.4	,,
,, ,, ,, III		7	,,	5.8	29.6	–	38.8	–	0.7	,,
,, ,, ,, IV		87	,,	9.7	43.9	–	58.7	–	1.1	,,
,, ,, ,, V		75	,,	10.3	49.0	–	64.9	–	0.97	,,

* Total DDT derived material.

† Hollingshead Social Classes I-V are correlated with decreasing affluence.

TABLE 8.3. Age and race comparisons of mean DDE, total DDT, and Dieldrin adipose residues (ppm) in Northern and Southern States grouped by mean monthly temperatures. (Human Monitoring and Florida Data, 1968.)

Age	White				Negro			
	No. of specimens	p,p'-DDE	Total DDT	Dieldrin	No. of specimens	p,p'-DDE	Total DDT	Dieldrin
Northern States ($<56°$F)								
0-5	39	2.74	4.04	0.15	9	4.27	6.53	0.19
6-10	8	1.40	2.48	0.08	2	3.00	5.98	0.17
11-15	14	1.64	2.35	0.03	2	1.37	2.27	0.0
16-20	36	2.01	2.98	0.07	8	2.88	4.24	0.08
21-25	35	2.32	3.47	0.08	10	5.23	8.74	0.11
26-30	58	2.29	3.26	0.07	7	3.67	5.21	0.12
31-40	141	2.90	4.12	0.10	27	5.31	7.35	0.13
41-50	328	3.11	4.40	0.10	50	4.97	7.28	0.12
51-60	584	3.46	5.07	0.12	55	5.24	7.91	0.17
61-70	729	3.45	5.23	0.11	67	5.66	8.38	0.12
71-80	615	3.43	4.97	0.11	41	6.36	9.46	0.16
81-90+	248	3.33	5.03	0.10	13	4.80	8.30	0.14
Total	2,835	3.29	4.84	0.11	391	5.26	7.86	0.14
Southern States ($>56°$F.)								
0-5	42	3.52	5.08	0.16	28	4.11	6.74	0.15
6-10	12	6.92	11.20	0.26	9	5.89	9.80	0.15
11-15	20	5.95	10.69	0.19	8	6.20	10.05	0.04
16-20	33	5.28	7.42	0.07	19	7.07	10.34	0.09
21-25	37	5.71	12.83	0.10	28	6.28	9.66	0.08
26-30	44	6.32	8.53	0.10	13	7.84	10.80	0.12
31-40	103	5.92	8.36	0.12	64	7.34	10.63	0.11
41-50	191	6.31	8.91	0.15	102	10.21	14.10	0.10
51-60	309	6.24	8.42	0.14	108	10.76	16.18	0.17
61-70	348	7.32	10.06	0.13	110	9.67	14.09	0.14
71-80	263	6.86	9.63	0.14	85	12.77	20.30	0.17
81-90+	104	6.54	6.19	0.10	28	12.89	18.61	0.16
Total	1,506	6.50	9.15	0.13	602	9.66	14.33	0.13

A quinquennial increase has occurred in whites up to the 21-25 year age group, with a levelling off thereafter in subsequent age groups. DDT residues in blacks, were in all age groups greater than whites and continued to rise with age. No such increases were observed for dieldrin residues. The same age trend was suggested when residue prevalences were compared between Northern and Southern States, although in the North, DDT residues continued to increase with age in both blacks and whites, but in the South, for blacks only after the 25 year age group.

Although the failure to associate residues with social class differences might be affecting these age trends, the concept of age dependency of the DDT fat residue is suggested by these data. If incidental DDT exposure

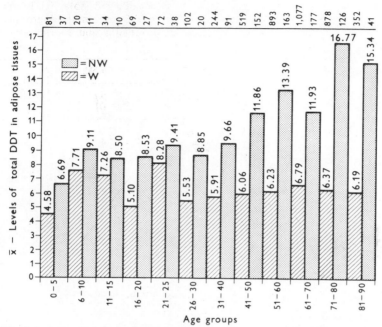

FIG. 8.1. x̄ adipose levels by age and race, of total DDT (ppm) in general population from 23 states of U.S.A., 1968. (Human monitoring and Florida data.)

were uniform, an increase in the average residues for each age group would be expected through the age when DDT residues level off (approximately 23 years), with residues from age groups over this age being the same thereafter. This pattern is suggested for the white human residue data in Fig. 8.1 though it is not seen in the blacks. The stabilisation or even decline with time in human DDT residues suggested by some, is not supported by these age distribution frequencies. The possible association of race differences with social class factors is discussed later.

Sex differences – Differences in DDT and dieldrin residue prevalences associated with sex are conflicting. A positive association of DDT residues with males was described by Wasserman *et al.* (1965), Dale *et al.* (1965), Robinson *et al.* (1965), Laug *et al.* (1951), Hayes *et al.* (1958) and Davies *et al.* (1969c). Most reports suggest higher DDT residue prevalences in males than females (Robinson *et al.*, 1965; Hunter *et al.*, 1963; Egan *et al.*, 1965; Abbott *et al.*, 1968).

Ethnic differences – In Israel, no ethnic differences of DDT-derived material were observed in comparison of residues from Ashekenazim, Sephardim, native Israelis, Yemen, and Indians (Wasserman *et al.*, 1967). Higher levels of serum residues of DDE were noted in Arizona in Yaki

TABLE 8.4 Concentrations of organochlorine pesticides in adipose tissue.*

Country	Year	No. of Samples	Analytical method	Storage level in body fat				Reference
				BHC-isomers	Dieldrin	Endrin	Heptachlor epoxide	
England	1965-67	248	GLC	0.31	0.21	—	0.04	Abbott et al., 1968
France	1961	10	GLC	1.19	—	—	—	Hayes et al., 1963
Holland	1965	11	GLC	0.11	0.20	—	0.01	Maes and Heyndrickx, 1966
Denmark	1965	18		—	0.20	—	—	Weihe, 1966
Hungary	1964	15		—	Tr.-0.16	—	—	Denes, 1966
Italy	1966	22		0.08	0.68	—	0.23	DelVecchio and Leoni, 1967
India	1964	35	GLC	1.43	0.04	nd**	nd**	Dale et al., 1965
Australia	1965-66	12		0.68	0.67	0.1	0.02	Wasserman et al., 1968
New Zealand		52		—	0.27	—	—	Brewerton and McGrath, 1967
Canada	1966	47		0.07	0.22	—	0.14	Brown, 1967
U.S. North-White	1968	2,835	GLC	—	0.11	—	—	Mrak Report, 1969
U.S. North-Negro	1968	291	GLC	—	0.14	—	—	,, ,, 1969
U.S. South-White	1968	1,476	GLC	—	0.13	—	—	,, ,, 1969
U.S. South-Negro	1968	588	GLC	—	0.13	—	—	,, ,, 1969

Indians when compared with the white population by Morgan (personal communication).

Occupation and drugs – Human pesticide residues are expectedly greater in persons with recent or remote occupational exposure to pesticides, and most residue profiles in such groups are reflective of multiple exposures to organochlorines. Interaction between pesticides, and also with certain commonly-used drugs, can modify residue concentrations of other persistent pesticides. Specific blood residue levels have been utilised as surveillance tools to prevent over-exposure. Brown *et al.* (1964) reported that a blood dieldrin level of 15 $\mu g/100$ ml appears to be the threshold level denoting the appearance of symptoms of dieldrin intoxication and the procedure was recommended both as a diagnostic and surveillance tool. Removal from occupational exposure, when levels of these orders of magnitude are reached in blood, has proven effective in the prevention of human dieldrin and aldrin toxicity in a manufacturing plant (Jager, 1970). Whole blood and serological surveys of DDT and DDE prevalences have also confirmed the distribution characteristics of these residues with person and place (Davies *et al.*, 1969c).

The influence of social class on human DDT residues – On almost all occasions where prevalences of DDT and its metabolites have been compared in blacks and whites, residues have been noted to be 25-50% greater in blacks. This race-associated correlation is one of the most challenging and one which begs interpretation. In another situation, Klemmer (personal communication) reported greater serum concentrations of p,p'-DDE in Hawaiian native males than in Hawaiian Caucasian males. The most plausible explanation would attribute such differences either to socio-economic factors, genetic factors, or a combination of both.

In a serological survey of total DDT and p,p'-DDE prevalences in 211 white males and 237 black males in U.S.A., residues were significantly lower in each race in the more affluent classes (Hollingshead Social Classes I and II) when compared with the poorer classes (Hollingshead Social Classes IV and V) (Davies and Edmundson, 1971) (Table 8.5). Within the individual social groups, residues of DDE were still significantly greater in blacks than whites. These differences in DDT and DDE residue prevalences are probably associated with such characteristics of social class as standards of hygiene, garbage collection, window screening, differences in pest control practices and differences in diet, and they emphasise an important variable in DDT epidemiology, particularly in tropical and sub-tropical countries, and one which may partly, if not totally, explain race-associated differences (Table 8.5).

8.3.3.2 *Geographical variables*

Tables 8.1 and 8.4 and Figs. 8.2 and 8.3 present adipose residue prevalences of DDT and its metabolites and dieldrin from various population surveys on both sides of the Atlantic. These suggest, not surprisingly, that DDT storage in man is greater in the warmer climates. Fig. 8.2 lists DDE, total

TABLE 8.5. Comparison of effects of socio-economic class (Hollingshead) on human serum pesticide residues.

Social class	n	$\bar{x}$ age	Total DDT (ppm$\times 10^3$)				DDE (ppm$\times 10^3$)			
			$\bar{x}$	range	S.D.	'p' value*	$\bar{x}$	range	S.D.	'p' value*
A. *White males only*										
I and II	65	35	32	9-89	16	<.01	24.1	8-59	12	<.01
IV and V	146	29	41	8-226	30					
B. *Black males only*										
I and II	12	27	43	27-63	13	<.01	32	18-49	11	<.01
IV and V	225	23	61	6-230	33		46	4-180	26	

* 2 sample *t* test.

DDT and dieldrin prevalences from 23 states of the United States, where the individual states are divided on the basis of the mean monthly temperature. In the warmer states (mean monthly temperature >13°C) the average levels of DDE and total DDT are almost twice as high as those in the cooler states. This geographical difference holds for blacks and whites. By contrast, dieldrin residues do not show this geographical difference. The same suggestion of temperature dependence is suggested from Fig. 8.3, which illustrates prevalences of organochlorines in European humans (reproduced by kind permission of J. Robinson). Not shown in this figure, but evidence further supporting this concept, is the total human DDT prevalences in Israel reported by Wasserman *et al.* (1967), and DDT prevalences for the New Delhi area reported by Dale *et al.* (1965). There are obviously several facets of these descriptive epidemiologic appraisals of the human DDT residue levels which make it difficult to totally support the concept that food is the main source of the human DDT residues in all areas of the world. Other factors are: the North-South residue prevalence differences in the United States, although there is extensive inter-state shipment of food, the occurrence of large DDE whole blood values in families (Mrak report, 1970), comparisons of amounts of DDE in blood with amounts of DDT in house dust, Mrak Report (1969), association of serum levels of DDT and DDE with social class, the report by Edmundson *et al.* (1970) of a tenfold increase in serum DDT in four months in a sentinel cat, in a home with large human DDE residues compared to small DDE changes observed in another similarly-fed cat in a home with occupants having lower serum DDE residues.

All these data suggest that conditions in the home, particularly the recent and past use of DDT in the home and its occurrence in house dust, may not infrequently, be a significant source of human DDT residues, especially in warmer climates.

The only demographic differences which negate this hypothesis are the greater levels of DDT residues usually noted in males than females who,

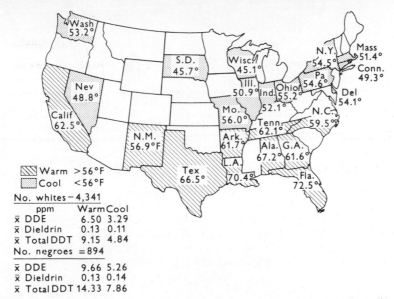

FIG. 8.2. Geographical comparisons of human adipose pesticide residues (ppm) in different mean monthly temperature zones of the U.S.A. (Human monitoring and Florida data, 1968.)

FIG. 8.3. Concentrations (in ppm) of DDT-type compound and HEOD in adipose tissue in Europe; concentrations of HEOD in italics.

for the most part, spend more of their time in the home environment than do the males.

Dieldrin residue data demonstrate neither person nor place variations, and the distribution characteristics of the residue in population surveys, entirely support the concept that food is the main source of the body burden of this pesticide.

8.4   FACTORS INFLUENCING THE PERSISTENCE OF PESTICIDES

Pesticide persistence is a term most frequently used to measure the relative biodegradable potential of the pesticide in the environment. Insofar as human tissues are concerned, the term is more frequently used in connection with the lipotrophic characteristics of the pesticide.

In contrast to the organophosphate and carbamate insecticides, which are much less soluble and usually do not present a residue problem, the organochlorine pesticides are all fat soluble and readily stored in all depots of the body. Highest concentrations are noted in the fat, with other tissues largely reflecting their respective adipose concentration. Thus, DDT concentrations in fat depots were ten times greater than those found in the liver, and one hundred times greater than those found in the brain, gonad and kidney (Fiserova-Bergerova et al., 1967). Organochlorine residue concentrations in different fat depots of the body were essentially the same (Hayes et al., 1958). Although several reports of the occurrence of DDT-toxicity in animals following starvation have been reported, the same has not been observed in man (Fitzhugh and Nelson, 1947; Dale et al., 1962; Brown, 1970).

DDT in human milk was first reported by Laug et al. (1951). Other countries which have reported organochlorine residue data in human milk have included England, Italy, Hungary, Germany, U.S.A., and Russia. Levels of DDT ranged from 0.05 to 0.26 ppm in most of these surveys.

8.5   HAZARDS CAUSED BY THE PESTICIDE RESIDUE

The clinical significance of the pesticide residue levels depends upon the concentration and toxicity of the individual pesticide, and the tissue in which the residue is measured. In acute poisonings and in conditions of occupational exposure, the residue is used to diagnose and delineate the toxicity, and as a surveillance tool. A threshold level for dieldrin intoxication of 0.15-0.20 μgml^{-1} has been used by Brown et al. (1969) and by Jager (1970). For 'Telodrin' and endrin, the blood threshold value below which no human intoxication was observed was 0.015 μgml^{-1} for 'Telodrin', and 0.05-0.100 μgml^{-1} for endrin. In Florida, symptoms of parathion intoxication occurred when blood concentrations reached 22×10^{-3} ppm (Davies, unpublished). In nine non-fatal parathion intoxications, the average admission urinary paranitrophenol was 10.8 ppm, and it was 40.3 ppm in 14 fatal parathion intoxications (Davies *et al.*, 1966). Symptoms of mercury intoxication may occur with levels of total mercury in excess of 500 ppm×10^3 in whole blood, 1,000 ppm×10^3 in red cells and 150 ppm in hair.

Incidental exposure is measured by the adipose or blood residue of the persistent pesticides and in the general population has not been causally associated with disease. Even in the occupationally exposed, fat levels of total DDT as high as 739 ppm and serum levels as high as 2,914×10^{-3} ppm have been recorded in individuals with no obvious disease at the time they were examined (Poland *et al.*, 1970). Cortisol and phenylbutazone metabolism were measured in 18 persons with a five year history of work exposure in a DDT manufacturing plant and compared with that of 18 matched control persons. The average serum *p,p'*-DDT and *p,p'*-DDE were 573 ppm×10^3 and 586 ppm×10^3 respectively in the DDT factory workers and 12 ppm×10^3 and 35 ppm×10^3 respectively in the control group. Their data indicated that under these prolonged and intensive periods of DDT exposure, the urinary β hydroxycortisol was increased by 57% and the serum phenylbutazone half-life was reduced by 19%. Drug and pesticide interactions have also been demonstrated by Kolmodin *et al.* (1969), who showed increased metabolism of antipyrene under conditions of occupational exposure to a variety of organochlorine pesticides (lindane, DDT and chlordane). The whole problem of drug and pesticide interactions is currently under intensive investigation and the pharmaco-clinical consequences of liver enzyme induction is just beginning to be observed.

Apart from drug interactions, several occupational studies of individuals with appreciable years of exposure to insecticides have failed to reveal serious disease under conditions of occupational exposure. The health profile of workers occupationally exposed to DDT, measured by both physical and biochemical surveys, was particularly reassuring with regard to the safety of these exposures (Laws *et al.*, 1967). Similar data for persons occupationally exposed to aldrin, dieldrin, Telodrin and endrin have been reported by Jager (1970). The results of these intensive occupational studies are particularly reassuring as to the possible occupational health hazard of these pesticide industries.

8.6   MINIMISING AND REMOVING RESIDUES

The occurrence of residues of the persistent pesticides in animal tissue in excess of acceptable tolerances can lead to serious economic hardship for the farmer. In the United States, food residues are monitored regularly by both Federal and State agencies. At the Federal level, the Food and Drug Administration is responsible for interstate surveillance, and at the State level, the State Department of Agriculture assumes the responsibility for intra-State monitoring, and the surveillance of imported foods coming from outside of the United States.

Occasionally, accidental contamination of food stock occurs, resulting in the occurrence of excessive pesticide residues in milk or meat. The acceptable tolerances are exceeded and the farmer is faced either with having to destroy his herd, or waiting until levels have dropped to an acceptable level. Under these circumstances, it is not surprising that efforts have been made in animals to lower tissue residues deliberately by pharmacological methods. Experimentally, the feasibility of this procedure has been suggested for several pesticides and in a variety of animal species. Thus, increased metabolism and diminished storage of benzene hexachloride (Karansky *et al.*, 1964), dieldrin (Street and Chadwick, 1967; Cueto and Hayes, 1967) and DDT (Alary *et al.*, 1968) have all been demonstrated. Applying this procedure practically, dairy scientist Robert M. Cook was reported to have lowered dieldrin residues in milk with phenobarbitone and charcoal from a herd of cows accidentally contaminated with aldrin (Reeder, 1969).

Although there are no clinical indications at present for the deliberate reduction of pesticide residues in man, the feasibility of this interaction was suggested from the reduction of fat storage residues of DDT and its metabolites and dieldrin, observed in fourteen volunteers given 100 mg of diphenylhydantoin daily for nine months. The median percent change for DDT, DDE and dieldrin was 75%, 61% and 73% respectively, after nine months medication (Davies *et al.*, 1971).

8.7 THE FUTURE OUTLOOK AND NEEDS FOR FURTHER RESEARCH

Human pesticide residues are but one example of the growing number of environmental pollutants, of which trace amounts can be detected in man. There is need to know the magnitude, distribution and secular trends of these traces in the population at large. This can be thought of as a field of chemical epidemiology, and the descriptive techniques that we have seen being applied to the DDT human residue could serve as a model for the study of these other environmental chemicals. The frequency distribution in the healthy population should substantiate toxicologic information as to environmental sources. Residue prevalences in disease should support animal toxicological data. It would appear that for DDT, even after twenty-two years of usage, we are only now beginning to understand national residue prevalences and the special demographic and geographical features of human pesticide residues. In this regard, as recently as 1969, that situation was summarised by the Mrak Report (1969) when it stated, 'the field of pesticide toxicology exemplifies the absurdity of a situation in which 200 million Americans are undergoing lifelong exposure, yet our knowledge of what is happening to them is at best fragmentary and for the most part indirect and inferential. While there is little ground for forebodings of disaster, there is even less for complacency. The proper study of mankind is man. It is to this study that we should address ourselves without delay.'

At present, large numbers of food analyses are conducted annually in the United States. During the period July 1, 1963 to June 30, 1966, the Food and Drug Administration examined 53,194 objective samples. Objective samples are those in which there is no reason to suspect residues. These samples included 3,836 imported samples. An additional 25,728 samples were examined because of suspected residues. During the fiscal years 1965 and 1966, the U.S. Department of Agriculture examined 5,036 objective samples of animal tissue collected from animals slaughtered in federally inspected establishments (Duggan, 1969). The largest human residue survey is the one reported in the Mrak Report (1969) which presents residue data from 4,969 persons. The goal of food monitoring programmes is presumably to prevent excessive intake of pesticides by man in his food. Insofar as man is concerned, is it not time that a decision should be reached as to whether the possibility exists that such traces are or could be harmful? If it is judged that there is no danger, then is there real need to maintain continuous monitoring of food for pesticide residues? If, on the other hand, the consensus of opinion is that there is indeed a potential hazard, then since, in many areas, non-dietary sources may be almost as important a source of residues as are dietary sources. should not man monitor himself as vigorously and efficiently as he currently analyses his food intake? It may well be that growing public concern, and the warning signals in nature, are such that they will not permit a protracted state of ambivalence on this issue.

REFERENCES

D. C. Abbott, R. Goulding and J. O'G. Tatton, *Brit. med. J.,* **3**, 146 (1968)
L. Acker and E. Schulte, *Naturwissenschaften,* **57**, 497 (1970)
J. G. Alary, J. Bordeur, M. G. Cote, J. C. Panisset, P. Lamothe and P. Guay, *Rev. Canad. Biol.,* **27**, 269 (1968)
A. Bevenue and H. Beckman, *Residue Reviews,* **19**, 83 (1967)
H. V. Brewton and H. J. W. McGrath, *New Zealand J. Sci.,* **10**, 486 (1967)
H. Bronisz, W. Rusiecki, J. Ochynski and E. Bernard, *Dissert. Pharm.,* **19**, 309 (1967)
J. R. Brown, *Canad. Med. Assoc. J.,* **97**, 367 (1967)
J. R. Brown, *Toxicol. Appl. Pharmacol.,* **17**, 504 (1970)
V. K. H. Brown, C. G. Hunter and A. Richardson, *Brit. J. Industr. med.,* **21**, 283 (1964)
J. E. Campbell, L. A. Richardson and M. L. Schafer, *Arch. Environ. Hlth.,* **10**, 831 (1965)
L. J. Casarette, G. C. Fryer, W. L. Yauger and H. W. Klemmer, *Arch. Environ. Hlth.,* **17**, 306 (1968)
C. Cueto, Jr. and F. Biros, *Toxicol. Appl. Pharmacol.,* **10**, 261 (1967)

J. C. Dacre and R. W. Jennings, *Toxicol. Appl. Pharm.*, **17**, 277 (1970)

W. E. Dale, T. B. Gaines and W. J. Hayes, Jr., *Toxicol. Appl. Pharm.*, **4**, 89 (1962)

W. E. Dale, M. F. Copeland and W. J. Hayes, Jr., *WHO Bull.*, **33**, 471 (1965)

W. E. Dale, A. Curley and C. Cueto, Jr., *Life Sci.*, **5**, 47 (1966)

W. E. Dale, A. Curley and W. J. Hayes, Jr., *Ind. med. Surg.*, **36**, 275 (1967)

J. E. Davies, J. H. Davis, D. E. Frazier, J. B. Mann and J. O. Welke. R. F. Gould (Editor), Organic Pesticides in the Environment. *Advances in Chemistry Series*, **60**, 67 (1966)

J. E. Davies, W. F. Edmundson, C. H. Carter and A. Barquet, *Lancet*, **ii**, 7 (1969a)

J. E. Davies, W. F. Edmundson, A. Maceo, A. Barquet and J. Cassady, *Amer. J. Pub. Hlth.*, **59**, 435 (1969b)

J. E. Davies, W. F. Edmundson, N. J. Schneider and J. C. Cassady, *Pest Monit. J.*, **2**, 80 (1969c)

J. E. Davies, W. F. Edmundson, A. Raffonelli and C. Morgade, *Amer. J. of Epid.* (1971)

J. E. Davies and W. F. Edmundson, *Symposia Enterprises* (in press)

V. DelVecchio and V. Leoni, *Nuovi Annali d'igine Microbiologia*, **18**, 107 (1967)

A. Denes, *Nahrung*, **6**, 48 (1962)

A. Denes, *Inst. of Nutrition* (1965)

R. E. Duggan, *Ann. N.Y. Acad. Sci.*, **160**, 173 (1969)

W. F. Durham, J. F. Armstrong, W. M. Upholt and C. Heller, *Science N.Y.*, **134**, 1880 (1961)

W. F. Durham, J. F. Armstrong and G. E. Quinby, *Arch. Environ. Hlth.*, **11**, 76 (1965)

W. F. Edmundson, J. E. Davies, M. Cranmer and G. A. Nachman, *Ind. Med. Surg.*, **38**, 145 (1969)

W. F. Edmundson, J. E. Davies, A. Maceo and C. Morgade, *So. Med. J.*, **63**, 1440 (1970)

H. R. Egan, R. Goulding, J. Roburn and J. O'G. Tatton, *Brit. med. J.*, **2**, 66 (1965)

R. Engst, R. Knoll and B. Nickel, *Pharmazie*, **22**, 654 (1967)

H. Enos. Personal Communication (1971)

V. Fiserova-Bergerova, J. L. Radomski and J. E. Davies, *Ind. Med. Surg.*, **36**, 65 (1967)

O. G. Fitzhugh and A. A. Nelson, *J. Pharmacol. Therap.*, **89**, 30 (1947)

K. Halacka, J. Hakl and F. Vymetal, *Cs. Hyg.*, **10**, 188 (1965)

W. J. Hayes Jr., G. E. Quinby, K. C. Walker, J. W. Walker, J. W. Elliott and W. W. Upholt, *Arch. Ind. Hlth.*, **18**, 39 (1958)

W. J. Hayes Jr., W. E. Dale and R. LeBreton, *Nature* (Lond.), **199**, 1189 (1963)

W. S. Hoffman, H. Adler, W. I. Fishbein and F. C. Bauer, *Arch. Environ. Hlth.*, **15**, 58 (1967)

A. B. Hollingshead, Yale University Press (1957)

C. G. Hunter, J. Robinson and A. Richardson, *Brit. med. J.*, **1**, 221 (1965)

C. G. Hunter and J. Robinson, *Arch. Environ. Hlth.*, **15**, 614 (1967)

C. G. Hunter, J. Robinson and M. Roberts, *Arch. Environ. Hlth.*, **18**, 21 (1969)

J. W. Jager, Elsevier Pub. Co. (1970)

W. Karansky, T. Portig and H. W. Vohland, *Arch. Exptl. Pathol. Pharmakol.*, **247**, 49 (1964)
H. W. Kleimmer. Personal Communication (1971)
B. Kolmodin, D. Azarnoff and F. Sjoquvist, *Clin. Pharmacol. Therap.*, **10**, 638 (1968)
H. F. Kraybill, *Canad. Med. Assn. J.*, **100**, 209 (1969)
D. S. Kwalick, *J. Amer. Med. Ass.*, **215**, 120 (1971)
E. P. Laug, F. M. Kunze and C. S. Prickett, *Arch. Indstr. Hyg. Occup. Med.*, **3**, 245 (1951)
E. R. Laws, A. Curley and F. J. Biros, *Arch. Environ.*, **15**, 766 (1967)
V. M. Llinares and M. Wasserman, Unpublished Data (1968)
R. Maes and A. Heyndrickx, *Meded. Rijksfaculteit Lanbouwetenschappen Gent.*, **31**, 1021 (1966)
H. Maier-Bode, *Med. Exp.*, **1**, 146 (1960)
D. P. Morgan. Personal Communication (1970)
D. P. Morgan and C. C. Roan, *Arch. Environ. Hlth.*, **22**, 301 (1971)
E. M. Mrak. Report of the Secretary's Commission on Pesticides and Their Relationship to Environmental Health (Parts I and II). U.S. Govt. Ptg. Off. (1969)
J. A. O'Leary, J. E. Davies, W. F. Edmundson and M. Feldman, *Amer. J. Obstet. Gynec.*, **106**, 939 (1970)
W. Perkow, Die Insektizide. Huther Verlage, p. 360. Heidelberg (1956)
A. Poland, D. Smith, R. Kuntzman, M. Jacobson and A. H. Conney, *Cl. Pharmacol. Therap.*, **11**, 724 (1970)
J. L. Radomski, W. B. Deichman, E. E. Clizer and A. Rey, *Fd. Cosmet. Toxicol.*, **6**, 209 (1968)
N. Reeder, *Farm J.* (1969)
C. Roan, D. Morgan and E. H. Pachall, *Arch. Environ. Hlth.*, **22**, 309 (1971)
J. Robinson, A. Richardson, C. G. Hunter, A. N. Crabtree and M. J. Rees, *Brit. J. Ind. Med.*, **22**, 220 (1965)
J. Robinson and C. G. Hunter, *Arch. Environ. Hlth.*, **13**, 558 (1966)
J. Robinson, *Ann. Rev. Pharm.*, **10**:
W. P. Schoor, *Lancet*, **ii**, 520 (1971)
J. C. Street and R. W. Chadwick, *Toxicol. Appl. Pharmacol.*, **11**, 68 (1967)
M. de Vlieger, J. Robinson, A.N. Crabtree and M.C. van Dijk, *Arch. Env. Hlth.*, **17**, 759 (1968)
K. C. Walker, M. B. Gvette and G. S. Batchelor, *J. Agr. Fd Chem.*, **2**, 1034 (1954)
M. Wasserman, M. Gon, D. Wasserman and L. Zellermayer, *Pest. Monit. J.*, **1**, 15 (1967)
M. Wasserman, D. H. Curnow, P. N. Forte and Y. Groner, *Ind. Med. Surg.*, **37**, 295 (1968)
M. Wasserman, D. Wasserman, I. Ivirani and Halos. Personal Communication (1970)
M. Wasserman, D. Wasserman, S. Lazarovici, A. M. Coetzee and L. Tomatis, *African Med. J.*, **44**, 646 (1970)
M. Watson, W. W. Benson and J. Gabica, *Pest. Monit. J.*, **4**, 47 (1970)
M. Weihe, *Ugeskr. Laeger*, **128**, 881 (1966)
H. R. Wolfe, W. F. Durham and J. F. Armstrong, *Arch. Environ. Hlth.*, **21**, 711 (1970)

Chapter 9

Pesticide Residues in Food

R. E. Duggan* and M. B. Duggan

*Department of Health, Education and Welfare,
F.D.A. Rockville, Maryland*

9.1 INTRODUCTION

9.1.1 Importance

Man suffers vicariously from the effects of pesticides on other animal species, but is considerably more directly concerned with threats to the health of his offspring. In particular, he will not tolerate a danger to them from the food intended for their nurture. Any prospects of birth defects or brain or nerve damage to them from knowable sources is unacceptable to him, yet equally untenable to him is human hunger. So he finds himself at the same time appreciative of the marvellous feats of the food producers, those masters of the green revolution, and anxious about unknown contaminants in his own food supply.

Many diligent and competent people spend their time analysing pesticides in foods, assessing the findings, and implementing their conclusions with regulations designed to guard the purity of their food supply. It has been no accident, but rather the result of just such endeavours, that the

* Present address: 6319 Anneliese Drive, Falls Church, Virginia.

level of pesticide residues in the food supply, in spite of the use of millions of pounds of pesticide chemicals each year, has remained within safe levels.

The discovery that chlorinated pesticide residues were present in most living creatures throughout the world, mobilised the attention of scientists and laymen alike. It was demonstrated to all in a one-world fashion that actions within one nation can affect all. No longer can it be assumed that the cycling of chemicals within this closed system, namely the earth, is a matter beyond the power of man to affect. Chlorine has been wrested by considerable force from its natural sink, the salt of the sea, and by what route, and how long before it returns, all have an interest.

9.1.2   History

This chapter will discuss the sources and amount of residues of pesticides in food and the factors that affect these residues. It is necessary, however, to know something of the history of the problem. In retrospect, in the period prior to World War II, before the discovery of the insecticidal effect of certain organochlorines, examination of foods for residues of pesticides was simple. There was then in use a limited list of lethal inorganic compounds and a handful of relatively non-toxic plant products, much less effective in the control of pests than present pesticides. Furthermore, the location of residues from these in the food supply was predictable; for example, one examined the crop upon which they were used. Translocation and biomagnification problems were yet to come. Wet chemistry methods, hardly more sophisticated than those used in an elementary course in quantitative analysis, were in general use. Chromatographic methods were not available for delicate separations. Spectrometers of various wave lengths were used mostly by professors and their graduate students and were not then, as now, the commonly available means of quantitative determination. The old gravimetric and volumetric methods had questionable accuracy levels in the milligram range. Now, routine instrumental methods can be reliable in analysing residues as small as picograms. This improvement in methodology, which is a continuing process, has had its own effect on regulations concerning pesticide levels in foods. Some noble ideals, such as the position that there should be positively no residues of anything whatsoever harmful, in milk or fish (zero residue), were caught in a pincers attack between the honest analyst who had to say that fractional parts per million of this or that were actually present, and the need of the whole world for milk. The milk or fish was not necessarily more polluted; it was simply now possible to tell exactly how polluted the milk or fish was and possibly had been all along. In the U.S.A., midway along in this history of a growing fear of contaminants and a growing capability for their detection, there was added to the body of law regulating food additives, the prohibition of any amounts of added carcinogens in food. Even though the prohibition was not directly applicable to pesticide regulation, for all practical purposes it may as well have been. Researchers produced different forms of cancer in various animals with massive doses of a variety of chemicals, and this was coincident with the new capability of finding very small amounts of these chemicals in some of man's oldest and most commonly used foods. Scientific findings outstripped scientific judgment of their significance. New dilemmas continually face regulatory officials and consumers alike. In other words, the history of coping with the problem of chemicals in the food supply, though it goes way back, is still being written.

The regulation of pesticides had its beginning in much simpler forms than the specific laws in effect throughout most of the world. The differing bases of regulation have resulted in misunderstandings between nations. Better understandings and more uniform standards may be the end result of the continued deliberations by the Food and Agriculture Organisation (F.A.O.) Working Party of Experts and the World Health Organisation (W.H.O.) Expert Committee on Pesticide Residues.

9.1.3 Perspective

The amount and kind of pesticide used on a crop is the farmer's choice, within two limitations: One of these is the availability of the material and the other is his expectation of the saleability of the crop. The pesticide regulations of his own country, as well as those of the country in which the food will be sold, determine these limitations. It is the choice, amount, and timing of pesticide applications that jeopardise the farmer's crop at harvest time. He can neither see nor measure these residues. He must rely on package instructions and advice of agricultural experts, that so much applied so many days before harvest will result in legal residues. Should anything go amiss, government agencies acting as the protector of the consumer, both judge and act to reduce the hazard by confiscation of the food or other means.

Development, testing, approval, and distribution of pesticide chemicals involve numerous specialised groups, e.g. research chemists, plant pathologists, toxicologists, farmers, public health and agricultural experts, etc., engaged in a common purpose to improve the yield and quality of food. More recently, however, a new group, the environmentalists, joined the debate. The members of this group exhibit all levels of expertise, ranging from concerned scientists, to those who feel that a potassium ion that arrives in the garden via a sack of fertiliser is a bad potassium ion. Emotional rejection of technology is a present factor shaping pesticide history. Many of the most vocal can contribute no information at all to assist in detection or regulation of real or potential dangers. Because the anxiety of the questioners is genuine, the worker in the field of pesticides would do well to control his impatience, and supply the facts that he has with full interpretations, explaining what should be their common concern.

9.2   AMOUNT OF PESTICIDE RESIDUES IN FOOD

Which pesticides contribute the residues found in foods? In what amounts are they present? How much residue is actually ingested? What changes have occurred and what further changes may be predicted in these residues? The answers to these questions should be matters of objective record.

9.2.1   Sources of data

The total amount of data on pesticide residues in food is so great that even bibliographies of the papers containing it are unmanageable in length. Carefully chosen descriptors and a computerised search will still result in a mixture of three types of data with consequent difficulties of comparison. These categories are research data, surveillance data, and survey data.

The research data relates the residues to known amounts of applications, known timing of applications, and known conditions of preparation and storage. The importance of this is quite evident, as it permits the interpretation of situations encountered in surveillance and surveys, in which the only facts known are the final residue figures. Should these figures be unacceptable, research information indicates changes in production conditions necessary to alter them.

Surveillance data are the findings for law enforcement of health inspectors looking for violations of regulations. These give a biassed picture of day-to-day hazards because it is the food that is suspected of violation that is sampled first. This data is important for its revelation of exactly how badly contaminated food can be. This data is not so easily found as the research data, because the latter is published in regular scientific channels. Substantial amounts of surveillance data is accumulated by various government agencies in the major food-producing countries. Too often the residue data, having served its immediate programme purpose, is consigned to permanent but little-consulted files.

Survey data is accumulated by monitoring studies, to determine what is the usual and average situation. Monitoring investigations are useful in regulatory or quality control processes. They provide data which can be studied statistically to estimate the percentage of samples exceeding any given level and to analyse annual changes in frequency of contamination with specific chemicals, which may represent trends or merely irregular behaviour.

Monitoring data are also less easily available than research data by ordinary literature search. It has been observed (Duggan, 1969) that there is a substantial body of literature showing the levels of residues remaining in foods under controlled experiments in the field and during processing. There is relatively little published information readily available on foods in market channels. This does not mean that the information is non-existent. On the contrary, various government agencies and food processing firms, in the major food-producing countries, have substantial banks of such data. Summaries of data from some continuing programmes in government may appear in little known official reports having limited distribution. To some extent, it is difficult and expensive to publish large quantities of data unless it is known to be of interest and use to other investigators.

The most important data available at present comes from a growing body of information collected as 'Total Diet Studies'. They were begun in the U.S.A. in 1961, and have been executed continuously since then (Mills, 1963), (Williams, 1964), (Cummings, 1965), (Duggan et al., 1966, 1967), (Martin and Duggan, 1968). These were preceded by 'total meal' studies (Walker et al., 1954) and (Durham, 1965). Researchers in Canada (Swackhamer, 1965; Smith, 1971), Hungary (Soos, 1969) and England (McGill and Robinson, 1968; McGill et al., 1969; Harries et al., 1969; Abbott et al., 1969; Abbott et al., 1970) have also conducted such studies. The American study poses the question: What is the greatest residue that might be consumed? The other studies ask: What is the average consumption? The dispassionate facts revealed by these studies are of especial value amid the emotional guesses recently current.

The series of monographs 'Evaluations of Some Pesticide Residues in Food', issued jointly by the Food and Agriculture Organisation of the United Nations and the World Health Organisation, are excellent continuing sources for summary information on levels of specific residues in foods. In their evaluations, summaries are given on levels of residues from supervised applications, surveillance programmes and total diet studies.

9.2.2 Difficulties encountered in comparison of data

It would seem possible to write each and every datum relating to pesticide residues into one computer and, thenceforth, formulate complex questions to it for the ultimate answers in pesticide wisdom. However, even when reports are available, there are hazards in correlating and comparing data from different investigations.

Surveys are variously structured according to the questions to be asked of its data, or the resources available to its planners. A survey of the pesticide residue consumed by an infant, on a normal diet of commercial infant foods (Lipscomb, 1968), differs greatly from one intended to measure residues consumed by a hearty adult eater with a normally diversified diet. There is little uniformity in sampling procedures among the various investigations having differing objectives, viz. quality control, regulatory control, monitoring and surveillance. Bias may be introduced deliberately or inadvertently by the relative importance of specific commodities to the dietary habits and economics of various countries, and by population groups within a country. The economics of sampling and designing analytical programmes require that judgmental factors involving the frequency of, or the potential for, significant (illegal or excessive) findings, be given consideration before initiating studies of their presence.

Choice of methods of analysis and choice of combinations of confirmatory methods would also have some effect on the residues actually found, as opposed to those potentially present. The potential presence of any one, or any combination, of over 800 different pesticidal compounds poses a very real problem in discussing residue levels in food. It is hardly possible to determine, even with the sophisticated multi-residue analytical methods currently available, the presence or absence of all of these compounds in any given sample of food, whether in the raw or ready-to-eat form (Burke, 1971). Relatively few pesticides, perhaps 100 or so, are used extensively enough for serious consideration in monitoring programmes. Schechter (1967 and 1971) initially listed 29 pesticide chemicals described as important for monitoring residue levels in the U.S. National Pesticide Monitoring Programme and increased the list to include 21 additional pesticides in the most recent evaluation.

Analytical procedures, although referenced to standard methods, as AOAC, are often modified by the examining laboratory. One valid reason for intra-laboratory method modifications is the difference in available instruments or even available attachments for them. The widespread use of non-specific methods of analysis, such as gas-liquid chromatography, requires expert interpretation of findings and confirmation of identity of the residue by another procedure (Schechter, 1968). Natural substances may give false positives (Frazier *et al.*, 1970), and industrial chemicals may interfere with the determination, for example, polychlorinated biphenyls have had an additive effect in analyses for DDT compounds (Eidelman, 1963; Harrison, 1966; Robinson *et al.*, 1967) (see chapter 1).

Despite the difficulties just catalogued, there are several studies on a national scale, designed to answer the question of the kinds and amounts of pesticide residues regularly occurring in food.

9.2.3 General findings of the kind and amount of pesticide residues in human food

Because total diet studies were structured to identify and to measure these residues, their results must be given particular attention.

The series of British studies are very valuable. McGill and Robinson (1968) estimated the average daily intake of organochlorine residues for persons in S.E. England, based upon a twelve-month study of completely prepared meals. They found that the DDT group of insecticides and dieldrin were consistently present in the whole diet, and that the rate of ingestion of the DDT group was about one-tenth of the acceptable daily intake recommended by W.H.O. Pesticide Committee in 1965.

Another study, quite extensive in scope, of the pesticide residues in the total diet of England and Wales, 1966-1967 (Harries *et al.*, 1969) has been reported in a series of articles (Abbott *et al.*, 1969; Abbott *et al.*, 1970). Sixty-six samples represented one-year diets of domestic and imported foods, subdivided into seven classes. They also found that about two-thirds of the dieldrin intake came from the meats and fats groups. In the portion of the study determining organophosphorus compounds (Abbott *et al.*, 1970), only 6 of 39 demonstrable compounds were found, and five of them only once. The sixth and most frequently occurring compound was malathion, which was found in 18% of the cereal samples.

These reports show that levels of certain pesticide chemicals exceeded the F.A.O.-W.H.O. acceptable daily intake during some periods. McGill and Robinson (1968) reported an ingestion rate of 21 mg HEOD (dieldrin) from completely prepared meals in England. Later reports by McGill *et al.* (1969) showing a decline in the dietary intake of HEOD (dieldrin) were confirmed by the investigations of Abbott *et al.* (1969). This decline was attributed to the restrictions governing the uses of aldrin and dieldrin in England.

Soos (1969) examined 79 meals representing 20 complete daily menus, to survey the content of organochlorines in Hungarian diets in 1967 and 1968. His comparison of DDT group compounds with British and American findings show levels two to ten times higher: 0.131 mgkg^{-1} (Hungary) compared with 0.046 mgkg^{-1} (British) and 0.013 (American). Soos also found very high lindane residues, with a mean value for the total diet of 0.052 mg/kg. He speculated that this higher level, exceeding the F.A.O.-W.H.O. acceptable daily intake, was caused by recent restrictions on DDT and prohibition of cyclodiene pesticides in Hungary, which contributed to a greater use of lindane.

With large dietary levels of DDT still persisting and exceeding the acceptable daily intake, Hungarian diets during the transitional period contained a double burden of organochlorine residues from both past and present usages.

The U.S. National Monitoring Programme for pesticide residues in Foods and Feeds (Duggan and McFarland, 1967; Duggan and Cook, 1971) is one of the most comprehensive continuing investigations of levels of residues in foods reported. The results of this investigation have been reported and evaluated on a regular basis as noted earlier. This investigation will be considered in some detail because of the large amount of data accumulated. There are three major elements in the study. First, the collection and examination of objective (monitoring) samples of raw agricultural commodities in wholesale distribution channels. Second, the collection and examination of objective samples of red meat and poultry at government-inspected slaughter houses and poultry processing plants. These elements are used for the enforcement of tolerances and are extended as necessary by the collection and examination of selective samples. Third, total diet studies on food collected in market basket samples from retail stores, prepared by trimming, washing and cooking as for human consumption then analysed for a measure of the dietary intake of pesticide chemicals.

The most recent report (Duggan *et al.*, 1971) summarising the investigations from July 1, 1963 to June 30, 1969, shows that residues of 83 different pesticide chemicals were found in the 111,296 samples of agricultural products examined during this period. Thirty of these pesticide chemicals were commonly found, i.e. in 1% or more of samples in any year, within broad food classes. Results are reported on the basis of broad food classes, such as leafy vegetables, legume vegetables, manufactured dairy products, etc., and related to total diet studies for the same period.

Not unexpectedly, residues of most of these chemicals were found at lower levels in the total diet samples. Twenty-two pesticide chemicals were found often enough to be considered commonly present in the U.S. diet.

Table 9.1 lists those organic pesticide residues (22) commonly found in the total diet for the period 1965-1970, and shows the range of incidences of positive findings, and the range of the calculated daily intake from all foods. The frequency of occurrence and the sensitivity of the analytical methods must be taken into consideration in attaching significance to this type of data. For example, even though the maximum daily intake for carbaryl is shown as 0.15 mg, and that for DDT as 0.041, the relatively insensitive analytical method for carbaryl, and a relatively few positive findings may affect the calculated value and leave false impressions of the

TABLE 9.1. Organic residues commonly found in total diet in the United States, 1965-1970.

	Incidence %	Daily intake range (mg)
DDT	37.3-55.6	0.015-0.041
DDE	31.1-50.6	0.010-0.028
TDE	19.4-32.8	0.004-0.018
Dieldrin	15.3-31.3	0.004-0.007
Lindane	10.6-15.8	0.001-0.005
Heptachlor epoxide	8.9-13.4	0.001-0.003
BHC	6.0-13.1	0.001-0.003
Malathion	1.9-11.1	0.003-0.013
Carbaryl	N.D.-7.4	N.D.-0.15
Aldrin	0.8- 5.6	T -0.002
2,4-D	0.3- 4.2	T -0.005
Diazinon	0.3- 5.8	T -0.001
Dicofol	0.5- 5.6	0.003-0.010
PCP	N.D.- 3.3	N.D.-0.006
Endrin	1.1- 3.3	T -0.001
Methoxychlor	N.D.- 1.9	N.D.-0.001
Heptachlor	N.D.- 1.9	N.D.-T
Camphechlor	N.D.- 3.6	N.D.-0.002
'Perthane®'	N.D.- 1.3	N.D.-0.004
Parathion	0.6- 5.0	T -0.001
Endosulfan	0.3- 5.3	T -0.001
Ethion	0.3- 4.4	T -0.004

N.D. = not detected.   T = <0.0005

relative amounts of these chemicals in the diet. Dicofol, methoxychlor, Perthane® and ethion were commonly found, but have not been included in the list of chemicals selected for national monitoring. All other chemicals in Table 9.1 have been included in the monitoring list.

It is not likely that the daily intake of carbaryl exceeds that of DDT, which can be readily seen when the annual values are reviewed.

The following eight additional pesticide chemicals were commonly found in the monitoring samples examined during the same period:

chlordane	carbophenothion
tetradifon	methyl parathion
PCNB	DCPA
chlorbenside	MCP

Of these, only chlordane, PCNB, methyl parathion and MCP are on the monitoring guide. Except for carbophenothion and DCPA, these chemicals have been found in one or more composites of the total diet sample. Other chemicals listed in the monitoring guide have also been found in one or more of the food class composites during the 6-year period of the investigation.

There are 22 additional organic pesticide chemicals and 4 element-based pesticide chemicals (arsenic, bromides, lead, mercury) listed in the monitoring guide. Some of these have been reported in one or more total diet composites during the investigation. Two of these, arsenic and bromides, have been included in the analytical programme since inception of the programme. No distinction is made between naturally-occurring forms of these elements in foods, and amounts that might result from pesticide applications. The daily intake of bromides, calculated on an annual basis, from both sources, ranges from about 16-28 mg, with lower intake levels in the past years. Grain and cereal foods account for most of the bromide intake. The average level of arsenic in the total diet samples, calculated on an annual basis, was 0.02 ppm during 1967-1969.

Mercury residues have been determined, by neutron activation, on 30 selected food classes in 1967 (Corneliussen, 1969). Levels found ranged from 0.002 - 0.01 ppm in dairy product composites; 0.01 - 0.05 ppm in meat, fish and poultry composites; 0.02 - 0.05 ppm in grain and cereal composites; 0.006 - 0.01 ppm in potatoes; 0.002 - 0.01 ppm in legume vegetable composites; and 0.002 - 0.006 ppm in beverages.

No major differences have been observed between imported and domestic (U.S.) foods in, either incidence, levels, or kinds of pesticide chemicals found (Duggan et al., 1971). No attempt was made to study the data by country of origin. Although no detailed comparison has been made, the residue findings do not appear to differ greatly between the various countries where surveillance is maintained.

The U.S. monitoring of residues in foods has shown a high incidence of residues; about 50% of the samples contained residues of one or more pesticide chemicals in foods produced in or imported into the United States. However, the levels of residues were generally low: 75% below 0.11 ppm and 95% below 0.51 ppm.

Table 9.2 shows the number of samples examined in broad commodity classes and the residues exceeding 2 ppm in two fiscal years of the investigation. Where there were enough positive findings for statistical testing, the statistical evaluation indicated that, in general, the trend of incidences of residues of chlorinated pesticides was about the same, or showed a decrease, during the last two or three years of the investigation. There are indications that the residues of organic phosphate compounds are increasing in raw products and in the total diet studies.

The U.S. investigation is of particular value because comparisons can be made over an extended period of time between residues in samples of raw foods reasonably representing a national food supply, and those in food ready for consumption. Furthermore, the data can be used in calculating the dietary intake of pesticides which might result from different food-consumption patterns.

Duggan (1967) observed that the organochlorine pesticide levels in milk and dairy products might serve to establish a relation between the two programmes. These chemicals are fat-soluble, and the normal processing of milk and dairy products does not change the residue levels in milk fat. Results of residues in milk and dairy products are usually reported on a fat basis. Table 9.3 shows the most recent comparison, adding confidence that this relation is reliable.

Duggan and Lipscomb (1971) noted that, although there was no direct relationship between the thousands of surveillance samples and the 134 market basket samples, a comparison of the data was of interest in providing more assurance, that the relatively few total diet samples were giving a reasonably accurate picture of the dietary intake of pesticides. Table 9.4 compares the relationship of DDT levels in raw and ready-to-eat foods for the period 1964-1970.

For dairy products, the averages and range of DDT levels are practically the same in the monitoring programme and the total diet programme. For other product classes, such as large fruits, grains and cereals, and root vegetables, the averages and ranges of DDT levels found in the two parallel investigations are separated approximately by a factor of 10, which adds to the confidence in the data from the limited number of total diet samples.

Figure 9.1 (Duggan and Corneliussen, 1972) shows the distribution of the major classes of organic chemicals, organochlorines, organophosphates, herbicides and carbamates in the dietary intake during four years of the study. Organochlorines, combined with organophosphorus compounds, account for 85-95% of the dietary intake of pesticide chemicals. The proportion of organophosphorus compounds increased from 5.5-26.9% of the total pesticide intake during the last three years. The proportion of herbicides and carbamates has decreased in each year. Herbicides

TABLE 9.2. U.S. food samples containing pesticide residues exceeding 2 ppm.

	Domestic					Imported			
Number samples examined	Number samples (>2.0 ppm)	Pesticide chemical	Number samples	Range of levels (ppm)		Number samples examined	Number samples (>2.0 ppm)	Pesticide chemical	Number samples	Range of levels (ppm)

July 1, 1967–June 30, 1968

Number samples examined	Number samples (>2.0 ppm)	Pesticide chemical	Number samples	Range of levels (ppm)	Number samples examined	Number samples (>2.0 ppm)
Large fruits						
1,551	10	Ethion	1	5.6	162	0
		Carbaryl	3	3.0-8.0		
		Dicloran	2	4.3-8.0		
		Zineb	1	3.6		
		Trithion	1	3.5		
		SOPP*	1	13.5		
		DDT	2	2.8-3.5		
Small fruits						
419	5	Dicloran	1	8.0	58	0
		Dicofol	2	2.1-2.5		
		Ethion	1	3.8		
		DDT	1	3.2		
Leaf and stem vegetables						
2,461	97	Camphechlor	59	2.1-84.0	122	0
		DDT	28	2.2-34.0		
		TDE	4	3.6-21.1		
		Zineb	12	2.4-6.2		
		Parathion	3	2.3-3.0		

Commodity	No.	Pesticide	n	Range		
		Sodium arsenite	1	5.4		
		Lead arsenate	1	4.2		
Root vegetables 1,954	6	Parathion	2	2.1-3.3	67	0
		DDT	2	3.5-3.9		
		Dicloran	1	3.0		
		Lindane	1	2.1		
Vine and ear vegetables 1,091	3	DDT	1	2.3	300	0
		Toxaphene	2	2.3-4.0		
Grains (human) 934	25	Malathion	20	2.1-15.1	8	0
		Methoxychlor	2	3.0-4.0		
		Camphechlor	1	2.3		
		Methyl bromide	1	31.4		
		Mercury	1	4.2		
Grains (animal) 371	1	Malathion	1	28.0	10	0
Dairy products 1,141	5	DDE	4	2.0-2.1	177	3
		DDT	1	2.8		
Fluid milk 1,552	10	DDE	10	2.0-3.9	0	0
		Heptachlor epoxide	1	2.3		
		Lindane	2	3.5-7.6		

* Sodium-o-Phenylphenate; Orthophenylphenol

(continued)

TABLE 9.2. (continued)

	Domestic					Imported				
	Number samples examined	Number samples (>2.0 ppm)	Pesticide chemical	Number samples	Range of levels (ppm)	Number samples examined	Number samples (>2.0 ppm)	Pesticide chemical	Number samples	Range of levels (ppm)
					July 1, 1968–June 30, 1969					
Large fruits	863	7	DDT	1	2.4	94	0			
			Methoxychlor	1	2.7					
			PCNB	1	4.1					
			Diphenyl	1	2.7					
			SOPP*	3	2.2-6.2					
Small fruits	410	9	Methoxychlor	1	2.3	144	2	Dicofol	2	2.1-3.1
			Ethion	4	2.3-3.2					
			Dicofol	2	2.2-3.9					
			Captan	2	5.1-5.9					
Beans	85	1	DDT	1	5.0	24				
Leaf and stem vegetables	1,920	139	Camphechlor	85	2.1-40.0	19	4	Camphechlor	2	2.9-183.0
			DDT	30	2.1-23.0			DDT	2	2.7-68.0
			TDE	1	2.5			Thiodan	1	4.7
			DDE	1	6.1					
			Zineb	6	2.1-20.1					
			Endosulfan	2	2.1-2.3					
			Malathion	1	2.9					
			Tedion	1	2.9					
			BHC	1	2.8					

Commodity (total)		Pesticide	No.	ppm			Pesticide	No.	ppm
		Methyl parathion	1	2.1					
Root vegetables 1,213	4	DDT	2	2.2-3.0	90	0			
		Chlordane	1	3.8					
		Camphechlor	1	2.3					
Vine and ear vegetables 856	2	DDT	2	2.0-7.5	280	0			
Grains (human) 590	14	Malathion	11	2.7-28.0	4	0			
		Methyl parathion	1	5.5					
		Camphechlor	1	2.9					
Grains (animal) 119	7	Malathion	3	2.1-3.8					
		Parathion	3	2.8-8.8					
		Calcium arsenite	1	3.7					
Dairy products 691	2	Heptachlor epoxide	1	4.1	620	11	BHC	7	2.1-6.1
		BHC	1	6.0			Lindane	2	3.7-7.2
							DDT	1	24.3
							DDE	2	2.0-2.3
							Tetradifon	1	7.9
Shell eggs 640	1	Chlordane	1	14.8	0				
Fluid milk 857	4	DDE	2	2.3-2.6	0				
		TDE	1	4.1					
		Methoxychlor	1	4.1					

Adapted from: W. R. Poage, Chrm. 1971. Fed. Pesticide Control Act of 1971. Hearings U.S. Gov't. Printing Office, #2.A.

TABLE 9.3.   Average amounts of pesticide chemicals in dairy products (fat basis).

Pesticide	Fiscal years			
	1964-1966	1967	1968	1969
Number objective	12,143	3,150	2,427	1,495
or total diet samples	44	30	30	30
	(ppm)	(ppm)	(ppm)	(ppm)
DDE				
Objective samples	0.066	0.087	0.069	0.058
Total diet	0.074	0.050	0.063	0.048
Dieldrin				
Objective samples	0.042	0.017	0.011	0.008
Total diet	0.016	0.019	0.012	0.019
DDT				
Objective samples	0.042	0.033	0.014	0.009
Total diet	0.037	0.032	0.030	0.023
Heptachlor epoxide				
Objective samples	0.036	0.005	0.003	0.005
Total diet	0.010	0.006	0.012	0.012
TDE				
Objective samples	0.026	0.017	0.011	0.007
Total diet	0.013	0.022	0.019	0.010
BHC				
Objective samples	0.007	0.018	0.006	0.013
Total diet	0.008	0.011	0.021	0.007
Lindane				
Objective samples	0.004	0.006	0.001	0.001
Total diet	0.005	0.003	T	0.001
Aldrin				
Objective samples	0.001	0.001	0.001	T
Total diet	0.001	0.001	. .	. .
Heptachlor				
Objective samples	0.002	T	T	T
Total diet	. .	0.001	. .	T
Methoxychlor				
Objective samples	0.001	0.001	0.002	0.005
Total diet	0.002	0.005	0.004	. .

(From: Duggan and Lipscomb (1971))

accounted for 5% of the residue load in 1968 and 1969, but declined to 0.8% in 1970.

Figure 9.2 (Duggan and Corneliussen, 1972) shows the distribution of the total chlorinated residues and DDT and its analogues among the different food classes for each year of the study. This figure also shows the percentage of the residues represented in each food class. The distribution

TABLE 9.4. DDT residue levels in raw and ready-to-eat foods (1964-1970).

Food class		Objective samples (ppm)	Total diet samples (ppm)
Large fruits	Average	0.10	0.010
	Range	0.02-0.25	0.006-0.018
Grains and cereals	Average	0.02	0.005
	Range	<0.005-0.04	0.003-0.008
Leaf and stem vegetables	Average	0.14	0.012
	Range	0.08-0.18	0.006-0.016
Root vegetables	Average	0.04	0.003
	Range	0.03-0.04	<0.001-0.007
Dairy products	Average	0.03	0.023
	Range	0.01-0.04	0.017-0.037

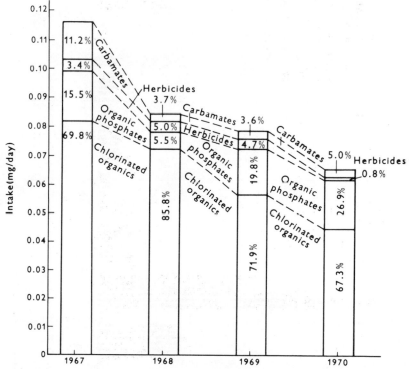

FIG. 9.1.   Distribution of residues by chemical class, 1967-1970. (From Duggan and Corneliussen, 1972.)

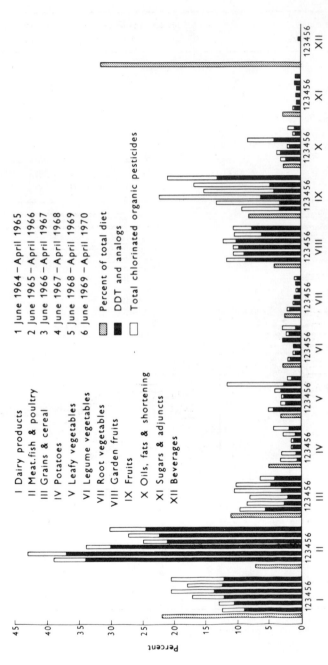

Percent

(From: Duggan and Corneliussen (1972))

FIG. 9.2. Distribution of total chlorinated organic pesticides and DDT and analogs among food classes compared to the percent of total diet represented by each food class.

I Dairy products
II Meat.fish & poultry
III Grains & cereal
IV Potatoes
V Leafy vegetables
VI Legume vegetables
VII Root vegetables
VIII Garden fruits
IX Fruits
X Oils, fats & shortening
XI Sugars & adjuncts
XII Beverages

1 June 1964 – April 1965
2 June 1965 – April 1966
3 June 1966 – April 1967
4 June 1967 – April 1968
5 June 1968 – April 1969
6 June 1969 – April 1970

Percent of total diet
DDT and analogs
Total chlorinated organic pesticides

of the chlorinated compounds and DDT compounds has been remarkably consistent among the food classes.

From such parallel data, the influence of certain types of food on the dietary intake can be assessed. Figure 9.3 shows the approximate proportionate contribution of foods of animal origin to the dietary intake of certain organochlorine pesticide residues, using data from both programmes. It is obvious, recognising that these values are approximations, that major contributors to the dietary intake of pesticide residues may be minor components of the diet. In this approximation, it was assumed that there were no significant losses in food preparation. This would not be a valid assumption for approximations derived from other types of foods, or for other than fat soluble pesticide chemicals in these classes of foods.

9.3 SOURCES OF RESIDUES

The sources of residues in foods are many. Agricultural procedures directly related to production of food are the most obvious sources of residues.

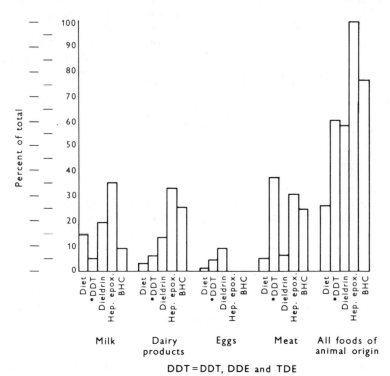

DDT = DDT, DDE and TDE

FIG. 9.3. Approximate proportionate contribution of residues in diet by foods of animal origin. Total diet studies, 1964-1969; Surveillance samples, 1963-1969.

Review of data in the U.S.A. on large residues (for examples, see Table 9.2) suggests that the major problem associated with the use of pesticide chemicals is not usually residues in foods to which the pesticides were applied. For instance, direct agricultural applications to other crops, and drift from aerial applications, have been associated with residues in milk. This is particularly true where the application is near a crop used for animal feed.

There are large amounts of pesticide chemicals used for purposes other than food production. Public health applications for disease control are very important in many areas of the world. Protection of forests and other fibre crops is desirable and important. Weed control on various kinds of roads or other rights of way requires herbicides of various kinds. Also, there are uses for insect control in homes, gardens and in the production of ornamental plants. Disposal of used pesticide containers is also an occasional source of residues in food. Pesticide chemicals used for these many different purposes are mobile. They may be transported by air, soil, and water movements (see chapter 10) and by incompletely-defined mechanisms into the food chain.

There are reservoirs of pesticide chemicals in the environment, resulting in residues in foods. There are the highly publicised sources such as the polluted lakes and rivers receiving the run-off from agricultural land, and disease control treatments or the by-products from production, formulation and industrial applications of pesticides. The recycling of residues in animal and vegetable wastes and by-products has not received much attention. The highly efficient recovery of valuable protein, carbohydrate and fat from these sources, for direct return to the food cycle must return residues of the parent pesticide or degradation products to human foods. Some of these food processes are efficient enough to allow speculation that the primary loss from a pesticide reservoir of this type is into human food. One could further speculate that the additions from direct and indirect sources may exceed the loss.

Changing patterns in types of pesticide chemicals and their uses can be expected to change the kind and amounts of residues in foods. Many of the indirect effects and routes of food contamination by the chlorinated pesticides have been studied, identified, and to some degree avoided. There is still much to be learned about the indirect effects due to the classes of newer chemicals being used in increasingly greater quantities to replace the long-lived organochlorine compounds. It is only logical to expect these types of chemicals to produce problems of indirect residues of parent compounds and their degradation products in the food supply. Perhaps these problems will not be as dramatic and lasting as with the chlorinated compounds.

Industrial contamination and natural sources complicate the evaluation of the elemental-based pesticides, bromides, mercury, lead and arsenic. Some of these compounds have played important roles in pest control of food crops. Arsenical compounds are important veterinary drugs. For the most part, permitted residues of these compounds in food are questioned rather severely. Health aspects of residues in foods appear to be more closely related to accidents, misuse and industrial contaminations than to sanctioned uses for food production. Commonly, statements are made attributing the residue levels in humans to the dietary intake of pesticide chemicals. Although this may be the case, the relative exposure of humans to pesticide chemicals from all sources is yet to be measured. Until such an investigation has been made, and the ambient levels established, especially in air, definitive statements of this nature are unsound. In addition to industrial elemental contaminants, there are organochlorines used for industrial purposes which mimic the organochlorine pesticides in their environmental behaviour and produce residues in food. The toxicity of these compounds has not been defined sufficiently for use in foods to be permitted.

Edwards (1970) gives an excellent review of the extent of contamination of fish and seafood, and discusses the migration from bottom sediment of streams through the different tropic levels to that of human food, which supports the hypothesis that indirect routes are most significant in evaluating sources of residues in foods. The earlier discussion concerning the approximate proportional dietary intake of chlorinated residues from food of animal sources, showed that the major intake of several of the more common pesticide chemicals were on foods in which there are no permissible direct use of pesticides.

9.4 FACTORS INFLUENCING THE PERSISTENCE OF RESIDUES IN FOODS

Pesticide chemicals differ in their persistence in the food chain depending largely on their chemical characteristics. Further complicating this aspect for some classes of pesticides, are the background levels of these chemicals naturally present, e.g. mercury, lead, arsenic and bromides, in some foods, as mentioned earlier. Many of the organochlorine chemicals were specifically developed for their continued activity for long periods, through one or more harvest seasons. The subsequent movement and long effective life of these compounds and their metabolites in other environmental substrates was not fully known at their inception.

Indeed, there is much yet to be learned about the transport of organochlorine compounds. There is no doubt that biological magnification in food chains, and particularly the aquatic system, of this class of pesticides is, and will continue to be, a persistent problem in human foods for some extended period of time. There has not been a dramatic rate of increase in the levels of these compounds, as measured by monitoring programmes in the U.S.A. for substrates such as water, soil, and foods, although billions of pounds have been used during the past 25 years. This suggests the existence of pesticide reservoirs not yet identified by existing monitoring programmes. Routes of degradation that are not fully identified and measured may exist. Data from the U.S.A., discussed previously (9.2), indicate a wider distribution of pesticide chemicals at lower levels throughout the food supply. Many wastes from food processing are recycled directly in the food chain by their use in animal feeds. Organo-chlorine chemicals soon reach a stable state with respect to animal metabolism, and one can speculate (see 9.3) that residues are being added to this cycle, with the major outlet being into human food. This view is supported by U.S. data on residues in poultry and red meat, but is not supported when data on residues in dairy products are examined.

The parent organophosphates have relatively short 'half-lives' compared with the organochlorine chemicals. This does not mean that the toxic effects may not be as persistent. Usually, much is known about the mammalian toxicity of the parent compounds and known metabolites prior to approval for use on foods. The chemical structure of many of these compounds is such, that the possibility of isomers and analogues not being detected by current analytical methodology is greater than for the organochlorines. It is significant that analytical schemes used in most monitoring programmes are limited to a relatively small number of the parent compounds of this class of pesticides.

The persistence of heavy metals, such as lead, arsenic, copper and mercury, is well known and need not be discussed here. It is likely that quantities of these elements have been generally present in man's food supply for many years, but at trace levels not routinely detectable until recently.

9.5 HAZARDS CAUSED BY RESIDUES

Discussion of hazards to man from pesticide residues ultimately become discussions of what is unknown about pesticides, especially their carcino-genic, mutagenic or teratogenic effects. Perhaps to approach the uncertain aspects from the certain ones will delineate real hazard from suspected hazard. The trend in amounts of residues in foods, and the trend in the location of these residues in various food classes, have been established by the work reviewed in Section 9.2 on amounts of residues currently occur-ring. The dietary intakes are within limits judged safe by the concurrent opinion of experts at national and international levels. The trend of the amounts of the most persistent types in food is downward. The known matters of concern consist of wider incidence (albeit at lower levels) and an increase of incidence in those foods which acquire contamination by indirect or unavoidable routes, as opposed to direct application. This implies that surveillance of levels in meat, dairy products, shellfish and fish must be continued. Trends toward higher incidences or levels, even though not of immediate concern, must be evaluated for cause, and steps taken to eliminate the source.

Data on residues of heavy metals, particularly lead, mercury, and arsenic in the food supply are not as extensive as data on organic pesticides. Especially on the border line of known facts are those about variation in background levels of these metals in foods in the pre-pesticide era. Whatever the source, whether by natural occurrence, from pesticide residue or by industrial contamination, the potential hazard is real, therefore the levels of these elements in foods are of concern. Continued investigation to identify sources and actions to limit, insofar as practical, any increase in the dietary intake of these metals is desirable.

There have been many epidemics of poisoning by pesticides in foods. Twenty-one incidents tabulated in the Mrak Report (1969) are given in Table 9.5. Eleven were ascribed to spillage during transportation or storage; five resulted from eating formulated chemicals; four were due to improper application; and two to other reasons. Fatalities were reported in 12 of these epidemics. While it may be argued that such accidents do not constitute residues in food in normal context, such hazards do exist and result from the consumption of food containing pesticides. The frequency of the epidemics of poisoning deserves mention in any discussion of hazards associated with residues in foods.

Manufacturing by-products, intermediates, and impurities in pesticide chemicals, as well as metabolites of pesticides and their by-products, are of as equal concern as the parent compound. Obvious impurities and known metabolic products are factors used in the judgment of safe tolerances. It is not always possible to predict or determine the presence of these substances. Limitations may be placed on known impurities, as with hexachlorocylopentadiene in chlordane. Others must be dealt with as they are identified, such as photo-dieldrin and the dioxins. Those organophosphates having chemical structures with potentials for dozens of metabolites are particularly worrisome. The safety factor used in toxicological recommendations is sufficiently large to ensure minimum hazard from these sources.

The dietary intake of specific pesticides, even in the extremely variable dietary patterns used in the U.S. investigation, is quite safe to those persons eating a well-balanced diet, but it is impossible to predict exposures from abnormal diets, and diets varying from this well-balanced norm are not uncommon in both the U.S.A. and Europe. National dietary habits and food preferences result in significant variations in dietary intake of pesticide chemicals. Exceedingly strange diets resulting from whim, fad, or ignorance might well result in undesirable increases of residues in the diet. The continuing evaluation of tolerances and actual levels of residues and the toxic effects of lesser known, as well as the more widely used, chemicals, by the F.A.O. Working Party of Experts and the W.H.O. Expert Committee on Pesticide Residues can only be reassuring. The international and national concern about pesticides must ultimately result in a compromise between the continued use of pesticides in food production with resulting residues in foods and minimal hazard to the consumer. The

TABLE 9.5. Epidemics of poisoning by pesticides reported in literature between 1952-1969.

Kind of accident	Pesticide	Material contaminated	Number of cases	Number of deaths	Location
Spillage during transportation or storage	Endrin	Flour	159	0	Wales
	Endrin	Flour	3	0	Egypt
	Endrin	Flour	691	24	Qatar
	Endrin	Flour	183	2	S. Arabia
	Dieldrin	Food	21	0	Shipboard
	Diazinon	Doughnut mix	20	0	U.S.A.
	Parathion	Wheat	360	102	India
	Parathion	Barley	38	9	Malaya
	Parathion	Flour	200	8	Egypt
	Parathion	Flour	600	88	Colombia
	Parathion	Sugar	300	17	Mexico
Consumption of treated soil	Hexachlorobenzene	Seed grain	>3,000	3-11%	Turkey
	Organic mercury	Seed grain	34	4	W. Pakistan
	Organic mercury	Seed grain	321	35	Iraq
	Organic mercury	Seed grain	45	20	Guatemala
Improper application	Toxaphene	Collards	4	0	U.S.A.
Miscellaneous	Toxaphene	Chard	3	0	U.S.A.
	Parathion	Crops	>400	0	U.S.A.

E. M. Mrak (Chairman), Report of the Secretary's Commission on Pesticides and their Relationship to Environmental Health, Part II, 311, U.S. Department of Health, Education, and Welfare (1969).

TABLE 9.6. Dietary intake of pesticide chemicals in the U.S.A. ($mgkg^{-1}$ of body weight/day).

Compound	W.H.O.-F.A.O. acceptable daily intake	Total diet studies						6-year average
		1965	1966	1967	1968	1969	1970	
Aldrin		0.00001	0.00004	0.00001	0.00001	0.0000001	0.0000006	0.00001
Dieldrin		0.00008	0.00009	0.00005	0.00005	0.00007	0.00007	0.00007
Total (Aldrin+dieldrin)	0.0001	0.00009	0.0001	0.00006	0.00006	0.00007	0.00007	0.00008
Carbaryl	0.02	0.002	0.0005	0.0001	–	0.00004	–	0.0005
DDT		0.0004	0.0005	0.0004	0.0003	0.0002	0.0002	0.0003
DDE		0.0003	0.0003	0.0002	0.0002	0.0002	0.0001	0.0002
TDE		0.0002	0.0002	0.0002	0.0002	0.0001	0.0001	0.0002
Total (DDT analogues)	0.005	0.0009	0.001	0.0008	0.0007	0.0005	0.0004	0.0007
Dichlorvos	0.04	N.D.						
Diphenyl	0.125	N.D.						
Gamma BHC (Lindane)	0.0125	0.00007	0.00006	0.00007	0.00004	0.00002	0.00002	0.00005
Bromide	1.0	0.39*	0.22*	0.29*	0.41*	0.24*	0.24*	0.30*
Heptachlor		0.000003		0.000001	0.000001	0.000001	0.0000001	0.000001
Heptachlor epoxide		0.00003	0.00005	0.00002	0.00003	0.00003	0.00002	0.00003
Total (Heptachlor+H. epoxide)	0.0005	0.00003	0.00005	0.00002	0.00003	0.00003	0.00002	0.00003
Malathion	0.02		0.0001	0.0002	0.00004	0.0002	0.0002	0.00015
Diazinon	0.002		0.00002	0.000001	0.000001	0.000004	0.00001	0.00001
Parathion	0.005		0.00001	0.00001	0.000001	0.000005	0.000003	0.00001
BHC		0.00003	0.00004	0.00003	0.00004	0.00002	0.00002	0.00003
Dicofol		0.00004	0.0002	0.0002	0.0001	0.00010	0.00005	0.00010
Endrin		0.000009	0.000004	0.000004	0.00001	0.00000	0.000005	0.000005
Total organochlorine pesticides		0.0012	0.0016	0.0012	0.0010	0.0008	0.0006	0.0011
Total organophosphates			0.00014	0.0003	0.00007	0.0002	0.0003	0.0002
Total herbicides		0.0001	0.00022	0.00005	0.00006	0.00005	0.000008	0.00008

* Total bromides present – includes naturally occurring bromides.

levels currently found in the food supply generally do not pose a hazard judged by current toxicological standards, as shown by data in Table 9.6, which compares the F.A.O.-W.H.O. acceptable daily intake with the U.S. dietary intake (Duggan, 1971). An acceptable daily intake is defined as 'the daily dosage of a chemical which, during an entire lifetime, appears to be without appreciable risk on the basis of all facts known at the time'. 'Without appreciable risk' is taken to mean the practical certainty that injury will not result even after a lifetime of exposure (F.A.O.-W.H.O., 1970).

The hazards to human health resulting from the dietary intake of pesticide chemicals used in the production of food are minimal. The regulatory constraints imposed by most countries on the use of toxic substances in food production are adequate, recognising the ability of toxicologists and pharmacologists to measure and extrapolate the results of the animal-testing protocols to human safety.

9.6   MINIMISING AND REMOVING RESIDUES

Sections 9.2 and 9.3 have discussed the levels of residues in foods and their sources. The levels applied to a crop or that present at the time of harvest, with few exceptions, bear little relation to the amounts in food as finally consumed. There is continual degradation and removal of pesticide chemicals through environmental exposure and food preparation processes, beginning with the application of the compound. Many factors influence the degree of disappearance from the food. These factors are related to the chemical structure of the pesticide and the chemical and physical characteristics of the food. The same chemical might disappear entirely from a root crop, yet be retained at practically the original level in milk or meat or vice versa. Generalisations concerned with minimising or removing residues have so many qualifications as to be impractical. A natural objective is to remove all toxic substances from the food supply insofar as may be practical, irrespective of how remote the hazard may be.

It is important to remember that most residues in food are the result of specific and controllable uses of pesticides, but not necessarily in production of the food. Identification of the source of some residues in food often requires sophisticated investigation extending over lengthy periods.

The most effective means of minimising residues lies in regulating the uses of these toxic compounds. For the most part, there is permitted and deliberately controlled use in the production of food. Furthermore, most countries do not permit the indiscriminate non-controlled use of pesticide chemicals, whether for agricultural, health or other purposes. The laws and regulations are in keeping with the national problems just as the enforcement of the regulations is in keeping with other national demands. At this time there seems to be a general attempt in many countries to improve the control of pesticide chemicals through additional legislation and through different organisational structures.

Some countries regulate pesticide residues exclusively through control of the importation or use of specific chemicals in the production of food. There are no specific tolerances established for residues in individual commodities. Other systems of regulations provide for tolerances applicable to groups of commodities such as leafy vegetables, dairy products, etc. Some countries, such as the U.S.A. and Canada, regulate both the use of the pesticide chemical and the amount of residue permissible in specific commodities or groups of similar foods.

There are two general systems used when tolerances are established. The first of these, employed by many countries, derives the residue tolerance from a safe acceptable daily intake based on toxicological considerations and normal national dietary patterns of food consumption. Tolerances established under this system are generally quite low and require careful definition of circumstances for sampling and enforcement. The second system derives the residue tolerance from the practical use of the chemical in food production, although requiring that the remaining residues are at safe consumption levels, also based on toxicological investigations. The second system usually results in higher tolerances which are enforced at the time of the initial introduction of food into distribution channels. A safe acceptable daily intake is considered in this system, but is not a primary factor in the establishment of the tolerance, since residue levels are ordinarily reduced from tolerance levels in food preparation procedures.

Both systems have merit. Selection of the high tolerance system may well be associated with countries having large geographic areas producing food under a wide range of climatic conditions with a wide variety of crops and pests. Lengthy intervals may be common between harvest and consumption, even for 'perishable' foods such as fresh vegetables. The first, or low tolerance, system is more commonly associated with countries having limited geographic range and production as well as a relatively short interval between harvest and consumption. It is of interest to observe that the levels of residues found on foods, prior to processing, are not greatly different irrespective of the control system.

Residues of pesticide chemicals for which there are no sanctioned uses or tolerances are often found in food. The great majority of these residues are from indirect sources, and do not result from direct application.

In the U.S.A., a number of agricultural uses were permitted on the basis that 'no residues' would remain on the food or zero tolerance were established. Later, improved analytical methods were capable of detecting residues resulting from such uses and from the wider distribution of pesticide chemicals in the environment generally. There is no legal standing for such residues which brings us to the concept of 'zero' tolerances. This legalistic concept is untenable for the practical control of residues in the food supply. Relatively satisfactory sytems of administrative guidelines and practical residue limits (F.A.O.-W.H.O.), based on toxicological safe levels, have been used to manage the problem. A more satisfactory solution of negligible residue tolerances (N.A.S.-N.R.C. Report Zero—No Residue) is being used to replace the present administrative guidelines.

9.7 FUTURE OUTLOOK

The expert study groups (Mrak Report, 1969; Anon, President's Science Advisory Committee, 1963; Anon, National Academy of Sciences, Pesticide Residue Committee, 1965; U.S.D.A., Report of Committee on Persistent Pesticides to U.S.D.A., 1969), who have examined the use and effects of pesticide chemicals in recent years, have not proposed the elimination of pesticide chemicals in food production. In fact, no identifiable human hazards have been associated with the levels commonly found in the general food supply. The most recent of these reports (Mrak Report) made extensive recommendations based on the following principles stated in the transmittal of the report to the Secretary of the Department of Health, Education, and Welfare.

1. Chemicals, including pesticides used to increase food production, are of such importance in modern life that we must learn to live with them.
2. In looking at their relative merits and hazards we must make individual judgments upon the value of each chemical, including the alternatives presented by the non-use of these chemicals. We must continue to accumulate scientific data about the effects of these chemicals on the total ecology.
3. The final decision regarding the usage of these chemicals must be made by those governmental agencies with the statutory responsibilities for the public health, and for pesticide registration.

Regulations, however, are of little value unless they are well known, understood and used by the food-producing industry. The consequences of failure to abide by regulations is a particularly important aspect in the understanding.

Education and training in the proper use of pesticide chemicals has been very successful in maintaining residues at minimal levels in several countries. The improper use or misuse of pesticides through lack of understanding or appreciation of the possibility of excessive residues has characterised the relatively few and infrequent serious problems with specific foods on a regional or national basis. There is also need for education of the consumer concerning the significance of residues in foods. This need extends to a proper information system to ensure proper perspective when new scientific evidence dictates more severe restrictions of a pesticide chemical to ensure safety.

Minimal residues in raw foods, except for fish and other foods not produced under man's direction, is a joint function of proper regulations and compliance with those regulations.

Normal food processing procedures, such as washing, trimming and cooking, remove residues because the largest amounts are usually in the skin or peel. However, these procedures are not effective in removing the fat-soluble residues (e.g. chlorinated organics) in many of the animal foods, as discussed in 9.2 (Tables 9.3 and 9.4).

Specific studies (Farrow *et al.*, 1969), on the removal of residues during commercial and home preparative procedures showed that residues of organochlorine and organophosphate pesticides on fruits and vegetables were reduced substantially, ranging from 20% by washing to more than 90% through peeling and cooking. Many years ago, apples were treated by washing with acid to remove excessive quantities of lead arsenate prior to marketing. Although investigations are being reported, there is no widespread use of specific processes for removing the organic pesticides as a routine food processing practice.

Reports of investigations concerned with the accumulation of residues from feeding various levels of the organochlorine compounds, are commonly available. Some of these were continued, in order to demonstrate the depletion after removal of the contaminated feeds. Organochlorine residues are retained and excreted by animals over long periods. Reduction of these residues, through normal depletion after removal of the source of the contamination, is very time-consuming and usually not profitable. Studies involving fat depletion, charcoal feedings and various drugs show some promise in removing residues from food animals but these are not yet in general use.

Table 9.7 compares the F.A.O.-W.H.O. Recommendations for tolerances or practical residue limits with the U.S. findings for specific food groups. This table clearly shows, based on U.S. data, that the actual levels of organochlorines and malathion in domestic or imported foods in the U.S.A. are substantially below F.A.O.-W.H.O. tolerances. It also shows that the levels are further reduced during food preparation procedures.

The world-wide (international) interest in residues in foods, having a primary base in health aspects, is of considerable concern, because of economic implications in international trade. The deliberations of the F.A.O.-W.H.O. Expert Committees on Pesticides have resulted in proposals for international tolerances. The adoption and enforcement of international tolerances based on scientifically established safety protocols, good agricultural practices and considerations of effect on other environment substrates appear to be goals within reach. Collection, analysis and evaluation of surveillance levels of pesticide chemicals in the ever-changing food supply are difficult to justify when the incidence of excessive residues is very low. There will be a continuing need for reassurance that levels are remaining within acceptable limits or that specific lots of foods are diverted to non-food uses when unacceptable levels are found. Continuing research is needed to perfect surveillance systems for these purposes. For example, selection and use of specific foods as 'markers',

TABLE 9.7. Comparison F.A.O.-W.H.O. recommendations with U.S. residues (ppm).

	Large fruits	Milk products (fat basis)	Meat (fat basis)	Eggs	Leafy vegetables	Root vegetables	Grains and cereals
DDT, DDE, TDE (singly or combined)							
F.A.O.-W.H.O. recommended tolerances or practical residue limits	7	1.25	7	0.5	7	1	—
U.S. domestic foods	0.11	0.11	0.33	0.06	0.16	0.06	0.02
U.S. imported foods	0.05	0.15	0.41	0.02	0.54	0.03	0.01
U.S. total diet	0.01	0.11	0.44	*	0.02	0.007	0.005
Aldrin-Dieldrin							
F.A.O.-W.H.O. recommended tolerances or practical residue limits	0.1	0.125	0.2	0.1	0.1	0.1	0.05
U.S. domestic foods	<0.005	0.02	0.05	0.01	<0.005	0.01	<0.005
U.S. imported foods	0.01	0.01	0.10	<0.005	<0.005	0.01	<0.005
U.S. total diet	<0.001	0.02	0.03	*	<0.001	0.001	0.003
Lindane							
F.A.O.-W.H.O. recommended tolerances or practical residue limits	3	0.1	2	0.2 (yolk)	3	3	0.5
U.S. domestic foods	<0.005	<0.005	0.01	<0.005	<0.005	<0.005	<0.005
U.S. imported foods	<0.005	0.02	0.01	<0.005	<0.005	<0.005	<0.005
U.S. total diet	<0.001	0.003	0.006	*	<0.001	<0.001	0.006
Heptachlor-Heptachlor epoxide							
F.A.O.-W.H.O. recommended tolerances or practical residue limits		1.25	0.2		0.05	0.05	0.02
U.S. domestic foods	<0.005	0.02	0.02	<0.005	<0.005	<0.005	<0.005
U.S. imported foods	<0.005	0.01	0.01	<0.005	0.01	<0.005	<0.005
U.S. total diet	<0.001	0.01	0.016	*	<0.001	<0.001	0.001
Malathion							
F.A.O.-W.H.O. recommended tolerances or practical residue limits							
U.S. domestic foods	*	*	*	*	*	*	0.56
U.S. imported foods	*	*	*	*	*	*	*
U.S. total diet	*	*	*	*	*	*	0.012

such as dairy products and carrots, within broad food classes, to detect trends toward unacceptable levels in time to institute corrective measures before significant parts of the food supply are jeopardised. Systematic statistical sampling of crops at harvest, supported by knowledge of agricultural uses in the growing area, might detect specific lots bearing excessive residues.

Improvement of the world residue situation can be predicted as the result of the trends already noted. More widespread use of new biological techniques for control of insects should begin to have a beneficial impact in the near future. New chemical pesticides will also continue to be developed in research laboratories. Inasmuch as introduction of these products is permitted only after careful evaluation this process can only result in lesser residues.

REFERENCES

D. C. Abbott, D. C. Holmes and J. O'G. Tatton, *J. Sci. Fd Agr.*, **20**, 245 (1969)

D. C. Abbott, S. Crisp, K. R. Tarrant and J. O'G. Tatton, *Pesticide Sci.*, **1**, 10 (1970)

Anon, 'Uses of Pesticides' The President's Science Advisory Committee (1963)

Anon, Report on 'No Residue' and 'Zero Tolerance' Pesticides, Pesticide Residue Committee, National Academy of Sciences — National Research Council (1965)

J. A. Burke, *Residue Rev.*, **34**, 59 (1971)

J. S. Cummings, *J. Ass. Off. Agric. Chem.*, **48**, 1177 (1965)

R. E. Duggan, H. C. Barry and L. Y. Johnson, *Science N. Y.*, **151**, 101 (1966)

R. E. Duggan, H. C. Barry and L. Y. Johnson, *Science N.Y.*, **1**, 2 (1967)

R. E. Duggan and P. E. Corneliussen, *Pest. Monit. J.*, **5**, 331 (1972)

R. E. Duggan and F. J. McFarland, *Pest. Monit. J.*, **1**, 1 (1967)

R. E. Duggan and J. R. Weatherwax, *Science N.Y.*, **175**, 1001 (1967)

R. E. Duggan, *Pest. Monit. J.*, **2**, 2 (1968)

R. E. Duggan, *Ann. N.Y. Acad. Sci.*, **160**, 173 (1969)

R. E. Duggan and G. Q. Lipscomb, *Pest. Monit. J.*, **2**, 4 (1969)

R. E. Duggan and G. Q. Lipscomb, *J. Dairy Sci.*, **54**, 695 (1971)

R. E. Duggan, G. Q. Lipscomb, E. L. Cox, R. E. Heatwole and R. C. King, *Pest. Monit. J.* (1971)

R. E. Duggan and H. R. Cook, *Pest. Monit. J.*, **5**, 37 (1971)

W. F. Durham, *Residue Rev.*, **4**, 34 (1965)

C. A. Edwards, *Critical Reviews in Environmental Control*, Chem. Rubber Co., Cleveland (1970)

M. Eidelman, *J.A.O.A.C.*, **46**, 182 (1963)

FAO/WHO Evaluations of Some Pesticide Residues in Food (1969). Food and Agriculture Organisations of the United Nations World Health Organisation, 240 (1970)

R. P. Farrow, E. R. Elkins, W. W. Rose, F. C. Lamb, J. W. Ralls and W. A. Mercer, *Residue Rev.*, **29**, 73 (1969)

B. E. Frazier, S. Chesters and S. B. Lee, *Pesticide Monit. J.*, **4**, 67 (1970)

J. M. Harries, C. M. Jones and J. O'G. Tatton, *J. Sci. Fd Agr.*, **20**, 242 (1969)

R. B. Harrison, *J. Sci. Fd Agr.*, **17**, 10 (1966)

G. Q. Lipscomb, *Pest. Monit. J.*, **2**, 104 (1968)

R. J. Martin and R. E. Duggan, *Pest. Monit. J.*, **1**, 11 (1968)

A. E. J. McGill and J. Robinson, *Fd. Cosmet, Toxicol.*, **6**, 45 (1968)

A. E. J. McGill, J. Robinson and M. Stein, *Nature* (Lond.), **221**, (5182), 761 (1969)

P. A. Mills, *J.A.O.A.C.*, **46**, 762 (1963)

E. M. Mrak, Report of the Secretary's Commission on Pesticides and their Relationship to Environmental Health, Part II. Dec. 1969. U.S. Dept of Health, Education and Welfare.

W. R. Poage, Chrm. 1971. Federal Pesticide Control Act of 1971. Hearings U.S. Govt. Printing Office, #92.A.

L. M. Reynolds, *Bull. Environ. Contam. Toxicol.*, **4**, 128 (1969)

J. Robinson, A. Richardson, A. N. Crabtree, J. C. Carlson and G. R. Potts, *Nature* (Lond.), **214**, 1307 (1967)

M. S. Schechter, *Pest. Monit. J.*, **1**, 20 (1967)

M. S. Schechter, *Pest. Monit. J.*, **2**, 1 (1967)

M. S. Schechter, *Pest. Monit. J.*, **5**, 68 (1971)

D. C. Smith, *Pest. Sci.*, **2**, 92 (1971)

K. Soos and Z. Lebensm-Unersuch, *Forsch.*, **141**, 219 (1969)

L. J. Stevens, C. W. Collier and D. W. Woodham, *Pest. Monit. J.*, **4**, 145 (1970)

A. B. Swackhamer, *Pest. Progr.*, **3**, 108 (1965)

U.S.D.A. 'Report of Committee on Persistent Pesticides to U.S. Department of Agriculture.' Division of Biology and Agriculture, National Research Council (1969)

K. C. Walker, M. B. Soette and S. S. J. Bachelor, *J. Agric. Fd Chem.*, **2**, 1034 (1954)

S. J. William, *J.A.O.A.C.*, **47**, 815 (1964)

Chapter 10

Pesticides in the Atmosphere

G. A. Wheatley

National Vegetable Research Station,
Wellesbourne, Warwick, England

10.1   INTRODUCTION

The word 'pesticide' has been gradually adopted as a collective term to describe a wide range of chemicals now used to manipulate biological systems in agriculture and other industries. The 'Review of Present Safety Arrangements for the Use of Toxic Chemicals in Agriculture and Food Storage' (Cook, 1967) defines a 'pesticide product' as including chemicals used 'to destroy any insect, fungus, bacterium, virus or rodent . . . or to attract, repel, sterilise, stupefy any pest, or to act as a plant growth regulator, defoliant, desiccant . . .,' and so on. For practical purposes, therefore, many chemicals with industrial uses and virtually all agricultural chemicals, with the exception of inorganic fertilisers, come within this definition and can be considered as pesticides. It is in this broad sense that the term is used to discuss pesticides in the atmosphere.

Pesticides are developed deliberately for a great many relatively restricted purposes. They may be intended to control unwanted weeds, pests or pathogens, to manipulate plant growth to man's advantage or to rid farm animals of parasites. Their performances as biologically-active agents have therefore been studied by scientists specialising in such disciplines as entomology, plant pathology, bacteriology, plant physiology, agronomy, weed control, veterinary science, food storage or public health fields. On the other hand, their physico-chemical properties and behaviour have mainly been investigated quite separately by different groups of specialists, the physicists, chemists and chemical engineers who have had the tasks of synthesizing, purifying and manufacturing the active ingredients and of formulating them into usable products. During the past decade, spectacular improvements have been made in analytical techniques for detecting and measuring very low concentrations of pesticide residues in food, wildlife and in the environment — soil, water and air, both for research and for regulatory purposes. It has been necessary to achieve a better understanding of the physico-chemical properties of pesticides in order to develop more sensitive and reliable analytical methods. In turn, the application of these improved methods has rapidly enlarged our knowledge of the physical and chemical behaviour of both pesticides and their residues. Because of the variety of scientific disciplines concerned with research on pesticides, this information is scattered widely in the scientific literature and a reviewer is soon forced to the conclusion that many studies undertaken empirically disregard very relevant findings in closely related fields. The behaviour of insecticides, fungicides and herbicides in soils is a good example of how disciplinary barriers have effectively inhibited the development of unifying concepts for what is clearly a single subject.

Like all other chemicals, natural or synthetic, pesticides obey the laws of science. The behaviour of each is determined by the interplay between its own physico-chemical properties and those of the environment in which it occurs. In this respect, it is of no consequence that one chemical may be put into soil to control an insect pest and another to control a weed. Their subsequent fates will depend on their physical and chemical properties, and those of the soil, and climate, not the purpose for which they are used. The most important, if not all, of the basic principles influencing the behaviour of pesticides in the environment are now probably recognised. Even so, we cannot predict the fate of chemicals released into the environment because the complex interactions of the variables cannot yet be expressed quantitatively. Mathematical models have now been used to assess the circulation of DDT in the biosphere (Woodwell *et al.*, 1971) and they should greatly assist in forecasting the course of future events as soon as there is a more accurate understanding of the processes involved. Already we have learnt that certain organic compounds, such as some organochlorine insecticides and polychlorinated biphenyls, can disperse from local areas of usage and become re-distributed on a global scale. We can predict that short-lived compounds such as tetraethyl pyrophosphate could not become such widespread contaminants. Between these extremes, there are several hundred pesticide chemicals with differing potentials in this respect. The problem is how to predict the potential of each to move uncontrollably away from sites of usage, and to identify the part(s) of the environmental complex which individual chemicals put at risk. The desirability of not having to rely on the empirical 'try-it-and-see' approach is obvious when the consequences could be an unpremeditated irreversible environmental catastrophe.

Only in recent years, has it become apparent that the atmosphere is a key transport medium, and a vast reservoir, for pesticides and their residues. Presumably, this also applies to other chemicals with appropriate characteristics. Pesticides get into the atmosphere in many ways, most of them known or at least predictable, and it is now possible to suggest which routes of entry are likely to be the most important. Public awareness of the problems of atmospheric pollution began about twenty years ago when it inspired a crusade against visible pollution of the atmosphere by black-smoke emission from domestic and industrial sources (MacKenzie, 1966). During the past decade, it has been increasingly realised that many gaseous (invisible) pollutants, arising from man's activities, seriously threaten health and the survival of living matter near intense sources of pollution. The effects of photochemical smog, sulphur dioxide and trioxide, carbon monoxide and other general atmospheric pollutants on health are now well-appreciated (Smith, 1966; Altshuller, 1967; Wolman, 1968; Epstein, 1970; Lave and Seskin, 1970), and the phytotoxic effects of sulphur dioxide, chlorine or ozone, at concentrations of less than 1 ppm, can seriously threaten forests and other vegetation (Altshuller, 1965; Katz and Drowley, 1966; Rich *et al.*, 1969; Reinert *et al.*, 1970). Information accruing from studies of these problems has direct relevance to the behaviour of pesticides in air, indicating how they move and become dispersed and diluted with time and distance from their source. The results from investigations following the accidental escape of [131]iodine (Chamberlain, 1959), or on the dispersal of pollen grains or fungal spores (Gregory, 1961), are also relevant. In the early 1960's, radionuclides released into the atmosphere by test explosions of nuclear devices provided some of the most precise results yet available on the mass movement and mixing potential of the atmosphere. Certain aspects of the behaviour of pesticides in the atmosphere, particularly those concerned with their physical movement and dispersal, are therefore largely predictable, by drawing intelligent parallels with some known information from a host of more intensively studied analogous situations.

10.2 THE ATMOSPHERE

The atmosphere is one of the two great transport systems of our planet, the other being water. The enormous mass of matter comprising the atmosphere ($5{\times}10^{15}$ tonnes) and the oceans and inland waters ($1.3{\times}10^{18}$ tonnes) have, until recent years, both been assumed to be so great that they can act as infinite sinks for man's waste products or effluents. Although the basic composition of the atmosphere is remarkably stable, traces of many alien chemicals arising from man's industrial activities can now be detected in air by highly sensitive analytical methods. The various chemical species present may occur in gaseous (vapour) or particulate (solid or liquid = aerosol) form, but most will exist simultaneously as vapour in equilibrium concentration with the solid phase. For practical purposes, the dispersion of pesticides in the atmosphere is controlled entirely by the mass movement of the atmosphere itself, the only exception being the downward loss of material by dry deposition (see below) and washout by rain. Consequently, some understanding of the general behaviour and movement of air is necessary to comprehend how pesticides enter the atmosphere, and how they become transported, dispersed and redistributed, both locally and globally.

The atmosphere is only a transport medium for pesticides. It is not a medium in which they are biologically active, even though they can react chemically or photochemically while airborne. Although substances may be effectively transferred from one site to another as vapour, biochemical processes occur only after they have become dissolved in liquids or deposited onto solid interfaces (Hartley, 1971). So-called vapour action is a misnomer, and Gerolt (1959) showed how initial observations could be misleading in this respect. He demonstrated how DDT could appear to have a fumigant action under inadequate test conditions, if minor temperature changes within an apparatus caused vapours to condense on downstream surfaces from which they could be picked-up mechanically by a test insect. By slightly increasing the temperature of an air-stream after it had flowed over a DDT source, the downstream-air was no longer saturated with DDT vapour. The insecticide did not then condense on internal surfaces in contact with the insects, and the 'vapour action' was no longer apparent. The term 'vapour transfer' is therefore a more apt description of this process.

10.2.1 Structure

The atmosphere is subdivided into several layers according to the changes in temperature which occur with increasing height above the Earth's surface. All phenomena recognisable as 'weather' occur in the troposphere (McIntosh and Thom, 1969), which is the layer of most interest in relation to pesticide distribution and transport. The upper limit of the troposphere, the tropopause, varies in height with latitude and the season, ranging from about 6-7 kilometres high over the winter pole, to about 18 kilometres over the Equator. Above the tropopause, the stratosphere extends up to the stratopause at a height of about 50 kilometres. Although there is evidence that air-masses move up through the tropopause into the stratosphere and *vice versa* (Volchok, 1965), the principal movements seem to be seasonal, and their relevance to pesticide redistribution is, at present, unknown. Our present knowledge of pesticides in the atmosphere is therefore confined to the phenomena occurring in the troposphere.

10.2.2 Meteorological scale

Meteorological phenomena and associated air movements can be considered on several different scales of magnitude, each with a particular significance to the entry and re-distribution of pesticides in the atmosphere.

In the very lowest layers of the atmosphere, air movement (wind velocity) falls quickly to zero with decreasing height above the Earth's surface. The actual height at which the air velocity is effectively zero varies with the overall roughness of the surface, and may range from a few centimetres over rough ground to a mere fraction of a millimetre over a smooth plane surface such as a pane of glass. Air-flow in the boundary layer next to surfaces is assumed to be laminar (non-turbulent) in a plane parallel to the surface and, within this layer, inward or outward fluxes of matter to or from the surface can only occur by diffusion, the net effect of random motions of molecules or suspended particles. Diffusion is primarily of importance in controlling the first stage of the loss of pesticides from deposits on treated surfaces. Other factors being equal, losses from similar effective areas of surface should tend to be faster from smooth than from rough surfaces with their associated thicker layer of non-turbulent air. The rate of diffusion is a function of the concentration gradient between the deposit surface and the interface of the laminar-flow and turbulent layers of air. It is therefore a maximum, when the concentration of chemical is zero in the outer turbulent air at its interface with the boundary layer. This tends to occur at high windspeeds, which should therefore enhance the losses of pesticides from deposits. In practice, however, rates of loss will tend towards a maximum at quite moderate windspeeds of perhaps 1-2 m/sec because the rate of diffusion through the boundary layer soon becomes the principal limiting factor of the loss process. In still-air conditions, losses should be minimal since they occur only by diffusion, which would be limited by the reduced and extended concentration gradient.

Up to a height of about 50 m, the approximate limit of the Earth's so-called 'surface boundary layer' (McIntosh and Thom, 1969), the wind velocity increases logarithmically, the air is usually very turbulent and its behaviour is noticeably affected by the magnitude and sign of the vertical temperature gradient, or lapse rate, for instance when inversions cause local mists or fog to develop. Micrometeorology is concerned with the processes in this layer where we live, work, grow our crops and release most of our pesticides and other aerial contaminants. The air movements are obvious to all, and are recognised as wind, and the associated turbulence as gusts. These vary in severity depending on both the wind speed and the terrain. Local topography has a considerable influence on the behaviour of air in the lower 50 m of the atmosphere, and thereby the extent of the movement, dispersal and fall-out of pesticides or other contaminants released into it.

From 50 m up to a height of about 500 m, the limit of the so-called 'planetary boundary layer', the horizontal wind velocity continues to increase logarithmically with increasing height, but it may change direction slightly (20-30°) as its movements become less influenced by 'friction' with the ground and surface obstructions. Air movements become relatively smooth above 500 m, implying that airborne contaminants should be less readily mixed, dispersed and diluted at high than at low altitudes. This would be countered by their greater chances of remaining airborne at high altitudes for long periods. They can then be transported great distances, entrained in the large-scale weather systems and moving with them along the complex paths dictated by temperature, pressure and other energy gradients in the atmosphere. In the major weather systems, enormous volumes of air move vertically when the atmosphere is unstable, so that air-masses contaminated at lower altitudes can become distributed up to the tropopause.

The atmosphere, as a whole, also moves relative to the Earth's surface, 'slipping' in a westerly to easterly direction and taking about 9-18 days for a complete revolution around the Earth, depending on the season and the latitude. Until recently, it has usually been assumed that very little mixing occurs between the atmospheres of the Northern and Southern Hemispheres. However, Sotabayashi et al. (1969) recorded transfer of radioactive debris between the two hemispheres at a rate of about 12 km/h, a rate of movement similar to that of typhoons. Kidson and Newell (1969) have indicated how transfers of relative angular momentum between the atmosphere and the Earth's surface, occur in such a way that a flux is created from the winter to the summer hemisphere, amounting to a substantial part of the total momentum budget. The transequatorial flux is minimal during April when the energy contents of the hemispheres are similar in this respect. It can also have a strong influence on the circulation of the atmosphere within each hemisphere.

10.2.3 Transport potential

Adequate mechanisms therefore exist in the atmosphere to redistribute entrained matter, not necessarily uniformly, on a local, continental, hemispherical or global scale, provided that it persists and remains airborne for sufficiently long periods. The behaviour and fate of vapours and small particles, released into the atmosphere at low altitudes, has been studied in many contexts other than that of pesticides. Since largely physical principles are involved, reliable analogies can be drawn to indicate how pesticides are likely to be transported from sites of origin to become dispersed and redistributed by the atmosphere.

Pollen grains and uredospores of cereal rust fungi can remain airborne for thousands of kilometres according to Gregory (1961), who provided tables enabling estimates to be made of the chances of particles remaining airborne for given distances when they are released at different heights. Once a particle has been airborne for a few minutes, it has a good chance of remaining aloft for several hours. According to the calculation on which Gregory's table is based, a 10 nm particle of unit density released 1 m above the ground has an 80% chance of travelling 1,000 km in dry weather. The mean airborne life of a smoke particle liberated in Britain is about 2 days, compared with a life of perhaps 12 hours for a molecule of sulphur dioxide vapour (Meetham, 1950) or about 1 day for [131]iodine (Chamberlain, 1959). At the other extreme, Junge (1963) has quoted the mean residence time for nuclear-bomb debris in the atmosphere to be about 30 days. This might perhaps be considered to be an upper limit for pesticide particles liberated nearer the ground and therefore likely to remain airborne for shorter periods and so have a shorter mean residence time. Even so, the mean residence times for pesticide particles would be sufficiently long to permit them to travel long distances in the major weather systems and to circumnavigate the Earth perhaps several times. Despite their capability for long-distance travel, studies have shown that most particles deposited in an area tend to have originated locally. This is a direct consequence of the dilution in concentration of the material by upwards and lateral spread in the atmosphere at increasing distances from a point source, the rate of decrease in concentration often being approximately proportional to the 1.75th power of the distance from the source.

It is evident that concentrations of stable airborne contaminants, arising intermittently from what are virtually point sources, will not become thoroughly dispersed and diluted for some time, perhaps weeks. Meanwhile, they will be circumnavigating the Earth as a contaminated air-mass, gradually becoming more diluted and less distinct with the passage of time. In addition to dilution by dispersion, the amount of airborne material will tend to decline steadily as the processes of dry deposition, sedimentation and wash-out by rain tend to deposit it on the land and into the oceans. Some molecules will also diffuse back into the air as part of the equilibration process responsible for maintaining a continuous low concentration of each constituent in the general circulation. Wide fluctuations must therefore be expected to occur in the concentrations of a relatively stable airborne contaminant passing a fixed point, or sampling station, on the Earth's surface. High transient concentrations would be indicative of recent or local release of pesticide into the air, and these would be superimposed on a more uniform background level determined by the equilibration between the mobile airborne phase and the static deposits.

Since eddy-turbulence and not diffusion is the most important factor in the redistribution of airborne contaminants, it follows that the atmosphere can most readily disperse them when it is unstable during windy conditions. Matter, released into the air while the temperature lapse rate is inverted, is effectively trapped within this lower layer and so is less readily dispersed or diluted. The atmospheric conditions prevailing while pesticides are being used have thus a marked influence on their immediate behaviour in the vicinity, and so determine the extent to which long-distance transport will occur. It is worth noting that conditions which tend to avoid high concentrations occurring locally while pesticides are being used will frequently be those most likely to promote their transport over long distances.

10.3 ENTRY OF PESTICIDES INTO THE ATMOSPHERE

Abundant evidence has been accumulated to show that the entry of pesticides into the atmosphere is an inevitable consequence of their use, but recognition of this fact has been acknowledged only within the past decade. Akesson and Yates (1964) stated that most agricultural problems arising from pesticide drift were confined to within a few kilometres of their application sites, but they also observed that 'larger-scale effects are surely present'. Middleton (1965) cited the meagre information available on the presence and persistence of insecticides in air, but he had insufficient data to assess the concentrations of fungicides and herbicides in air. Even in 1966, a progress report by Wilson, on a research plan to study the fate of pesticides in the environment, showed plants, soil and water as possible routes followed by the chemicals after application but it made no reference to air! By 1970, Akesson et al. had reported studies showing the extent of the transport of pesticides in air and confirmed the earlier hypothesis that the total contamination of an air-shed can occur when large areas (4,000-20,000 ha) are treated. They were emphatic that this was 'a problem that can no longer be ignored'.

Pesticides can enter the atmosphere either intermittently or more continuously, depending on the source. Intermittent release occurs during their application, at certain stages of batch manufacture, while concentrates are being handled, or when enclosed fumigated spaces are subsequently ventilated. Relatively high concentrations of the chemicals are released from what are virtually point sources, but for only brief periods. A more continuous release into the atmosphere occurs as part of the normal weathering processes of pesticide deposits on or in treated substrates, the rates of loss normally diminishing rapidly as the deposits age. Eventually, the loss becomes part of the equilibrating exchange of pesticide molecules between the atmosphere and every substrate/air interface. This is the least appreciated and most insidious of the entry and redistribution mechanisms. The dynamic equilibration processes follow the basic laws of physical chemistry, controlling and directing the redistribution of molecules from foci of relatively high concentrations, such as treated areas. Although the rates of exchange may appear to be slow, appreciable quantities can nevertheless be transferred between deposits in and on substrates and the atmosphere, because of the extensive and continuous nature of the phenomenon, as the calculations of Woodwell *et al.* (1971) make clear.

Pesticide molecules may enter the atmosphere either in particulate form or as vapour. The particulates may be either liquids (aerosols) or solids and the latter may be wholly comprised of the pesticide active ingredient or they may be mostly a carrier matrix in which the pesticide molecules are entrained, absorbed or adsorbed onto the interstitial surfaces. Mineral particles, waxy debris from plant cuticles or carbonaceous particles generated during incomplete combustion of pesticide-containing organic matter are examples of carrier substrates. Normally, both vapour and particulate forms co-exist, and there are considerable technical difficulties in ascribing the exact quantitative importance of each as entry or transport mechanisms. Few, if any, direct measurements of pesticide losses to the atmosphere are claimed to be complete assessments.

10.3.1 Intermittent entry from point sources
The escape of pesticides while they are being applied outdoors is probably the most important way that high concentrations enter the atmosphere intermittently. The physical drift of solid particles or small droplets away from the site of application has long been recognised as a factor reducing the effectiveness of pesticide applications, creating hazards to operators or to other persons or animals nearby, and being a serious source of contamination of nearby vegetation (Edwards and Ripper, 1953). It is now being seen in its wider context as an important route for the entry of pesticides into the general atmospheric circulation.

10.3.1.1 *Entry of particulates*
Particulate drift is essentially a problem of fine particles with such low settling velocities that they are at the mercy of even small upward components of air movement or turbulence. At the same time their mass, and consequently their momentum, is so low that they have little power to

impact on target surfaces, and must deposit primarily by sedimentation. Yeo (1959), Hoskins (1962), Ebeling (1963) and Akesson and Yates (1964) reviewed and discussed the deposition of pesticides in general terms. It is largely a physical and mechanical problem outside the scope of this review. To ensure satisfactory biological performance, pesticides must be distributed uniformly over a target area and hence they must first be finely divided, the function of the application equipment. The mechanics of achieving this inevitably result in the production of particles or droplets embracing a range of sizes and hence a range of impaction and settling characteristics. The upper limit of particle size constituting the drift-prone 'fines' depends on the density of the material and the circumstances in which it is applied.

Droplets of aqueous or oil sprays usually have a density of about 0.8-1, and for technical pesticide chemicals it is between 1 and about 1.4 g/cm^3. Akesson and Yates (1964) showed that less than 1% of spray particles detected 217 m downwind exceeded 30 μm diameter. According to Yeo (1959), the greatest deposition on the ground within a few kilometres of a release point is usually caused by droplets of 10-50 μm diameter. Droplets of 100 μm diameter do not drift much unless winds are very strong, whereas 5 μm droplets show little inclination to deposit, and can drift for many miles. Aqueous spray droplets quickly evaporate when released into air, so losing mass and becoming more drift-prone. Amsden (1962) reported that 50 μm droplets of pure water had lifetimes of only 3.5 sec in air at 30°C and 50% relative humidity, during which time they fell only about 3 cm. Droplets of a 30% by weight aqueous suspension, 150 μm in diameter, evaporated to dryness while falling 6.5 m through air at 40°C and 20% relative humidity, so that the solid material deposited as a dust. The sizes of aqueous droplets, occurring downwind of a release-point, will therefore normally be very much smaller than when they were released, and the non-volatile oil phase or the technical pesticide liquid or solid may be the main constituent remaining.

Carriers for pesticide dusts normally have particle densities of from about 2.2 to 3 g/cm^3, so that particles of less than about 10 μm are particularly drift-prone. Dust formulations usually have an appreciable proportion of their total weight comprised of 10 μm or smaller particles, and they are notorious for drifting beyond the target area. Gerhadt and Witt (1965) found that aerial applications of DDT, camphechlor (Toxaphene®) and tetradifon gave six times more drift deposit downwind when dust formulations were used than when the chemicals were applied as sprays. MacCollom et al. (1968) found that tetradifon dust particles of less than 10 μm in diameter could remain airborne for 20-60 min and that half of the particles taken 0.4 km downwind down a slope, and all of those recaptured 1.6 km away, were less than 10 μm in diameter when the windspeed was 0.4-3.2 km/h and during a temperature inversion of 1°C. Dust formulations are thus not a very acceptable means for dispensing highly potent active ingredients, and their use is declining. Granular products having relatively large particles have now tended to succeed dusts or sprays for situations where drift has been an acute problem (Luckman and Decker, 1960; Akesson and Yates, 1964). Pesticides dispensed out-of-doors as aerosols, fogs or smokes, are probably the ultimate means of promoting the drift-entry of particulates into the atmosphere because of the minute particle sizes involved.

10.3.1.2   Entry as vapour

Compared with studies on particulate drift, vapour-phase losses during application have received little attention, and most of the evidence available is circumstantial. It is, nonetheless, impressive, even though it is not possible to differentiate fully the respective contributions of vapour and particulate losses. Their relative importance must obviously vary with the properties of the chemical, formulation, the mode of application and the prevailing weather. Generalisations are therefore of only limited value. Even under the best application conditions in the field, substantial proportions of many chemicals have been reported as 'missing' from the target areas. Decker et al. (1950) sprayed DDT, camphechlor, dieldrin, aldrin and chlordane insecticides onto alfalfa, but recovered only 85, 84, 56 and 45% of the four last-named compounds relative to the amount of DDT recovered. The losses were in accordance with the known volatilities of the compounds. Even when they collected DDT at the nozzle outlet, only 88% could be recovered, and this declined to 74% at a distance of 1.2 m from the nozzle. When chemicals were applied to soil, and samples were taken immediately afterwards, Bohn (1964) reported a 16% loss of dimethoate and Menzer et al. (1970) an 18% loss of disulfoton but apparently a full recovery (103%) of phorate. Hindin et al. (1966) recovered only 22-30% of aerially-applied DDT on targets placed at the level of maize tassels 2.5 m above the ground, and only 35 and 49% respectively of diazinon and ethion. Kiigemagi and Terriere (1971) found that 40% of thionazin and 24% of fonofos failed to reach the soil surface when they were applied as diluted emulsion sprays and thionazin is known to be the more volatile (v.p. 3×10^{-3} compared with 3.8×10^{-4} mm Hg at $30°C$). Since the amounts caught in dishes were similar to those calculated as being present in the soil samples, the losses had most probably occurred in the brief passage through the air between the spray leaving the nozzle and contacting the soil or dishes. More fonofos was lost when it was applied at a high, than at a low, prevailing temperature, again implicating loss by volatilisation. Bearing in mind the quantities that Decker et al. (1950) reported missing, namely from 15 to 55% for DDT and aldrin respectively, and allowing for imperfect recovery of DDT directly from the spray nozzle (12% missing), it seems possible to account for only 3 to 12% of the loss as particulate drift. This would agree with the findings of Akesson and Yates (1964) that only a small proportion, probably less than 5%, of the volume of a conventional spray would be present as drift-prone droplets less than 30 to 40 μm diameter. The 55% loss of aldrin must have occurred mostly as vapour, as

Decker *et al.* themselves reasoned, and many pesticides are more volatile than aldrin. MacCuaig and Watts (1963) estimated, for instance, that half of a 100 µm diameter dichlorvos droplet would evaporate while falling about 50 m through air at 30°C as during aerial application.

Thus, it is evident that losses of up to half of the applied amounts of pesticides frequently occur during application, as a result of direct loss of vapour to the atmosphere. It is significant that, where comparisons have been made, losses have invariably been greatest with the more volatile chemicals. Even DDT, one of the least volatile pesticides, escapes in part by this route which may be the most important means whereby pesticides become entrained in vast air-sheds.

10.3.1.3 *Drift during application*

Numerous incidents have been attributed to pesticide drift, and these indicate the proportions that can stray from target areas during application. Some of the most striking visible evidence has resulted from the drift of hormone-type herbicides, of which 2,4-D and its ester formulations seem to have been most frequently indicted for affecting susceptible crops such as grapes, cotton or tomatoes. Akesson and Yates (1964) describe how this problem played an important part in promoting legislation to control 2,4-D and similar herbicides in the United States during the early 1950's. The need to avoid unexpected residues of organochlorine insecticides in stockfeed, dairy products and crops for human consumption soon led to other regulations aimed at limiting the consequences of pesticide drift.

Except in a few special circumstances, the direct consequences of drift have been confined to within a few kilometres of the source. In Australia, Mears (1964) considered that it was risky to use hormone-type herbicides within 3.2 km of cotton, even in calm weather, especially if ester formulations were used, and he also stated that they should not be applied by aircraft within 6.4 km. At the high temperatures of the Punjab, Bakshi (1967) observed severe damage to American cotton by 2,4-D vapour drifting from neighbouring treated fields. Gerhadt and Witt (1965) considered that forage crops for stockfeed would have to be at least 0.9 or 2.4 km downwind of a DDT spray or dust application respectively to avoid deposits on the foliage, exceeding levels likely to cause excessive residues in the butterfat of grazing animals. MacCollom (1962) found up to 0.45 ppm of tetradifon in forage, 60 m from an orchard where the insecticides had been applied as a dust, and investigations showed that drift could occur by air-drainage down a slope during calm, non-turbulent, inversion conditions. Luckman and Decker (1960) reported that drift from dieldrin sprays, applied during a Japanese beetle control programme, had killed sheep on nearby pastures and affected others for up to a week. They also found that both the drift and the residues on the forage could be reduced by using granular formulations. Akesson and Yates (1964) tabulated a range of separation distances applicable to seven organochlorine insecticides, to keep contamination of nearby forage to acceptable levels. Taking into account rates of application, persistence of the chemicals and the intervals between treatment and cutting the crop, they considered that a separation of about 0.9 km during 'good weather' would enable aramite to be applied and downwind forage to be cut immediately afterwards, the most extreme case cited. They also calculated that 10 μm diameter particles could theoretically travel 1.4 km while falling about 3 m in a wind of 4.8 km/h and that a 2 μm aerosol droplet could remain airborne for about 34 km in the same conditions. In practice, however, it seems likely that some particles would have stayed airborne much longer. Akesson and Yates also recorded a three- to tenfold increase in drift deposition in a 13-16 km/h wind, compared with that found in a 6.4-11 km/h wind. Presumably, the amount of material more permanently suspended must have increased similarly. Pesticide applications in the field are therefore limited to periods of low windspeeds, usually less than 10-16 km/h, to ensure deposition on the target area and this fortunately minimises drift-entry into the atmosphere.

Marvin (1965) considered that the four main factors responsible for drift incidents were: (i) direct transport of particulates by wind; (ii) up-draughts or 'thermals' initiated by thin patches in crop-stands (less important when vegetation is dense); (iii) vapour emanations, particularly from ester formulations of herbicides of the plant growth-regulating type (which he considered was an overrated phenomenon); and (iv) convective drift, often associated with evening applications on a hilltop from which entrained particles drift with cool air down a slope, following ravines or gullies in a manner analogous to mist movement and tending to stay close to the soil, so that drift of hormone-type herbicides affect plant roots more than when primarily wind-borne.

Methods of controlling particulate drift fall into two categories: those concerned with the chemical, its formulation and the application machinery, and those based on the environmental factors such as the meteorological conditions and the topography of the area being treated. Examples are seen in the development of precision low-pressure nozzles and atomising devices to restrict droplet size-range (Anon., 1966; Roth, 1965), special jets imparting high speeds to the emerging liquid to improve impaction (Little et al., 1966), the use of additives to reduce droplet evaporation and the development of invert emulsions (Akesson and Yates, 1964; Kirch and Esposito, 1967), and the application of undiluted pesti-cide concentrates (Argauer et al., 1968). Applying heavy oil concentrates of DDT, with Micronair equipment in Tanzania, Lee (1966) obtained a deposit on seed beans and cotton two to three times greater than with a conventional sprayer. Even aerial application of herbicides has become sufficiently precise for dalapon to be sprayed along ditches to control *Phragmites* spp., without drift damage occurring, and using lower doses and less spray than for ground equipment (Robson et al., 1966). The environmental conditions most likely to promote drift are now well-recognised and have already been discussed. Presumably, development aimed at reducing the local consequences of pesticide drift and increasing deposition on the target should at the same time, reduce the amounts entering the atmospheric circulation. An exception would seem to be the avoidance of inversion conditions in favour of unstable conditions associated with normal temperature lapse rates in the atmosphere. This must promote upward drift of particles and vapour to high altitudes and hence increase the contamination of the circulatory air-shed.

10.3.1.4 *Entry by wind erosion*

Intermittent entry of pesticides into the atmosphere can also arise by wind erosion of dry soil. An extreme instance has been described by Cohen and Pinkerton (1966) who collected and analysed an unusually heavy deposit of atmospheric dust at Cincinnati, Ohio on 26 January 1965. The dust had originated largely from the High Plains area of the southern United States as a result of a severe dust storm on the previous day. They detected five organochlorine compounds, DDT, DDE, chlordane, heptachlor epoxide and dieldrin, the organophosphorus insecticide fenchlorphos (Ronnel®), and the phenoxyacetic acid herbicide 2,4,5-T at concentrations totalling about 1.5 ppm in the dust. This was not unexpected, in view of the known heavy pesticide usage in the area of origin and the well-established fact that residues of many pesticides persist for months or years in arable soils (Edwards, 1970). Cohen and Pinkerton reasoned from this and other data that pesticide-contaminated soil is constantly being taken-up, transported considerable distances at high altitudes by wind, and then deposited either by sedimentation or by rain far from its origin. The cultivation of dry soils, during even mildly unstable (cylconic) meteorological conditions, and the natural 'blowing' of poorly-structured soils are obvious ways for pesticide-bearing particulates to become airborne and freely-dispersed. The importance of this contribution to the intermittent entry of pesticides into the atmosphere is not known, but arable soils in areas of intensive pesticide usage can contain up to a few ppm of persistent chemicals, the concentrations present being in equilibrium with the frequency and rates of usage (Wheatley *et al.*, 1960; Decker *et al.*, 1965; Hamaker, 1966).

10.3.1.5 *Entry from other sources*

As well as from applications to agricultural crops and soils, other uses of pesticides provide local sources for their intermittent entry into the atmosphere. The amounts of different pesticides used for veterinary purposes, for moth-proofing fabrics, for food storage or domestic hygiene (Wilson, 1969) and the relative contributions that these uses make to the pesticide content of the atmosphere are not known. When they are applied by methods similar to those used to treat crops and soils, the same principles govern their escape. For example, fogging techniques, to treat large open areas for public health purposes, must clearly contribute and may be an important source for certain compounds such as DDT or malathion. Brief periods must also occur when pesticides escape at high concentrations into the atmosphere when food stores, ships or aircraft are ventilated after fumigation. Similarly, the highly controversial use of thermal vaporisers to disseminate insecticides such as gamma-BHC or DDT will introduce small amounts in the ventilating air-changes. In contrast, where pesticides are used as bulk liquids to dip sheep, impregnate timber or moth-proof fabrics, significant amounts should not escape during the treatment process. Viewed as a whole, intermittent release of pesticides from these sources seems unlikely to be very important, except where very large quantities are used, or occasionally, when local inversion conditions limit dispersal and dilution, or where very large quantities are being manufactured. For example, the escape of chlorphenols during the manufacture of batches of certain phenoxyalkanoic acid herbicides has occasionally been closely associated with taints in food baked a few kilometres distant from the source.

For centuries, burning has been an accepted means for disposing of waste organic materials. Widespread open-burning of tree prunings, straw, stubble or the inefficient incineration of waste citrus peel can cause general atmospheric pollution for a few weeks at certain times of the year (Osterli, 1967; Cook, 1966). This may be another route for the entry of pesticides into the atmosphere, more important perhaps, because of the extent of the practices than because of the amounts involved at any one time. Reynolds (1969), and Risebrough et al. (1969) suggested that polychlorinated biphenyls may be introduced into the atmosphere by combustion of waste material. Small pesticide residues must occur very widely, if not ubiquitously, in organic waste and at least part will escape as vapour or entrained in carbonaceous smoke particles when combustion is incomplete. Several million tonnes of straw or stubble are probably burned each year in the U.K., in a few weeks during late summer. Assuming an average gamma-BHC content of 0.005 ppm, and that half of this would escape unpyrolysed, then only about 2 kg would be released from this source annually. Even if released all at the same moment, it would provide only 20 fg/m^{-3} (2×10^{-14} g/m^{-3}) to a height of 300 m over the U.K. (3×10^5 km^2), a seemingly insignificant concentration which nevertheless could be important in transporting the compound between sources and sinks in the environment.

10.3.2 Continuous entry

All pesticide-treated surfaces or substrates are potential sources from which pesticides enter the atmosphere more or less continuously, either by wind erosion of particulates or by vapour transfer. Descriptions of the decline of pesticide residues make frequent use of the words 'disappear' and 'loss' to refer to that part of the amount initially applied which cannot subsequently be recovered. Some will have been chemically degraded to compounds that are not detected by the analytical method(s) used, some may have been physically removed by leaching or by irreversible adsorption onto the substrate, but some will almost certainly have escaped into the atmosphere. That this *can* occur with many pesticides is incontroversial. What matters is the *extent* to which the route can account for 'missing' pesticide, and work in recent years indicates that it is probably a major contributor to the entry of many chemicals into the atmosphere, including those most deeply involved in present problems of environmental contamination. The relative proportions of pesticides degraded or physically removed from application sites can, at present, only be surmised from a knowledge of the properties and behaviour of the different compounds.

10.3.2.1   *Entry from treated surfaces*

Hoskins (1962) discussed the physical nature of deposits on surfaces, arising from the use of different insecticide formulations, and showed how the substrate influences their subsequent behaviour, but he referred to losses into the atmosphere only in general terms. Ebeling (1963) reviewed the influences of pesticide formulations, rainfall, humidity, volatilization, wind and temperature on the persistence of deposits. Deposits from aqueous suspensions are more susceptible to loss by weathering processes than are deposits from oil or solvent-based formulations, and loose, fluffy dust deposits are the most readily removed of all formulations. The kinetics of residue decline have been discussed by Gunther and Blinn (1955), Hoskins (1961), and Ebeling (1963) with particular reference to losses from plant surfaces, and by Edwards (1966), and Hamaker (1966), in relation to losses from soil. Often, but not invariably, residues decline very rapidly for a brief period, minutes or a few hours, while physical removal losses dominate immediately after deposition. The loss-rate gradually declines, so that many residues ultimately persist longer than would be predicted by first-order reaction kinetics. With the passage of time they seem to become more intimately 'bound' to the substrate. David (1959) pointed out that decomposition and volatilization losses would eventually predominate over the removal of dust particles by wind, rain or vibration. Hence, the direct loss of deposited particulate pesticides into the atmosphere will only be important for a short period until the physical form of the deposit stabilises with age. Simulated wind and rain transfer the infective stages of the bacterium, *Erwinia amylovora,* that causes fireblight of pears, from tree to tree (Bauske, 1968) and there is no reason to doubt that comparable particles of other matter could be transported in the same way. Compared with losses during application, however, this is not likely to be an important source, even though it could be of greater duration, especially since dust formulations have become less popular as a means of distributing pesticides.

The volatilization of deposited pesticides has been recognised for many years as probably being an important loss mechanism. Decker *et al.* (1950) and Decker (1957) observed that insecticide residue losses from deciduous fruit and forage crops were closely correlated with the vapour pressures of the chemicals, but Gunther and Blinn (1955) were not able to correlate this property with the disappearance of residues from citrus fruits. They pointed out that the effective volatility of the chemicals would be influenced by factors such as the presence of plant waxes and oils, stickers, and the physical form of the deposit. The *in vitro* saturation vapour pressure of a pesticide is only one of the properties controlling its vaporisation. Other properties include its solubility in, and its adsorption on the substrate, its molecular weight and the saturation deficit of the atmosphere over the deposit with respect to the chemical in question, which, in turn, is affected by the prevailing wind-speed, local microturbulence and temperature.

The importance of vapour losses from surface deposits in ideal conditions can be judged from calculations by Hartley (1971), who stressed that the importance of apparently negligible volatility of an organic compound should not be discounted when spread over a large interfacial surface. Choosing dichlorbenil, lindane and simazine to represent a range of volatilities spanning that of most pesticides, he calculated that the lifetimes of freely suspended 10 μm spheres of each would be 20 min, 14 h and 100 days respectively. Hemispherical particles resting on a surface would have half of the surface area of the spheres and would consequently last twice as long. The lifetimes of compounds such as dichlorvos, demeton, parathion or DNOC free acid would lie between those of dichlorbenil and demeton, and those of most other pesticides, including DDT, would be in the 14 h to 100 day range. Particle size is important, since evaporation is a function of the particle radius; a 10 μm diameter dichlorbenil particle weighing 0.5 ng would evaporate completely in 20 min, one weighing 0.5 μg would lose 0.1% min^{-1}, and a 0.5 mg particle 1.4% in 24 h. The saturated atmospheric hemispheres around 10 μm particles of the above three chemicals, lying on impervious flat surfaces, would have radii of 0.8, 2.7 and 33 cm respectively, indicating the degree of separation necessary between particles to prevent mutual interference with volatilisation. Alternatively, Hartley calculated that a 10 m depth of air over 1 ha would be saturated by 500, 12 and 0.01 g of dichlorbenil, lindane and simazine, respectively, in the vapour phase. Under completely calm conditions, it would take several months for this volume of air to become saturated by molecular diffusion alone, whereas with normal wind and turbulence, saturation could occur in minutes. The surface areas over which pesticides are distributed, in normal use, are frequently very great. The total surface area of foliage of a mature crop will usually be many times (10-20) the nominal area of ground on which the crop is growing. Hartley and West (1969) calculated that 1.12 kg active ingredient ha^{-1} would form a continuous deposit of 11 μg cm^{-2} in a layer 0.11 μm thick (density = 1), or if all were present as spherical particles 100 μm diameter, there would be 16 particles cm^{-2}. Taking the rate of volatilization of DDT, as observed by Lloyd Jones (1971) to be about 1 ngcm^{-2}h^{-1}, this deposit would take about 1,100 h, or about 46 days, to approach zero if no other factors were operating, an order of persistence agreeing with measured rates of loss of DDT from plant surfaces.

Temperature and air-movement both affect the rate of evaporation of pesticides from deposits. Temperature has a profound effect on the volatility of chemicals in general, a several-fold increase usually occurring if the temperature is raised by 10°C. For this reason, deposits evaporate more rapidly in the tropics than in more temperate climates, and losses tend to be greatest during the few hours of maximum temperature each day. Wind is important not only for redistributing the vapours. It also enhances losses, by maintaining a high saturation deficit next to the surface boundary layer through which the chemicals can only move by diffusion. In still air, all movement further afield from surface deposit must be diffusive. The layer saturated with the vapour soon deepens and so restricts the process. Increasing wind-speeds at first have a very marked effect in enhancing the volatilization loss, but it then proceeds, at a somewhat lower almost constant rate, once the air movement exceeds a few metres per second, because the rate of diffusion of the molecules through the surface boundary layer becomes the limiting factor. Air movement within a crop canopy will, therefore, greatly promote volatilization losses of deposited pesticides. An example of the magnitude of this effect was given by Mistric and Gaines (1953) who found that camphechlor residues declined in effectiveness by 85% when subjected to a simulated wind, but by only 9% in the absence of wind. Walton and Howell (1954) showed how TEPP depended on vapour transfer for its action in controlling aphids. At wind-speeds of 40-49 kmh^{-1}, the insecticide killed only 29% of the aphids compared with 70% when the wind-speeds were 5-8 kmh^{-1}.

10.3.2.2   Entry from treated soil

Soils are one of the principal environmental depositories for many commonly used pesticides. The subject has been well-reviewed by Edwards (1966 and 1970), and Marth (1965). Residues at concentrations of up to several parts per million are present widely in the cultivated layer of arable soils (Wheatley et al., 1962; Decker et al., 1965; Harris et al., 1966; Edwards, 1969), and trace residues must be universally distributed over the land-masses of the globe, since they cannot avoid participating in the physical transfer and redistribution process of the environment. The concentrations present represent the net balance between local rates of pesticide input from all sources and the rates of decline by all routes, including continual vapour phase transfer into the atmosphere.

Soil particles have an enormous surface area over which pesticide molecules can become distributed, providing an excellent opportunity for volatilization to occur when circumstances permit. However, many factors interact to determine the extent to which vapour phase transfer into the atmosphere occurs. The most relevant properties of the pesticide are its volatility, its solubility in soil moisture (note: not water per se) and its tendency to be adsorbed by the soil particles. The physico-chemical nature of the soil mineral structure, the organic matter content, its moisture status and its texture, also combine to influence volatilization of different chemicals to differing degrees. Environmental factors such as temperature, moisture saturation deficit (not relative humidity), and most especially, wind velocity over the soil surface, also play a determining role.

In general, the most rapid losses of pesticides occur for only a short time immediately after treating soil, the chemicals soon becoming 'bound' to the soil particles. Harris (1961) and Harris and Lichtenstein (1961) observed a rapid initial vapour loss of several insecticides, and Getzin (1958) found that the amounts of radioactive phorate lost within an hour of treatment were 25, 20 and 18% of the amount applied to a sandy soil, a silt loam and a muck soil respectively. The thiolcarbamate herbicide EPTC is more volatile than phorate (about 2×10^{-1} compared with 8.4×10^{-4} mm Hg at $20°C$), and Gray and Weierich (1965) found 20, 27 and 44% of EPTC lost from dry, moist and wet soil, respectively, within 15 min of spraying it onto the soil surface. After one day, 23, 49 and 69% had been lost and this increased to 44, 68 and 90% six days after application. When EPTC was applied as a granular formulation, initial losses were negligible when the soil was dry or contained 4-6% moisture, but 60% was lost from soil with 15% moisture, and they concluded that there was a critical moisture level for each soil, above which large losses occurred when the chemical was left on the soil surface or only shallowly incorporated. The second most important factor was the depth of incorporation, which had to be at least 5 -7.5 cm to prevent large losses when the treatment was followed by light rain, or sprinkler irrigation, amounting to about 2.5 cm in 3 days. Increasing the temperature from $0°$ to $20°C$ increased volatilisation of EPTC from moist, but not from dry soil. Vernolate and pebulate are much less volatile thiolcarbamates than EPTC, and Horowitz (1966) did not consider volatilisation of pebulate from dry soil to be important, its activity as a herbicide persisting for 1-2 months when it was applied to dry soil and not incorporated. Fang (1969) observed that the amount of pebulate lost from wet soil was correlated with the amount of soil organic matter present, indicating that adsorption probably played an important part in limiting the loss of thiolcarbamates from soil. Harris and Lichtenstein (1961) investigated the volatilization of aldrin in detail and found that vapour losses of this insecticide also decreased in dry soil and in relation to the clay and organic matter content. These findings contradict the popular belief that rapid vapour losses of chemicals occur if they are left on a dry soil surface. In fact, they are most likely to be strongly adsorbed when the soil particles are dry, and this will be an important controlling factor in the loss of vapours into the atmosphere. When moisture is present, the water molecules displace some pesticide molecules from the adsorption sites and loss from the moisture/air interface can proceed. Not all chemicals, however, interact strongly with the soil particles, and others may be sufficiently polar to compete favourably with the water molecules and resist desorption. Some chemicals can be so firmly bound to adsorption sites that their volatilisation behaviour is little affected by the presence of moisture, for instance, the base-exchange process which so effectively inactivates the bipyridyl herbicides diquat and paraquat. In contrast, the chlorinated aliphatic acid herbicides show little tendency to interact physically with the soil matrix and their tendencies to volatilise will therefore be determined primarily by their physical properties, vapour pressure in soil and air-water partition ratios (a function of vapour pressure and water solubility) in the case of moist or wet soils. Extreme volatilization tendencies may exist within a main group of compounds, for instance the benzoic acid herbicides, amiben and dichlorbenil. The former seems not to be lost as vapour to any measurable extent over a period of 4 weeks at 15-35°C and relative humidities of 0-100% (Donaldson and Foy, 1965). In contrast, the need to incorporate dichlorbenil into soil to minimise volatilization loss is well-recognized (Swanson, 1969), and rainfall can apparently *increase* its persistence by *reducing* volatilization (Barnsley and Rosher, 1961). Casely (1968), discussing volatilization of chloronitrobenzene fungicides, noted that tecnazene was about four times more volatile than quintozene (v.p. 1.33×10^{-4} mm Hg at 25°C), and he concluded that a considerable proportion of the 2×10^4 kg of quintozene, applied annually in the San Joaquin Valley in California would volatilise, and thereby lead to environmental contamination. He found, as with many herbicides and insecticides, that the presence of water in soil enhanced loss of tecnazene by volatilization, although it seemed not to codistil with the water vapour in the manner proposed by Acree *et al.* (1963) for DDT.

There is universal agreement that the losses of pesticides in vapour form are much reduced by incorporating them into the soil, and indeed, the physico-chemical processes involved dictate that this should be so. Molecules on the soil surface need only diffuse through a millimetre or two of air to escape and be transported further afield by turbulence and wind. Those in the soil have first to make their way to the surface either by diffusion, or be carried upwards by mass-flow of air or liquid (soil moisture) through the tortuous air channels. Diffusion through water is about ten thousand times slower than through air, and so diffusion upwards through water-filled (undrained) pores can be discounted. In free-drained or dry soil, gaseous diffusion at linear speeds of a few millimetres an hour are theoretically possible, but the fine capillary channels and the strong adsorptive power of most soils effectively limit diffusion to rates well below the theoretical maxima attainable. Even when only 2.5 cm deep, the rates of decline of gamma-BHC, aldrin, dieldrin and *p,p'*-DDT seem to be the same as at greater depths and, depending on the chemical, are 5-20 times less than when the insecticides are left on the soil surface (Wheatley, 1965).

The water solubilities of organic pesticides vary from about 1 μg litre^{-1} for DDT to about 700 gl^{-1} for salts of diquat or paraquat. The extents to which different pesticide chemicals can be carried upwards, by the mass-flow of soil water through the capillary channels, to the evaporating surface of the soil have not been determined. However, it cannot be a very important transport mechanism for compounds of extremely low water solubility or very high sorptive potentials, although it could be significant for some water-soluble compounds, depositing them exposed on the surface. Transport by mass-flow of air through the soil may also assist pesticide molecules to reach the atmosphere. Diurnal temperature fluxes will cause air in the upper few centimetres of soil to expand and contract in volume cyclically by perhaps 5% (Abbott *et al.,* 1966) or more, so creating a periodic mass-flow of air carrying pesticide molecules as vapour into the atmosphere. Like diffusion, this process is unlikely to be a rapid loss mechanism. If 50% of the volume of a soil is 'voids' (a loose soil), then the upper 20 cm of soil will hold about 1,000 m^3 air ha^{-1}, of which not more than 100 m^{-3} would be exchanged daily by the temperature-induced diurnal flux. This volume could hold only about 20 μg of *p,p'*-DDT vapour at $20°C$, so that only about 90 mg ha^{-1} could be lost in this way each year, an insignificant amount compared with observed losses even for this very persistent compound. Probably of greater significance, is the mass-flow induced within the surface layers of the soil by minute air pressure gradients, the result of air moving over the irregular soil surface and of general atmospheric turbulence. Spencer and Cliath (1969) found that rates of loss of dieldrin residues by volatilization were a function of air movement through soils. Partial changes of soil air must therefore be occurring continuously in interstitial channels communicating directly with the atmosphere. The soil air will therefore tend not to be saturated with pesticide vapour in the upper layer, encouraging losses by volatilization. This must be an erratic process, very dependent on the texture and porosity of the surface layers of the soil and on the vagaries of air movement.

10.3.2.3 *Entry from other localised sources*

There are many other sources from which pesticides are released more or less continuously as vapour. Fortunately, only minor uses are involved. Methods of application have been devised to control vapour-release for prolonged periods. The use of thermal vaporisers for gamma-BHC or DDT, and of resin-impregnated strips for dichlorvos, are two examples which have been and are the subject of both controversy and restrictions. A number of casual uses, such as the inclusion of gamma-BHC, in polymer emulsion paints, chlordane in lacquers, or DDT in window-cleaning fluids have become either obsolete or restricted in use. Certain industrial uses of persistent compounds, such as dieldrin for moth-proofing clothes and carpets, have also been reviewed elsewhere (Wilson, 1969). Vapour transfer is a vital mechanism in the action of many fungicides or fungistats, examples being the use of biphenyl-impregnated wrapping paper for inhibiting the development of *Penicillium* moulds on oranges (Anon, 1965), and the vapour treatment of seed to control seed-borne fungi (Crosier and

Natti, 1965). Bevenue (1969) reported the uptake of chlordane vapour by flour and rice exposed to it for 90 days and showed that when baking flour was heated for 30 min at 180° about 47% of the adsorbed chlordane was eliminated (presumably into the atmosphere).

None of these uses can be major sources in themselves but each must make a small contribution to the pesticide content of the atmosphere and some may have particular significance in indoor situations, such as transfer of pesticide into foodstuffs.

10.3.2.4 *Entry from ubiquitous sources*

All surfaces, solid or liquid, in contact with freely circulating air, must be continuously contaminated by pesticides present in the atmosphere, either by deposition of particulates or by vapour sorption. Deposited material is, of course, free to re-volatilize and so some will return to the atmosphere. The exchange of pesticide molecules between gaseous and solid or liquid phases occurs in both directions simultaneously, but at rates which depend, among other factors, on the magnitude and direction of the concentration gradients at any given instant. The process is always moving towards a state of equilibrium where the number of molecules depositing equals those escaping, a situation likely to exist only momentarily from time to time in practice. Whether there is a net loss or gain of material to the atmosphere will, therefore, depend on the concentrations of the chemical in the air and in the adjacent substrate. Where the concentrations in the substrate are high, as in areas of pesticide usage, there will normally be a net loss to the atmosphere, as already discussed. In other land areas, the passage of contaminated air-masses will lead to periods of deposition, followed by some re-evaporation into less-contaminated air once the peak concentrations have passed. This will gradually smooth-out fluctuations in the pesticide content as the atmosphere moves further afield over land-masses, an effect analogous to peak-broadening during chromatography.

The enormous mass of even the mixed (75-100 m depth) volume of the oceans, ensures that concentrations of pesticides in the water will normally be so low, relative to those in air, that the net vapour phase exchange will tend to be from air to water and not *vice versa*. However, the entrainment of sea-spray into the air and the subsequent evaporation of the water droplets will leave solutes as particulates suspended in the atmosphere. Pesticides present in the oceans could thereby enter the atmosphere only as minute traces. However, the phenomenon is so extensive that it may be an important mechanism regulating the exceedingly low background concentrations of some pesticides in the atmosphere.

The continuous entry of pesticides into the atmosphere from universal sources, cannot be considered sensibly in detail as isolated processes. Each is only one part of the complex of dynamic equilibrium mechanism determining the dispersion and fate of pesticides in the atmosphere.

10.4   ASSESSMENTS OF CONCENTRATIONS OF PESTICIDES IN THE ATMOSPHERE

The foregoing discussion provides abundant circumstantial evidence that many, if not all, pesticides have ample opportunity to enter the atmosphere and become entrained in the general circulatory systems of the air-masses. No new scientific principles have to be invoked to explain their presence and, with hindsight, it is surprising that the significance of the escape of pesticides into the atmosphere from all sources was not realised until the mid-1960's.

The obvious hazards of working in close proximity to high concentrations of certain very toxic pesticides had been recognised for many years and appropriate safety precautions had been implemented in many countries to safeguard workers. Factual evidence that significant quantities of pesticides 'disappear' during application had been available for at least 15 years (Decker et al., 1950). During 1963, pesticides were demonstrated in air over several cities in California (West, 1964) and near to a treated area in Ohio (Cohen et al., 1964). Harris and Lichtenstein (1961) had shown clearly that aldrin and certain other insecticides were lost from soil even in vapour phase and, taking all sources into consideration, Wheatley and Hardman (1965) realised that certain organochlorine insecticides may be escaping into the atmosphere at sufficient rates for them to be continually present at very low concentrations on a world-wide basis. They explored the possibility that this might account for low concentrations of gamma-BHC, dieldrin and DDT, in soil which was believed never to have been treated with these chemicals. Fortunately, ultra-sensitive analytical techniques were available by this time and, when these were applied to the problem, compounds with identical chromatographic behaviour to gamma-BHC, dieldrin and p,p'-DDT were found in central England in rain, the natural sampling mechanism for atmospheric pollutants. The amounts did not seem to be sufficient to account for the residues observed in the untreated soil, but the results were of wider implication. Abbott et al. (1965) quickly confirmed and extended these findings, and subsequent studies by Wheatley and Hardman (Figs. 10.1 and 10.2) and by Tarrant and Tatton (1968) showed that rain falling over the British Isles invariably contained very low concentrations of a range of organochlorine pesticides, despite the limited local seasonal usage outdoors. Atmospheric contamination on a world-wide basis seemed to be implicated. Meanwhile, Abbott et al. (1966) demonstrated that the insecticides could be recovered directly from London air. Similar investigations, in the United States from 1963 to 1966, showed that pesticides were usually present in air and in rain over and near treated zones, both rural and urban (Antommaria et al., 1966; Bamesberger and Adams, 1966; Cohen and Pinkerton, 1966; Tabor, 1966; Weibel et al., 1966). Finally, direct evidence of long-distance transport of pesticides by air was obtained by Cohen and Pinkerton (1966) and by Risebrough et al. (1968). The latter demonstrated organochlorine pesticides in dust collected over Barbados. Presumably it had been transported there over several thousand miles of ocean by north-east trade winds.

Are the concentrations in air sufficient to constitute toxicity hazards to life in any form?

Jegier (1969) dealt with the significance of pesticides in the atmosphere under three main levels of 'population exposure'. Firstly, high exposure situations applied to workers in pesticide manufacturing or formulating plants, operators engaged professionally in the applications of pesticides or living or working in premises where aerosol (or vapour-dispensers) are used. Secondly, were people in communities near areas where pesticides are used extensively for agricultural or public health purposes and likely to be exposed to lower but significant concentrations. Populations living far from sites of usage who would be exposed to minimum (background) concentrations, formed a third category. Jegier's subdivisions (1969) are helpful for discussing the concentrations of pesticides in the atmosphere, but they also imply differing durations of exposure, a very important toxicological consideration. The general background exposure is obviously continuous with relatively little variation; the exposure of the intermediate group includes the background exposure, but additionally has periodic marked increases, during and for some time after, pesticides have been applied; the first group experience additional acute short-term exposures, the frequency of which are also important.

10.4.1 High concentrations

The high-exposure situations were the first to be investigated, not only because they clearly represented the most obvious hazards to man or animals, but because the relatively high concentrations of pesticides could be determined by analytical methods of only modest sensitivities. The concentrations present are often in terms of μg or mg (10^{16}-10^{13} g) per cubic metre (m^3) of air, so that it is not necessary to sample large volumes to recover measurable quantities of the chemicals. Middleton (1965) reviewed examples of measurements of insecticide concentrations determined in a variety of high exposure conditions, and Jegier (1969) tabulated insecticide concentrations as determined during spraying. Examples of their data are listed in Table 10.1, together with the findings of Lloyd and Tweddle (1964), who reported an investigation of dimefox concentrations in air, when diluting the concentrated chemical into the tank of a sprayer. Most vapour arose from spillage of the concentrate, and turbulence caused a five-fold variation in the dimefox concentration around the tank. The vapour is heavier than air so that greater concentrations occurred below, rather than above, the level of the filler cap, and the worker filling the tank was less-exposed than others standing nearby. Concentrations upwind were lower than those downwind, as would be expected. Soil treated with this hazardous chemical emitted detectable concentrations of dimefox vapour for three days following treatment.

From the know saturation vapour pressures and molecular weights, it is

TABLE 10.1. Examples of insecticide concentrations in air in the vicinity of sites of application.

| Insecticides | Operation/site | Concentrations in air | | Ref. |
		Range $(\mu g\ m^{-3})$	Mean	
Parathion	Loading and mixing	0-5,530	–	a
	Tank-filling	0-2,430	530	b
	Orchard, California	40-290	130	a, b
	Orchard, Florida	360	360	a, b
	Orchard, Canada	2,000-15,000	–	a
	Apple orchard, Canada	50-260	150	b
Azinphos-methyl	Apple orchard	50-2,550	670	b
	Apple orchards, tank-filling	260-6,200	2,770	b
Malathion	Apple orchard, vegetable fields	410-760	590	b
	Mosquito control	70-90	–	a
Carbaryl	Apple orchards	180-810	600	b
Endrin	Potato fields	0	0	b
	Vegetable fields, aerial spraying	20-90	50	b
BHC	Forests, ground-treated	2,600-12,500	–	a
	Forests, aerially-treated	540-4,100	–	a
	Vaporised in restaurants, stores, etc.	40-780	–	b
DDT	Forests, ground-treated	2,600-4,600	–	a
	Forest, aerially-treated	1,900-171,000	–	a
Dimefox	Hopyards, tank-filling	210-610	120	c
	Hopyards, 0-108 h after soil application	50-20	–	c
Chlordane	Apartments, stores, cafeterias	800-920	440	b
Diazinon	Cloakrooms	260-10,200	2,680	b
Methoxychlor	Barns, interiors, cattle spraying	500-4,500	7,680	b
Demeton	Greenhouses	4,220-19,600	9,150	b

References: a = Middleton (1965); b = Jegier (1969); c = Lloyd and Tweddle (1964)

apparent that many of the higher concentrations of the chemicals listed in Table 10.1 greatly exceeded those possible if the chemicals were present solely as vapour, thereby implying that much of the recovered chemicals were present in particulate form. BHC is one of the more volatile of those listed. Its molecular weight is 291 and the saturation vapour pressure of the gamma-isomer at 20°C has recently been redetermined by Atkins and Eggleton (1971) as 3.1×10^{-5} mm Hg. The maximum concentration of gamma-BHC in air at 20°C is thus

$$\frac{291}{22.4} \times \frac{3.1 \times 10^{-5}}{760} \times \frac{273}{293} = 5 \times 10^{-6}\ gl^{-1}, \text{ or } 5\ mgm^{-3}.$$

Concentrations in excess of this imply that some was present in particulate form, but lower concentrations do not necessarily mean that the chemical was all present as vapour. It is most unlikely that the ambient air can become fully saturated with vapour under normal outdoor conditions. DDT is one of the least volatile of the chemicals represented in Table 10.1 and the concentrations cited all exceeded 5 $\mu g/m^{-3}$, implying that some of each sample was particulate.

10.4.2 Intermediate concentrations

Pesticides occur in air at intermediate concentrations immediately down-wind of zones of usage, and within the zones during the early phases of residue loss. The areas concerned are essentially those discussed in relation to the drift-entry of pesticides into the atmosphere during application, and their release from residual deposits before they are more widely dispersed into air-sheds. Jegier (1969) cited instances of the insecticides azinphos-methyl, carbaryl, malathion and parathion occurring at concentrations of up to 500 μgm^{-3} in the air of residential areas in Quebec, 300-600 m from apple orchards where they were being applied by air-blast equipment. Breidenbach (1965) was also able to detect pesticides in the air over peach-growing areas in Georgia and South Carolina.

Some of the most detailed studies of atmospheric pesticide concentrations associated with nearby usage, were described by Tabor (1966), who determined the concentrations in the air over urban and rural communities. The sampling method which he used, was primarily for particulates, and took no account of the chemicals in the vapour phase, or of the loss of vapour from the trapped particles. His methods were especially suitable for detecting and assessing the organochlorine pesticide-content of the air, but he also noted that numerous unidentified organo-thiophosphates could be detected in many samples, although he only presented results for sites where malathion had been specifically used. Tabor's results, therefore, probably underestimated not only the amounts, but also the range of types of pesticides present in air near large areas of usage. Nevertheless, they demonstrated correlations between usage and the concentrations occurring in associated air, showing that they were mainly present at concentrations of ngm^{-3}, although mgm^{-3} concentrations sometimes arise during periods of usage (Table 10.2). In urban neighbour-hoods, the wind direction was a determining factor in DDT dispersal. There was reasonable agreement between the chemicals expected and those actually found, although he was unable to detect chlorbenzilate in one area where it was expected, or endrin at two sites near to areas of usage, but this may have been due to technical analytical difficulties.

Bamesberger and Adams (1966) described an investigation in which they collected 24-hour air samples to determine the concentrations of gaseous (ester) 2,4-D herbicides, at two sites, during the 100-day period of main summer usage in Washington State, U.S.A. during 1964. They used a rotating disc impactor to collect large particles ($>$3 μm) and a midget impinger to trap smaller particles and the gaseous fraction. Results were obtained for four 2,4-D esters and one of the methyl esters of 2,4,5-T was trapped on both the impactor and impinger parts of the sampling device. The results (adapted in Table 10.1) show air-concentrations comparable to those found by Tabor for insecticides in communities with pest-control programmes, or near rural areas of usage. Bamesberger and Adams' data also reveal the irregular occurrence of high concentrations, obviously corresponding to the frequency of usage in areas upwind of their sampling

TABLE 10.2. Ranges of pesticide concentrations in air associated with nearby usage in various areas of North America.

Locality	Characteristics	Pesticides	Concentrations (ng m^{-3})	Fraction of samples in which detected	References
Rougemont, Quebec	300-600 m from application	Azinphos-methyl, Carbaryl, Malathion, Parathion	up to 500,000	–	Jegier (1969)
Florida	6 small communities in usage areas	DDT	0.1-22	88/103	Tabor (1966)
		Chlordane	0.1-6	66/103	
		Camphechlor	1.2-15	12/103	
		Aldrin	0.14	9/103	
		Thiophosphates	traces	?	
	1 rural area during usage nearby	DDT	0.3-8,500	11/11	
		Chlordane	1-31	11/11	
Chesapeake Bay township	Samples before, during and after fogging for mosquito control	Malathion	before 0.2	4/4	
			during 8-22	4/4	
			after 0.2-2.3	4/4	
Atlantic coast resort	Samples during and after fogging	DDT	during 100-8,000	6/6	
			after 2-11	6/6	
		Malathion	during 1-30	6/6	
			after <0.1-1	6/6	
Eastern City	Pest control programme, 3-week period	DDT	230-300	2/4	
		Malathion	6-25	3/4	
		Mixture: DDT+	3-430	3/3	
		Malathion	0.5-140	3/3	
Pullman, Wash.	–	2,4-D esters impactor	<1,960	–	Bamesberger and Adams (1966)
		impinger	1,040		
		2,4,5-T esters impactor	<3,380 (av. 45)	9/99	
		impinger	0	0/99	
Kennewick, Wash.	Hotter and drier than Pullman	2,4-D esters impactor	<820	–	
		impinger	<5,120		
		2,4,5-T ester impactor	<630 (av. 12)	14/102	
		impinger	<780 (av. 13)	7/102	

points. The maximum concentrations shown in the table occurred infrequently, the average for 24-hour periods usually being less than 1 μgm^{-3}. Their data for the four 2,4-D esters indicated that one or other, or all, of the chemicals were detectable in air at Pullman on about half of the days in a 100-day sampling period, and on almost three days out of four at Kennewick. The distributions of the observed concentrations were typically skewed, almost certainly reflecting the proximity of the usage areas to the sampling points and the prevalence of low-dispersion atmospheric conditions. In the hotter and drier climate of Kennewick, the impinger samples collected more of the herbicides as fine particulates (aerosols) and/or vapour than at the Pullman site. The authors also made reference to the presence of peaks of other compounds on their chromatograms, most probably organochlorine and/or organophosphate insecticides.

Some of the most extensive uses for pesticides occur in forest protection practices. Yule et al. (1971) monitored the phosphorus content of the atmosphere daily for 5 weeks in 1969, covering a period when the insecticide fenitrothion was being used in New Brunswick for spruce budworm control. About 300 tonnes of fenitrothion were applied by aircraft to approximately 1.2×10^6 ha of pulpwood forest within 4 weeks. They analysed air samples collected five feet above ground at five stations distributed around the treated zones and found average daily phosphorus concentrations ranging up to about 3 μgm^{-3}, although they were mainly between 0.5 and 1.5 μgm^{-3}, as compared with the background level which was estimated to be about 0.5 μgm^{-3}. The peak concentrations occurred for only a few days towards the beginning and again at the end of the treatment period and, although they were not able to correlate the peaks with particular meteorological conditions, they showed that relatively more fenitrothion was in vapour form while concentrations were high, the amounts present as particulates being much less variable. The atmospheric contamination was partly attributable to local application and partly to down-wind drift of material applied further afield. They concluded that forest spraying contributed relatively small amounts of insecticide to the atmosphere and to other parts of the environment outside the target area. The results of this exercise are obviously very similar to those of other investigations described above.

There is therefore a striking measure of agreement as to the concentrations of pesticides likely to occur in air in and around zones of usage. In areas near to those where pesticides are being used, the concentrations commonly range up to about 10 μgm^{-3}, although higher concentrations may occur occasionally, as shown by the data from Quebec cited by Jegier (1969). Given adequate analytical methods, it is not unreasonable to assume that virtually any pesticide should be demonstrable in zones stretching a few kilometres down-wind of an area where it is being applied by conventional methods, or where it is being 'lost' from deposits by removal mechanisms.

10.4.3 Background concentrations

Since it was first realised, about 1963-1964, that some pesticides may be sufficiently stable for them to exist in the atmosphere for periods of days or weeks, several investigations have been undertaken, firstly, to demonstrate them in the atmosphere in areas remote from areas of usage, and then to determine the concentrations present and so assess their significance. It was obvious from the outset, that it would be difficult to find reliable sampling techniques, and that ultra-sensitive analytical methods would have to be used, embodying stringent precautions to avoid inadvertent contamination of samples. The sampling methods adopted have involved either the collection of rainwater or direct air-sampling. Either method is valid for merely demonstrating the presence of the pesticides, but neither is entirely above criticism for quantitative results.

The scrubbing action of raindrops falling through the air is a well-known natural sampling method, and the collection of rain is technically simple. However, the relationships between the concentrations present in rain and in the rain-free atmosphere are complicated. Particulates provide a nucleus around which raindrops can form and they are also readily washed-out or scrubbed from the air by impacting with the falling droplets. Fine droplets should do this more efficiently than large droplets, and the body of air cleared of particles or sampled by a prolonged, slow-moving, frontal drizzle is likely to be larger than that swept by a brief, intense shower originating from a rapidly moving, but locally-formed, cumulus cloud. The concentration of pesticide in solution in raindrops will also tend to equilibrate with concentrations in the air through which they fall, according to the air-water partition ratios of the compounds concerned. Smaller droplets will equilibrate more quickly than large droplets, and so the pesticide content of the former will tend to be more representative of that in the lower few tens of metres of the atmosphere, whereas the latter will tend to be representative of a greater depth of atmosphere, but they may not have achieved the equilibrium concentration during their more rapid fall. The first raindrops falling through a given air-mass should contain higher concentrations than those falling later. Variations in concentrations in the rain between the beginning and the end of a period of rainfall are thus to be expected, but they could be either slight or extreme depending on the circumstances. The quality and history of the rainfall sampled will thus greatly influence the extent to which it reflects the general pesticide content of the air-mass, and consequently, rainwater samples have usually been accumulated over periods of weeks in order to damp-out the short-term variability and provide a reasonable average.

The collection of air-samples is obviously a more direct method for determining the concentrations of contaminants, but it is difficult to devise a sampling device which will truly represent the concentrations present. The very low concentrations present in the general atmosphere make it necessary to sample large volumes of air, to get even the minimal quantities needed for highly sensitive methods of analysis. Several forms of apparatus have been used. Particulate matter has been filtered out, with the knowledge that some vapour escapes, and that prolonged passage of air over the trapped particles may lead to a net loss or gain of contaminants as vapour. There is uncertainty as to whether the concentrations reported from the particles are real or artefacts. Alternatively, the air may be passed through solvent bubblers, or through an adsorbent such as charcoal. They, too, may become secondarily equilibrated or lose liquid in aerosol form. A combination of filter and solvent methods has also been used. In contrast to rainwater sampling, the physical limitations of air-sampling systems have usually restricted collections to periods of only a few hours. Consequently, the results tend to be variable because they record relatively short-term fluctuations in the pesticide content of the atmosphere. Many samples must therefore be analysed to obtain meaningful results.

10.4.3.1   *Presence in rain*

The continual presence of organochlorine insecticides in rainwater, and hence in the general atmosphere was demonstrated, apparently simultaneously, in the British Isles and in the U.S.A., from 1964 to 1966 by several teams of workers, and the results obtained are very substantially in agreement. In an agricultural area in central England (Wellesbourne in Warwickshire) Wheatley and Hardman (1965) collected rainfall samples for successive monthly periods from April 1964 until February 1965. The first seven monthly samples were analysed when the analytical methods were being refined to improve sensitivity, but nonetheless, they revealed the presence of compounds, tentatively identified as gamma-BHC and dieldrin, at concentrations of from 77-120, and 19-36 ngl^{-1} respectively. Subsequently, improvement in the analytical method enabled p,p'-DDT to be detected at concentrations of 2-4 ngl^{-1} and the collection of additional samples during periods of rain, confirmed that the chemicals were in the rain itself and not simply stray contamination of the collecting system during lengthy non-rain periods. This study continued until July 1966, and the results are given in Fig. 10.1 as the monthly mean concentrations (ngl^{-1}) of the three insecticides in the rainwater.

During 1964 and 1965, the gamma-BHC concentration of the rain fluctuated considerably from month to month, but the results suggested a downward trend from an initial level of about 100 to a minimum of about 20 ngl^{-1} by mid-1965. Thereafter, there was an increase during the autumn, a further decline, and then, in common with other compounds, a trend towards a very large peak during March 1966, a time when BHC would not be used outdoors in England. The results for dieldrin were similar. It declined from about 30 to about 10 ngl^{-1} by autumn of 1966, showing a progressive increase between December 1965 and March 1966, the concentration then falling rapidly to about the original level. Although the p,p'-DDT content appeared to change relatively little during 1965, the proportional changes in its concentration, month by month, were nevertheless appreciable, as with the other compounds, and it also increased to an unprecedented degree in the March 1966 samples. No entirely satisfactory explanation has been forthcoming for the exceptionally high pesticide concentration of the March 1966 sample, except that it was a month of low rainfall (22 mm; Fig. 10.2). The two preceding samples

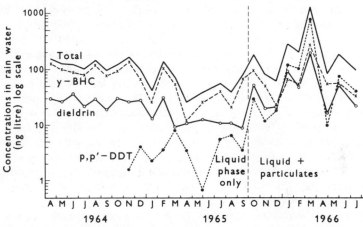

FIG. 10.1. Concentrations of organochlorine insecticides measured in cumulative monthly samples of rainwater falling at Wellesbourne from April 1964 to July 1966. (Wheatley and Hardman, unpublished data.)

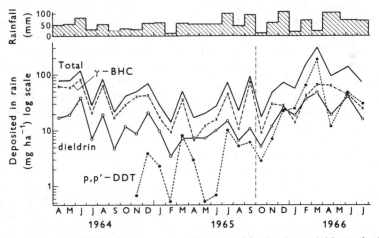

FIG. 10.2. Rainfall and amounts of organochlorine insecticides calculated to have been deposited per month at Wellesbourne from April 1964 to July 1966. Prior to October 1965 (vertical dashed line) the amounts present on particulates were estimated. (Wheatley and Hardman, unpublished data.)

taken during January and February also showed a progressive increase in p,p'-DDT and dieldrin content, suggesting that it was not caused by inadvertent contamination at the rain-collection site or in the laboratory.

During the early part of the study, the rainwater was filtered through a small glass-wool pad to remove any large particles of solid matter. The material trapped on the filter pad was a tarry, sooty, amorphous mass present in variable amounts. Analysis of this matter from October 1965 onwards showed that it contained substantial amounts of pesticides. Consequently, the concentrations shown in Fig. 10.1 prior to this date are underestimates of the total pesticide content of the rain. For gamma-BHC, an average of 8% was usually associated with the solid matter, and so the difference between the observed and the total present would be small. About 22% of the total dieldrin and 39% of the total p,p'-DDT content of the rain were recovered from the particulate matter, indicating more serious underestimates of these insecticides prior to October 1966. In deriving the data for Fig. 10.2, which shows the amounts calculated to have been deposited by rain at Wellesbourne each month, the concentrations of gamma-BHC, dieldrin and p,p'-DDT observed during the period were multiplied by 1.08, 1.28 and 1.65 respectively. Thereafter, the amounts shown are as measured.

Abbot *et al.* (1965) confirmed the presence of organochlorine insecticides in rain falling on different parts of England. They reported the occurrence of the alpha-, beta- and gamma-isomers of BHC, HEOD (the principal constituent of dieldrin), and the p,p'-isomers of DDE, TDE, and DDT, in samples from two sites in central London. They also noted that similar results had been obtained from the analysis of samples of snow collected within 25 km of London, and in rain falling at sites as far apart as Camborne in south-western England, and Stornoway in the Outer Hebrides. The concentrations of gamma-BHC were similar to those recorded in central England by Wheatley and Hardman (up to 155 ngl^{-1}), but they found rather higher maximum concentrations of HEOD (up to 95 ngl^{-1}), and very much higher concentrations of p,p'-DDT (up to 400 ngl^{-1}) in their London samples. These differences are exactly what would be expected from the much greater amount of particulate matter observed in the rain from the London area, compared with that from rural Wellesbourne, and the differing distributions of the compounds between the solid and liquid phases. The results were further amplified by Tarrant and Tatton (1968). They reported a study of the organochlorine pesticide content of rainwater from seven widely separated sites in the British Isles, including London and Wellesbourne, from August 1966 to July 1967. Alpha- and gamma-BHC, dieldrin, p,p'-DDT, p,p'-TDE and p,p'-DDE were detected in rain at all seven sites, at concentrations of 10's or 100's of ng/I^{1}. They also detected polychlorinated biphenyls (PCB's), which had by this time become implicated as components of environmental contamination (Jensen, 1966), but the amounts present were not estimated. The compounds, aldrin, chlordane, endosulfan and heptachlor were not detected, implying that any present were at concentrations of less than about 1 ngl^{-1}. This study confirmed the presence of organochlorine insecticides in rainwater throughout the year, and it also showed that the concentrations in duplicate samples, taken in London and at Wellesbourne, could be determined with adequate precision, usually to within a few ngl^{-1}, so giving confidence in the reliability of the sampling and analytical procedures used for this and the earlier studies. Tarrant and Tatton's results also indicated an apparent decline in the concentrations present at Wellesbourne and London, compared with those reported during 1964-1966 which, they suggested, might be associated with the reduced usage of dieldrin in Britain and, in the case of DDT, to the marked effect of the Clean Air Act in reducing the smoke content of the London atmosphere by 1967 (Brazell, 1970). In part, however, lower maximum concentrations of the compounds were to be expected as a result of analysing three-monthly, pooled samples from the seven sites, a procedure which itself would have had an averaging-effect on the data.

Meanwhile, a parallel series of investigations in North America were revealing a very similar picture. Weibel et al. (1966) reported that the organochlorine content of rainwater collected in Ohio between August 1963 and March 1964, contained averages of 280 and 220 ng of organo-chlorines $litre^{-1}$ respectively. The first definite report indicating the presence of pesticide in air, other than as a direct consequence of loss during application, seems to be that of Cohen et al. (1964). Atrazine and polypropylene glycol butyl ether ester of 2,4-D had been applied to an experimental agricultural plot. Three weeks later, rainwater was found to contain organochlorine compounds which produced a gas chromatogram 'strikingly similar to those obtained on samples of the treated soil located more than a mile away' (Cohen and Pinkerton, 1966). The organochlorine content of the rain was 300 ngl^{-1}, the two herbicides each contributing about 100 ngl^{-1}. The soil contained atrazine at a concentration of 6 μgg^{-1} (6 ppm), and they calculated that 16-17 mg of the soil in a litre of rainwater would have caused the observed herbicide concentrations in the rain. This led them to postulate that pesticides could exist in the atmosphere adsorbed on soil particles which would then be redeposited by wash-out onto land areas remote from the site of the original deposit. Their subsequent analysis of dust washed-out of the atmosphere by a trace-rainfall, in January 1965, amply confirmed the long-distance transport of pesticide-contaminated soil particulates in the atmosphere as already discussed.

10.4.3.2 *Presence in air*
The discovery of organochlorine pesticides in rain in the British Isles arose as a direct corollary of studies, particularly by Harris and Lichtenstein (1961), that several commonly used pesticides were lost from soil as vapour and should therefore be present in air, which was being 'sampled' naturally by the rain. Direct confirmation of their presence in air was

obtained in a careful study by Abbott *et al.* (1966) who drew air first through a 0.22 µm filter and thence to a series of four bubbler bottles containing dimethylformamide, a solvent of relatively low volatility. By this means, they were able to trap quantitatively the compounds present in 8,000-litre samples of air in central London during August 1965. They found a similar spectrum of compounds to those already detected in rainwater samples at their site, namely alpha and gamma-BHC, HEOD, and the *p,p'*-isomers of DDE, TDE and DDT, at concentrations ranging from approximately 1 to 21 ngm^{-3} (parts in 10^{12} air, w/w). They also noted on gas chromatograms, a number of unidentified peaks with short retention times, which they tentatively ascribed to polynuclear hydrocarbons known to be associated with atmospheric dust, and, of greater historical interest, three peaks with long retention times, which they thought might be break-down products of organochlorine pesticides, but which, viewed in retrospect, were most probably polychlorinated biphenyls, not at that time recognised as general atmospheric contaminants.

One of the earliest studies of pesticides in air seems to have been that undertaken by the State of California Department of Public Health during November and December 1963. DDT was detected in all but 2 out of 18 air samples from four Californian cities, at concentrations ranging from <0.3-19 ngm^{-3} (West, 1964; Jegier, 1969), similar to those recorded in London air by Abbott *et al.* (1966). Attempts to trace sources of DDT contamination of remote areas, by sampling from aircraft, are also reported to have revealed DDT in amounts ranging from a trace in 25 hours flying-time over Canada, to 45 µg in one half-hour over New Mexico (Sheldon *et al.,* 1964). Also, according to Newsom (1967), when Huddleston *et al.* (1966) collected air over 24-hour sampling periods at nine different stations they recorded DDT concentrations of from 0.2-340 µgm^{-3} in filtered air.

Antommaria *et al.* (1965) sampled air on a roof in Pittsburgh, Pennsylvania, on ten occasions from June to December 1966 for 15-day periods. They, too, analysed only particulate matter trapped on filters, and, consequently, their results represent only minima, because of potential vapour loss during sampling. However, they were able to detect and identify several of the compounds normally associated with DDT residues, but assessed only the *p,p'*-DDT content, which totalled 2.36 µgm^{-3}. Their results indicated that more DDT was adsorbed on finer than on coarser particulates, which would be expected by the greater surface area of the finer particulates providing more adsorption sites; mean specific surfaces of about 3m^{2}g^{-1} having been determined for particulates in Pittsburgh air (Corn and Montgomery, 1968).

Risebrough *et al.* (1968) attempted to estimate the relative importance of air and river effluents, as sources of marine contamination by pesticides. They used a nylon mesh (50% efficient for particles >1 µm) to filter dust from air. Dust collected off La Jolla pier in California averaged

70 pgm^{-3} of organochlorine pesticides, compared with 0.078 pgm^{-3} in air over Barbados in the Caribbean. Over the Pacific Ocean, they were unable to detect organochlorine compounds, although they reported finding unnaturally high concentrations of talc particles, which they presumed indicated the probable presence of pesticides, at concentrations too low to detect. The average pesticide content of the dust collected at Barbados corresponded to 41 ngg^{-1} of dust, from which they calculated an annual deposition of about 600 kg of pesticide between the Equator and 30°N latitude. They also detected PCB's in their samples, and their results implied that these compounds tended to be present more in the vapour phase than did the organochlorine pesticides, although they were similarly dispersed, depending on the wind pattern, circulation and fall-out, a subject subsequently investigated in greater detail (Risebrough *et al.*, 1969).

10.5   FATE IN THE ATMOSPHERE

The fate of pesticides after they have escaped into the atmosphere, has to be deduced by the application of basic physical and chemical principles, from analogies with the behaviour of other atmospheric contaminants, from the observations already made on the presence of pesticides in air, and from knowledge of their redistribution to remote sites. The probable pattern of events is not difficult to outline qualitatively. Few meaningful quantitative estimates of the process have been attempted to date, how-ever, and invariably these have been based on assumptions which grossly over-simplify the complexity of the dynamic interactions between the variables in the system. A recent model proposed by Woodwell *et al.* (1971) may well have indicated at least the order of magnitude of certain of the events in the redistribution process.

The fate of pesticides in the atmosphere can be considered under four main headings, dealing with their vapour phase/particulate behaviour, chemical changes, removal processes and residence times, and the extent of their redistribution. Much of the discussion is of necessity speculative and the evidence circumstantial, because there have been few scientific measurements made in the atmosphere itself that are fully relevant to pesticide behaviour.

10.5.1   Vapour phase/particulate behaviour

Whether a pesticide becomes entrained in the atmosphere as vapour or on particulates is largely immaterial. It will not remain entirely in either state for long. The molecules will immediately begin to interchange between the two phases, endeavouring to establish appropriate equilibrium concentra-tions of vapour and particulate forms. The equilibrium concentrations depend on the nature of the particulate matter, as well as on the chemical properties of the pesticide, interacting according to the basic laws governing gas solid adsorption processes. Amorphous carbon from black-smoke emissions, for instance, is a powerful chemical adsorbent whereas fine quartz sand particles are not. Differences between chemicals can be illustrated by reference to gamma-BHC and p,p'-DDT. Gamma-BHC seems to be largely present in the atmosphere as vapour, and consequently, the concentrations in air over rural areas should be similar to those in air over smoke-contaminated city air. In contrast, p,p'-DDT tends to be more adsorbed onto particulates, often less than about one-half being present as vapour, and much less when the concentration of particulates is high. In general, therefore, pesticide escaping initially as vapour will tend to be adsorbed onto uncontaminated suspended particulates, whereas contaminated particulates will tend to relinquish pesticide molecules to the air until equilibrium is approached. If however, more pesticide enters a system already in equilibrium, then the concentrations will automatically adjust themselves to re-establish the equilibrium ratios between the phases once again. Although the time taken for this to occur is not likely to be long (perhaps less than 1 h in a well-stirred closed system in the laboratory) the highly dynamic natural systems will probably never attain a steady state of equilibrium. It will tend to fluctuate as the air moves from a pesticide source to a sink or *vice versa*, for example, from treated to untreated zones or from land to ocean. This will tend to progressively damp-out the more violent fluctuations in pesticide content of the atmosphere, caused by the intermittent escape of relatively high concentrations in limited volumes of an airmass. If all further entry were to be prevented, a relatively uniform background concentration would eventually prevail, steadily declining as material is depleted from the atmosphere by other mechanisms.

10.5.2 Removal processes

The principal removal processes are probably chemical degradation or change, and fall-out. The latter can be sub-divided into wash-out by rain, and dry-deposition, which includes both sedimentation and trapping or filtering-out of particulates, and adsorption of vapours onto solid or liquid surfaces. The contributions of the separate aspects of the dry-deposition process are not usually determined.

The importance of the chemical reactions that can take place in the atmosphere, is well-illustrated by the formation of photochemical smog by the action of sunlight on exhaust gases from internal combustion engines. Many pesticides in solution or as deposits, can also undergo photochemical changes when exposed to sunlight, or irradiated artificially with ultraviolet light. The photochemical decomposition of herbicides and plant growth-regulating substances, has been reviewed by Crosby and Ming-Yu Li (1969) who concluded that most classes of the chemicals are liable to be broken-down by ultraviolet light. For example, the phytotoxicity of certain substituted-urea herbicides is reduced when deposits on a soil surface are exposed to sunlight and kept at high temperatures (Comes and Timmins, 1965). Complex changes can also be induced in cyclodiene and certain other organochlorine insecticides, if they are irradiated in the laboratory or exposed to sunlight either in solution or as dry deposits (Robinson et al., 1966; Henderson and Crosby, 1967, 1968; Rosen and Carey, 1968; Zabik et al., 1971). The transformation products may differ from the parent compounds both physico-chemically and toxicologically. Rosen et al. (1966) found that exposure of dieldrin to sunlight for 3 weeks or 2 months caused 7 and 25% respectively, to change to a closely allied compound, compared with a 66% conversion when the insecticides were irradiated for 48 hours in the laboratory, with a 2537 nm wavelength ultraviolet light source. At high altitudes, therefore, it is not unreasonable to assume that photoconversion or decomposition will be more pronounced than at sea-level, where the ultraviolet radiation from the sun is less intense. Unfortunately, not all light-induced changes are degradative in the sense that they convert toxic to non-toxic compounds; some photoconversion products are more toxic to certain organisms than the parent compounds (Rosen et al., 1966), and some may also be more persistent, although seemingly of lesser toxicological importance (Robinson et al., 1966). The capacity of the atmosphere to provide a medium, within which large amounts of pesticides may be converted or degraded photochemically, is quite unknown, both within and above the troposphere.

Other chemical transformations can also occur, changing the nature of the compounds and their toxicities. Among insecticides, the epoxidation of aldrin to dieldrin, and of heptachlor to heptachlor epoxide, occurs spontaneously when either of the parent compounds is used, the epoxides always being recoverable subsequently as residues. Although aldrin has been widely used in many parts of the world, it has not been specifically identified as a general contaminant of the atmosphere, nor has heptachlor. It is virtually certain that both will rapidly change to their respective epoxides once they become finely divided in air and, hence, this is probably the origin of much of the dieldrin and presumably most, if not all, of any heptachlor epoxide detected in the atmosphere, although some may escape as vapour from treated soil.

Wash-out of chemicals in rain occurs in two ways. Firstly, rain-drops impact with, and 'scrub' from the atmosphere much of the particulate matter present, and with it any adsorbed pesticide as already discussed. Secondly, pesticide vapour will partition into the raindrops as they fall and so be removed in solution. The ratios of the equilibrium concentrations in the air and the water droplets will depend upon the water solubility, the saturation vapour pressure and molecular weight of the pesticide and on temperature. The behaviour of the pesticide molecules will be complex, because air, solid and liquid substrates will be present simultaneously, and the pesticide molecules will partition between all three phases. Atkins and Eggleton (1971) have shown that gamma-BHC, dieldrin and p,p'-DDT, at very low concentrations, closely follow Raoult's Law. After re-determining the solubilities of these insecticides in water they were therefore able to calculate the water air partition ratios as 11,500, 1,900 and 540 parts of gamma-BHC, p,p'-DDT and dieldrin by weight, respectively, in the water, to 1 part in the air. They calculated wash-out ratios $= \left(\dfrac{\text{mass/kg rain}}{\text{mass/kg air}}\right)$ of these compounds in the same manner as that used for other atmospheric contaminants, and found them to be 14, 2.3 and 0.65 respectively, on the assumption that all of the chemical present in the atmosphere would be in the vapour phase. (The wash-out ratio for particulates at ground level would probably exceed 100.) A comparison of these calculated wash-out ratios with those obtained from independent measurements of the con-centrations of these three pesticides, in air and rainwater, showed good agreement in the case of gamma-BHC, but the calculated ratios were low for dieldrin and very low for p,p'-DDT. Atkins and Eggleton concluded therefore, that gamma-BHC was present in the London atmosphere largely in gaseous form, dieldrin almost entirely so, but that an appreciable pro-portion of the p,p'-DDT was present in particulates, so corroborating the inferences drawn from other work already discussed. The calculated and observed values of the wash-out ratios were in much better agreement when the comparisons were made on the basis of the pesticides present in rain in rural central England, where particulate matter would be much less. They therefore speculated that there was a higher pproportion of DDT adsorbed into particulates in London air than in central England. They also pointed out that the levels of 'particulate DDT' found in Barbados by Risebrough *et al.* (1968) were four orders of magnitude smaller than the total DDT measured in London air. These explanations seem to accord well with all known facts.

Atkins and Eggleton (1971) also measured the deposition velocities (Vg) of these insecticides onto grass in a wind tunnel, where $Vg = \left[\dfrac{\text{rate of deposition per unit area}}{\text{volumetric concentration}}\right]$. The values for all three chemicals were similar, namely 0.04×10^{-2} cm sec^{-1}. After making certain assump-tions concerning the mean rainfall in latitudes 30-60°N (73 cm yr^{-1}), and the mass of air over 1 m^2 of the earth's atmosphere up to the tropopause (8.25×10^3 kg), they calculated the fraction of the insecticides washed from the atmosphere on average each week, to be approximately 2.5, 0.4 and 0.11% for gamma-BHC, p,p'-DDT and dieldrin respectively, compared with the dry deposition rate of approximately 3 to 3½% for all three compounds. The combined rates of removal of the pesticides from the atmosphere were therefore estimated as 5.8, 3.8 and 3.6% per week respectively, indicating average lifetimes in the atmosphere of about 17, 26 and 28 weeks. Chamberlain (1971) calculated the deposition rate of these three compounds onto a water surface, such as an ocean, to be about 0.3, 0.12 and 0.05 cm sec^{-1}, respectively, from the vapour phase. These results are high, compared with those suggested by Atkins and Eggleton, the discrepancies emphasising the experimental difficulties involved in this type of study and the risks inherent in over-simplification of complex situations. However, taking these results as a whole, it seems that the average lifetimes of the three chemicals in the atmosphere ought not to exceed Atkins and Eggleton's estimates and, particularly in the case of gamma-BHC, they could be appreciably shorter. Woodwell *et al.* (1971), however, calculated from gross usage of DDT and the known average residue concentrations in the environment, that its removal from the atmosphere by rain would have a time constant of about 3.3 years. By analogy with the transfer of carbon dioxide into the oceans, they suggested that the time constant for the downward transport might be up to about 7 years, but, in conclusion, they assumed that its mean residence time in the atmosphere was probably not longer than 4 years. This estimate is almost an order of magnitude greater than the residence time suggested from Atkins and Eggleton's study. Woodwell *et al.* made the point that 'residues deposited on the ground are, of course, available for re-evaporation'. It seems probable that the discrepancy is indicative of a very considerable recycling of the residues between the earth's surface and the atmosphere, so that the actual mean residence time is, in practice, greatly extended beyond that indicated by direct measurements of the deposition rates.

10.5.3 Redistribution
The demonstration of pesticides in the air over Barbados by Risebrough *et al.* (1968) implied transport over 5,000 km of Atlantic Ocean by the north-east trade winds from West Africa. Evidence of movement of this scale carries with it, the virtual certainty that contamination of the atmosphere is global.

Interesting circumstantial evidence for ad-polar movement of residues, was suggested by the study of Koeman and Pennings (1970), who reported that residues of organochlorine pesticides were surprisingly low in wild-life in areas of central Africa in the vicinities of Lakes Chad and Victoria, compared with those found in Atlantic and Pacific biota. They pointed out that DDT is widely used in neighbouring countries for vector control, and that endrin was extensively used in cotton and yet was not detectable in samples (<0.008 μgg^{-1}). Therefore, they reasoned that insecticide used in hot, arid countries must be lost to the atmosphere by evaporation and co-distillation. Arid parts of the globe would thus be sources for the relatively high concentrations found in temperate zones, to which the residues were transported by air and precipitated by rain. The inference, therefore, is that certain residues may distil from tropical zones and re-condense at higher latitudes, accumulating in polar regions where they are likely to be exceptionally stable at the low prevailing temperatures. From residues present in snow-melt (40 pgg^{-1}), Peterle (1969) calculated that up 2.4×10^6 kg of DDT residues may have accumulated in Antarctic snow; if this is correct, the atmosphere is the only feasible transport system which would be capable of redistributing pesticides extensively over such a remote area of the globe.

The possible speed of atmospheric redistribution of particulate matter

J. H. Brazell, *Nature, Lond.*, **226**, 694 (1970)

A. W. Breidenbach, *Arch. Environ. Health*, **10**, 827 (1965)

J. C. Casely, *Bull. Environ. Contam. Toxicol.*, **3**, 180 (1968)

A. C. Chamberlain, *Q. Jl R. met. Soc.*, **85**, 350 (1959)

A. C. Chamberlain, Private communication (1971)

J. M. Cohen and C. Pinkerton. R. F. Gould (Editor), Organic Pesticides in the Environment, p. 163, Adv. Chem. Ser. No. 60., *American Chem. Soc.*, Washington, D.C. (1966)

J. M. Cohen, F. L. Evans and C. Pinkerton, Report to the 27th Annual Conference of the Water Pollution Control Federation, Bal Harbour, Fla., 30 Sept. (1964)

Sir J. Cook, Review of the present safety arrangements for the use of toxic chemicals in agriculture and food storage. London: H.M.S.O., pp. 72 (1967)

R. W. Cook, *Proc. Fla State hort. Soc.*, **78**, 260 (1966)

R. D. Comes and F. L. Timmons, *Weeds*, **13**, 81 (1965)

M. Corn and T. L. Montgomery, *Science, N.Y.*, **159**, 1350 (1968)

D. G. Crosby and Ming-Yu Li. P. P. Kearney and D. D. Kaufman (Editors), Degradation of Herbicides, p. 321. Marcel Dekker, Inc. New York (1969)

W. F. Crosier and J. J. Natti, *Phytopathology*, **55**, 128 (1965)

W. A. L. David, *Outl. Agric.*, **2**, 127 (1959)

G. C. Decker, *Agric. Chem.*, **12**, 39 (1957)

G. C. Decker, W. N. Bruce and J. H. Bigger, *J. econ. Ent.*, **58**, 266 (1965)

G. C. Decker, C. J. Weinman and J. M. Bann, *J. econ. Ent.*, **43**, 919 (1950)

T. W. Donaldson and C. L. Foy, *Weeds*, **13**, 195 (1965)

W. Ebeling, *Residue Rev.*, **3**, 35 (1963)

C. A. Edwards, *Residue Rev.*, **13**, 83 (1966)

C. A. Edwards, *N.A.A.S. quart. Rev.*, **86**, 47 (1969)

C. A. Edwards, Critical Reviews in Environmental Control, Chem. Rubber Co. Cleveland (1970)

C. J. Edwards and W. E. Ripper, *Proc. Br. Weed Control Conf.*, 348 (1953)

S. S. Epstein, *Nature, Lond.*, **228**, 816 (1970)

S. C. Fang. P. C. Kearney and D. D. Kaufman (Editors), Degradation of Herbicides, p. 147. Marcel Dekker, Inc. New York (1969)

P. D. Gerhadt and J. M. Witt, *Proc. 12th Int. Congr. Entomol., London, 1964*, p. 565 (1965)

L. W. Getzin, *Diss. Abstr.*, **19**, 625 (1958)

P. Gerolt, *Nature, Lond.*, **183**, 1121 (1959)

R. A. Gray and A. J. Weierich, *Weeds*, **13**, 141 (1965)

P. H. Gregory, The microbiology of the atmosphere. Leonard Hill, London (1961)

R. A. Gunther and R. C. Blinn, Analysis of insecticides and acaricides. Interscience Publishers, London (1955)

J. W. Hamaker. R. F. Gould (Editor), Organic Pesticides in the Environment, p. 122. Adv. Chem. Ser. No. 60, *American Chem. Soc.*, Washington, D.C. (1966)

C. R. Harris, *Diss. Abstr.*, **22**, 375 (1961)

C. R. Harris and E. P. Lichtenstein, *J. econ. Ent.*, **54**, 1038 (1961)

C. R. Harris, W. W. Sans and J. R. W. Miles, *J. Agric. Fd. Chem.*, **14**, 398 (1966)

G. S. Hartley, Vapour transfer of pesticides. Proc. Soc. Chem. Ind., Pesticides Group, Biophysical and Biochemical Panel, London (Unpublished, private communication) (1971)

G. S. Hartley and T. F. West, Chemicals for Pest Control. Pergamon Press, London (1969)

G. L. Henderson and D. G. Crosby, *J. Agr. Fd. Chem.*, **15**, 888 (1967)

G. L. Henderson and D. G. Crosby, *Bull. Environ. Contam. Toxicol.*, **3**, 131 (1968)

E. Hindin, D. S. May and G. H. Dunstan. R. F. Gould (Editor), Organic Pesticides in the Environment, p. 132. Adv. Chem. Ser. No. 60, *American Chem. Soc.*, Washington, D.C. (1966)

M. Horowitz, *Weed Res.*, **6**, 22 (1966)

W. M. Hoskins, *F.A.O. Plant Protection Bull.*, **9**, 163 (1961)

W. M. Hoskins, *Residue Rev.*, **1**, 66 (1962)

E. W. Huddleston, D. Ashdown and D. C. A. Herzog, *Texas Tech. Coll. Entomol. Dept.*, 66-1, 13 pp. (1966)

Z. Jegier, *Ann. New York Acad. Sci.*, **160**, 143 (1969)

S. Jensen, *New Scientist*, **32**, 612 (1966)

C. E. Junge, Air Chemistry and Radioactivity. Academic Press, New York (1963)

M. Katz and W. B. Drowley, *Canad. J. Public Health*, **57**, 71 (1966)

J. W. Kidson and R. E. Newell, *Nature, Lond.*, **221**, 352 (1969)

U. Kiigemagi and L. C. Terriere, *Bull. Environ. Contam. Toxicol.*, **6**, 336 (1971)

J. H. Kirch and J. E. Esposito, *Proc. 21st Ann. Northeast Weed Control Conf.*, 442 (1967)

J. H. Koeman and J. H. Pennings, *Bull. Environ. Contam. Toxicol.*, **5**, 164 (1970)

L. B. Lave and E. P. Seskin, *Science, N.Y.*, **169**, 723 (1970)

C. W. Lee, *Agric. Aviat.*, **8**, 54 (1966)

E. C. S. Little, T. O. Robson and D. R. Johnstone, *Proc. 7th Br. Weed Control Conf.*, **2**, 920 (1965)

G. A. Lloyd and J. C. Tweddle, *Sci. Fd Agric.*, **15**, 169 (1964)

C. P. Lloyd-Jones, *Nature, Lond.*, **229**, 65 (1971)

W. H. Luckman and G. C. Decker, *J. econ. Ent.*, **53**, 821 (1960)

G. B. MacCollom, D. B. Johnstone and B. L. Parker, *Bull. Environ. Contam. Toxicol.*, **3**, 368 (1968)

R. D. MacCuaig and W. S. Watts, *J. econ. Ent.*, **56**, 850 (1963)

V. G. MacKenzie, *Canad. J. Public Health*, **57**, 65 (1966)

E. H. Marth, *Residue Rev.*, **9**, 1 (1965)

P. Marvin, *Agric. Chem.*, **20**, 69 (1965)

D. H. McIntosh and A. S. Thom, Essentials of Meteorology. Wykeham Publications Ltd., London (1969)

A. D. Mears, *Agric. Gaz. N.S.W.*, **75**, 1399 (1964)

A. R. Meetham, *Q. Jl R. met. Soc.*, **76**, 359 (1950)

R. E. Menzer, E. L. Fontanilla and L. P. Ditman, *Bull. Environ. Contam. Toxicol.*, **5**, 1 (1970)

J. T. Middleton. C. O. Chichester (Editor), Research in Pesticides, p. 191, Academic Press, New York (1965)

W. J. Mistric and J. C. Gaines, *J. econ. Ent.*, **46**, 341 (1953)

L. D. Newsom, *A. Rev. Entomol.*, **12**, 257 (1967)

V. P. Osterli, *Calif. Dept. Agric. Bull.*, **56**, 1 (1967)

D. H. Peirson and R. S. Cambray, *Nature, Lond.*, **216**, 755 (1967)

T. J. Peterle, *Nature, Lond.*, **224**, 620 (1969)

R. A. Reinert, A. S. Heagle, J. R. Miller and W. R. Geckerler, *Pl. Dis. Rep.*, **54**, 8 (1970)

L. M. Reynolds, *Bull. Environ. Contam. Toxicol.*, **4**, 128 (1969)

S. Rich, G. S. Taylor and H. Tomlinson, *Pl. Dis. Rep.*, **53**, 969 (1969)

R. W. Risebrough, R. J. Huggett, J. J. Griffin and E. D. Goldberg, *Science N.Y.*, **159**, 1233 (1968)

R. W. Risebrough, P. Reich and H. S. Olcott, *Bull. Environ. Contam. Toxicol.*, **4**, 192 (1969)

J. Robinson, A. Richardson, B. Bush and K. E. Elgar, *Bull. Environ. Contam. Toxicol.*, **1** 127 (1966)

T. O. Robson, E. C. S. Little and D. R. Johnstone, *Weed Res.*, **6**, 254 (1966)

J. D. Rosen and W. F. Carey, *J. agric. Fd. Chem.*, **16**, 536 (1968)

J. D. Rosen D. J. Sutherland and G. R. Lipton, *Bull. Environ. Contam. Toxicol.*, **1**, 133 (1966)

L. D. Roth, *Weeds*, **13**, 326 (1965)

M. G. Sheldon, M. H. Mohn, G. A. Ise and R. A. Wilson. *U.S. Fish Wildlife Circ.*, **199**, 55 (1964)

A. R. Smith *S.C.I. Monogr. No. 22.* Pergamon Press, London (1966)

T. Sotobayahsi, T. Suzuki and A. Furusawa, *Nature, Lond.*, **224**, 1096 (1969)

W. F. Spencer and M. M. Cliath (1969). *Environ. Sci. Technol.*, **3**, 670 (1969)

N. G. Stewart, R. N. Crooks and E. M. R. Fisher, *U.K.A.E.A. Rep. A.E.R.E. HP/R 1701* (1965)

C. R. Swanson. P. C. Kearney and D. D. Kaufman (Editors), Degradation of Herbicides, p. 229. Marcel Dekker, Inc., New York (1969)

E. C. Tabor, *Trans. N.Y. Acad. Sci. Ser. 2*, **28**, 569 (1966)

K. R. Tarrant and J. O'G. Tatton, *Nature, Lond.*, **219**, 725 (1968)

H. L. Volchok, *Nature, Lond.*, **206**, 1031 (1965)

R. R. Walton and D. E. Howell, *J. econ. Ent.*, **47**, 780 (1954)

S. R. Weibel, R. B. Weidner, J. M. Cohen and A. G. Christianson, *J. Amer. Water Works Assoc.*, **58**, 1075 (1966)

I. West, *Arch. Environ. Health*, **9**, 626 (1964)

G. A. Wheatley, *Ann. appl. Biol.*, **55**, 325 (1965)

G. A. Wheatley and J. A. Hardman, *Nature, Lond.*, **207**, 486 (1965)

G. A. Wheatley, J. A. Hardman and A. H. Strickland, *Pl. Path.*, **11**, 81 (1962)

G. A. Wheatley, J. A. Hardman and D. W. Wright, *Pl. Path.*, **9**, 146 (1960)

A. Wilson, Further review of certain persistent organochlorine pesticides used in Great Britain. London: H.M.S.O. (1969)

B. R. Wilson, *Trans. N.Y. Acad. Sci. Ser. 2*, **28**, 694 (1966)

A. Wolman, *Science, N.Y.*, **159**, 1437 (1968)

G. W. Woodwell, P. P. Craig and A. Johnson, *Science, N.Y.*, **174**, 1101 (1971)

D. Yeo, *Rep. 1st Int. Agric. Conf., 1959*, 112 (1959)

W. N. Yule, A. E. W. Cole and I. Hoffman, *Bull. Environ. Contam. Toxicol.*, **6**, 289 (1971)

M. J. Zabik, R. D. Schuetz, W. L. Burton and B. E. Pape, *J. Agric. Fd. Chem.*, **19**, 308 (1971)

Chapter 11

Pesticide Residues in Soil and Water

C. A. Edwards

*Rothamsted Experimental Station,
Harpenden, Herts.*

409

11.1   RESIDUES IN SOIL

Large amounts of pesticides reach the soil, either as direct applications, from fall-out from aerial spraying, in rain or dust or from plant or animal remains which become incorporated with the soil. Thus, the soil is an environmental reservoir for these residues from which they move into the atmosphere, water or living organisms.

11.1.1   Amounts occurring in soil

With few exceptions, the only pesticide residues reported in surveys have been either persistent inorganic chemicals, such as arsenic, copper or lead, that were used as insecticides or fungicides prior to the second world war (Taschenberg *et al.,* 1961; Miles, 1968; Woolson *et al.,* 1971; Wiersma *et al.,* 1971), or persistent organochlorine insecticides. Residues of a very few organophosphate insecticides have been reported from areas of intensive use (Wiersma *et al.*, 1972; Stevens *et al.*, 1970; Harris and Sans, 1971) (Table 11.4). Unfortunately, hardly any of the surveys have been random, but instead have been associated with particular crops or patterns of pesticide usage.

In the few surveys for arsenic residues that have been made (Table 11.1), the amounts found have varied very much, the largest amount being 830.0 ppm (Woolson *et al.*, 1971). All soils contain some arsenic (up to about 10 ppm), but the amounts reported from many orchards are excessive and would certainly harm susceptible crops. There are many reports in the literature of damage to crops due to arsenic and other inorganic pesticides (Linder, 1943; Morris and Swingle, 1927; Vincent, 1944). Recently, several decades since arsenical compounds have been used, Bishop and Chisholm (1962) found potentially harmful residues in 22 out of 25 orchards and Miles (1968) reported that in seven out of 58 soils sampled, there were sufficient residues to damage crops.

In almost all of the soils surveyed for pesticide residues, the commonest chemical, and the one that has occurred in the largest amounts, has been DDT, with the next most common dieldrin. However, dieldrin has been reported in much smaller quantities than DDT, although possibly more frequently. This is surprising, because the calculations made by Edwards (1966) from all the published data available, showed that although DDT was the most persistent insecticide residue in soil, it was only slightly more persistent than dieldrin. When it is considered that during 1962-1964, according to Strickland (1965), 262 tons of DDT (and related compounds were used on 262,000 acres of arable land, compared with the 163 tons of aldrin and dieldrin (aldrin breaks down to dieldrin) were used on 715,000 acres in Great Britain. Similar figures from the U.S.D.A. show that 23,660 tons of DDT were used in 1964-1965, compared with 36,000 tons of the aldrin and dieldrin group of insecticides in U.S.A. The most likely explanations for the larger residues of DDT in soil seem to be the longer period of usage of DDT than aldrin or dieldrin. Currently, DDT seems to present a greater potential hazard in the environment than any of the other persistent pesticides, because of its greater persistence and affinity for fatty tissues.

In Great Britain, very few other pesticide residues have been found in soil, and even these have been little more than traces; this is certainly because of the very small quantities of other kinds of organochlorine insecticides that have been used in this country. Small quantities of chlordane have been used for killing earthworms, heptachlor has been used as seed dressings and endrin as a spray to kill blackcurrant gall mite, but the total quantities of all these insecticides used have been negligible compared with the amounts of DDT and dieldrin that have been used. Table 11.3 shows quite large residues of chlordane and heptachlor in agricultural soils in the United States, and probably, when thorough soil surveys are made in the southern United States, residues of endrin will also

TABLE 11.1. Residues of arsenic and copper in soil.

Location	Reference	No sites	Crop	Arsenic Maximum	Arsenic Mean	Copper Maximum	Copper Mean
Orchards and vineyards							
U.S.A.	Taschenberg et al., 1961	4	Grapes		—	130.0	118.0
,,	Stevens et al., 1970	30	Orchard fruits	219.20	110.24*		
Canada	Miles, 1968	8	,, ,,	121.0	53.5		
,,	Bishop and Chisholm, 1962	25	,, ,,	124.4	61.95		
Horticultural							
Canada	Miles, 1968	4	Greenhouse crops	6.90	5.10		
,,	,, ,,	12	Vegetables	26.60	5.40		
Agricultural							
U.S.A.	Wiersma et al., 1971	6	Mixed	16.98	5.30		
,,	Woolson et al., 1971	58	,,	830.0	165.0		
Canada	Miles, 1968	2	Alfalfa	5.50	4.20		
,,	,, ,,	9	Corn	8.60	3.90		
,,	,, ,,	8	Cereals	6.90	3.90		
,,	,, ,,	3	Sugar beet	8.60	3.70		
,,	,, ,,	5	Tobacco	6.90	4.70		
Pasture and Grassland							
Canada	Miles, 1968	1	Pasture	N.A.†	3.70†		
Non-cropland							
U.S.A.	Wiersma et al., 1971	6		7.0	4.24		

Residues in ppm

* Mean estimated from total range. † Residues found only at one site. N.A. = Figure not available.

be found to be common in these areas, because large amounts of endrin are used on cotton crops. Quite large endrin residues have been found in soils of the Mississippi River Delta (U.S.D.A., 1966).

There is little doubt that the large residues of DDT and dieldrin remaining in agricultural and forest soils have been partly due to misuse of these insecticides. Aerial spraying and annual 'insurance' treatment of large areas, irrespective of whether pests were present or not, have contributed greatly to these residues, and the practice of combining persistent insecticides with fertilisers has often resulted in use of insecticides when they were not really needed.

11.1.1.1 *In orchards*
Most orchards contain very large amounts of DDT with residues up to 245 ppm (approx. 500 lb/acre or 550 kg/ha). All orchards surveyed in the 14 investigations reported in Table 11.2 had appreciable residues of DDT; this is hardly surprising when most orchard management programmes require about six separate sprays of DDT during the growing season, and as much as half of these fall on to the soil surface. Amounts of other organochlorine insecticides in orchard soils were much smaller, and differed little from those in other cultivated soils.

11.1.1.2 *In horticultural soils*
Residues of DDT were generally smaller than in orchard soils although a few contained very large residues. Of the other organochlorine insecticides, only aldrin and dieldrin were reported in significant amounts, other than in isolated sites.

11.1.1.3 *In agricultural soils*
Most agricultural soils contained less DDT residues than orchard or horticultural soils, although large amounts were reported from a survey of onion-growing soils. These soils also contained comparatively large amounts of other organochlorine insecticides, particularly chlordane and dieldrin. Otherwise, there were few instances where large organochlorine residues were found in agricultural soils.

11.1.1.4 *In other soils*
Quite large amounts of DDT have been reported from pasture soils (Fahey *et al.,* 1965), but from the sparse information available, amounts were usually smaller than those from arable land. It is interesting that residues of DDT have been reported from all of the few areas surveyed where insecticide treatments would be rare or non-exisitent, such as forests, tundra and desert, although there were very few reports of other organochlorine residues from these areas. In view of the very considerable debate and speculation on environmental contamination by DDT, it is imperative that more information be available on the amounts of persistent insecticide residues in woodlands, grasslands and other natural areas that have

TABLE 11.2.   Residues of DDT and related compounds in soil.

Location	Reference	No. sites	Crop	Residues in ppm	
				Maximum	Mean
Orchards and vineyards					
Great Britain	Edwards, 1969	5	Orchard fruit	131.10	61.80
U.S.A.	Ackley et al., 1950	3	,,	32.10	24.90
,,	Chisholm et al., 1951	31	,,	58.0*	18.55*
,,	Ginsburg and Reed, 1954	11	,,	106.0·	29.40
,,	Lichtenstein, 1957	14	,,	116.0	37.10
,,	Murphy et al., 1964	35	,,	181.0*	74.0*
,,	Terriere et al., 1966	2	,,	80.09†	57.52†
,,	Stevens et al., 1970	30	,,	245.41	122.67‡
,,	Mullins et al., 1971	6	,,	41.10†	18.12†
Canada	Harris et al., 1966	8	,,	3.60	2.40
,,	Duffy and Wong, 1967	6	,,	73.0	42.80
,,	Harris and Sans, 1969	1	,,	147.37†	109.85†
,,	,, 1971	2	,,	111.76†	84.41†
U.S.A.	Taschenberg et al., 1961	4	Vineyard grapes	42.0	36.8
Horticultural					
U.S.A.	Seal et al., 1967	19	Carrots	12.80	3.70
,,	Deubert and Zuckerman, 1969	6	Cranberries	4.24	3.57
,,	U.S.D.A., 1968	27	Soybeans	7.50	1.70
,,	McCaskill et al., 1970	12	,,	3.58	0.19
,,	Mullins et al., 1971	13	Vegetables	22.27†	3.06†
,,	Saha and Sumner, 1971	41	,,	6.75	0.72
Canada	Harris et al., 1966	4	Greenhouse crops	2.60	1.50
,,	Harris and Sans, 1969	5	Mixed	78.52	18.36
,,	Duffy and Wong, 1967	3	Strawberries	1.81	1.0
,,	Harris et al., 1966	11	Vegetables	47.60	9.50
,,	Duffy and Wong, 1967	11	,,	6.70	2.0
,,	Harris and Sans, 1971	5	,,	33.45	24.26
Agricultural					
Great Britain	Wheatley et al., 1962	21	Potatoes	0.96	0.20
,,	Davis, 1968	10	Mixed	0.80	0.30

Location	Reference	n	Habitat/Crop		
	W(i)erma et al., 19??		"	1.?3	0.58
	Mullins et al., 1971	6	"	1.72	1.26
	" , " , 1971	12	Forage	0.23	0.02
	Ware et al., 1968	12	Alfalfa	4.90	1.40
	Ware et al., 1971	30	"	5.19	1.58
	Ginsburg and Reed, 1954	10	Corn	6.50	4.0
	Mullins et al., 1971	13	Grain	0.31	0.03
	Stevens et al., 1970	33	Small grains and roots	9.23	4.61‡
	Lahser and Applegate, 1966	3	Cotton	2.60	2.40
	Stevens et al., 1970	25	Cotton and/or vegetables	15.63	7.67‡
	Seal et al., 1957	5	Peanuts	0.70	0.30
Canada	Harris and Sans, 1971	4	Mixed	0.74	0.48
	Harris et al., 1966	6	Corn	3.70	1.20
	" , " , 1966	9	Cereals	5.10	1.40
	Saha et al., 1968	20	Cereal and legume	0	0
	Duffy and Wong, 1967	24	Roots	17.10	1.70
	Harris et al., 1966	4	Sugar beet	1.50	0.40
	" , " , 1966	5	Tobacco	5.10	3.20
	Harris and Sans, 1971	4	"	4.56	3.67
East Pakistan	Nasim et al., 1971	6	Rice	4.10	1.21

Pasture and Grassland

Location	Reference	n	Habitat/Crop		
U.S.A.	Fahey et al., 1965	227	Turf and cultivated	87.30	2.90
Canada	Harris et al., 1966	5	Pasture	2.60	0.50

Non-cropland

Location	Reference	n	Habitat/Crop		
U.S.A.	Wiersma et al., 1967	6		4.46	0.81
	Laubscher, 1971	13		1.24	0.14

Forest

Location	Reference	n	Habitat/Crop		
Canada	Woodwell, 1960	6		N.A.	0.50
	Yule, 1970	2		0.42	0.39
	Yule and Smith, 1971	3		21.33	15.21

Tundra

Location	Reference	n	Habitat/Crop		
Canada	Brown and Brown, 1970	5		0.15	0.09

Desert

Location	Reference	n	Habitat/Crop		
U.S.A.	Lahser and Applegate, 1966	5	Desert and prairie	2.30	1.60
	Ware et al., 1971	12	Desert	2.92	0.48

* Data given as lb/acre and converted to approximate ppm.
‡ Mean estimated from total range.

† Includes Dicofol residues.
N.A. = Figure not available.

TABLE 11.3.   Residues of cyclodiene and other organochlorine insecticides

Location	Reference	No. sites	Crop	Aldrin Max.
Orchards				
U.S.A.	Stevens et al., 1970	30	Orchard fruits	–
,,	Mullins et al., 1971	2	,, ,, –	–
Canada	Duffy and Wong, 1967	6	,, ,,	0.15
Horticultural				
Great Britain	Edwards, 1969	12	Carrots	–
U.S.A.	Seal et al., 1967	19	,,	–
,,	Deubert and Zuckerman, 1969	6	Cranberries	–
,,	U.S.D.A., 1968	27	Soybeans	0.18
,,	McCaskill et al., 1970	12	,,	N.A.*
,,	Mullins et al., 1971	6	Vegetables	–
,,	Saha and Sumner, 1971	41	,,	0.28
Canada	Harris et al., 1966	4	Greenhouse crops	–
,,	Duffy and Wong, 1967	3	Strawberries	0.01
,,	Harris et al., 1966	11	Vegetables	2.10
,,	Duffy and Wong, 1967	11	,,	2.50
,,	Harris and Sans, 1969	3	,,	1.84
,,	,, ,, ,,	1	Crucifers	N.A.*
Agricultural				
Great Britain	Wheatley et al., 1962	21	Potatoes	0.12
,, ,,	Davis, 1968	10	Mixed	0.70
,, ,,	Edwards, 1969	5	Cereal	–
East Germany	Heinisch et al., 1968	222	Mixed	–
U.S.A.	Seal et al., 1967	25	Potatoes	–
,,	Sand et al., 1972	92	Sweet potatoes	0.11
,,	Wiersma et al., 1972	71	Onions	0.96
,,	Trautman, 1968	41	Mixed	0.15
,,	Mullins et al., 1971	10	,,	–
,,	Wiersma et al., 1971	6	,,	–
,,	Mullins et al., 1971	5	Forage	–
,,	Decker et al., 1965	35	Corn	–
,,	Mullins et al., 1971	11	Grain	0.61
,,	Lahser and Applegate, 1966	3	Cotton	–
,,	Seal et al., 1967	5	Peanuts	–
Canada	Harris and Sans, 1969	1	Potatoes and tobacco	N.A.*
,,	Harris, 1969	1	Mixed	N.A.*
,,	Harris and Sans, 1971	16	,,	2.33
,,	Harris et al., 1966	6	Corn	0.50
,,	,, ,, ,, ,,	9	Cereals	0.50
,,	Saha et al., 1968	20	Cereal and legume	0.05
,,	Duffy and Wong, 1967	24	Roots	2.13
,,	Harris et al., 1966	5	Tobacco	0.20
East Pakistan	Nasim et al., 1971	6	Rice	
Pasture and grassland				
U.S.A.	Fahey et al., 1965	227	Turf and cultivated	–
Canada	Harris et al., 1966	5	Pasture	0.20
Non-cropland				
U.S.A.	Wiersma et al., 1971	6		–
,,	Laubscher et al., 1971	13		–
Desert				
U.S.A.	Lahser and Applegate, 1966	5		–

* Residues found at only one site.   † Mean estimated from total range.
N.A. = Figure not available.   T = Trace.

IC	Chlordane		Dieldrin		Endosulfan		Endrin		Heptachlor and heptachlor epoxide		Camphechlor	
Mean	Max.	Mean	Max.	Mean	Max.	Mean	Max.	Mean	Max.	Mean	Max.	Mean
–	N.A.*	0.10*	2.84	1.41†	4.63	2.30†	12.61	6.30†	–	–	N.A.*	7.72*
0.05	–	–			–	–	–	–	–	–	–	–
–	–	–	0.16	0.04	–	–	–	–	0.02	T	–	–
–	–	–	1.47	0.67	–	–	–	–	–	–	–	
–	–	–	0.26	0.20	–	–	–	–	0.26	0.16	–	–
.	–	–	3.15	2.08	–	–	–	–	–	–	–	–
–	1.11	0.24	0.31	0.08	–	–	1.17	0.10	0.16	0.02	–	–
0.12	–	–			–	–	–	–	–	–	–	–
0.07	–	–			N.A.*	0.04*	T	T	T	T	N.A.*	1.0*
0.001	3.91	0.12	0.77	0.06	–	–	0.48	0.01	0.34	0.03	–	–
–	–	–	0.40	0.10	–	–	–	–	–	–	–	–
–	–	–	0.02	0.01	–	–	–	–	–	–	–	–
–	0.60	0.08	1.60	0.80	–	–	–	–	0.20	0.02	–	–
–	0.86	0.12	1.14	0.25	–	–	–	–	1.39	0.16	–	–
–	0.11	0.04	2.64	1.55	–	–	5.11	1.70	0.17	0.06	–	–
–	–	–	N.A.*	0.81*	–	–	–	–	–	–	–	–
–	–	–	0.41	0.09	–	–	–	–	–	–	–	–
T	–	–	0.70	0.15	–	–	–	–	–	–	–	–
N.A.	–	–	0.40	0.02	–	–	–	–	–	–	–	–
N.A.												
–	–	–	0.20	0.10	–	–	–	–	0.10	0.08	–	–
–	5.07	0.28	2.18	0.17	–	–	–	–	0.39	0.02	–	–
–	23.84	1.63	16.72	0.79	0.38	0.0007	2.05	0.06	2.24	0.09	7.77	0.55
T	–	–	1.52	0.06	–	–	–	–	0.01	T	–	–
–	–	–	–	–	–	–	–	–	N.A.	T	–	–
–	0.02	0.01	0.02	0.02							0.55	0.13
–	0.05	0.03	N.A.*	0.02*					0.07	0.05	–	–
–	–	–	1.22	0.50							–	–
–	–	–	0.44	0.24	–	–	–		N.A.*	0.02*	–	–
0.26	–	–	–	–	–	–	–	–			–	–
–	–	–	0.20	0.15	–	–	–	–			–	–
–	–	–	N.A.*	0.47*	–		N.A.*	0.09*	–	–	–	–
			N.A.*	1.08*								
–	0.63	0.06	3.33	0.67	–	–	6.55	0.34	0.50	0.06	–	–
–	0.10	0.02	0.90	0.30							–	–
			1.10	0.10								
–	0.02	T	0.30	0.05	–	–	–	–	0.05	T	–	–
–	0.48	0.03	4.04	0.47	–	–	–	–	0.73	0.04	–	–
–	0.20	0.06	0.50	0.30	–	–	–	–	0.20	0.06	–	–
0.23	–	–	0.53	0.46	–	–	–	–	–	–	–	–
0.04	120.0	1.20	2.20	0.03	–	–	–	–	1.60	0.03	–	–
–	–	–	1.10	0.20	–	–	–	–	–	–	–	–
–	0.02	0.01	–	–	–	–	–	–	–	–	–	–
			0.0013	0.0003	–	–	–	–	–	–	–	–
0.20	–	–	–	–	–	–	–	–	–	–	–	–

never been treated with any insecticides. This would enable more satisfactory conclusions to be reached as to the amounts of persistent pesticides transported intercontinentally in the atmosphere.

11.1.2   Sources of residues in soil
Only a proportion of the insecticide residues found in soil result from direct application. Many soils that have never been treated with any pesticide contain small amounts of DDT; presumably these come from spray drift or atmospheric fall-out.

11.1.2.1   *From soil treatments*
In the years after organochlorine insecticides were first discovered, large amounts were applied to soil as broadcast treatments and thoroughly incorporated. Because these chemicals were so cheap, and a single treatment lasted for at least the whole of season, often, larger amounts were used than necessary, and annual treatments used as insurance measures whether needed or not. The results of these policies are that many of the more intensively cropped soils now have quite large residues of persistent insecticides (Tables 11.2 and 11.3). Usually, it was only after it was realised that organochlorine insecticides might be harmful to wildlife that more care was taken with their use. Now, most treatments take the form of seed dressings, spot treatments, or in-row band applications, and the amounts of chemical used per acre are very much smaller. Whereas in the early days 2-4 pounds of active ingredient of organochlorines per acre (2.2-4.5 kg/ha) was not considered to be an excessive annual soil treatment, currently ½-1 pound of active material per acre would be much more usual.

11.1.2.2   *From spray fall-out*
Sprays applied to crop foliage do not all reach their target, and it has been estimated that as much as 50% of the insecticide applied to crop foliage reaches the soil, either as spray drift or run-off from the leaves or on leaves which fall to the ground. For instance, Cope and Bridges (1963) reported that an average of 43% of the methoxychlor applied to an alfalfa crop eventually reached the ground, and Terriere *et al.* (1966) found that waste land, adjacent to orchards that were regularly sprayed with DDT, had accumulated as much as 5 pounds active ingredient per acre.

Insecticides that reach the soil from foliar sprays are often left undisturbed, and there is evidence (Lichtenstein and Schulz, 1961; Wheatley and Hardman, 1965; Edwards, 1966) that they break down much faster on the soil surface than when cultivated into the soil. The use of herbicides instead of cultivation to keep down weeds means that the insecticide is not mixed into the soil and its persistence is lessened. The largest residues of organochlorine insecticides that have been reported from soil during surveys have been in orchards and vegetable fields (Table 11.2 and 11.3). All the insecticides used in orchards, are applied to the foliage, so the soil residues are due to foliar spraying and not from insecticides applied directly to the soil.

Another important source of residues of organochlorine insecticides in soil, particularly in the United States and Canada, is from aerial spraying of crops and forests. The overall residues in soil from such treatments may not be large, unless they are from an intensive spraying campaign (Yule, 1970); but since they occur over large areas and contaminate forest soils they may well adversely affect the soil microflora and fauna that are important in soil formation and fertility. Insecticides can drift considerable distances when applied as aerial sprays (Akesson and Yates, 1964); for instance, in Connecticut, when 0.5 pound active ingredient of DDT per acre (0.56 kg/ha) was applied to trees, 0.01-0.24 pound per acre (0.011-0.27 kg/ha) was recovered from soil under the trees and 0.06-0.21 pound per acre (0.066-0.24 kg/ha) from soil in the open (Turner, 1963). Hoffman and Surber (1948) reported an average of 0.4 pound active ingredient of DDT per acre (0.45 kg/ha) on the soil, after 1 pound per acre was applied to trees, and Pillmore and Finlay (1963) estimated that as much as 1 pound per acre (1.12 kg/ha) of DDT had to be sprayed to control spruce budworm although only 0.2 pound per acre at the site of action was sufficient to control the pest, so the rest must have missed its target. Clearly, a large proportion of the residues in soils originate from such spray fall-out.

11.1.2.3 *In rain and dust*
In recent years, evidence has accumulated that the atmosphere contains residues of organochlorine insecticides (see Chapters 10 and 12). Residues have been reported in rain (Wheatley and Handman, 1965; Abbott *et al.,* 1965; Cohen and Pinkerton, 1966; Weibel *et al.,* 1966; Tarrant and Tatton, 1968, in air (West, 1964; Abbott *et al.,* 1966; Tabor, 1966; Huddleston *et al.,* 1966; Jegier, 1969) and dust (Antommaria *et al.,* 1965; Riseborough *et al.,* 1968, 1969). It is assumed that these originate from spray-drift or by volatilization from soil to water. Probably the residues become concentrated on to particulate matter or in moisture drops and fall on to soil either with dust or rain, the latter being the more important route (Woodwell *et al.,* 1971). The amounts that reach soil in this way are unlikely to be large.

The maximum amounts of DDT likely to be added to soil with rainfall can be calculated. The greatest concentration of DDT reported from Great Britain was $210 \text{ ppm} \times 10^6$ (Tarrant and Tatton, 1968) and similar amounts have been reported from U.S.A. (Cohen and Pinkerton, 1966). Assuming a moderate rainfall of 50 in (125 cm) per acre per annum, 0.0023 lb a.i. per acre (0.0027 kg/ha) of DDT would reach the soil annually. However, this is a maximum amount, and the mean amount would be likely to be less than one-third of this, so that it would be reasonable to expect that no more than 0.01 lb a.i. per acre (0.011 kg/ha) of DDT would reach soils annually in rainfall. This agrees with the conclusions of Wheatley and

Hardman (1965) who stated that no significant increase in the contamination of agricultural land seems likely to arise from the amounts of organochlorine insecticide residues they found in rain water, and as amounts reported by other workers do not differ greatly, their conclusions remain valid.

11.1.2.4 *From crop and animal remains*

Small quantities of organochlorine insecticide residues are taken up into the tissues of plants (Lichtenstein and Schultz, 1965; Hurtig and Harris, 1966; Burrage and Saha, 1967; Lichtenstein *et al.,* 1968; Ware *et al.,* 1968) (see Chapter 2), most invertebrates (Newson, 1967; Edwards, 1970) (Chapter 3) and vertebrates (Stickel, 1968; Edwards, 1970) (Chapter 7).

The amounts vary considerably but sufficient data is available to enable calculations to be made of the contribution of these channels to addition or removal of residues from soil. In general, they are more likely to remove residues than to add to them.

Amounts of organochlorine residues reported from cereal crops have averaged about 0.02 ppm (Edwards, 1970). The average yield of grain from a cereal crop is about 20 cwt per acre (2,500 kg/ha), and the corresponding yield of straw about 50 cwt (6,250 kg/ha), a total of about 7,840 lb/acre (8,750 kg/ha). This portion of the crop that is removed, would contain only about 0.0001 lb per acre of insecticide, an extremely small amount. The crop would have about 3,000 lb per acre (3,360 kg/ha) of roots and stems containing a further 0.00006 lb per acre but this would remain in the soil, although this insecticide may still be locked up in the organic matter.

Root crops, such as potatoes, turnips, mangels, carrots, or sugar beet, may be more important in removing insecticide residues not only because they contain more residues, averaging about 0.2 ppm, but also because yields from these crops are much larger (8-20 tons per acre). Based on a similar calculation to that for cereals, a yield of 20 tons per acre (5,000 kg/ha) could remove on average about 0.01 lb per acre (0.011 kg/ha) of insecticide, and in optimal conditions this might be as high as 0.1 lb per acre (0.11 kg/ha). However, this is still a relatively small proportion of an average dose of insecticide (about 2.0 lb per acre (2.24 kg/ha)). However, Onsager *et al.* (1970) reported much larger residues of dieldrin, DDT and chlordane in sugar beet. These averaged 8.4% of the amount in the soil in which they were grown for dieldrin, 5.5% for DDT and 9.6% for chlordane. Residues in the sugar beet were up to 0.96 ppm for dieldrin, up to 0.33 ppm for DDT and up to 1.12 ppm for chlordane. This could account for considerable losses from soil. These amounts are from larger than usual treatments and results from two other workers support my calculations. Beestman *et al.* (1969) reported that less than 3% of dieldrin applied to soil was translocated into the aerial parts of corn. Caro and Taylor (1971) found that after applying two doses of 5 lb a.i. per acre (6.7 kg/ha) to soil, 0.01 to 0.02 ppm (0.03% of the dose applied) was found in corn grown in this soil, although as much as 0.75 ppm occurred in the leaves.

Insecticides are concentrated into the bodies of invertebrates and microorganisms that live in soil. Stöckli (1950) calculated that the top 6 in (15 cm) of an average soil could contain about 25 metric tons of these organisms. If it is taken that the soil fauna and flora contain a mean of 1.0 ppm of organochlorine residues (under some conditions they can have much more than this, but this is a reasonable average), then this is equivalent to 0.02 lb per acre (0.022 kg/ha) of insecticide residues that are locked up in this way. Some of this total may be removed by migration either by the animals that crawl over the surface of the soil or by flight of winged adults that develop from larvae that live in soil. However, it is doubtful if more than about 20-30% would be removed from an area by this route, although the rest may be locked up in the invertebrate tissues. If anything, this is an underestimate of the amount of insecticides bound up in this way, because Stöckli's figure for biomass of soil organisms is almost certainly too low, and many invertebrates that are killed by the residues contain much more than 1 ppm. Furthermore, there is some evidence that microorganisms can take even more insecticide than invertebrates. For instance, Ko and Lockwood (1968) reported that fungal mycelia could take up as much as 10% of the dieldrin or DDT in a soil.

The amounts of residues removed by vertebrates that feed on grass from contaminated pasture, or other food containing pesticides, is much more difficult to calculate, but it is very unlikely that it would be more than that removed with a root crop. If all these amounts are added it is unlikely that the total loss of residues from soil via crops or animals would be much more than 0.1 ppm.

11.1.3   Factors influencing the persistence of insecticides in soil

11.1.3.1   *Chemical nature of insecticide*

These include the chemical stability of the insecticide, its volatility, solubility, concentration and formulation.

Most organochlorine insecticides are non-volatile but there is some correlation between the vapour pressure and persistence of these insecticides in soil. Harris (1961) reported that the rate of volatilization of some organochlorine insecticides was in the order of aldrin < heptachlor < heptachlor, epoxide < dieldrin < DDT. Harris and Lichtenstein (1961) found that the volatilisation of insecticides from soil increased with the concentration of the insecticide, the relative humidity of the air over the soil, the soil temperature, the movement of air over the soil surface, and the amount of moisture in the soil.

Chisholm and Koblitsky (1959) first suggested that organochlorine insecticides may be lost into the atmosphere by codistillation with water vapour escaping from the soil, and Acree *et al.* (1963) demonstrated that as much as 50% of the DDT in a 10 ppm$\times 10^3$ suspension in water could disappear in 24 hours, probably because most of the DDT in the water was concentrated near the surface. Very large amounts of water disappear from the soil by volatilization, and it has been suggested that considerable amounts of insecticide may be lost with this, but this theory has now been largely discredited, and probably the amounts lost by codistillation are very small.

Solubility of the insecticide is also an important factor influencing its persistence. Most organochlorine insecticides are relatively insoluble in water but although the more soluble insecticides are generally leached much more readily from soils, leachability and solubility are not always associated. The water solubility of an insecticide is often inversely related to its degree of absorption onto soil fractions (Bailey and White, 1964), but this may not be true for some insecticides. Certainly, when the water solubilities of insecticides are compared with their persistence in soil they seem strongly correlated (Edwards, 1966).

Concentrations of insecticide residues in soil differ greatly, and there is now evidence that large doses break down proportionately more slowly than small ones. Lichtenstein and Schultz (1959) found that twice the percentage of DDT remained from a dose of 100 pounds active ingredient per acre than from one of 10 pounds active ingredient per acre (11.2 kg/ha), 3½ years after treatment, and they confirmed this in laboratory experiments. They concluded that those reactions within the soil responsible for the breakdown of the insecticide in soil depended on the concentration of the insecticide.

Nash and Woolson (1967), and Wiese and Basson (1966) in South Africa, confirmed that large doses disappeared proportionately more slowly than small ones. These conclusions were confirmed by Edwards (1966) who plotted all available published results and calculated regressions of percentage of insecticide remaining against time. These regressions showed clearly that a larger percentage disappeared from a small dose than a large one per unit time. Nevertheless, greater gross quantities may disappear from soil containing the larger dose, i.e. the rate of breakdown was not constant, irrespective of dose, but diminished logarithmically with it.

Insecticides can be formulated as dusts, wettable powders, granules, microcapsules, seed dressings, emulsions, miscible liquids or solutions, and the formulation may greatly affect the persistence. For instance, there is little doubt that the water-soluble formulations leach faster from soil than the oil-soluble water-miscible formulations. The size of particles of insecticide in a formulation is also important; for instance, Barlow and Hadaway (1955) reported that the rate at which particles of lindane were adsorbed onto soil immediately after application was inversely proportional to their size. Hence, at first, the effectiveness of the insecticide was directly proportional to the size of particles in the formulation because the larger particles were adsorbed more slowly than the smaller ones. However, when all the particles had disappeared from the soil surface and insecticidal effectiveness depended mostly on fumigant action, the smaller particle treatments were more effective, probably because the insecticide was more thoroughly distributed through the soil. There is also evidence that emulsions persist longer than wettable powders and as granular formulations are designed to increase the persistence of insecticides in soil they can be expected to persist longer than other formulations. To summarise, insecticides usually persist longer in soils as granules than as emulsions, which in turn persist longer than miscible liquids; they disappear fastest from wettable powders and dusts.

11.1.3.2   *Type of soil*

In general, insecticides are retained longer in heavier soils and those with much organic matter, also insecticides are much less toxic to insects in heavier and organic soils. Hadaway and Barlow (1949, 1951, 1952, 1957, 1963, 1964) studied the influence of the type of soil to which an insecticide is applied, on its persistence, adsorption, and inactivation. They showed that solid particles of organochlorine insecticides, deposited on the surface of soil blocks, disappeared rapidly into the interior of the blocks and they believed this was a physical process of adsorption because it was not restricted to a particular chemical structure or group of insecticides. The rates of loss of insecticides from the surface of mud were increased with greater vapour pressures.

The soil type greatly influences the adsorption of insecticides. Edwards *et al.* (1957) showed that aldrin and lindane were adsorbed least in a sand, and by increasing amounts in silty clay loam, light sandy clay loam, coarse silt, silty clay, sandy loam, clay loam, and muck. Up to thirty times as much insecticide was needed in a muck soil as in a sand to kill test insects.

Harris (1966) suggested that insecticides may be bound to soils in several different ways depending on the structure of the insecticide. Heptachlor and DDT seemed to be inactivated by the clay fraction, diazinon and parathion by the sand and silt fraction, and dichlofenthion by both fractions. All of them were inactivated to some extent by the organic matter. Wiese (1964) found that organochlorine insecticides were only half as active against termites in a loam soil and one-sixth as active in a clay compared with a sandy soil, and Roberts (1963) stated that much more dieldrin was needed to kill test insects in muck or loam soils than in sands.

The soil type influences not only the persistence and activity of insecticides in soils, but also the rate at which they are converted into other chemicals. For instance, Lichtenstein and Schultz (1960) found that aldrin changed into dieldrin sooner in a loam than in a muck soil.

The ways in which soil structure affects the persistence of insecticides is intimately linked with such features as hydrogen ion concentration, organic matter and clay content. For instance, soil structure influences the porosity of soil, and the persistence of lindane in soil has been related to porosity (Swanson *et al.,* 1954). These workers concluded that the movement of moisture was retarded in soils with high capillary porosity, and probably such soils would also retard volatilization and movement of lindane. Roberts (1963) considered that the increasing persistence of dieldrin in sand, loam and muck, could be correlated with the increase in adsorption surface area associated with a decrease in size of soil particles. Thus, it seems that the main influence of soil structure on insecticide persistence is the mechanical composition of the soil.

11.1.3.3 *Organic matter content*

Soil organic matter seems to be the most important single factor influencing the persistence of insecticides. The amount of organic matter in soils can range from less than 1% in some sands to as much as 50% in muck or peat soils. All the evidence indicates that the more organic matter is in a soil, the longer an insecticide persists in it, for instance, Edwards *et al.* (1957) found a highly significant correlation between the persistence of aldrin and lindane, and the amount of organic matter in ten greatly differing soils. They postulated that the relationship between the persistence of insecticides and the amount of organic matter in soils was curvilinear, so that the persistence in muck soils could not be predicted by a simple linear equation, and this was confirmed by Hermanson and Forbes (1966).

These conclusions were also supported by Lichtenstein and Schultz (1960), Roberts (1963), Wheatley *et al.* (1960), Menn *et al.* (1960), Yaron *et al.* (1967) and Konrad *et al.* (1967). Thus, pesticides are likely to persist much longer in a soil with much organic matter, for instance, in forest litter. This is clearly undesirable, because in this situation it may affect the soil animals that are responsible for breaking down dead plant material and returning it to the soil.

11.1.3.4 *Clay content*

A factor that is almost as important as the organic matter is the amount of colloidal material a soil contains. Soils which contain much clay have a much larger internal surface area than sandy soils and so might be expected to retain insecticides longer because there is more area for adsorption. Although there is good evidence that soils with much clay retain insecticides longer, it is often difficult to be sure that the clay content is the principal factor influencing the persistence of insecticides in soils, because the clay content and organic matter content are themselves correlated. Swanson *et al.* (1954) showed that clay/sand mixtures retained lindane more than mixtures of silt/sand, and both much more than sand alone. The adsorption of pesticides onto soil colloids was reviewed by Bailey and White (1964) and they calculated this was a phenomenon that greatly affected pesticide peristence.

Hermanson and Forbes (1966) concluded that the persistence of dieldrin in soils was closely related to their colloidal properties and the specific surface area of the minerals they contained. Edwards *et al.* (1957) compared the correlations between persistence of insecticides and moisture-holding capacity ($r = 0.8468$ for aldrin, $r = 0.8204$ for lindane) with those between persistence and organic matter content ($r = 0.8935$ for aldrin, $r = 0.9141$ for lindane), and concluded that, although moisture-holding capacity which in turn correlated with clay content, was important, it was less so than the amount of organic matter. Harris and Lichtenstein (1961) found that less aldrin volatilised from soils with much clay, and assumed that aldrin was adsorbed onto the clay particles. Abdellatif *et al.* (1967) claimed that 90% of the adsorption was due to the soil clay content and only after this was the organic matter content important. The insecticide was adsorbed by both fractions but degraded faster in the soils with much organic matter. Porter and Beard (1968) reported that adding organic colloids to soils decreased the rate of loss of insecticides. Thus, most of the evidence available shows that clay content is important in influencing insecticide persistence.

11.1.3.5 *Soil acidity*
The hydrogen ion concentration may influence the breakdown of insecticides in soil in several ways. It can affect the stability of clay minerals, the ion exchange capacity, or the rate at which both chemical and bacterial decomposition occurs. However, there is not much direct evidence that pH affects the persistence of organochlorine insecticides in soils. Downs (1951), Swanson *et al.* (1954), Fleming and Maines (1953), and Bollen *et al.* (1958) all reported that pH did not affect the persistence of insecticides. However, Chawla and Chopra (1967) reported that benzene hexachlorine and DDT broke down more rapidly in alkaline soil (pH 9.5) than in what they termed normal soil, and Hermanson and Forbes (1966) believed pH influenced dieldrin activity in soils. Champion and Olsen (1971) claimed that their experiments showed that soils with a low pH, and hence a high anion exchange capacity, would tend to adsorb more DDT than soils of higher pH.

Organophosphate insecticides seem much more sensitive to pH than the organochlorines; for instance, in experiments by Griffiths (1966), they persisted longer in acid soils, Konrad *et al.* (1967) found that diazinon decomposed much faster in acid soils, and Getzin (1968) confirmed this. Konrad and Chesters (1969), and Konrad *et al.* (1969) reported that both 'Ciodrin®' and malathion broke down more rapidly under alkaline conditions. Clearly, the influence of pH on rate of breakdown of insecticides in soil, depends on the main routes of breakdown and whether they are mainly chemical or microbiological; the degradation of some insecticides is affected by pH whereas that of others is not.

11.1.3.6 *Mineral ion content*
The amounts and kinds of minerals in a soil influence both its type and structure and, hence, how long insecticides persist in it. Adsorption is probably the first stage in the process of catalytic decomposition, and adsorption occurs very rapidly in soils with much iron. Barlow and Hadaway (1955) found that activated iron oxide gels adsorbed DDT, and many lateritic soils contain large amounts of iron oxide. Downs (1951)

showed that those soils which catalysed the decomposition of DDT most effectively had most iron and aluminium. Swanson *et al.* (1954) found that lindane in soil was retained longer in soils with much magnesium. Gallaher and Evans (1961) examined a range of New Zealand soils for catalytic effects of the clay minerals and found that some soils derived from volcanic ash favoured decomposition of DDT. Catalytic decomposition of DDT in the presence of inorganic salts, notably those of iron, aluminium and chromium, was reported by Fowkes *et al.* (1960). It seems clear that minerals are important in influencing breakdown of insecticides in soil, but we do not yet have enough evidence to assess this importance fully.

11.1.3.7   *Temperature*

Insecticides are lost from soil mainly by chemical degradation, bacterial decomposition and volatilisation, and all these processes are influenced by temperature, so that at low temperatures these processes slow down and little insecticide is lost. For instance, Lichtenstein and Schulz (1959) reported that no aldrin or heptachlor was lost from frozen soils, but, at $6°C$, 16-27% of the dose applied to soil was lost in 56 days, at $26°C$, 51-55% disappeared, and at $46°C$ 86-98% was lost, Patterson (1962) recorded that phorate disappeared faster at $100°F$ than at $40°F$, and Harris and Lichtenstein (1961) found that increased temperatures increased the rate of volatilisation of aldrin from soil. Conversion to other compounds is also accelerated by higher temperatures; for instance, Lichtenstein and Schulz (1959) found that more aldrin was converted to dieldrin at $26°C$ and $46°C$ than at $7°C$, and Getzin (1968) found that diazinon and thionazin both degraded faster at higher temperatures than at low ones.

Temperature also influences the adsorption of insecticides in soils because sorption tends to be exothermic, so that increased temperatures decrease adsorption and release insecticides. Furthermore, the solubilities of insecticides usually depend on temperature, so that more insecticide becomes dissolved in the soil moisture as temperatures increase, and the amounts of insecticides that are leached from soils may increase.

Obviously, the influence of temperature on the persistence of insecticides in soils is not simple because, although increased temperatures increase the rate of conversion to other compounds, volatilisation, desorption and leaching, warm soils are usually also dry ones, and dry soils hold pesticides much more firmly than wet soils.

11.1.3.8   *Soil moisture*

The main influence of soil moisture on the persistence of insecticides in soil is by its effect on the adsorption of the insecticide onto various soil fractions. Water can compete for adsorption sites with insecticides because it is a very polar molecule, strongly adsorbed by the soil colloids, and in drier soils there are fewer water molecules to compete with the insecticide molecules for adsorption sites. For most insecticides the adsorption seems to be reversible.

Soil humidity, which is directly related to soil moisture, can influence the persistence of insecticides in three ways. Firstly, it can influence the adsorption of an insecticide, secondly it can affect the rate at which it diffuses into soil, and finally it can affect the availability of the adsorbed toxicant. The rate of adsorption from the particulate state decreases as humidity increases, particularly in soils with small adsorptive capacity. The rate of inward diffusion increases with greater relative humidity so that a 10% increase in humidity can double the toxicity.

Most insecticides do seem to be affected by soil moisture, for instance, Roberts (1963) found that soil moisture influenced the effectiveness of dieldrin, and Harris and Lichtenstein (1961) showed that aldrin volatilised much faster from moist sand than from dry sand. Harris (1964) showed that parathion and diazinon were adsorbed by dry soil, so that diazinon was 134 times more toxic and parathion 28 times more toxic in moist soil than in dry soil. In other experiments, they estimated that within the range of normal soil moisture, the toxicity of heptachlor and DDT may differ between two or three times. Harris also showed that heptachlor was 7.8, DDT 9.9, parathion 24.4, diazinon 132.1, and dichlofenthion 188.6 times more toxic in moist sand than in dry sand, but these insecticides were not activated to the same extent in dry muck soils. He also estimated that trichlorfon was 20.1 times more toxic in wet than in dry soil, dieldrin 16.4 times, 'Zectra®' 16.1 times, DDT 15.9 times, heptachlor 12.7 times, and mevinphos 1.4 times.

Harris and Mazurek (1966) tested the persistence of 34 insecticides in soil and found that they were adsorbed equally in dry mineral soil and dry muck soil, but were released from the mineral soil when it was moistened, although not from the muck soil. On the basis of their results they concluded that not all insecticides were affected the same way by soil moisture, and divided the ones they had tested into four categories; (i) aldrin, lindane, and phorate, which were not strongly adsorbed by dry soil and did not compete with water for active sites on the soil particles so that they controlled insects consistently well; (ii) fenitrothion and methyl parathion, which were only moderately adsorbed by dry soil and moderately competitive with water for active sites on soil fractions and gave fairly good results as soil insecticides; (iii) thionazin, diazinon, and dichlofenthion which were strongly adsorbed by dry mineral soil, but were so non-competitive with water that they were highly active in moist mineral soil and, hence, their toxicity to insects was very erratic; and (iv) materials such as azinphos-methyl, which were so strongly adsorbed on the soil and so competitive with water that they remained adsorbed even in the moist state, and were quite ineffective against insects in soil.

Clearly, in soils that are consistently wet, insecticides will persist for a much shorter time than in dry soils. There is an obvious interaction with soil temperature because dry soils are usually in hotter climates and wet soils in colder areas. Thus, temperature and moisture may act in opposition in their effects on the persistence of insecticides.

11.1.3.9   Effect of crops
Open or fallow soil is exposed to much more wind, sun, and rain than the soil under a growing crop. This shading effect may greatly influence the persistence of insecticides in soil. For instance, when soil was artificially shaded after treatment with heptachlor, more remained in the shaded than in the unshaded soil, although the differences were not significant (Young and Rawlins, 1958); this was probably a temperature effect. Lichtenstein *et al.* (1962) reported that a dense cover of alfalfa considerably increased the persistence of aldrin and heptachlor in soil. After they applied 25 pounds active ingredient of aldrin per acre (28 kg/ha) to plots, after 3 years, 9.1 ppm remained in plots cropped with alfalfa compared with only 4.8 ppm in plots with no crop.

Another factor is that insecticides are taken up from soil into crops, and this contributes to their disappearance from soils. For instance, Lichtenstein (1964), Lichtenstein *et al.* (1965), and Glasser (1958) reported that organochlorine insecticides were taken up into carrots, and there have been many similar reports for other crops (Hurtig and Harris, 1966; Wingo, 1966; Lichtenstein *et al.,* 1965; Ware *et al.,* 1968). Amounts taken up are small and probably do not contribute greatly to the disappearance of persistent insecticides from soil.

11.1.3.10   Cultivation
Insecticides are usually cultivated into the soil; but those which reach soil accidentally, as spray drift or run-off from foliage, may remain undisturbed on the surface for a long time. The persistence of insecticides differs greatly with the degree of cultivation. For instance, Wheatley and Hardman (1962) reported that 95% of a dose of 1 pound of aldrin per acre (1.12 kg/ha), applied to the surface of a sandy loam, disappeared in 6 weeks; but if it was cultivated into the soil, 40% of the aldrin still remained 20 weeks after treatment. Lichtenstein and Polivka (1959) reported that aldrin persisted longer when cultivated into bare soil than when applied to turf and left uncultivated, and Lichtenstein *et al.* (1962) stated that plots rototilled to a depth of 5 inches, after aldrin or heptachlor had been applied, still had from 26-50% of the amount applied remaining 4 months after treatment, compared with 2.7-5.3% in untilled soil. Lichtenstein (1964) reported that 44-62% of the aldrin applied to the top 5-inch layer of Carrington silt loam remained after 1 year, compared with 6.5-13% if left uncultivated. Saha and Stewart (1967) recovered 3% of the heptachlor which they applied to the surface and left undisturbed, compared with 15% when the soil was cultivated. Similar results were obtained by Mulla (1960), Parker and Dewey (1965), and Nash and Woolson (1968).

There have been attempts to accelerate the loss of persistent insecticides from soil by regular cultivation. For instance, when Lichtenstein and Schulz (1961) disked a loam soil containing DDT or aldrin daily for 3 months, there was 25% less DDT and 38% less aldrin than in plots cultivated only once. Thus, it is well established that insecticides persist longer when thoroughly mixed into the soil than when not, although continued cultivation tends to decrease the persistence.

11.1.4   Hazards caused by residues in soil

Persistent insecticides in soil or forest litter may have a variety of effects on the living organisms in these media; (i) they may be directly toxic to elements of the soil fauna or microflora; (ii) they may affect these organisms genetically so that resistant populations develop; (iii) they may have a range of sublethal effects on activity, behaviour, reproduction or metabolism of soil organisms; and (iv) they may be taken up into the body tissues of the soil fauna and flora.

11.1.4.1   *To soil micro-organisms*

Serious disturbance of microbial activity in soil by persistent pesticides could have very adverse effects on soil fertility. Fortunately, most of the evidence is that, although microbial activity may be disturbed by excessively large residues of organochlorine insecticides, the effects of the amounts likely to occur in practice are insufficient to cause any significant decrease in soil fertility (Eno, 1958; Bollen, 1961; Alexander, 1969; Harris, 1970). In fact, a number of workers have reported increases in numbers of micro-organisms due to organochlorine insecticides, possibly because microbes can use pesticides carbon sources (Verona, 1952; Fletcher and Bollen, 1954; Roberts and Bollen, 1955). Several workers have confirmed that none of the organochlorine insecticides affected micro-organisms (Martin *et al.,* 1959; Eno and Everett, 1958; Wegorek, 1957; Shaw and Robinson, 1960). Fungi seem more susceptible than bacteria, which are almost immune (Bollen *et al.,* 1954). The effects of organochlorine insecticides on nitrification and ammonification seem to be very small (Eno, 1958), although, if anything, the nitrifiers are more sensitive. Of all the organochlorine insecticides, BHC seems to have the greatest influence on micro-organisms and this is one of the least persistent of these insecticides. In general, where effects of persistent pesticides on micro-organisms have been reported they are only transient and microbial populations soon recover.

11.1.4.2   *To terrestrial invertebrates*

Some soil invertebrates may be pests, in which case it is desirable that they be killed or harmed in some way. Unfortunately, the pests are greatly outnumbered by the beneficial organisms such as predatory mites, centipedes and carabid beetles that prey on pests, and the myriads of invertebrates that contribute to the breakdown of dead plant and animal organic matter, and thus ultimately to soil fertility. In particular, earthworms, enchytraeid worms, Collembola, Diptera larvae and some Acarina all feed on decaying material and help to incorporate it into soil. Harmful effects due to killing these beneficial animals are more likely to occur in uncultivated forests and woodlands than in arable land, where tillage and the use of fertilisers can compensate for the activities of soil invertebrates.

The effects of persistent pesticides on soil invertebrates are complex and can be both direct and indirect. It is not possible to do more than mention them here so the reader is referred to reviews by Edwards (1970) and Edwards and Thompson (1973) that summarize their effects on the whole of the soil ecosystem. In general, the more active animals, many of which are predators, are more susceptible to pesticides than more sluggish forms. The overall effect of organochlorine pesticides on the total numbers of soil animals is not usually drastic, numbers seldom being decreased by more than 50% of those in untreated soil.

Since the commonest residues in soil are those of DDT and dieldrin, these are the chemicals most likely to affect the soil biota. DDT is by no means equitoxic to all soil organisms. Many mites, especially predatory species, are killed by DDT; this often results in very considerable increases in numbers of Collembola, which are unaffected by this insecticide (Sheals, 1955; Edwards and Dennis, 1960; Edwards, 1963, 1965, 1970; Edwards et al., 1967; Bund, 1965). DDT does not kill earthworms and is not very toxic to the larger arthropods in soil. Dieldrin is not often applied to soil, the residues reported being usually due to the breakdown of aldrin, which is commonly used as a soil insecticide. Residues of dieldrin and aldrin are more toxic than DDT to most soil invertebrates, except predatory mites and earthworms. The other organochlorine insecticides, except for endrin, are less toxic to most species of soil invertebrates than DDT, aldrin and dieldrin. However, chlordane, endrin and heptachlor are all very toxic to earthworms, and as these animals contribute to soil fertility, this is certainly undesirable.

To summarize, there is still no evidence that the effects of persistent pesticides on soil invertebrate populations influence soil fertility, although if residues become large and widespread in woodland or pasture habitats, there is the possibility of long-term effects on soil structure.

11.1.4.3 *Entry into food chains*
The first report of organochlorine residues in the tissues of soil invertebrates was when Barker (1958) reported that he had found large residues of DDT, in earthworms from soil under elm trees that had been sprayed with DDT. Subsequently, many other workers reported organochlorine insecticide residues in earthworms (Stringer and Pickard, 1963; Cramp and Conder, 1965; Hunt, 1965; Davis, 1966; Davis and Harrison, 1966; Dustman and Stickel, 1966; Davis, 1968; Dimond et al., 1970; Korschgen, 1970; Edwards, 1970; Gish, 1970; Edwards and Thompson, 1972) and many of these reports showed that the amounts in the earthworm tissues

were greater than in the surrounding soil (see Chapter 3).

Similar concentration of organochlorine insecticides from soil into invertebrates have been reported for molluscs (Cramp and Conder, 1965; Stringer and Pickard, 1965; Davis and Harrison, 1966; Gish, 1970; Edwards and Thompson, 1972) for beetles (Cramp and Conder, 1965; Davis and Harrison, 1966; el Sayed *et al.*, 1967; Davis, 1968; Korschgen, 1970; Gish, 1970; Newsom, 1967; Edwards and Thompson, 1972) and for caterpillars (Newsom, 1967). It seems likely that many of the small invertebrates that inhabit soil contain organochlorine residues, because Butcher *et al.* (1972) showed that not only could mites contain DDT residues, but they were able to convert DDT to DDE.

Thus, it is not unreasonable to suppose that the majority of soil invertebrates living in contaminated soil contain at least small amounts of organochlorine residues. These need not be of much importance, because animals may not be affected by the residues they contain. However, many of these small invertebrates are a major source of food for some birds and mammals, which, if they eat sufficient of them, may gradually acquire a lethal dose. Furthermore, it is now well-established that vertebrates can concentrate some pesticides from their food into certain of their body tissues (Edwards, 1970). If some of these animals are in turn eaten by other vertebrates in higher trophic levels, there may be further concentration into these animals, and potential hazards to them. These problems are more fully discussed in Chapter 7, but this tendency for organochlorine residues to concentrate into the tissues of higher organisms seems one of their greatest hazards.

11.1.4.4 *Sublethal effects on soil invertebrates*

The most important sublethal effect of organochlorine insecticides on soil invertebrates is the development of resistance to organochlorine insecticides, by some of the species exposed to them for a number of generations. Such exposure selects genetically for resistant individuals, with the result that ever-increasing amounts of insecticides are required to kill those species that are pests. This in turn creates greater potential for pollution by the insecticide residues.

Another sublethal effect of residues, is that there is evidence that they may alter the reproductive potential of some invertebrates, and thus greatly affect their numbers. If the invertebrate is a pest this is to be preferred, but if it is beneficial, then it is clearly undesirable.

Finally, pesticide residues in invertebrates may completely change the behaviour and rate of feeding of these animals. Any, or all, of these effects exert pressure on the soil ecosystem, and tend to upset predator/prey relationships; at present the overall effects are unknown and they may well be undesirable, even if they are not one of the major hazards caused by insecticide residues in soil.

11.1.4.5   *Phytotoxicity*

There have been a very large number of investigations into the effects of organochlorine insecticide residues on the growth of plants (Foster, 1951; Stone, 1953; Cox and Lilly, 1952; Allen *et al.*, 1954; Eno, 1958; Dennis and Edwards, 1961, 1963, 1964; Edwards, 1965).

There is little doubt that excessive amounts of organochlorine insecticide residues in soil can be harmful to plants. The symptoms differ greatly between crops, and include poor growth or germination, chlorosis or wilting, distortion or stunting, and off-flavours or taints in the edible parts of plants. However, the amounts of most of the organochlorine insecticides reported in surveys are insufficient to damage most crops, except in orchard soils that often contain excessive residues.

BHC is the most phytotoxic of the organochlorines, moderate residues in soil sometimes stunting growth, and, more important, producing distinct taints or off-flavours in many edible crops. BHC seed dressings often affect the germination of seeds treated a long time before sowing. There are also a few reports of BHC stimulating plant growth.

DDT residues are only slightly toxic to a very few crops, except at excessive dosages, tomatoes, cucumbers and beans being the most susceptible plants. Amounts greater than 20 lb a.i. per acre (22.4 kg/ha) have affected the growth of cereals and beet. There are a few reports that the growth of carrots, parsnips and turnips was stimulated by DDT residues.

Aldrin is rather more phytotoxic than DDT, and moderate dosages (more than 8-16 lb a.i. per acre (8.9-17.9 kg/ha)) have damaged cucumbers, tomatoes, beans, beet and cereals. Chlordane, endrin and heptachlor are rather similar to aldrin in their effects on plants, and dieldrin is the least phytotoxic of all the organochlorines.

Thus, it seems unlikely that the levels of residues currently reported from soils (Edwards, 1970) (see Tables 11.2 and 11.3 and Chapter 2) seriously affect the growth of crops.

11.1.5   The disappearance of residues from soil

Organochlorine insecticides are all very stable chemicals that are relatively non-volatile and poorly-soluble. It is these characteristics that make them so very persistent, but nevertheless, even DDT gradually disappears from soil. It is still not absolutely clear which are the major pathways of loss, although as evidence accumulates, it is becoming easier to assess the relative importance of the various factors.

11.1.5.1   *Volatilization*

It has been assumed that the low volatilities of organochlorine insecticides (Table 11.4) mean that only a small proportion of them are lost through volatilization from the soil surface. But considerable evidence is accumulating that considerable amounts of even a relatively non-volatile compound, exposed to the atmosphere for long periods, can volatilize. For instance when Nash and Woolson (1968) studied the profiles of distribution of

TABLE 11.4.   Vapour pressure of organochlorine insecticides.

Insecticide	Vapour pressure at 20°C (mm Hg)	
Dieldrin	1.8×10^{-7}	
DDT	1.9×10^{-7}	Slightly volatile
Endrin	2.0×10^{-7}	
Camphechlor	1.0×10^{-6}	Moderately volatile
Lindane	9.4×10^{-6}	
Chlordane	1.0×10^{-5}	
Aldrin	2.3×10^{-5}	Volatile
Heptachlor	3.0×10^{-4}	

aldrin, dieldrin, heptachlor, camphechlor, isodrin, endrin, chlordane and benzene hexachloride in soil, 13 years after they were applied, they found less residues in the top 7.6 cm of soil than in the 7.6-23 cm depth, although the insecticides had been applied to the surface. They concluded that this was evidence for the action of volatility and photodecomposition, although there are other possible explanations of this distribution.

The amount of incorporation of residues into soil greatly influences the importance of volatilization as a factor in their loss; if the residues are on the soil surface they disappear much faster than if cultivated into the soil. Also, in soil, the vapour pressure of an insecticide is not constant, but changes with its concentration, because the insecticide is dissolved in the soil water and adsorbed on to soil fractions (Graham-Bryce, 1972). Hence, the amount of volatilization occurring may depend upon the concentration of the insecticide. There is also much less volatilization from very dry soils than from wet ones, because unless they are displaced by water molecules, the insecticides remain tightly bound to adsorptive surfaces in the soil. When the insecticide is cultivated into the soil, volatilization becomes limited by the rate at which the insecticide can be transported to the soil surface. Thus, transport occurs either by capillary movement in the soil solution or by molecular diffusion, and both these processes may be slow.

As might be expected from their vapour pressures, organochlorine insecticides differ in their rates of loss due to volatilization Guenzi and Beard (1970) concluded that considerable quantities of lindane volatilize from soil, and even with DDT, volatilization could be responsible for considerable losses, and Cliath and Spencer (1971) supported this conclusion. Harris (1970) stated that volatilization was a major factor in the loss of heptachlor from soil. Caro and Taylor (1971) made a careful study of possible routes of loss of dieldrin from soil and reported that a minimum of 2.9% was lost by volatilization; but because of the crude nature of their traps it is likely that the amount lost is actually much more than this. Farmer and Jensen (1970) gave experimental data that supported the conclusion that a large proportion of dieldrin in soil could be lost by volatilization; they calculated that in ideal conditions all the dieldrin applied could volatilize from soil within one year. However, this did not take account of the need for dieldrin to be transported to the soil surface.

Even DDT can volatilise from soil in quite considerable amounts, and Lloyd-Jones (1971) calculated that half the amount of DDT applied to soil may volatilize, but Spencer and Cliath (1972) showed that o,p'-DDT and p,p'-DDE volatilized much more than p,p'-DDT, so some degradation is necessary before much volatilization occurs. Moreover, vapour pressure is not constant but varies with temperature, so more volatilizes from warm soils than from cold soils.

There seems little doubt that volatilization is a major pathway of disappearance of organochlorine insecticides from soil, and this is probably the source of the residues reported from atmospheric air and rainwater (see Chapter 10).

11.1.5.2   Leaching and run-off

The movement of insecticides through soil with water can really be distinguished into two components: the lateral movement of the chemical over sloping areas of the soil surface, and the downward movement into the soil with drainage water. Lichtenstein (1958) reported that aldrin and lindane moved down slopes, and Edwards (1966) also reported the movement of DDT down slopes. Edwards et al. (1971) found that small quantities of dieldrin were carried over the surface of soil that sloped toward ponds.

Several workers have studied the leaching of insecticides down columns of soil. Kirk (1952) applied DDT to the surface of 1.5 cm diameter columns of soil and ran approximately 30 ml of water through them. Only 1-1.5% of the amount applied leached through to the third 2.5 cm layer. Lichtenstein (1958) found that some lindane leached down columns, but was unable to detect any lindane in the leachate. Guenzi and Beard (1967) followed the leaching of lindane and DDT down 10 cm diameter soil columns, and found that all the DDT remained in the 0-3 cm layer after 25 cm of water had been poured through the column, but small amounts of lindane penetrated to the 6-9 cm layer. Edwards et al. (1971) studied the movement of dieldrin down columns, 30 cm diameter and 35 cm deep, filled with 3 types of soil that was either broken up or left as intact soil profiles. The amounts of dieldrin which appeared in the leachates differed with the type of soil, but usually about ten times as much insecticide leached through intact soil columns as through those containing broken soil; nevertheless, the greatest amount of dieldrin that leached through any column after about 1 year's rainfall was only 2% of the amount applied, and usually it was much less.

There is evidence that progressively less insecticide is leached with successive leachings after application to soil (Bowman et al., 1965). Thompson et al. (1971) found that when they passed a year's rainfall down columns, much less dieldrin was leached with the second half of the water passing through the column than in the first half. They compared the amounts of dieldrin that were leached vertically downward into soil, with those which ran over a sloping surface, by building a trough with vertical transverse partitions that were progressively lower toward one end. The trough was filled with soil to a height of 5 cm above each partition and the soil in the first compartment was treated with dieldrin, then the box was left outdoors exposed to rainfall for 17 weeks. Only 0.02% of the applied dieldrin appeared in the leachate, and 99% of this was collected during the first 9 weeks, mostly from the first compartment or the last one. Caro and Taylor (1971) found that much more dieldrin occurred in run-off water from a water shed during the first few weeks than the rest of the year. Probably, these results are due to progressive adsorption of the insecticide.

The degree of leaching of an insecticide from soil is correlated with its water solubility, but this is not a simple relationship because it is also affected by the capacity of the insecticide for adsorption onto soil fractions. In the simplest form of adsorption, water molecules compete with insecticide for sites on the soil particles so that, in wet soil, the insecticide becomes released. King and McCarty (1968) developed a chromatographic model to account for the movement of pesticides through soils by elution. They estimated the amounts of insecticides that would be leached by different quantities of water and found good agreement between calculated and actual amounts if allowance was made for degradation of the pesticide.

It seems clear from available data, that for most insecticides, leaching accounts for only a small proportion of total losses although rather more may be removed by water flow over the surface during periods of heavy rainfall.

11.1.5.3 *Microbial breakdown*

Much evidence has accumulated that micro-organisms are important in breaking down persistent pesticides (see Chapter 13), although it is very difficult to differentiate between biological and non-biological breakdown. The simplest way of differentiating between these two methods is by comparing the rates of breakdown in sterile and non-sterile soils (Lichtenstein and Schulz, 1960; Menn *et al.,* 1965), but the method of sterilization can affect the validity of the results. For instance, Getzin and Rosefield (1968) found that some pesticide decomposed faster in soils that were sterilised by gamma-irradiation than in the same soils sterilised by autoclaving.

Korte *et al.* (1962) discovered that several micro-organisms slowly degraded cyclodiene insecticides other than dieldrin. They studied the fate of these insecticides in micro-organisms, and found that the moulds *Aspergillus flavus, A. niger,* and *Penicillium notatum* could not metabolize ^{14}C-dieldrin, but converted significant amounts of ^{14}C-aldrin, isobenzan, chlordane, and heptachlor into hydrophilic degradation products. Matsumura *et al.* (1968) found that a *Pseudomonas* sp. could break down aldrin to dieldrin and four other metabolites. Matsumura and Boush (1967) tested more than 500 microbial isolates from soil containing various insecticides, and found some that actively degraded dieldrin to various metabolites. Tu *et al.* (1968) screened 92 soil micro-organism cultures, for activity in breaking down insecticides, and found that most had some ability to convert aldrin to dieldrin, especially *Fusarium* sp., which converted 9.2% of added aldrin to dieldrin in 6 weeks. However, *Trichoderma* spp. were most efficient of the fungi, with *Penicillium* next and *Rhizopus* and *Mucor* of minor importance. The most active actinomycetes and bacteria were *Nocardia* sp., *Streptomyces* sp., and *Micromonospora* sp. Chacko *et al.* (1966) demonstrated that several actinomycetes could degrade DDT, but none of the micro-organisms they tested were able to degrade dieldrin. Yule *et al.* (1967) isolated two species of bacteria from soil, *Bacillus cereus* and *Bacillus* sp., which were able to dehydrochlorinate lindane *in vitro*.

To summarize, there is ever-accumulating evidence that micro-organisms play an important part in degrading persistent insecticides, but we still have little evidence that they can completely break them down.

11.1.5.4 *Other losses*
Organochlorine insecticides are taken up into the plants growing on soil containing residues and, if these are crop plants, residues are removed when the crop is harvested. This has been discussed more fully in 11.1.2.5. They also become locked up and inactivated in organic material, soil colloids and the tissues of plants and animals living in soil.

11.1.6 Minimizing residues in soil
The organochlorine insecticides have so many valuable attributes and there is still a lack of suitable alternatives for controlling some pests. Hence, large quantities of these chemicals are still being used and it is important to keep their residues to a minimum. This can be done by more refined methods of application and better regulation of their use. There has also been considerable interest in possible ways of decontaminating soil.

11.1.6.1 *Methods of application*
Many of the problems that have arisen through the use of persistent pesticides are due to careless use of these chemicals. Often, large areas of land have been sprayed indiscriminately and such sprays fall on all parts of an ecosystem, so it is not surprising that residues occur throughout the environment after such treatments. Haphazard spraying operations should be severely restricted, even when non-persistent pesticides are used. Fortunately, the efficiency of methods of spraying insecticides has increased greatly in recent years, for instance, less and less water is required, and this has made it possible to use less chemicals. For instance,

it is now possible, using sophisticated spraying equipment, to apply ½ lb of active ingredient of insecticide per acre (0.56 kg/ha) undiluted with water.

When organochlorine insecticides were first developed they were broadcast over the surface of agricultural soils every year, then thoroughly incorporated with a harrow, plough or rotovator. Such treatments use much more insecticide than is really necessary to control a particular pest. For pests where treatment of the whole soil is essential, it is much better to calculate a 'topping-up' dose, that will bring existing residues back to a sufficient level to control pests. Decker et al. (1965) described how the root pests of corn could be controlled by addition of aldrin in amounts calculated in this way. However, a much better method is to use localized treatments that place the insecticide exactly where it is required, such as by using a seed-dressing which places a very small quantity of insecticide around the seed. The development of greatly improved 'stickers', or other methods of ensuring a uniform cover of insecticide over the seed, in recent years should increase the efficiency and use of seed-dressings. Another method is to apply small doses of insecticides close to they seeds as they are sown, or narrow bands of insecticide in the rows of the crop. Alternatively, some pests can be controlled by dipping the roots of transplanted crops in insecticides before planting out.

Insecticides which persist only a short time in soil, can be made to remain for a whole growing season, by applying them absorbed on the surface of inert granules which gradually release the insecticide; different granules release insecticides at different rates and the most suitable ones for a particular pest can be chosen. There is also the use of insecticides in micro-capsules, which gradually break down to release the insecticide into the soil.

Finally, eradication programmes should be considered for suitable pests and areas. However, such programmes are usually only successful against pests introduced to countries other than that of their origin, or in delimited geographical areas.

11.1.6.2 *Legislation*
In the United States, Great Britain and many other countries, there is governmental legislation that controls the marketing of pesticides. In the United States, there are stringent regulations and requirements before a pesticide manufacturer can market any pesticide and there are tolerance limits on the amounts of pesticides that are permitted in crops to be used as foods (see Chapter 9).

So far, there are few outright bans on insecticides, but many countries have restricted the use of DDT, aldrin and dieldrin. The education of insecticide users and legislation controlling the labelling and instructions on insecticide containers can considerably decrease residue problems. Possibily, legislation restricting the methods of application, such as aerial spraying, might also be of value.

11.1.6.3   *Cultivation*

In some situations where the persistent pesticides lie on the soil surface, it is advisable to avoid cultivation and use herbicides to keep down weeds; this should greatly accelerate the disappearance of the chemicals both by volatilization and by photodecomposition. On the other hand, once the insecticides become thoroughly incorporated into the soil it may be possible to accelerate their disappearance by frequent cultivation. Lichtenstein *et al.* (1961) investigated this possibility, but although very frequent cultivations did increase the loss of the insecticides, it certainly would not be economic to cultivate as often as they did — every three days. Cultivation does not seem to be a practical proposition, for decreasing residues.

11.1.6.4   *Flooding or irrigation*

Heavy irrigation has been suggested as a way of increasing losses of insecticides from soil. Unfortunately, the degree of adsorption of the insecticides on to soil means that the amounts of water required would be very large and might well remove nutrients in the process. However, there is evidence that the disappearance of persistent insecticides from soil can be accelerated by flooding, which creates anaerobic conditions in the soil and favours the activity of certain micro-organisms. For instance, Kearney *et al.* (1969) were able to accelerate greatly losses of DDT from soil by flooding and inoculating with a bacterium (*Aerobacter aerogenes*). Guenzi and Beard (1968) also reported that DDT broke down faster under anaerobic conditions, and they thought that flooding soil might be a practical proposition. More work is required before such methods could be recommended on a field scale.

11.1.6.5   *Addition of materials*

Various workers have attempted to decrease the activity of residues by addition of adsorbent materials. A practical example of this was where Kring and Ahrens (1968) showed that addition of 100 - 400 lb of activated carbon per acre (112 - 448 kg/ha) could detoxify chlordane (1.5 - 12.0 lb a.i. per acre (1.7 - 13.7 kg/ha)) and heptachlor (1.0 - 4.0 lb a.i. per acre (1.12 - 4.5 kg/ha)), and Lichtenstein *et al.* (1971) showed that a single treatment of soil with 1,000 to 4,000 ppm of activated carbon, could substantially reduce the uptake of heptachlor and aldrin from soil into crops. The amount of carbon required depends on the amount of insecticide involved, the soil type and the crops to be grown. Although the method works, it is expensive and probably uneconomic for many crops. Furthermore, addition of large amounts of adsorbent means that very large amounts of chemicals will be required to control pests on future crops.

There remains the possibility that addition of localized treatments of activated carbon around individual crop plants or in rows would allow crops to be grown in contaminated soil.

11.1.6.6   *Stimulating microbial activity*

Micro-organisms are important in degrading pesticides, and there have been several attempts to accelerate degradation by introducing micro-organisms to soil, encouraging their growth by adding suitable substrates, such as composts to soil, or by flooding in order to create anaerobic conditions suitable for microbial activity. Some success has been reported, for instance, Guenzi and Beard (1968) reported that DDT degraded to DDE faster in anaerobic flooded conditions, especially when chopped alfalfa was added as an energy source. Kearney *et al.* (1969) used similar methods but added cultures of bacterium, *Aerobacter aerogenes* to soils. Flooding accelerated DDT decomposition, but it occurred even faster in the inoculated soils. Such experiments demonstrate the possibility that such techniques may be used, but much more work is needed before it becomes a practical possibility.

11.1.6.7   *Alternative chemicals*

Many non-persistent substitutes for organochlorine insecticides have been developed and are in common use, but it is still not possible to control some pests with alternative chemicals. Metcalfe *et al.* (1972) have suggested that it is quite feasible to develop much less persistent and biodegradable analogues of DDT and other organochlorine insecticides. They demonstrated, in a model ecosystem, that such analogues are not concentrated into the upper trophic levels of food chains. There have also been reports of attempts to produce 'self-destructing' forms of DDT containing catalysts that promote degradation. The difficulty of such methods is that the large bulk of soil might buffer the activity of reactive chemicals added. However, such studies merit further serious attention.

11.2   RESIDUES IN WATER

Water, and the mud at the bottom of rivers, streams, lakes, ponds or the seabed are other major reservoirs for persistent pesticide residues. There are many ways in which pesticides can reach water: (i) they may be directly applied as aerial sprays or granules to control water-inhabitating pests; (ii) they may fall on to the surface of water when forests or agricultural land are sprayed from the air; (iii) spray drift from normal agricultural operations may reach the water; (iv) residues may reach water as surface run-off from treated soil; (v) insecticides may be discharged into rivers with factory or sewage effluents; (vi) insecticide containers may be split or washed into rivers; and (vii) there may be uptake from the atmosphere or residues may be carried down with rain or dust. It is difficult to assess the relative importance of the various sources of insecticide residues in rivers, but probably small background amounts originate in run-off from agricultural land and rain, whereas massive amounts sometimes reported, come from industrial effluents, emptying sheep dips, and emptying or washing of spraying equipment.

11.2.1   Amounts occurring in water

11.2.1.1   In freshwater

Reports of insecticide residues from streams and rivers in the United States started about ten years after the introduction of organochlorine insecticides, It soon became obvious that small amounts of these insecticides occurred in many waterways in the United States and, because of this, extensive surveys of most of the major waterways in the U.S.A. have been made in recent years. A National Pesticide's Water-Monitoring Network was set up with 161 sampling sites spread all over the country. Each site is sampled quarterly for water and semi-annually for bed materials (Feltz *et al.,* 1971). There have also been more local surveys of U.S. rivers (Green *et al.,* 1967; Lichtenberg *et al.,* 1970; Johnson and Morris, 1971), streams (Brown and Nishioka, 1967; Manigold and Schulze, 1969), ponds (Weatherholz *et al.,* 1967) and marshes (Kolipinski *et al.,* 1971) (see Table 11.5). Schafer *et al.* (1969) surveyed residues in drinking water sources. In the United States all of the different organochlorine insecticides have been

TABLE 11.5.   Residues of organochlorine insecticides in water.

Location	Reference	No. sites	Types of water	A
				Max
Great Britain	Croll, 1969	76	British rivers (two-monthly samples)	—
,,	,, ,,	15	British rivers (single samples)	—
,,	Lowden *et al.,* 1969	9	British rivers	—
,,	,, ,, ,, ,,	9	Yorkshire rivers	—
,,	,, ,, ,, ,,	21	Sewage effluents	—
,,	Holden and Marsden, 1966	—	,, ,,	—
West Germany	Herzel, 1970	51	Major rivers	—
U.S.A.	Weaver *et al.,* 1965	97	Major river basins	85.0
,,	Anon. U.S.D.A. A.R.S. 81-13, 1966	10	Mississippi river delta	30.0
,,	Keith and Hunt, 1966	82	Water areas (California)	—
,,	Warnick *et al.,* 1966	48	Water areas (Utah)	—
,,	Breidenbach *et al.,* 1967	99	Major river basins	—
,,	Brown and Nishioka, 1967	11	Major rivers (Western U.S.A.)	5.0
,,	Green *et al.,* 1967	109	Major rivers	—
,,	Weatherholtz *et al.,* 1967	35	Ponds (S.W. Virginia)	—
,,	Manigold and Schulze, 1969	20	Streams (Western U.S.A.)	40.0
,,	Lichtenberg *et al.,* 1970	110 (1967)	Surface waters	—
,,	,, ,, ,, ,,	114 (1968)	,, ,,	—
,,	Johnson and Morris, 1971	6 (1968)	Iowa rivers	—
,,	,, ,, ,,	10 (1969)	,, ,,	—
,,	,, ,, ,,	10 (1970)	,, ,,	—
,,	Kolipinski *et al.,* 1971	11	Water areas (S. Florida)	—
,,	Zabik *et al.,* 1971	12	Major rivers (Michigan)	—
Canada	Miles and Harris, 1971	4	Water areas (Ontario)	—
Hawaii	Bevenue *et al.,* 1971	101	River and drinking water	—

* These high means are due to large residues in few samples.   † Residues found at only one site.
‡ Mean estimated from total range.   T = Trace.   N.A. = Figure not available.

reported from rivers, sometimes in large quantities; Green *et al.* (1966) reported that dieldrin occurred in almost all U.S. rivers in 1964, with little consistency in its geographical distribution. Residues of lindane, heptachlor, and endrin have all been reported from a number of U.S. rivers although the amounts of residues reported have been quite variable. DDT has been found in the largest amounts but it is not always the most common residue, although some rivers have contained it in all the surveys. Dieldrin also occurs very commonly in smaller quantities, but there are few reports of residues of chlordane, endosulfan and toxaphene in U.S. waterways. In British rivers, only DDT, BHC, aldrin and dieldrin have been reported; this is not surprising because these are the insecticides used most in that country.

There have been only two surveys of British rivers, the first being a study of residues in 8 south-east rivers and 18 rivers from other parts of the country (Croll, 1969). In this survey, the amounts reported were relatively small, especially since samples were taken from areas that might

		Concentration of residues ppmx10^6 (ng/1)										
HC	Chlordane		DDT R		Dieldrin		Endosulfan		Endrin		Heptachlor and heptachlor epoxide	
Mean	Max.	Mean	Max.	Mean	Max.	Mean	Max.	Mean	Max.	Mean	Max.	Mean
25.82	—	—	—	—	423.0	25.16	—	—	—	—	—	—
53.60*	—	—	43.0†	8.67	2840.0	291.60*	—	—	—	—	—	—
18.7	—	—	15.0	1.6	40.0	3.3	—	—	—	—	—	—
38.6	—	—	908.0	64.6	630.0	114.0	—	—	—	—	—	—
92.5	—	—	800.0	130.9	1900.0	145.0	—	—	—	—	—	—
—	—	—	130.0	36.0	300.0	200.0	—	—	—	—	—	—
138.2	—	—	300.0	18.9	165.0	3.2	3400.0	126.7	—	—	2000.0	39.4
—	—	—	102.0	9.3	118.0	7.5	—	—	94.0	5.5	—	—
28.0	—	—	720.0	112.0	60.0	10.0	—	—	4230.0	541.0	10.0	2.0
0.01	—	—	22.0	0.6	—	—	—	—	—	—	T	T
—	—	—	32.4	9.7	—	—	—	—	—	—	—	—
T	—	—	149.0	8.2	68.0	6.9	—	—	116.0	2.4	155.0	6.3
2.8	—	—	120.0	10.3	15.0	2.3	—	—	40.0	1.4	90.0	2.6
2.2	75.0	0.1	127.0	8.3	167.0	5.9	—	—	69.0	3.6	19.0	0.1
—	—	—	<1000.0	N.A.	—	—	—	—	—	—	15800.0	N.A.
0.5	—	—	180.0	9.3	70.0	1.1	—	—	70.0	0.3	60.0	1.4
0.9	—	—	840.0	18.5	87.0	5.0	—	—	133.0	2.0	—	—
3.0	—	—	417.0	7.8	407.0	8.2	—	—	—	—	—	—
—	—	—	12.0	2.3	10.0	1.8	—	—	—	—	—	—
—	—	—	16.0	1.8	63.0	8.5	—	—	—	—	—	—
—	—	—	23.0	3.9	65.0	8.7	—	—	—	—	—	—
—	—	—	40.0	26.4	—	—	—	—	—	—	—	—
—	—	—	160.0	111.0	—	—	—	—	—	—	—	—
—	—	—	397.0	64.0	110.0	7.1	187.0	19.6	—	—	—	—
1.6‡	13.0	7.0‡	83.0	44.0‡	19.0	9.4‡	—	—	—	—	—	—

be expected to receive drainage from insecticide-treated land; in general, residues were less than 50 ppm×10^6 (ng/1), the occasional large residues being attributed to industrial effluents. There were no residues in 12 different underground waters sampled. The other survey (Lowden *et al.*, 1969) was concentrated on Yorkshire and Lancashire, and sampled rivers that received considerable amounts of industrial effluents, so the results are very biassed, and it would be unwise to conclude as a result of this survey that British rivers contain more residues than those in the United States.

Lowden *et al.* (1969) also reported preliminary results, from 17 British rivers, that approximated much more closely to the data given by Croll (1969). The only survey available from the rest of Europe was that by Herzel (1970) who assessed residues in the main West German rivers. The only pesticide that was present commonly, and in large amounts, was lindane (39 out of 51 of the waters sampled). DDT was found in 6 rivers, endosulfan in 5, heptachlor in 2 and dieldrin in 1. However, there were isolated instances of very large amounts of some residues.

There have been few other surveys but in Hawaii, Bevenue, *et al.* (1971) found relatively small amounts in 46 water samples taken from over the State. There are very few reports of wells or ground water becoming contaminated with appreciable quantities of pesticides.

To summarize, the amounts of organochlorine residues in freshwater are relatively small, except where there is some obvious cause such as industrial effluent, direct application or accidental addition.

11.2.1.2   *In estuaries*

Since rivers contain appreciable amounts of pesticides, it would be expected that they would carry these chemicals down to the sea so that there might be large amounts of residues near the mouths of rivers. However, there is surprisingly meagre evidence of the amounts of pesticide residues near river mouths and in estuaries. In the U.S.A., it was estimated that 1.9 metric tons of pesticides are carried into the San Francisco Bay annually by the Sacramento and San Joaquin rivers (Risebrough, *et al.*, 1968) and 10 metric tons reach the Gulf of Mexico each year from the Mississippi. Although these may seem very large amounts, they are small relative to the pesticide usage in the vicinity of these rivers. These residues represent no more than 0.1% of the amounts of pesticides used in the area supplying water to the estuaries, but some may be rapidly taken out to sea because there is a strong tidal exchange in San Francisco Bay (Frost, 1969). In a study of an estuary in Louisiana, only small amounts of endrin and dieldrin were reported (Rowe *et al.*, 1971), and these were less than in the same area in 1964-1966 (Hammerstrom *et al.*, 1967). Small DDT residues have been reported from a Washington estuary (Butler, 1969). There are two probable reasons for these low residues. Not only are pesticides strongly adsorbed on to soil fractions and do not reach rivers in large amounts, but once they reach the water, they become adsorbed onto particulate matter in suspension which eventually sinks to the bottom, and the residues may remain inactivated until stirred up by strong currents, the thermocline or taken up into the tissues of mud-living organisms. For instance, in a study of dieldrin and endrin residues in a Louisiana estuary less than 1 ppm×10^3 was found in the water, with about four times as much in the bottom mud (Rowe *et al.*, 1971). In pesticide surveys such as that in the Mississippi River Delta (U.S.D.A., 1966), much larger quantities of organochlorines were reported from the mud and sediment than from the water. There is some evidence that pesticide residues are greater in estuaries that receive run-off from large agricultural and urban areas, than in those that are geographically isolated from extensive agricultural lands (Modin, 1969).

There is another source of pesticide residues in estuaries. Many U.S. estuarine salt marshes have been treated with large quantities of DDT annually to control mosquitoes, etc. (Butler, 1969); as much as 13 lb/acre (14.6 kg/ha) has been reported from one marsh near New York (Woodwell *et al.*, 1967).

It is important to remember that there are distinct seasonal changes in insecticide residue input to estuaries, with peak amounts in spring and to a lesser extent in autumn (Butler, 1969). There has been so much anxiety about estuarine contamination by pesticides in the U.S.A. that a monitoring programme has been set which analyses residues at 30-day intervals from about 175 stations along the Pacific Gulf and Atlantic Coasts.

To summarize, it seems that the extent of estuarine pesticide pollution is less than might be expected, but as stated by Butler (1966), 'Despite two decades of research, the extent and importance of pesticide pollution in estuaries is poorly understood.'

11.2.1.3   *In sea-water*
We do not have much information on the amounts of pesticides in sea-water. Since the amounts of residues in estuaries are relatively small, it seems unlikely that these could be a major source of residues in sea-water. It seems much more likely that sea-water becomes contaminated with pesticides by atmospheric fall-out of rain or dust or even by surface interchange (see Chapter 10). Goldberg *et al.* (1971) agreed with this conclusion, and calculated that as much as one-quarter of the total annual production of DDT was passed from the atmosphere into the oceans of the world. Once in sea-water, most of the pesticide residues probably become bound onto organic or inorganic particles or held in the biota. It is for this reason that most data we have on marine contamination by pesticides has been based on residue analyses made on fish, plankton and aquatic invertebrates. Amounts in sea-water have either been so small that either they have not been quoted, or analyses have not been made.

Probably, much of the residues held in the inanimate particulate matter eventually falls into the abyss below the surface water after circulating for some time in the top 75-100 metres of water. Unfortunately, we have no data as to how much sedimentation of this kind occurs although studies of carbon dioxide movement indicate a mean time of about 4 years before it is completely transferred to the abyss (Woodwell *et al.*, 1971). The fate of residues within the abyss is impossible to forecast due to the vast size of this reservoir.

The residues held within the biota may be transported considerable distances, but the amounts moved in this way must be only a minute fraction of the total oceanic contamination. Some residues are concentrated into the wide variety of non-living organic matter in the surface waters (Garrett, 1967; Williams, 1967). For instance, surface slicks of organic matter have been shown to be effective concentrators of organochlorine insecticides; when these are moved by wind the pesticides may be important agents in the transport of pesticides in the marine environment (Seba and Corcoran, 1969). Possible mechanisms for the uptake and distribution of persistent pesticides in water are discussed more fully in Chapter 4.

11.2.2   Sources of residue in water
Many of the sources of organochlorine residues in water have been discussed in relation to soil (see Section 11.1.2). These include accidental fall-out of spray from agricultural treatments, fall-out from the atmosphere and surface run-off from agricultural land. The movements of pesticides from air to water are also discussed in Chapters 4, 10 and 12 in some detail.

11.2.2.1   *Run-off from agricultural land*
In a careful study of dieldrin movements from soil to water in two watersheds, no more than 0.07% of the soil dose appeared in run-off water (Caro and Taylor, 1971). Another study monitored the amounts of organochlorine insecticides in a creek and a controlled drainage system draining 280 square miles and leading to L. Erie. The largest amount of DDT was 68 ppm$\times 10^6$ in water and 441 ppm$\times 10^3$ in mud. There was a correlation between rainfall and concentration of insecticides in the creek water (Miles and Harris, 1971). There have been several other studies, such as that by Edwards *et al.* (1971) who showed that only minute amounts of dieldrin moved from treated agricultural land. Close to ponds, the small amounts that did reach the ponds, got there in surface run-off. There now seems enough evidence to be certain that only a small proportion of the pesticide residues found in freshwater gets there via drainage or run-off from agricultural land. Nevertheless, amounts from this source may occasionally be large and many workers have reported seasonal increases in residues that coincide with peak periods of agricultural usage. Irrigation or flooding of agricultural land may be a source of contamination of water. For instance, water from rice paddies contained aldrin, and water from a cotton field contained endrin (Sparr *et al.*, 1966).

11.2.2.2   *From spraying*

Contamination of water by persistent pesticides from aerial spraying is important, and may be due to accidental spray fall-out from large-scale aerial spraying of forest areas to control pests such as spruce budworm (Dimond *et al.*, 1968; Yule and Tomlin, 1971), or from aerial spraying of large-scale crops such as cotton (Sparr *et al.*, 1966). Alternatively, pollution may be due to direct spraying of water; probably the greatest single source of water pollution is the tens of thousands of tons of DDT that are applied to water annually to control mosquitoes (Westlake and Gunther, 1966).

Such applications are seasonal, and the water is contaminated for only short periods. For instance, when water was sampled from a forest stream over a period of two years the DDT concentration exceeded 0.5 ppm$\times 10^3$ for only a few hours after treatment of the surrounding forest (Yule and Tomlin, 1971). Nevertheless, residues persist in the bottom mud for a long time. For instance, Dimond *et al.* (1971) showed that DDT persisted in small streams for 10 years after a single application.

Sprays falling onto water are responsible for maintaining a low level of pesticide pollution in many lakes, rivers and streams, but usually the amounts present, are too small to be toxic to many aquatic organisms. For instance, even when comparatively large residues of BHC, dieldrin and DDT were reported from rivers in Yorkshire and Lancashire in England, it was calculated that at only 2% of stations sampled were there enough total insecticide residues to be lethal to fish (Lowden *et al.*, 1969). Woodwell *et al.* (1971) calculated, from the annual precipitation in the U.S.A. and the estimated run-off, that only about 0.1% of the DDT produced annually reached natural waters via run-off from agricultural land. Although the continual replenishment of natural waters with persistent pesticides is a hazard that cannot be ignored, it does not seem to be of major importance as a source of pollution.

11.2.2.3   *Industrial effluent*

Many industries use insecticides in their processes, and the effluents from such factories may contain large amounts of persistent and other pesticides. For instance, DDT and dieldrin are used in large quantities for moth-proofing by woollen and carpet manufacturers and are also used by dry cleaners. Many industrial plants treat their waste before it passes out into water but these processes do not always remove pesticides completely. For example, Wilroy (1963) reported 12-65 mg/l of dieldrin in effluent from a wood factory in the U.S.A. and Holden (1966) found that an effluent from a wood mill contained 0.005 mg/l of dieldrin, several months after the insecticide had last been used. The large residues in Yorkshire and Lancashire rivers (Lowden *et al.*, 1969) have been attributed to industrial effluents of this kind.

The waste from pesticide factories is another source of contaminated effluent. Some instances where there have been serious contaminations from this source have been summarized by Nicholson (1967) and a more recent incident was the large quantities of endosulfan reported from the River Rhine (Greve and Wit, 1971). There seems little doubt that such instances are more likely to cause large-scale mortality of fish than the comparatively small but widespread amounts of residues originating from agricultural applications of pesticides.

11.2.2.4 *Sewage*

Material heavily contaminated with persistent pesticides often reaches sewage-treatment plants particularly from industrial sources, but also from domestic sewage effluent which may contain as much as 300 ppm$\times 10^6$ of dieldrin and 100 ppm$\times 10^6$ of DDT (Holden and Marsden, 1966). This would require a dilution factor of $\times 20$ before it could be discharged into rivers without hazard to fish. Fortunately, this is well within the capabilities of sewage plants. Lowden *et al.* (1969) reported decreases in concentration during treatment of 100:1 for BHC, 1,000:1 for dieldrin and 2,000:1 for DDT. Sewage effluents containing discharge from industries usually contain more pesticides than those with only domestic sewage. Residues of 1,000-1,000 ppm$\times 10^6$ of dieldrin and up to 500 ppm$\times 10^6$ of DDT have been reported from English sewage works. In another survey, pesticide residues of up to 390 ppm of lindane, up to 1,900 ppm of dieldrin and up to 800 ppm of DDT have been reported. The problem of removing residues from water is discussed more fully later in this chapter.

11.2.2.5 *Sheep dips*

There have been many reports of streams receiving the drainage from inefficient soakaway pits at sheep-dipping sites (Holden, 1966). In one such instance, there was 1,600 ppm$\times 10^6$ of BHC and 1,000 ppm$\times 10^6$ of dieldrin in a stream for a day after the sheep dip had been discharged. Fortunately, less persistent pesticides are now more commonly used as sheep dips.

11.2.2.6 *From dust and rain*

This seems to be the major source of contamination of freshwater and an important source of pesticides in larger areas of freshwater. For instance, Risebrough *et al.* (1968) estimated that two-thirds of a ton of pesticides fell out with dust into the Atlantic Ocean between the Equator and 30°N latitude, annually (see also sections 11.1.2.3).

There have been numerous other reports of pesticides in rainwater (Wheatley and Hardman, 1965; Weibel *et al.,* 1966; Tarrant and Tatton, 1968) and dust (Antomommaria *et al.,* 1965; Cohen and Pinkerton, 1966; Tabor, 1966; and Södergren *et al.,* 1972) and the amounts reported are often large enough to account for considerable contamination of water. This route of pesticide input to water is discussed more fully in sections 11.10 and 11.12.

11.2.3 Factors influencing persistence of insecticides in water

Even the more persistent insecticides do not remain long in water unless they are carried in suspension adsorbed on to particulate matter. Within hours or days of contamination, the amounts in water fall to low background levels (see Figs. 11.1 a and b). There are several factors that influence the persistence of pesticides in water.

11.2.3.1 *Solubility*

The organochlorine insecticides are all relatively insoluble, but they differ considerably in their water solubilities, which are very much temperature-dependent. For instance, over a temperature range 60-95°F, the solubility of lindane increased as much as five times, three to five times for aldrin, dieldrin and endrin and two times for p,p'-DDT.

TABLE 11.6. Water solubilities of organochlorine insecticides.

Insecticide	Water solubility at 20-30°C (ppm)
DDT	0.0002
Aldrin	0.027
Heptachlor	0.06
Chlordane	0.1
Dieldrin	0.186
Endrin	0.1
Camphechlor (toxaphene)	3
Lindane	10

Even the more soluble insecticides disappear from water quite rapidly and most of the evidence is that the pesticides are gradually taken up by the bottom mud and organic matter.

11.2.3.2 *Mud bottom*

When a persistent insecticide reaches water, a large proportion disappears rapidly, with little remaining longer than one week. In a laboratory experiment, when water containing DDT was kept above soil covered with filter paper, after six hours, 56% of the insecticide was in the soil, and after 24 hours a further 22% had moved into the soil (Weidhass *et al.,* 1960). In a study of the distribution of toxaphene in a lake in New Mexico, after several days there was a concentration of 0.01-0.28 ppm in the water, 0.04-0.13 in the bottom sediment, 0.4-18.3 in aquatic plants and 2.5-15.2 in fish (Kallman *et al.,* 1962). In a similar study of the distribution of DDT in a pond, it was found that it was quickly concentrated into the bottom mud and vegetation (Bridges *et al.,* 1963). In a study of the distribution of pesticides in a lake, Keith and Hunt (1966) found that very little DDT, BHC or camphechlor were in the water, but very much more was in the suspended particulate matter and bottom sediment. Edwards *et al.* (1971) and Beynon *et al.* (1971) reported that insecticides virtually disappeared from water within 24 hours. Several other workers have reported much larger residues in mud than in the water above it (Simmons, 1945; Dimond *et al.*, 1971; Miles and Harris, 1971; Rowe *et al.*, 1971). There seems little doubt that just as some soils can adsorb more pesticides than others, so some types of mud bottom can bind more pesticide. Bailey and Hannum (1967) showed that pesticide concentrations in water are related to the particle size of the sediments, the largest concentrations appearing in the sediments composed of smaller particles, and Lotse *et al.* (1968) supported this conclusion. We still need much more investigation into the binding of pesticides in different types of bottom mud.

11.2.3.3   *Organic matter*
It seems well established that pesticides show an affinity for both living and dead organic matter particularly the lipoid portion of such material. Bailey and Hannum (1967) and Lotse *et al.* (1968) both confirmed this conclusion. If the organic matter is floating, the pesticide tends to remain in suspension in the water, whereas if it is in the bottom mud it tends to remove the pesticide from the water above it. The relative importance of organic matter in adsorbing and inactivating pesticides in water and also in transporting them still remains to be fully defined.

11.2.3.4   *Temperature and pH*
The evidence that these factors influence the persistence of pesticides in natural waters is purely inferential. It is well-established that temperature affects both solubility and volatility of pesticides, and there is good evidence that they are much more stable at some hydrogen ion concentrations than at others. There is need for much more field data on how temperature and pH affect pollution of water by pesticides; both have been shown to be important in soils but parallel investigations in water are still lacking.

11.2.4   Hazards caused by residues in water
It is not proposed to discuss the effects of pesticides in water or aquatic invertebrates, fish and mammals, because these are discussed at length in Chapters 4, 5, 6 and 7. The only topics discussed here will be those not dealt with elsewhere in this book.

11.2.4.1   *Human drinking water*
Man is exposed to uptake of pesticides from the air he breathes, the food he eats and the water he drinks. River water and lakes all contain appreciable amounts of persistent organochlorine insecticides; it is important that these do not reach the sources of human drinking water. In one of the earlier studies in U.S.A. (Nicholson, 1959), it was concluded that the amount of DDT in drinking water was only a small proportion of

the total exposure to DDT encountered by most of the population. In an intensive study of pesticides in drinking water from the Mississippi and Missouri Rivers (Schafer *et al.*, 1969), 500 grab samples of finished drinking water and raw water were assayed. Only endrin (23 samples), chlordane (5 samples) and dieldrin (1 sample) were found in concentrations that exceeded the suggested maximum permissible amount. In a large-scale survey of pesticides in surface waters of the United States, 1964-1968 (Lichtenberg *et al.*, 1970), there were no samples that exceeded the maximum permissible limits for human intake.

Nicholson (1969) quoted the suggested permissible limits for amounts of pesticides in public water supplies that had been decided by the National Technical Advisory Committee on Water Quality and submitted as recommendations to the U.S. Secretary of the Interior (Table 11.7). Although these have not been officially adopted they have been used as guide-lines.

TABLE 11.7. Surface water criteria for pesticides in public water supplies (mg/l).

Pesticide	Permissible criteria	Desirable criteria
Aldrin	0.017	Absent
Chlordane	0.003	,,
DDT	0.042	,,
Dieldrin	0.017	,,
Endrin	0.001	,,
Heptachlor	0.018	,,
Heptachlor epoxide	0.018	,,
Lindane	0.056	,,
Methoxychlor	0.035	,,
Camphechlor	0.005	,,

Adapted from Water Quality Criteria, Report of the National Technical Advisory Committee to the Secretary of the Interior, April 1968, by Nicholson, 1969.

11.2.4.2 *Recycling into the atmosphere*

The sea constitutes a vast reservoir for chemicals, and since there is continual evaporation of water to the air, pesticides may be carried back into the atmosphere from the oceans (Goldberg *et al.*, 1971). The movement of micro-organisms from the oceans into the atmosphere has been demonstrated (Blanchard and Syzdek, 1970) and there seems no reason that pesticides should not be discharged back into the atmosphere in a similar manner. With the vast surface area of the oceans, this could well represent a continual source of replenishment for atmospheric pollution (see Chapter 12). Much more work is needed to establish the importance of this interchange between air and water.

11.2.5 Minimizing and removing residues from water
It is possible to greatly decrease the hazards due to contamination of water by persistent pesticides. This can be achieved either by using smaller quantities of pesticides or by using alternative control measures, and currently, both these trends are being followed.

11.2.5.1 *Use of alternative control measures*
If less persistent pesticides can be used, then there should be much less long-term contamination of water and mud, because these chemicals break down rapidly in water. Nevertheless, some are even more toxic to aquatic organisms than organochlorine insecticides, and the short-term hazards may be greater. For instance, large numbers of fish have been reported to have been killed by less persistent pesticides, such as zinophos in England (Muirhead-Thompson, 1971), parathion in U.S.A. (Nicholson, 1969), and endosulfan in Germany (Greve and Wit, 1971). Hence, such alternative chemicals may not be the full answer to the problem.

Even if persistent pesticides continue to be used, the way they are employed can greatly diminish their ecological impact. Such techniques include precise placement in soil and the use of seed dressings.

Alternative methods of pest control, include biological control, using parasites, predators or pathogens, development of resistant crops or animals, release of sterile males of the pest, the use of attractants or repellents, the use of biological pesticides, and various physical and cultural techniques.

11.2.5.2 *Water treatment*
Most industrial installations that produce waste containing considerable amounts of pesticides have facilities for decreasing the amounts of residues before the effluent is discharged into water. Such methods include treatment by chemical and biological processes, the use of settling basins or incinerators. Nevertheless, large amounts of pesticides do sometimes reach water, and it is essential that these are removed before the water is reprocessed for drinking or other use. Fortunately, the normal treatment of water seems to remove a large proportion of the pesticide residues, but its effectiveness varies considerably with the pesticide and concentration. It seems to be much more difficult to remove continuous small amounts of pesticides than the large amounts that occur occasionally. In one study, all of the conventional water treatment processes, involving coagulation sedimentation and filtration removed from 80 to 90% of the DDT in water containing 0.1–10.0 ppm (Carollo, 1945). However, Robeck *et al.* (1965) tested the effectiveness of various filtering methods in removal of dieldrin, endrin, lindane and DDT from water and found that conventional treatment using coagulation techniques followed by sand filtration varied in

effectiveness. Almost all of the DDT (10 ppm×10^3) was removed by this treatment, but less than 20% of the lindane and parathion. The use of lime and soda ash as a softening agent with an iron salt as coagulant, did not improve on the results obtained with alum coagulation alone, in fact the results were poorer. Treatment with chlorine or potassium permanganate did not decrease the insecticide residues either, but large and impractical concentrations of ozone did decrease amounts of organochlorine insecticides, although the by-products were not identified.

In an experimental study of oxidation of pesticide residues in water, lindane concentrations were readily decreased by ozonation and partially affected by potassium permanganate, but chlorination, peroxides or aeration had no measurable effects. Aldrin could be attacked by chlorination, potassium permanganate, ozone or aeration but not by peroxides, whereas dieldrin was only affected by ozonation and aeration (Buescher *et al.*, 1964). As much as 82-85% of endosulfan residues in water could be removed by filtration or centrifugation (Greve and Wit, 1971).

A number of workers have studied the use of activated carbon in the filtration system as a means of removing pesticides from water (Kawahara and Breidenbach, 1966). Nicholson *et al.* (1966) were able to remove DDT from water by carbon adsorption, but they were unsuccessful with low concentrations of BHC and toxaphene. However, Robeck *et al.* (1965) found that large amounts of carbon would remove lindane, but it required 29 ppm of activated charcoal to decrease a concentration of lindane from 10 ppm×10^6 to 1 ppm×10^6, and Cohen *et al.* (1960) obtained some diminution of toxaphene residues with between 1 and 9 ppm of activated carbon.

Thus, with the amounts of persistent pesticides currently occurring in water there seems to be little cause for anxiety, but should they increase, action would be necessary to ensure their more efficient removal.

REFERENCES

D. C. Abbott, R. B. Harrison, J. O'G. Tatton and J. Thompson, *Nature,* (Lond.), **208**, 1317 (1965)

D. C. Abbott, R. B. Harrison, J. O'G. Tatton and J. Thomson, *Nature,* (Lond.), **211**, 259 (1966)

M. A. Abdellatif, H. P. Hermanson and H. T. Reynolds, *J. econ. Ent.,* **60**, 1445 (1967)

F. Acree, M. Beroza and M. C. Bowman, *J. agric. Fd Chem.,* **11**, 278 (1963)

N. B. Akesson and W. E. Yates, *A. Rev. Ent.,* **9**, 285 (1964)

M. Alexander, Soil Biology, p. 209. U.N.E.S.C.O., Paris (1969)

N. Allen, R. L. Walker and L. C. Fife, *Tech. Bull. U.S.D.A.,* **1090**, 19 (1954)

P. Antommaria, M. Corn and L. Demaio, *Science, N.Y.,* **150**, 1476 (1965)

T. E. Bailey and J. R. Hannum, *J. Sanit. Eng. Div. Am. Soc. Civil Engrs.,* **93**, 27 (1967)

G. W. Bailey and J. L. White, *J. agric. Fd. Chem.*, **12**, 324 (1964)

R. J. Barker, *J. Wildl. Mgmt.*, **22**, 269 (1958)

F. Barlow and A. B. Hadaway, *Bull. ent. Res.*, **46**, 547 (1955)

G. B. Beestman, D. R. Keeney and G. Chesters, *Agron. J.*, **61**, 247 (1969)

A. Bevenue, J. W. Hylin and T. W. Kelley, *Hawaii Farm Sci.*, **20**, 1 (1971)

K. I. Beynon, M. J. Edwards, A. P. Thompson and C. A. Edwards, *Pestic. Sci.*, **2**, 5 (1971)

R. F. Bishop and D. Chisholm, *Canad. J. Soil Sci.*, **42**, 77 (1962)

D. C. Blanchard and L. Syzdek, *Science N. Y.*, **170**, 626 (1970)

W. B. Bollen, *Ann. Rev. Microbiol.*, **15**, 69 (1961)

W. B. Bollen, H. E. Morrison and H. H. Cravell, *J. econ. Ent.*, **47**, 302 (1954)

W. B. Bollen, J. E. Roberts and H. E. Morrison, *J. econ. Ent.*, **51**, 214 (1958)

M. C. Bowman, M. S. Schechter and R. L. Carter, *J. agric. Fd. Chem.*, **13**, 360 (1965)

W. R. Bridges, B. J. Kallman and A. K. Andrews, *Am. Fish. Soc. Trans.*, **92**, 421 (1963)

E. Brown and Y. A. Nishioka, *Pestic. Monit. J.*, **1**, 38 (1967)

C. A. Buescher, J. H. Dougherty and R. T. Skrinde, *Proc. 36th Water Pollut. Contr. Fed. Conf.*, 1963, 27 pp. (1964)

R. H. Burrage and J. G. Saha, *Can. J. Pl. Sci.*, **47**, 114 (1967)

C. F. Van De Bund, *Boll. Zool. Agr. Bachic.*, **7**, 185 (1965)

J. W. Butcher, R. M. Snider and J. L. Aucamp, *Colloquium Pedobiologiae*, p. 207, I.N.R.A., Paris (1972)

P. A. Butler, *J. appl. Ecol.*, **3** (Suppl.), 253 (1966)

P. A. Butler, *BioScience*, **19**, 889 (1969)

J. H. Caro and A. W. Taylor, *J. agric. Fd. Chem.*, **19**, 379 (1971)

J. A. Carolla, *J. Am. Wat. Wks. Ass.*, **37**, 1310 (1945)

C. I. Chacko, J. L. Lockwood and M. Zabik, *Science, N. Y.*, **154**, 893 (1966)

D. F. Champion and S. R. Olsen, *Soil Sci. Soc. Am. Proc.*, **35**, 887 (1971)

R. P. Chawla and S. L. Chopra, *J. Res. Punjab. agric. Univ.*, **4**, 96 (1967)

R. D. Chisholm and L. Koblitsky, *Trans. 24th N. Am. Wildl. Conf.*, 118 (1959)

M. M. Cliath and W. F. Spencer, *Soil Sci. Soc. Am. Proc.*, **35** (5), 791 (1971)

J. M. Cohen, L. J. Kamphake, A. E. Lempke, C. Henderson and R. L. Woodward, *J. Am. Wat. Wks. Ass.*, **52**, 1551 (1960)

J. M. Cohen, C. Pinkerton. R. F. Gould (Editor), Organic pesticides in the Environment, p. 309. Advances in Chemistry, **60**, American Chemical Society, Washington, D.C. (1966)

O. B. Cope and W. R. Bridges, *U.S. Fish Wildl. Circ.*, **167**, 32 (1963)

H. C. Cox and J. H. Lilley, *J. econ. Ent.*, **45**, 421 (1952)

S. Cramp and P. J. Conder, *R. Soc. Prot. Birds Rept.*, 20 (1965)

B. T. Croll, *Water Treat. Exam.*, **18**, 255 (1969)

B. N. K. Davis, *J. appl. Ecol., (Suppl.)*, **3**, 133 (1966)

B. N. K. Davis, *Ann. appl. Biol.*, **61**, 29 (1968)

B. N. K. Davis and R. B. Harrison, *Nature* (Lond.), **211**, 1424 (1966)

G. C. Decker, W. N. Bruce and J. H. Bigger, *J. econ. Ent.*, **59**, 266 (1965)

J. B. Dimond, G. Y. Belyea, R. E. Kaduce, S. A. Getchell and J. A. Blease, *Can. J. Ent.*, **102**, 1122 (1970)

J. B. Dimond, A. S. Getchell and J. A. Blease, *J. Fish. Res. Bd. Can.*, **28**, 1877 (1971)

J. B. Dimond, R. E. Kadunce, A. S. Getchell and J. A. Blease, *Bull. Env. Contam. Toxicol.*, **3**, 194 (1968)

W. G. Downs, *Science, N.Y.*, **114**, 259 (1951)

E. H. Dustman and L. F. Stickel, Pesticides and Their Effects on Soils and Water, *Am. Soc. Agron. Spec. Publ.*, **8**, 109 (1966)

C. A. Edwards, *New Scientist*, **19**, 282 (1963)

C. A. Edwards, *5th Symp. Brit. Ecol. Soc.* (Blackwell, Oxford) 239 (1965)

C. A. Edwards, *Res. Rev.*, **13**, 83 (1966)

C. A. Edwards, *Scient. Am.*, **220**, 88 (1969)

C. A. Edwards, Critical Reviews in Environmental Control, **1**, Chemical Rubber Co., Cleveland Ohio, U.S.A. (1970)

C. A. Edwards, S. D. Beck and E. P. Lichtenstein, *J. econ. Ent.*, **50**, 622 (1957)

C. A. Edwards and E. B. Dennis, *Nature* (Lond.), **188**, 4752 (1960)

C. A. Edwards and E. B. Dennis, *Pl. Path.*, **10**, 54 (1961)

C. A. Edwards and E. B. Dennis, *Pl. Path.*, **12**, 27 (1963)

C. A. Edwards and E. B. Dennis, *Pl. Path.*, **13**, 173 (1964)

C. A. Edwards, E. B. Dennis and D. W. Empson, *Ann. appl. Biol.*, **59**, 11 (1967)

C. A. Edwards and A. R. Thompson, *Res. Rev.*, **45** (1973)

C. A. Edwards, A. R. Thompson, K. I. Beynon and M. J. Edwards, *Pestic. Sci.*, **1**, 169 (1971)

C. F. Eno, *J. agric. Fd. Chem.*, **6**, 348 (1958)

C. F. Eno and P. H. Everett, *Soil Sci. Soc. Am. Proc.*, **22**, 235 (1958)

J. E. Fahey, J. W. Butcher and R. T. Murphy, *J. econ. Ent.*, **58**, 1026 (1955)

W. J. Farmer and C. R. Jensen, *Proc. Soil Soc. Am.*, **34**, 28 (1970)

H. R. Feltz, W. T. Sayers and H. P. Nicholson, *Pestic. Monit. J.*, **5**, 54 (1971)

W. E. Fleming and W. W. Maines, *J. econ. Ent.*, **46**, 445 (1953)

D. W. Fletcher and W. B. Bollen, *Appl. Microbiol.*, **2**, 349 (1954)

A. C. Foster, *U.S. Dept. Agric. Circ.*, **862**, 41 (1951)

J. Frost, *Environment*, **11**, 14 (1969)

F. M. Fowkes, H. A. Benesi, L. B. Ryland, W. M. Sawyer and K. D. Detling, *J. agric. Fd. Chem.*, **8**, 204 (1960)

P. J. Gallagher and L. Evans, *N.Z. Jl. Agric. Res.*, **4**, 466 (1961)

W. A. Garrett, *Deep-Sea Res.*, **14**, 221 (1967)

L. W. Getzin, *J. econ. Ent.*, **61**, 1560 (1968)

L. W. Getzin and I. Rosefield, *J. agric. Fd. Chem.*, **16**, 598 (1968)

C. D. Gish, *Pestic. Monit. J.*, **3**, 241 (1970)

E. D. Goldberg, P. Butler, P. Meier, D. Menzel, G. Paulik, R. Risebrough and L. F. Stickel, National Academy of Sciences, Washington, D.C., 42 pp. (1971)

I. J. Graham-Bryce, *Proc. 1st. Int. Congr. Pollut. Engnr. Sci. Sol.*, Tel Aviv. Plenum Press, London and New York (1972)

R. S. Green, C. G. Gunnerson, J. J. Lichtenberg and N. C. Brady (Editor), Agriculture and the Quality of our Environment, 137. A.A.A.S. Publication 85 (1967)

P. A. Greve and S. L. Wit, *J. Water Pollut. Contr. Fed.*, **43**, 2338 (1971)

D. C. Griffiths, *Rep. Rothamsted Exp. Sta. for 1965*, 170 (1966)
W. D. Guenzi and W. E. Beard, *Proc. Soil. Sci. Soc. Am.*, **31**, 644 (1967)
W. D. Guenzi and W. E. Beard, *Soil Sci. Soc. Am. Proc.*, **34**, 443 (1970)
A. B. Hadaway and F. Barlow, *Bull. ent. Res.*, **40**, 323 (1949)
A. B. Hadaway and F. Barlow, *Nature* (Lond.), **167**, 854 (1951)
A. B. Hadaway and F. Barlow, *Bull. ent. Res.*, **43**, 281 (1952)
A. B. Hadaway and F. Barlow, *Ann. Trop. Med. Parasit.*, **51**, 187 (1957)
A. B. Hadaway and F. Barlow, *Bull. ent. Res.*, **54**, 329 (1963)
A. B. Hadaway and F. Barlow, *Bull. Wld. Hlth. Org.*, **30**, 146 (1964)
R. J. Hammerstrom *et al.*, Public Health Serv. Publ. 999–UIH–2 26 pp. (1967)
C. R. Harris, *J. econ. Ent.*, **57**, 946 (1964)
C. R. Harris, *J. econ. Ent.*, **59**, 1221 (1966)
C. R. Harris, *J. econ. Ent.*, **63**, 782 (1970)
C. R. Harris and E. P. Lichtenstein, *J. econ. Ent.*, **54**, 1038 (1961)
C. R. Harris and J. H. Mazurek, *J. econ. Ent.*, **59**, 1215 (1966)
C. R. Harris and W. W. Sans, *Pestic. Monit. J.*, **5**, 259 (1971)
H. P. Hermanson and C. Forbes, *Soil Sci. Soc. Amer. Proc.*, **30**, 748 (1966)
F. Herzel, *Bundesgesundheitsblatt*, **13**, 49 (1970)
C. H. Hoffman and E. W. Surber, *Trans. Am. Fish Soc.*, **75**, 48 (1948)
A. V. Holden, *J. appl. Ecol. (Suppl.)*, **3**, 45 (1966)
A. V. Holden and K. Marsden, *J. Proc. Inst. Serv. Purif.*, **4**, 295 (1966)
E. W. Huddleston, D. Ashdown and D. C. Herzog, *Texas Tech. Coll. Ent. Dept.*, 13 (1966)
L. B. Hunt, *U.S.D.I. Wildl. Serv. Circ.*, **226**, 12 (1965)
H. Hurtig and C. R. Harris, Conference on Pollution and Our Environment, Montreal, 16 (1966)
Z. Jegier, *Ann. N.Y. Acad. Sci.*, **160**, 143 (1969)
L. G. Johnson and R. L. Morris, *Pestic. Monit. J.*, **4**, 216 (1971)
B. J. Kallman, O. B. Cope and R. J. Navare, *Trans. Am. Fish Soc.*, **91**, 14 (1962)
F. K. Kawahara and A. W. Breidenbach, *Soil Sci. Soc. Am. Spec. Publ.*, **8**, 122 (1966)
J. O. Keith and E. G. Hunt, *Trans. 31st N. Am. Wildl. Nat. Res. Conf.*, 150 (1966)
P. H. King and P. L. McCarty, *Soil Sci.*, **106**, 248 (1968)
V. M. Kirk, *Cornell Univ. Agric. Exp. Sta. Mem.*, **312**, 48 (1952)
W. H. Ko and J. L. Lockwood, *Can. J. Microbiol.*, **14**, 1075 (1968)
M. C. Kolipinski, A. L. Higer and M. L. Yates, *Pestic. Monit. J.*, **5**, 281 (1971)
J. G. Konrad, D. E. Armstrong and G. Chesters, *Agron. J.*, **59**, 591 (1967)
J. G. Konrad and G. Chesters, *J. agric. Fd. Chem.*, **17**, 226 (1969)
J. G. Konrad, G. Chesters and D. E. Armstrong, *Soil Sci. Soc. Am. Proc.*, **33**, 259 (1969)
L. J. Korschgen, *J. Wildl. Mgmt.*, **34**, 186 (1970)
F. Korte, G. Ludwig and J. Vogel, *Liebigs Ann. Chem.*, **656**, 135 (1962)
J. B. Kring and J. F. Ahrens, *Am. Potato J.*, **45**, 405 (1968)
R. C. Linder, *Proc. Am. Soc. Hort. Sci.*, **42**, 275 (1943)
J. J. Lichtenberg, J. W. Eichelberger, R. C. Dressman and J. E. Longbottom, *Pestic. Monit. J.*, **4**, 71 (1970)

E. P. Lichtenstein, *J. econ. Ent.*, **51**, 380 (1958)

E. P. Lichtenstein, *J. econ. Ent.*, **57**, 133 (1964)

E. P. Lichtenstein, T. W. Fuhremann and K. R. Schulz, *J. agric. Fd. Chem.*, **16**, 348 (1968)

E. P. Lichtenstein, T. W. Fuhremann and K. R. Schulz, *J. agric. Fd. Chem.*, **19**, 718 (1971)

E. P. Lichtenstein, C. H. Mueller, G. R. Mydral and K. R. Schulz, *J. econ. Ent.*, **55**, 215 (1962)

E. P. Lichtenstein, G. R. Mydral and K. R. Schulz, *J. econ. Ent.*, **57**, 133 (1964)

E. P. Lichtenstein and J. B. Polivka, *J. econ. Ent.*, **52**, 289 (1959)

E. P. Lichtenstein and K. R. Schulz, *J. econ. Ent.*, **52**, 124 (1959)

E. P. Lichtenstein and K. R. Schulz, *J. econ. Ent.*, **53**, 192 (1960)

E. P. Lichtenstein and K. R. Schulz, *J. econ. Ent.*, **54**, 517 (1961)

E. P. Lichtenstein and K. R. Schulz, *J. agric. Fd. Chem.*, **13**, 57 (1965)

E. P. Lichtenstein, K. R. Schulz, T. W. Fuhremann and T. T. Liang, *J. agric. Fd. Chem.*, **18**, 100 (1970)

E. P. Lichtenstein, K. R. Schulz, R. F. Skrentney and P. A. Stitt, *J. econ. Ent.*, **58**, 742 (1965)

C. P. Lloyd-Jones, *Nature* (Lond.), **229**, 65 (1971)

E. G. Lotse, D. A. Graetz, G. Chesters, G. B. Lee and L. W. Newland, *Environ. Sci. Technol.*, **2**, 353 (1968)

G. F. Lowden, C. L. Saunders and R. W. Edwards, *Water. Treat. Exam.*, **18**, 275 (1969)

D. B. Manigold and J. A. Schulze, *Pestic. Monit. J.*, **3**, 124 (1969)

J. P. Martin, R. B. Harding, G. H. Carmell and L. D. Anderson, *Soil Sci.*, **87**, 334 (1959)

F. Matsumura and G. M. Boush, *Science, N.Y.*, **156**, 959 (1967)

F. Matsumura, G. M. Boush and A. Tai, *Nature* (Lond.), **219**, 965 (1968)

J. J. Menn, J. B. McBain, J. B. Adelson and G. G. Patchett, *J. econ. Ent.*, **58**, 875 (1965)

J. J. Menn, G. G. Patchett and G. H. Batchelder, *J. econ. Ent.*, **53**, 1080 (1960)

R. L. Metcalf, I. P. Kapoor and A. S. Hirwe, *Chemtech.*, Feb., 105 (1972)

J. R. W. Miles, *J. agric. Fd. Chem.*, **16**, 620 (1968)

J. R. W. Miles and C. R. Harris, *Pestic. Monit. J.*, **5**, 289 (1971)

J. C. Modin, *Pestic. Monit. J.*, **3**, 1 (1969)

H. E. Morris and D. B. Swingle, *J. agric. Res.*, **34**, 59 (1927)

E. M. Mrak, Report Secretary's Commission on Pesticides and Their Relationship to Environmental Health, U.S. Dept. Hlth. Educ. Wlfr. (1969)

R. C. Muirhead-Thomson, Pesticides and Freshwater Fauna, 248 pp. Academic Press, London and New York (1971)

M. S. Mulla, *J. econ. Ent.*, **53**, 650 (1960)

R. G. Nash and E. A. Woolson, *Science, N.Y.*, **157**, 924 (1967)

R. G. Nash and E. A. Woolson, *Soil Sci. Soc. Am. Proc.*, **32**, 525 (1968)

L. D. Newsom, *A. Rev. Ent.*, **12**, 257 (1967)

H. P. Nicholson, *J. Am. Wat. Wks. Ass.*, **51**, 981 (1959)

H. P. Nicholson, *Science, N.Y.*, **158**, 871 (1967)

H. P. Nicholson, *Wash. Acad. Sci.*, **59**, 77 (1969)

H. P. Nicholson, A. R. Grzenda and J. J. Teasley, *Proc. Symp. Agric. Waste Water*, 132 (1966)

J. A. Onsager, H. W. Rusk and L. I. Butler, *J. econ. Ent.*, **63**, 1143 (1970)

B. L. Parker and J. E. Dewey, *J. econ. Ent.*, **59**, 106 (1965)

R. S. Patterson, *Diss. Abstr.*, **23**, 1471 (1962)

R. E. Pillmore and R. B. Finley, *Trans. 28th N. Am. Wildl. Conf.*, 409 (1963)

K. Porter and W. E. Beard, *J. agric. Fd. Chem.*, **16**, 344 (1968)

R. W. Risebrough, R. J. Huggett, J. J. Griffin and E. D. Goldberg, *Science N.Y.*, **159**, 1233 (1968)

R. W. Risebrough, P. Reich and H. S. Olcott, *Bull. Environ. Contam. Toxicol.*, **4**, 192 (1969)

G. G. Robeck, *J. Am. Wat. Wks. Ass.*, **57**, 181 (1965)

G. G. Robeck, K. A. Dostal, J. M. Cohen and J. F. Kreissel, *J. Am. Wat. Wks. Ass.*, **57**, 181 (1965)

R. J. Roberts, *J. econ. Ent.*, **56**, 781 (1963)

J. E. Roberts and W. B. Bollen, *Appl. Microbiol.*, **3**, 190 (1955)

D. R. Rowe, L. W. Canter, P. J. Snyder and J. W. Mason, *Pestic. Monit. J.*, **4**, 177 (1971)

J. G. Saha and W. W. A. Stewart, *Can. J. Sci.*, **47**, 79 (1967)

E. I. El Sayed, J. B. Graves and F. L. Bonner, *J. agric. Fd. Chem.*, **15**, 1014 (1967)

M. L. Schafer, J. T. Peeler, W. S. Gardner and J. E. Campbell, *Environ. Sci. Technol.*, **3**, 1261 (1969)

D. B. Seba and E. F. Corcoran, *Pestic. Monit. J.*, **3**, 190 (1969)

W. M. Shaw and B. Robinson, *Soil Sci.*, **90**, 320 (1960)

J. G. Sheals. D. K. McE. Kevan (Editor), Soil Zoology, p. 241. Butterworth, London (1955)

S. W. Simmons, *Publ. Hlth. Rep.*, **60**, 917 (1945)

A. Södergren, B. Svensson and S. Ulfstrand, *Environ. Pollut.*, **3**, 25 (1972)

B. I. Sparr, W. G. Appleby, D. M. DeVries, J. V. Osmun, J. M. Mc Bride, G. L. Foster and R. F. Gould (Editor), Organic pesticides in the Environment, Advances in Chemistry, Series **60**, 146. *American Chemical Society*, Washington D.C. (1966)

W. F. Spencer and M. M. Cliath, *Soil Sci. Soc. Am. Proc.*, **34**, 574 (1970)

L. J. Stevens, C. W. Collier and D. W. Woodham, *Pestic. Monit. J.*, **4**, 145 (1970)

L. F. Stickel, *U.S.D.I. Bur. Sport Fish Wildlife Rept.*, **119**, 1 (1968)

A. Stöckli, *Z. Pflernähr. Dung.*, **48**, 264 (1950)

M. W. Stone, *J. econ. Ent.*, **46**, 1100 (1953)

M. W. Stone, F. B. Foley and D. H. Bixby, U.S. Dept. Agric. Circ., 926 (1953)

A. H. Strickland, *Ann. appl. Biol.*, **55**, 319 (1965)

A. Stringer and J. A. Pickard, *Rept. Agric. Hort. Res. Sta., Univ. Bristol, 1962*, 127 (1963)

A. Stringer and J. A. Pickard, *Rept. Agric. Hort. Res. Sta., Univ. Bristol, 1964*, 172 (1965)

C. L. W. Swanson, F. C. Thorp and R. B. Friend, *Soil Sci.*, **78**, 379 (1954)

E. C. Tabor, *Trans. N.Y. Acad. Sci.*, **28**, 569 (1966)

K. R. Tarrant and J. O'G. Tatton, *Nature, Lond.*, **219**, 725 (1968)

E. F. Taschenberg, G. L. Mack and F. L. Gambrell, *J. agric. Fd. Chem.*, **9**, 207 (1961)

L. C. Terriere, U. Kiigemagi, R. A. Zwich and P. H. Westigard. R. F. Gould (Editor), Organic Pesticides in the Environment, 263. Advances in Chemistry Ser. **60**, *Am. Chem. Soc.* (1966)

A. R. Thompson, C. A. Edwards, M. J. Edwards and K. I. Beynon, *Pestic. Sci.*, **1**, 174 (1971)

G. M. Tu, J. R. W. Miles and C. R. Harris, *Life Sciences*, **7**, 311 (1968)

N. Turner, *Conn. Agric. Exp. Sta. Bull.*, **655**, 1 (1963)

U.S.D.A. Publ. ARS 81-13, 53 pp. (1966)

O. Verona, *Soils and Fertilizers*, **20**, 1435 (1957)

C. L. Vincent, *Wash. Agric. Exp. Sta.*, 437 (1944)

G. W. Ware, B. J. Estesen and W. P. Cahill, *Pestic. Monit. J.*, **2**, 129 (1968)

W. M. Weatherholz, G. W. Cornell, R. W. Young and R. E. Webb, *J. agric. Fd. Chem.*, **15**, 667 (1967)

W. Wegorek, *Soils and Fertilizers*, **20**, 1435 (1957)

S. R. Weibel, R. B. Weidner, J. M. Cohen and A. G. Christianson, *J. Am. Wat. Wks. Ass.*, **58**, 1075 (1966)

D. E. Weidhaas, C. H. Schmidt and M. C. Bowman, *J. econ. Ent.*, **53**, 121 (1960)

I. West, *Arch. Environ. Health*, **9**, 626 (1964)

W. E. Westlake, F. A. Gunther. R. F. Gould (Editor), Organic Pesticides in the Environment, 110. Advances in Chemistry, Ser. **60**, *Am. Chem. Soc.*, Washington, D.C. (1966)

G. A. Wheatley and J. A. Hardman, *Rep. Nat. Veg. Res. Sta. Wellesbourne, 1961*, 52 (1962)

G. A. Wheatley and J. A. Hardman, *Nature* (Lond.), **207**, 486 (1965)

G. A. Wheatley, J. A. Hardman and A. H. Strickland, *Pl. Path.*, **11**, 81 (1962)

G. A. Wheatley, D. W. Wright and J. A. Hardman, *Pl. Path.*, **9**, 146 (1960)

I. H. Wiese, *S. Afr. J. agric. Sci.*, **7**, 823 (1964)

I. H. Wiese and N. C. J. Basson, *S. Afr. J. agric. Sci.*, **9**, 945 (1966)

G. B. Wiersma, W. G. Mitchell and C. L. Stanford, *Pestic. Monit. J.*, **5**, 345 (1972)

G. B. Wiersma, P. F. Sand and R. L. Schutzmann, *Pestic. Monit. J.*, **5**, 223 (1971)

P. M. Williams, *Deep-Sea Res.*, **14**, 791 (1967)

R. D. Wilroy, *Proc. 18th Ind. Waste Conf. Purdue Univ. Eng. Ext. Ser.* No. 115, 413 (1963)

C. W. Wingo, *Res. Bull. Missouri Agric. Exp. Sta.*, **914**, 27 (1966)

G. M. Woodwell, P. P. Craig and H. A. Johnson, *Science N.Y.*, **174**, 1101 (1971)

G. M. Woodwell, C. F. Wurster Jr. and P. A. Isaacson, *Science N.Y.*, **156**, 821 (1967)

E. A. Woolson, J. H. Axley and P. C. Kearney, *Soil Sci. Soc. Am. Proc.*, **35**, 938 (1971)

B. Yaron, A. N. R. Swoboda and G. W. Thomas, *J. agric. Fd. Chem.*, **15**, 671 (1967)

W. R. Young and W. A. Rawlins, *J. econ. Ent.*, **51**, 11 (1958)

W. N. Yule, *Bull. Environ. Contam. Toxicol.*, **5**, 139 (1970)

W. N. Yule, M. Chiba and V. H. Morley, *J. agric. Fd. Chem.*, **15**, 1000 (1967)

W. N. Yule and A. D. Tomlin, *Bull. Environ. Contam. Toxicol.*, **5**, 479 (1971)

Chapter 12

Dynamics of
Pesticide Residues in the Environment

J. Robinson

Shell Research,
Woodstock Agricultural Research Center,
Sittingbourne, Kent

12.1   INTRODUCTION

Residues of pesticides or their degradation products have been reported in a wide variety of environmental samples. The concentrations of compounds of one particular group of pesticides, the organochlorine insecticides, have been extensively and intensively studied in recent years, and the results are summarised in other chapters in this book. The apparent ubiquity of the organochlorine compounds has rightly caused concern about any effects that may be occurring currently, or future effects that may arise from their continued use.

Pesticides are used to control the populations of certain organisms which, at a particular time and in a particular place, are considered to be inimicable to man's welfare. Excluding uses on crops grown indoors, or on animals reared intensively inside buildings, the attempt to control pests by chemical means, necessitates the deliberate introduction of the pesticides into some part or other of the open environment. This aspect of the use of pesticides differentiates them from the great majority of chemicals used by man. Nevertheless, other compounds do, of course, enter the environment, either naturally (e.g. terpenes that volatilise from trees, sulphur compounds from volcanoes, etc.), accidentally, or in waste material arising from man's activities (stack gases, industrial effluents, domestic waste and sewage). Moreover, these compounds may be locally of greater ecological importance than pesticides.

Furthermore, compounds which are not deliberately introduced into the environment as an integral part of their use may also be found to be dispersed widely in the environment. Perhaps the most outstanding example of compounds that appear to be as ubiquitous as the organochlorine insecticides, despite the fact that they are not introduced deliberately (large amounts are even used in closed systems), is provided by the polychlorinated biphenyls. Finally, it cannot be assumed that the reported ubiquity of residues of organochlorine insecticides is necessarily, or even predominantly, the result of their deliberate introduction into the open (mainly agricultural) environment; industrial effluents and waste disposal may also be involved.

All pesticides have measurable vapour pressures and solubilities in water; some examples are given in Table 12.1. Consequently, like all other compounds with measurable vapour pressures and solubilities in water, we may expect that use of these compounds will result in a proportion of the applied pesticide being dispersed from the point (or area) of application. The proportion dispersed, and the rate of dispersion, depend upon a

TABLE 12.1. Vapour pressure and solubility in water of some pesticides.

Compound	Vapour pressure (mm Hg)	Solubility in water (ppm)
Simazine	6.1×10^{-9} (20)*	230 (25)*
2,4-D	0.4 (160)	620 (25)
Parathion	3.78×10^{-5} (20)	24
p,p'-DDT	1.9×10^{-7} (20)	0.037 (25)
HHDN	6×10^{-6} (25)	0.2 (25)
HEOD	7.78×10^{-7} (25)	0.5 (30)

* Temperature ($^{\circ}$C) in parentheses.

number of factors, some being related to the intrinsic properties of the compound and others to extraneous (physical) factors. Furthermore, the entry of the pesticide, at the point of application, into a mobile organism, will result in the transport of the absorbed material away from the point of application, as a result of the movement of the host.

We may express the situation in general and simple terms as follows. A quantity of a compound, containing N molecules, is applied to a particular point. At a given instant, time t_1 after the application, a proportion (p_{1i}) of the molecules is present in a unit volume situated some distance from the point of application. The number of molecules in the specified unit volume is n_{1i} $(=p_{1i}N)$, and the total number of molecules in the biosphere originating from the point of application is $\Sigma p_{1i}N$, where $\Sigma p_{1i}N$ is equal to or less than N. The latter case corresponds to loss from the biosphere as a result of degradation (losses by escape into space being regarded as negligible). This is an explicit statement of one made by Tarrant and Tatton (1968), 'it seems logical to assume that these organochlorine insecticides now have world-wide distribution'. At time t_2 the number of molecules in the specified unit volume is n_{2i} $(=p_{2i}N)$, where the difference $(n_{1i}-n_{2i})$ is equal to the difference between the number of molecules entering during the period (t_2-t), and the number of molecules leaving the unit volume, and may be zero (corresponding to a steady state). If $n_{1i}>n_{2i}$ then the losses exceed the gains; the losses may be by degradation, transfer to adjacent volumes, or by scavenging (e.g. washout in rain). The problems to be considered are: (1) can we predict the number of molecules in the ith volume at time t_1; and (2) can we estimate the change in the numbers between t_1 and t_2? It is convenient to divide the mechanisms involved into two groups, one corresponding to physical mechanisms in the abiotic components, the other to biological mechanisms.

12.2   DISPERSION, DEGRADATION AND SCAVENGING
MECHANISMS IN ABIOTIC COMPONENTS

12.2.1   Atmosphere

In this discussion, attention will be concentrated on the troposphere, that layer of gases, predominantly nitrogen and oxygen, extending from the earth's surface to the tropopause. The temperature in the troposphere declines steadily with the height above the earth's surface until the tropopause is reached. The average height of the tropopause varies from about 7,000 m at the polar areas to about 16,000 m at the equator. Above the tropopause is the stratosphere, which is almost isothermal.

The composition of the atmosphere at ground level (excluding the major components: nitrogen, 78.08%; oxygen, 20.95% by volume) is given in Table 12.2. The concentrations of water vapour and the other com-

TABLE 12.2.   Composition of the atmosphere (other than nitrogen and oxygen) at ground level.*

Element or compound	Concentration ($\mu g/m^3$ at STP)	Residence time (years)
Argon	1.6×10^7	–
Helium	920	$\sim 2 \times 10^8$
Water vapour	$(3\text{-}3,000) \times 10^4$	~ 0.03
CO_2	$(4\text{-}8) \times 10^5$	4
CH_4	$(8.5\text{-}11) \times 10^2$	~ 100
N_2O	$(5\text{-}12) \times 10^2$	~ 4
CO	1-20	~ 0.3
SO_2	0-50	~ 0.015
H_2S	3-30	~ 0.1

* Derived from Junge (1963).

ponents of the atmosphere show considerable variations, both with regard to time and location. The variations in the concentrations of some components have been studied in detail, and as these variations are the result (sometimes wholly) of man's activities, these investigations are of great interest in attempting to understand the factors controlling the concentration of pesticides in the atmosphere

Carbon dioxide, for example, is a normal constituent of the atmosphere, and the concentration until recent times was virtually independent of man's activities. The main routes of input into the atmosphere include oxidation of organic compounds (including the combustion of fossil fuels), volcanic emissions, and transfer from the oceans; the main routes of removal from the atmosphere include reaction with silicates (Urey, 1952), photosynthesis, and transport into the oceans (oceanic sedimentation of carbonates plays a role in this). There have been a number of studies that indicate a small, but statistically significant, increase of the concentration of carbon dioxide in the atmosphere (Revelle, 1965; Pales and Keeling, 1965; Keeling, 1970). This change of the concentration of CO_2 in the atmosphere, indicates that the former steady state has been disturbed and that the rate of input of CO_2 into the atmosphere exceeds the rate of removal. This example of a state of imbalance of a 'natural' constituent, is of interest in relation to the subject of this chapter. The sources of carbon dioxide are so widespread that it is difficult to discriminate between variations of local origin and those operating over long distances. However, Bischof and Bolin (1966) have published some interesting results on the variations in the CO_2 content of the atmosphere, from sea-level to about 1,800 m, up to a distance of about 100 km from the coast of South Carolina; the concentrations of CO_2 were a factor of both height and distance from the coast, and this was interpreted as indicating the importance of two sources of CO_2 : release from soil and human activity.

Studies of sulphur compounds (SO_2 and H_2S) are perhaps more illuminating in the study of factors controlling dispersion. Sulphur compounds occur in the atmosphere both as a result of natural processes (decomposition of organic matter in swamps, tidal flats, etc.), volcanic activity, and transfer of aerosols containing sulphate ions from the oceans. Factors involved in the removal of SO_2 from the atmosphere include oxidation to SO_3 and subsequent formation of hygroscopic nuclei of sulphuric acid, and hence of mists that are washed out by rain, washout of SO_2 by rain as a result of absorption from the air (Beilke and Georgii, 1968), dry deposition of dust particles containing adsorbed SO_2 and so on. The variations in the concentration of SO_2 in the atmosphere around cities, indicates the importance of these processes of removal, but there are also reports of movements over longer distances. Reiquam (1970a) has discussed the deposition in Sweden, of sulphate ions that originated from SO_2 emissions in Britain and West Germany, some hundreds of kilometres away.

An important study of the deposition of a chemical over a large area, after emission from a point source, was made following an accident at the Windscale (North-West England) Works of the United Kingdom Atomic Energy Authority. A quantity of radioactive material was released into the atmosphere from a stack 120 m high. The effluent was widely dispersed as a result of the prevailing meteorological conditions, and three publications on one of the radioactive components, ^{131}I, are particularly pertinent (Chamberlain, 1959; Crabtree, 1959; Chamberlain and Chadwick, 1966). Deposition of ^{131}I occurred by three main mechanisms: deposition in particulate form, washout from the atmosphere, and vapour diffusion to the grass surface. The latter process was considered the more important in this particular episode. The meteorological factors involved were discussed by Crabtree (1959), and other factors, such as the physiological state of plant leaves, by Chamberlain and Chadwick (1966).

Another example of a chemical which has been extensively studied, in regard to its behaviour in the atmosphere, is that of lead and its compounds. Sources of lead in the atmosphere include industrial effluents, and aerosols containing lead emitted from the exhausts of automobiles. Scavenging processes that remove lead from the atmosphere, include deposition of particulate matter and washout by rain. Variations in the concentration of lead in the atmosphere, in relation to traffic density, have been studied intensively (Lehman and Möller, 1967; Motto *et al.*, 1970). The distances involved in these studies are small, 10-100 m, but transport over long distances has also been investigated; Patterson and his co-workers concluded that dispersion of lead particulates is occurring over large distances (Patterson, 1965; Tatsumoto and Patterson, 1963a, 1963b).

These brief summaries of the variations of the concentrations of some chemicals in the atmosphere, and the factors causing them, are sufficient to indicate the complexity of the variables affecting the dispersion and scavenging of chemicals in the atmosphere. Chemical degradation, chiefly by photolytic processes, is also important in the case of many organic compounds. Meteorological factors are also important, and some comment upon these is appropriate at this point.

12.2.1.1   *Meteorological factors involved in the processes of the dispersion and scavenging of chemicals in the atmosphere*

Factors that are important in relation to the dispersion of chemicals from the source, are wind speed, turbulence and temperature gradients (which may be inter-related). Probably, the most important scavenging process is precipitation. A number of theoretical approaches have been made to the dispersion of chemicals in the atmosphere, but in spite of their subtlety and great theoretical interest, they are not, unfortunately, of great practical importance.

Mathematical models of the atmosphere have been developed for meteorological and climatological forecasting. The models are necessarily complex; they involve seven parameters (temperature, pressure, density, moisture content, and the three orthogonol velocity components), four postulates (the first law of thermodynamics, conservation of mass, Newton's second law of motion, and the equations of state for a gas), boundary conditions, and certain external restraints (e.g. solar radiation energy, etc.). Models of a similar type (not necessarily including the same set of parameters) have been developed to describe the dispersion of chemicals in the atmosphere. Bosanquet and Pearson (1936) developed relationships based on classical diffusion theory and dimensional analysis; diffusion coefficients which are indices of atmospheric turbulence are included in the equations. Another relationship, using the statistical correlation theory of turbulence, combined with a classical diffusion equation, was proposed by Sutton (1953).

The relationships are of the form given in equations (1) and (2) respectively:

$$C = \frac{aQ}{ux^2} \exp -\left(\frac{h}{px} - \frac{y^2}{2(qx)^2} \right) \qquad (12.1)$$

$$C = \frac{bQ}{ux^{2-n}} \exp -\left(\frac{1}{x^{2-n}} \frac{y^2}{C_y{}^2} \times \frac{h^2}{C_z{}^2} \right) \qquad (12.2)$$

where a and b are constants for the particular set of meteorological conditions prevailing, Q is the rate of emission of the chemical, u is the mean wind speed, h is the height of the source above ground, x is the distance downwind of the source, y is the distance crosswind, p and q are diffusion coefficients (dimension-less) C_y and C_z are generalised eddy-diffusion coefficients, n is a turbulence parameter. The maximum concentrations at ground level predicted by both these equations are of the form:

$$C_{\max} \propto \frac{\text{source strength}}{\text{wind velocity} \times (\text{height of source})^2} \qquad (12.3)$$

The distances downwind at which these maximum concentrations occur are $h/2p$ (Bosanquet and Pearson, 1936) and approximately h/C_z (Sutton, 1953).

Several other relationships have been prepared (Scorer, 1959; Pasquill, 1962), and Reiquam (1970b) has also published an interesting treatment of the problems.

12.2.1.2 *Pesticides in the atmosphere*
From a consideration of the modes of use of pesticides, their measurable vapour pressure, and the evidence briefly summarised above, it would not be surprising if pesticides, particularly those used in the open environment, were being dispersed into the atmosphere from the areas of application. However, the distances over which dispersal will recur are difficult to predict, because our knowledge of the rates of entry into the atmosphere, from the various sources and rates of removal, is meagre. Processes leading to the removal of pesticides from the atmosphere include chemical reactions (probably photochemical), and scavenging, either by dry deposition or washout in rain.

There are few studies of the relationship between meteorological factors and the dispersion of pesticides in the atmosphere, and such studies as have been made were mainly concerned with aerosol drift during application of pesticides by aircraft. Yeo and Thompson (1953) investigated the effects of atmospheric turbulence upon the deposition of DDT applied as an aerosol (mass median diameter, 70-80 microns) from an aircraft flying at a height of 30 feet. It was concluded that the aerosol was deposited more readily during inversions than during conditions of super-adiabatic lapse-rate, and that simple practical relationships could probably be found between aerosol behaviour and meteorological factors.

Yeo *et al.* (1959) obtained some very interesting results on the dosage deposited on the ground at varying distances at right angles from the flight path of a plane flying at a height of 10 feet; airborne concentrations were also determined. They concluded that toxicological hazards (to crops, livestock or man) may occur at distances up to one mile downwind from the sprayed area. Akesson et al. (1964) studied the drift of aerosols sprayed from aircraft in California, and gave estimates of the distances downwind of the area of application, at which the residues of six insecticides on alfalfa (to be used as dairy feed) were considered acceptable. The hazard arising from drift was considered to be greater under inversion conditions, since vertical diffusion or upward movement of the chemical-laden air does not take place. A detailed review of residues arising from the drift of aerosols was prepared by Akesson and Yates (1964). Yates et al. (1966) derived relationships between the amounts deposited and the distance downwind of the flight path, using a dyestuff to estimate the amounts deposited.

Other studies have been made of the amounts of pesticides in the air, in which the interest was primarily that of determining the exposure of workers to such compounds (Durham and Wolfe, 1962), and of the exposure of the general population in agricultural areas (Tabor, 1966).

The above investigations were concerned with air concentrations and ground deposits up to about 2 miles from the area of application, but there are suggestions that aeolian transport of pesticides over much greater distances may be occurring (see Chapter 10). The evidence adduced in support of this hypothesis consists of analyses of air (Abbott et al., 1966), rain (Wheatley and Hardman, 1965; Abbott et al., 1965; Weibel et al., 1965; Tarrant and Tatton, 1968), airborne particulate matter (Antommaria et al., 1965; Cohen and Pinkerton, 1966; Risebrough et al., 1968) and Antarctic snow (Peterle, 1969). The inferences drawn from these investigations require careful consideration. Firstly, the concentrations found in most instances are extremely small. Thus, Abbott et al. (1966) reported a concentration of about 0.004 $\mu g/m^3$ p,p'-DDT in London air; Risebrough et al. (1968) found a mean concentration of 0.002 μg HEOD/g in dust deposited from the air in Barbados, corresponding to a mean atmospheric concentration of 5.4×10^{-9} $\mu g/m^3$. The analytical procedures, by which such small concentrations are determined, require critical evaluation as regards both their qualitative and quantitative reliability before it can be accepted that the presence of the specified compounds has been established beyond reasonable doubt. Analytical problems are discussed elsewhere (Chapter 1). Secondly, prescinding from the analytical difficulties, it is not clear that the reported results have demonstrated unequivocally that the pesticides found had been transported over long distances in the atmosphere. There are three investigations that may be regarded as crucial and these are discussed below.

Cohen and Pinkerton (1966) analysed particulate matter deposited in Cincinnati; this particulate matter was plausibly associated with a dust storm that had occurred in Western Texas and South-eastern New Mexico.

The precipitation of this particular particulate matter over Cincinnati was brought about by a trace of rain, e.g. the air over Cincinnati was scavenged by the dust and rain precipitated. In attempting to relate the presence of a particular pesticide in the dust deposited in Cincinnati with the presence of that pesticide in the dust that became airborne in Texas, it is necessary to consider the transport process. If we consider a unit mass of dust sample collected in Cincinnati, the quantity (Q_c) of, say, Ronnel, in that sample is given by:

$$Q_c = Q_0 + Q_{in}^a - Q_{out}^a + Q_{in}^c - Q_{out}^c$$

where Q_0 is the amount in the dust at the source, Q_{in}^a and Q_{out}^a are the amounts adsorbed and lost respectively during the period of aeolian transport, Q_{in}^c and Q_{out}^c are the amounts adsorbed or lost during the precipitation process over Cincinnati. Some rainfall had occurred a few hours before the dust cloud appeared over Cincinnati, so it may be concluded that Q_{in}^c was zero. The following inferences may be drawn:

(i) if $Q_c = Q_0$, then $Q_{in}^a - Q_{out}^a - Q_{out}^c = 0$

    (a limiting case being $Q_{in}^a = Q_{out}^a = Q_{out}^c$)

(ii) if $Q_c > Q_0$, then $Q_{in}^a > (Q_{out}^a + Q_{out}^c)$

(iii) if $Q_c < Q_0$, then $Q_{in}^a < (Q_{out}^a + Q_{out}^c)$

No information is available concerning Q_{in}^a, Q_{out}^a or Q_{out}^c (although the statement that the dust came from an area that had received heavy annual treatment with pesticides for many years, taken in conjunction with the small residues in the dust, probably implies that $Q_c < Q_0$). The dust had passed over areas in which agricultural usage of pesticides was customary, and accretion of pesticides by the dust cannot be excluded. No valid inferences can therefore be drawn concerning the proportion of any particular pesticide in the dust that was originally present in the dust in Texas.

The interpretation of the results of this investigation has been discussed in a detailed and simple manner, since the lack of knowledge of Q_{in}^a etc., as a result of which we can draw no valid inferences concerning the origin of the pesticides, also applies to other studies of aeolian transport of pesticides in which samples of rain, air or airborne dust are analysed. These difficulties of interpretation are particularly acute when the transport occurs over land areas in which pesticides are used extensively in agriculture: volatilisation, which is presumably occurring continuously (even if at a variable rate from day to day), makes it impossible to assign the source of pesticides in a particular sample.

There are two investigations in which contamination of dust particles, etc., during transport, is probably negligible, namely, the analyses by

Risebrough *et al.* (1968) of airborne dust in Barbados, and of Antarctic snow by Peterle (1969).

The dust samples in Barbados had been collected on screens facing the prevailing wind at Kitridge Point (Delany *et al.*, 1967). Extreme care was taken during the collection of the samples and subsequent operations with them (one can only express admiration for the rigorous precautions and scrupulous work of these investigations). Consequently, the origin of the dust was assigned, with some confidence, to the European-African continents, using minerological and biological criteria for this identification. The total concentration of organochlorine compounds in the various samples collected was in the range <1.2-164 $\mu g/kg$ of dust, and the equivalent concentrations in the air were in the range 6.5×10^{-9} to 242×10^{-9} $\mu g/m^3$. These concentrations are extremely small, smaller than those found in London air for example, by a factor of 10^5-10^6. One hesitates to raise the possibility of contamination of the samples during collection, examination, transport and analysis in view of the rigorous precautions taken, but the amounts present in the samples are so small that this possibility cannot be completely excluded. However, the difference between the estimated concentrations in the atmosphere at Kitridge Point and those in London (see Table 12.3) are so large that it is

TABLE 12.3.   Concentrations of organochlorine insecticides in air.

Geographical Area	Concentration ($\mu g/m^3$) in the air				
	p,p'-DDT	p,p'-DDE	o,p'-DDT	p,p'-DDD	Dieldrin
London (Abbott *et al.*, 1966)	3.8×10^{-3}	7.1×10^{-3}	—	3.8×10^{-3}	2.6×10^{-2}
Barbados (Risebrough *et al.*, 1967)	4.5×10^{-8}	1.4×10^{-8}	5×10^{-9}	4.3×10^{-9}	5.4×10^{-9}

plausible to suggest the existence of an efficient scavenging process during the airborne transport from the European-African continents to Barbados:

$$Q_k = Q_{ea} + Q_{in}^a - Q_{out}^a$$

where Q_k is the amount of a pesticide in unit mass of dust deposited at Kitridge Point, Q_{ea} is the amount present in unit mass of dust at the western edge of the European-African continents, Q_{in}^a and Q_{out}^a are the amounts adsorbed and lost during aeolian transport over the Atlantic ocean. The scavenging processes, presumably, include dust deposition, washout by rain, photochemical decomposition, and partition of pesticides (in the vapour form) between the atmosphere and the surface of the ocean. By contrast, spray from the oceanic surface may contain traces of pesticides and thus contribute to Q_{in}^a.

Peterle (1969) analysed samples of melted Antarctic snow and reported the presence of p,p'-DDT at concentrations of the order of 0.04 µg/kg, similar to those in rain in Britain and Ohio. However, the latter samples also contained other organochlorine insecticides and the absence from the Antarctic snow of analytically detectable amounts of the other organochlorine compounds is puzzling. Furthermore, the reported concentration of p,p'-DDT, presumably scavenged from the Antarctic atmosphere by snow, appears to indicate atmospheric concentrations of a similar order to those in London and Ohio, which would be surprising, in view of the low concentrations in the atmosphere at Kitridge Point — it would appear to indicate that the scavenging processes in the southern polar seas are much less efficient than those in the northern Atlantic ocean. It is pertinent to refer to the conclusion of Lorius et al. (1969) that the concentration of salts in Antarctic ice declines as the distance from the coast increases.

Risebrough et al. (1968) stress the large difference they found between the concentration of pesticides in the atmosphere at Barbados, and that at La Jolla in California; the latter concentration was some 1,000 times greater than that in Barbados. They concluded that this showed the difference in pesticide load between marine air adjacent to agricultural areas, and marine air remote from sites of application. In other words, pesticides in the atmosphere tend to be localised. This conclusion is consistent with the results obtained by Windom et al. (1967) who investigated the occurrence of talc in dusts recovered directly from the atmosphere or from rain and snow. They concluded that the talc arose from the use of pesticide formulations based on this mineral, and that the distribution of talc in atmospheric dust indicated local introductions rather than generalised global transport.

The absence of talc from the dust samples collected in Barbados is noteworthy, but its significance is difficult to assess: was talc not used in the pesticide formulations in Europe and Africa? If it was used, then its absence may imply that the pesticides in the dust did not originate in the European-African continents. Alternatively, that all the dust particles containing talc, derived from pesticide formulations in Europe and Africa, had been scavenged from the atmosphere and consequently the pesticides in the dust collected in Barbados had their origin from some other source (unless transfer had occurred in the atmosphere between airborne talc formulations and other dust particles).

Scavenging processes have been mentioned frequently above. Detailed investigations of the scavenging of pesticides from the environment have not been made, but it seems appropriate to refer to a predictive theoretical treatment by Makhon'ko (1967), of the removal of contaminants from the atmosphere. Photochemical reactions may be also important but studies of vapour phase reactions do not appear to have been published, although several papers are available on reactions in solution or on surface films.

A summary of the tentative conclusions that may be drawn concerning the dynamics of pesticides in the atmosphere is as follows: agricultural use of pesticides is an important source of entry of pesticides into the atmosphere; the rate of entry and the distances over which they move are dependent upon the vapour pressure of the pesticides, and the meteorological conditions. The majority of the pesticides entering the atmosphere in a particular area is deposited locally, and aeolian transport over long distances appears to be a minor phenomenon. The tendency for localised deposition is related to the scavenging processes that remove pesticides from the atmosphere, deposition of dust, photochemical decomposition, and washout by rain (which appears to be the most important).

12.2.2   Hydrosphere

The oceans cover an area of $3.61 \times 10^8 \, km^2$ (about 70.8% of the earth's surface), of which 2.06×10^8 km^2 are in the southern hemisphere. The estimated total volume of the oceans is 1.37×10^9 km^3, with an average depth of 3,795 m (Sverdrup, 1954). The oceans have been described as the ultimate sink for the world's natural wastes and human artefacts. Many of these wastes enter the oceans via rivers, e.g. industrial effluents, sewage, etc., but a proportion of some wastes is transported to the oceans by the atmosphere. Fresh water (rivers, lakes and ground water) cover an estimated area of 5×10^5 km^2, and the continental ice in Antarctica and Greenland, 2.3×10^7 km^2 (Goldschmidt, 1962).

The oceans are not a homogeneous phase, because there are both suspended mineral particles and living organisms in them, neither is it a mere continuum of water in turbulent motion as a result of surface wind, thermal motion, etc. There is, for example, a thermohaline circulation, and this accounts for the variations in the distribution of salinity and temperature in the deep water of the oceans; water at the greatest depth has the greatest density, consequently there is a stable stratification. There is an appreciable deep water circulation southwards from Iceland and Greenland; this crosses the Equator and rises around the Antarctic continent. Conversely, there is a deep circulation flowing northward from Antarctica. This crosses the Equator, mixes with the south-flowing deep water and returns to the Antarctic. This particular type of circulation is confined almost completely to the Atlantic ocean; no meridional deep water circulation is found to the north of the Equator in either the Pacific or Indian oceans.

Sub-surface currents, 'pure wind' currents, and semi-permanent surface currents have also been identified. There is a striking agreement between the directions of the prevailing winds and the prevailing surface and sub-surface currents, and taking these observations in conjunction with the oceanographic equations of motion, it has been concluded that these ocean currents are in fact maintained by the prevailing winds.

Given this knowledge of water motions, and a knowledge of the concentrations of a chemical in the various parts of the oceans, it should be possible to delineate the sources and sinks of that chemical (and *vice versa*). Such an approach has been made by Bolin and Stommel (1961) and

Keeling and Bolin (1967, 1968). These studies involve the use of models in which the behaviour of chemicals in the oceans is simulated by regarding them as consisting of reservoirs or boxes (these models have some analogies with the compartmental models used to simulate the pharmacokinetics of drugs in animals). Reservoir-type models, involving both the atmosphere and hydrosphere, have also been proposed (Craig, 1957, 1963; Lal, 1963; Manabe and Bryan, 1969).

Superimposed upon this 'hydrological' heterogeneity of the oceans is that caused by localised abiotic phases, e.g. petroleum oil, or lipids of biotic origin (Jarvis *et al.*, 1967; Garrett, 1967; Seba and Corcoran, 1969). In view of the lipophilic character of most pesticides, the existence of these phases, particularly as surface films, may be of great importance in relation to their dispersion and fate in the hydrosphere.

In recent years, there has been a growing interest in the distribution of heavy metals in the oceans. The occurrence of lead, for example, has been investigated. On the basis of variations in the concentration of lead in relation to depth, it has been suggested that the higher concentrations found in some surface waters are the result of washout of lead aerosols from the atmosphere (Tatsumoto and Patterson, 1963a, 1963b; Chow and Patterson, 1966). These investigations illustrate the use of studies of the distribution of a chemical in the oceans, in the elucidation of sources of a contaminant and of scavenging processes (the removal of lead, for example, from the upper mixed layer was attributed to sedimentation, either of inorganic particles or of the detritus of dead organisms that had absorbed lead).

The entry of pesticides into the hydrosphere arises from many sources including direct application (for pest control), industrial effluents (including disposal of agricultural waste), sewage, leaching and run-off from soil, aerosol and particulate deposition, rainfall, and absorption from the vapour phase. Direct application of pesticides to lakes and rivers is probably declining. Industrial effluents, sewage, and leaching and run-off from soil, usually enter rivers and are then transported to the oceans; large dilution occurs on entering the oceans, so there are localised areas of relatively high concentration in the rivers and estuaries and, as expected, estuarine and inshore organisms tend to contain higher residues than deep-sea fish. Effluents can be controlled and, consequently, this source of contamination of the hydrosphere could be reduced if the necessary steps are taken; conditions have certainly improved locally in recent years in this regard, but there is no doubt that further improvements should be made. Movement of pesticides from agricultural land by leaching or by run-off (caused either by flooding or irrigation) has been investigated; leaching depends upon the partition of the pesticide between the soil constituents (organic and inorganic) and the water percolating through the soil, and varies according to the properties of the pesticide and the type of soil. For the organochlorine insecticides, leaching from soils does not appear important (Lichtenstein *et al.*, 1962; Park and McCone, 1966; Sparr *et al.*,

1966; Hindin, May and Dunstan, 1966; Eye, 1968; Harris, 1969; Edwards, 1970; Edwards *et al.*, 1970; Thompson *et al.*, 1970).

Run-off varies according to the rate of flow of surface water and the soil type. Although there are several reports of riverine pollution by this mechanism (e.g. Holden *et al.*, 1968; Weidner *et al.*, 1969), it is probable that this is not a generally important source of contamination of waterways, and hence of the oceans (Moubry *et al.*, 1968; Edwards, 1970; Caro and Taylor, 1971); probably industrial effluents, etc., are of much greater importance.

There is little information on the amounts of pesticides entering the oceans by deposition of aerosols or dusts (see section 12.2.1), nor on that entering by rainfall. However, it has been plausibly suggested by Goldberg *et al.* (1971) that washout of residues from the atmosphere is a major route of entry of DDT-type residues into the oceans. These authors calculated the annual input of DDT-type compounds as follows:

$$
\begin{array}{cccc}
\text{annual quantity of} & \text{average annual} & \text{concentration} \\
\text{DDT-type compounds} = \text{precipitation} \times \text{in proportion} = 2.4 \times 10^7 \text{ kg} \\
\text{in rainfall} & (3 \times 10^{20} \text{ cc}) & (80 \times 10^{-12})
\end{array}
$$

The value used for the average concentration of DDT-type compounds in rain can only be regarded as tentative, as even the little evidence available indicates there are large variations from place to place and from time to time.

Removal of pesticides from the hydrosphere may occur by several routes: volatilisation (i.e. return to the atmosphere), absorption by aquatic organisms, and settling of particles to which pesticides are adsorbed. Goldberg *et al.* (1971), in a consideration of the upper mixed zone of the ocean, made some enlightening calculations on the input and loss of DDT-type compounds, but these calculations are probably best regarded as an outline of an appropriate procedure, rather than as an accurate estimation of the real situation. More sophisticated models can be constructed, but as these entail increasing numbers of assumptions, it is doubtful if the predictions of such a model will have any greater validity than those derived from the simple model of Goldberg *et al.* (1971).

12.2.3   Lithosphere

The behaviour of pesticides in the third major abiotic component of the biosphere differs from that in the atmosphere and hydrosphere in one essential respect, namely, that the latter are mobile phases, whereas the lithosphere is essentially an immobile matrix. Movement of pesticides in the lithosphere, therefore, involves the processes already discussed, i.e. volatilisation, transport in dusts, flow of water in waterways, etc. These processes may be modified, of course, by their occurrence in or near the lithosphere; for example, the partial vapour pressure of a pesticide in the soil will differ, as a result of adsorption, from the vapour pressure of the pure compound.

The term lithosphere has several meanings, but in this article is taken to be the iron core of the earth (magna), plus the fairly homogeneous mantle, and the heterogeneous silicate crust. The latter extends to the Mohorovičić discontinuity at a depth of about 33 km below the surface. So far as chemical contaminants are concerned, we are interested in a thin outer shell, a few metres deep (except for refuse disposed in deep pits), around the earth's surface; greater depths may be of interest in some other instances, such as the transfer of chemicals into ground water.

Routes of entry of chemicals into this outer shell include direct application (e.g. fertilisers and soil insecticides), waste disposal, flooding of land by rivers, rainfall, volcanic activity, etc. Routes of loss include volatilisation, leaching and run-off, and microbial degradation.

Studies of the dynamics of pesticides in the soil have been almost solely concerned with the rates of their disappearance after application, the major interest being in the influence of meteorological factors and soil type upon the change in the residues with time. Convenient summaries of residues in soil have been given by Edwards (1970). The effects of soil micro-organisms upon pesticides have been studied quite intensively (Korte, 1963; Matsumura and Boush, 1967; Tu et al., 1968; Matsumura et al., 1969; Edwards, 1970). Mathematical models of the accumulation of pesticides in soil, as a function of annual applications, has been prepared by Hamaker (1966) and Lin et al. (1971).

12.3 TRANSPORT AND DEGRADATION OF PESTICIDES BY ORGANISMS

The methods of use of pesticides result, either directly or indirectly, in the exposure of many types of organisms to their residues, and, for certain organochlorine insecticides, it may be postulated that virtually every type of living organism is potentially exposed to those compounds. Thus, soil micro-organisms come into contact with compounds used to control soil pests; the entry of pesticides into the hydrosphere results in the exposure of aquatic organisms to those pesticides; and insects come into contact with pesticidal formulations, aerosols, dusts, etc., or with deposits on plant surfaces. The exposure of the higher animals, particularly birds and mammals (including man) is adventitious, except for a few species which are regarded as pests in some areas. The entry of pesticides into these higher organisms occurs by inhalation, percutaneous absorption, and ingestion of residues present in food. The latter mode of entry is probably the most important for most vertebrates.

12.3.1 The use of surveys of residues in organisms to demonstrate the transport of pesticides by organisms

We may consider the transport of pesticides in the biosphere under two main headings: movement between organisms (i.e. along food-chains or through food-webs), and movement within a particular host over large distances. The two types of movement are not independent, nevertheless it is convenient to consider them separately.

The movement of residues through food-chains has received particular attention in the case of organochlorine insecticides. The concept of biological magnification has been derived from such studies (Rudd, 1964; Woodwell *et al.*, 1967; Robinson *et al.*, 1967; Macek, 1970), but the need for clarifying the use of this term has been pointed out (Robinson, 1970). For our discussion, we are particularly interested in the potential and actual transport of residues over long distances, and detailed discussion will be confined to this particular topic.

Studies of the concentration of organochlorine insecticides in a wide variety of organisms collected from different areas have demonstrated the ubiquity of residues of these compounds. This ubiquity does not necessarily demonstrate that large-scale movements are occurring throughout the whole of the biosphere. Use of these compounds is so widespread geographically that the occurrence of many of these residues is probably of fairly localised origin (movements of the order, say, of 200 km). There are a few surveys, however, that are at least consistent with the postulate of transport over long distances. Four surveys, for example, have been made of the residues of organochlorine insecticides in samples collected in the Antarctic (Sladen *et al.*, 1966; George and Frear, 1966; Tatton and Ruzicka, 1967; Brewerton, 1969). Although there are difficulties in the interpretation of these results (Robinson, 1970), nevertheless, they do constitute plausible grounds for tentatively concluding that most, if not all, of the residues in the Antarctic samples are the consequence of movement of residues over long distances, either in the mobile abiotic phases (atmosphere and hydrosphere), in the host species analysed, or in the fish, etc., eaten by them.

The plausibility of the postulated transport over long distances by organisms is strengthened by the following considerations:

The transport and dispersal of pesticides by organisms is dependent upon two factors; the mobility of the host and the rate of degradation within the host. The large and increasing amount of information available on the metabolism of pesticides is adequately summarised in several detailed reviews (Korte, 1967; Brooks, 1969; Kearney and Kaufman, 1969; Dauterman, 1971; Hollingworth, 1971); it is apparent that most pesticides are metabolised by a variety of organisms, although the rate of metabolism of organochlorine insecticides is slow in some cases. If the host organism travels long distances, e.g. during migration, and if the rate of metabolism of a pesticide in that host is slow, then movement of the pesticide over large distances becomes feasible. Swallows (*Hirundo rustica*), for example, migrate from Britain and other European countries to South Africa and *vice versa*, and short-tailed shearwaters (*Puffinus tenuirostris*) migrate from Australia to the Bering Sea and back again. Insects may be carried large distances by winds and thus act involuntarily as mobile hosts for pesticides.

A series of events such as the following may therefore be envisaged. Swallows (number, N_0) in Britain eat insects contaminated with pesticide X immediately prior to migration to South Africa. If the average number of molecules present in each swallow is $\dot{n}_0$, then the total number of molecules available for transport is $N_0 \bar{n}_0$. Some weeks later, some of the swallows (number, N) arrive in South Africa; the average number of molecules per bird is $\bar{n}$, where $\bar{n} < n_0$ (it is assumed, for simplicity, that no exposure to X occurred during the period of migration). A predator in South Africa eats N_i of these swallows, and thus ingests $N_i\bar{n}$ molecules of a pesticide that had originally been used some 10,000 km distance from the territory of the predator. The contamination of the predator, relative to a similar one in Britain (again, for simplicity, we assume that a comparable predator does so exist), is given by the ratio $N_i\bar{n}/N_0 \bar{n}_0$, a ratio which is dependent upon the predation rates in the two areas and also upon $\bar{n}/\bar{n}_0$, which is related to the rate of the concentration of X during the mobile period during the period of migration (t). If we assume that the rate of change, of the concentration of X in the swallow, corresponds to a pseudo-first-order reaction (the factors limiting the validity of this assumption are discussed below), then the biological half-life, t^*, is a convenient index of the rate of change. The proportion of X retained in the swallows at the end of the period of migration is given by $\exp(-t \ln 2/t^*)$. The calculated proportions retained for different ratios of t/t^* $(=\eta)$ are given in Table 12.4.

TABLE 12.4. Fraction of a pesticide retained, expressed as a function of its biological half-life.

$\eta = \dfrac{\text{lapsed time}}{\text{biological half-life}}$	Fraction retained (%)
0	100
0.01	99.3
0.1	93
0.5	70.7
1	50
5	3.1
10	0.1
20	0.001

The results of the calculations given in Table 12.4 suggest that the term 'persistent', as commonly used, is ambiguous and may be misleading in some circumstances: persistence is a function of the particular set of circumstances, as defined by η, being considered. It is proposed that if $\eta \leqslant 0.01$ then, as the change in concentration is very small, the compound may be termed persistent; if $0.01 < \eta < 0.1$, the term quasi-persistent appears appropriate; if $1 < \eta < 5$, quasi non-persistent; and if $5 < \eta < 10$, non-persistent seems appropriate.

The study of the rate of change of the concentration of pesticides is usually called pharmacokinetics, and, as the values of η for many organochlorine insecticides, for periods of t of the order of weeks, correspond to persistence or quasi-persistence in vertebrates, it is necessary to consider the dynamics of these compounds in vertebrates.

12.3.2 Pharmacokinetics of organochlorine insecticides

The dynamics of organochlorine insecticides in vertebrates is a topic which often appears to be misunderstood, and as an understanding of pharmacokinetics is important in relation to one of the possible mechanisms of the transport and dispersal of these compounds in the biosphere, a fairly detailed discussion of this topic is warranted.

A phenomenological approach is adopted in an attempt to consider the available evidence, free from preconceptions or presuppositions that may arise from the various inferences drawn from the environmental surveys. Presuppositionless knowledge, as proposed by philosophers such as Husserl (1901), is perhaps an ideal to which we strive, and, as regards a particular science, it must be appreciated that the presuppositionless attitude is 'limited' and restricted to that particular form of knowledge; it is unnecessary to adopt a presuppositionless attitude to the problem of perception, validation of logical or mathematical procedures, etc., for the phenomenological analysis of a specific scientific or technological problem.

In order to survey the available experimental results succinctly, they will be examined from four different aspects (examining the data in this manner does not conflict with the presuppositionless approach, since the validation of the type of analyses used is subject of other sciences, and not of pharmacokinetics as such). Subsequently, the implications of such empirical relationships as are found will be considered: firstly, the theories or models that may be proposed to account for the relationships, secondly, the inferences that may be drawn concerning the dispersal of organochlorine compounds in the environment by 'biotransport'.

The four aspects under which the data is examined are: what is the relationship, if any, between the intensity of the exposure, the duration of the exposure and the concentration found in the tissues; what relationships, if any, are there between the concentrations of a pesticide in the various body tissues; and what changes, if any, occur in the concentrations in the tissues when the exposure is terminated.

12.3.2.1 *Relationship between the intensity of exposure and the concentrations of organochlorine insecticides in the tissue of vertebrates*

Entry of organochlorine insecticides into vertebrates occurs either by ingestion (of residues in food), percutaneous absorption, or inhalation. Percutaneous absorption and inhalation, except in the case of vertebrates present in an area at the time of use of a pesticide or shortly thereafter, are probably of minor importance. Quantitative studies of the entry of pesticides by these routes are, anyway, few in number. Oral ingestion, however, has been studied quite intensively, although there are deficiencies in both the design and commission of some studies, which reduce their value in assessing the relationship, if any, between the intensity of the exposure and the concentrations of a pesticide found in the body tissues. Hayes (1959) concluded that, when other factors are left constant, the peak storage of DDT in each tissue varies directly with the daily dose. This conclusion was based mainly upon studies with rats, but the experimental results obtained with DDT in the rhesus monkey (Durham *et al.,* 1963) and man (Hayes *et al.,* 1956; Durham *et al.,* 1965; Hayes *et al.,* 1971; Morgan and Roan, 1971) are also consistent with this conclusion. A similar conclusion can be drawn from studies of the concentration of HEOD in the body tissues of rats (Walker *et al.,* 1969a), steers, hogs and sheep (Gannon *et al.,* 1959), the tissues and eggs of chickens (Robinson, 1970b), eggs of quail (Walker *et al.,* 1969b), and the whole blood and adipose tissue of man (Hunter and Robinson, 1967; Hunter *et al.,* 1969; Robinson and Roberts, 1969).

Significant correlations (or regression equations) have been derived in several instances, and it is concluded that such correlations between the intensity of the exposure and the concentrations in the tissues are general. The intensity of the exposure may be expressed as concentration ($\mu g/g$) in the diet, or the intake per unit, body weight (mg/kg), and the concentration in the tissue is usually expressed as $\mu g/g$ (or $\mu g/cm$ in the case of blood).

A statistically significant correlation between two variables implies that they are stochastically dependent; this may be regarded as the weakest form of association between two variables. If, as in many of the investigations of dieldrin for example, one of the variables is controlled (the dietary concentration) and is thus non-stochastic, and the concentrations of HEOD (a stochastic variable) in the tissues are determined, then a regression analysis is appropriate. One of the major differences between a regression analysis and a correlation analysis is that the latter can never be used as proof of causal relation.

In the case of HEOD, explicit regressions have been derived for rats, dogs and man, and we may represent the relationship in the following general form:

$$c_{ijk} = a_{ijk}(C_{jk}) \qquad (12.4)$$

where C_{jk} is the concentration of the kth insecticide in the diet of the jth species, and c_{ijk} is the concentration in the ith tissue of that species.

12.3.2.2 *Relationships between the concentration of an organochlorine insecticide in the tissues of vertebrates*

Significant correlations are found between the concentrations of HEOD in the various tissues of experimental animals, including rats (Walker *et al.,* 1969a), dogs (Walker *et al.,* 1969a; Keane and Zavon, 1969), and between the concentrations of HEOD in the whole blood and adipose tissue of man (Hunter and Robinson, 1967). The results for DDT are also consistent with this conclusion (Hayes, 1959), and we may symbolise the relationship as follows:

$$c_{ijk} = b\,(c_{njk}) \qquad\qquad (12.5)$$

where c_{ijk} and c_{njk} are the concentrations of the kth insecticide in the ith and nth tissues respectively of the jth species.

This type of relationship is implicit in equation (12.4) above.

12.3.2.3   Relationship between the concentration of organochlorine insecticides in tissues and the duration of continuing exposure

Hayes (1959) concluded that the degree of storage of DDT in the adipose tissue of rats varied directly with the number of doses (or days of continuing dietary ingestion), until a peak or plateau was reached, but that the relationships for other tissues are less clear. The results obtained during the feeding of diets containing DDT to rhesus monkeys (Durham et al., 1963) are consistent with this conclusion, although there was an apparent tendency for the concentration in the adipose tissue to decline after about 3 years (a similar tendency is apparent in the results obtained by Laug and Fitzhugh (1946) in their study of rats). The results of studies of the dynamics of DDT in man are also consistent with this conclusion (Hayes et al., 1956, 1971; Morgan and Roan, 1971).

A similar tendency was also found for the concentration of HEOD in the tissues of rats to approach an upper limit of storage, which is characteristic of the daily intake (Walker et al., 1969a; Deichmann et al., 1968), and the whole blood and adipose tissue of man (Hunter et al., 1969). The concentration of HEOD in the blood of dogs given daily doses of 0.1 mg/kg for 128 days was correlated with the time of exposure (Richardson et al., 1967). These investigators derived a significant relationship between the logarithim of the concentrations in the blood and the logarithm of the time of treatment. Keane and Zavon (1969) determined the concentration of HEOD in the blood of dogs given 1 mg HEOD kg/day for 5 days, followed by 0.2 mg HEOD kg/day for 54 days, and concluded that the concentration of HEOD in the blood remained very constant for the last 53 days of the trial. (This conclusion appears to be invalid as the concentration of HEOD in the blood increased slightly, but significantly, from days 7 to 59 of the trial:

$$C_{blood} = 0.00078t\,(\pm 0.00016) + 0.097$$

The concentration of HEOD in the eggs of quail (Walker et al., 1969b) and chickens (Robinson, 1970b), also appears to approach an upper limit, characteristic of the concentration of HEOD in the diet, as the exposure continues.

The dependence of the concentration in the tissues upon the duration of the exposure may be written as:

$$C_{ijk} = d(t) \qquad\qquad (12.6)$$

where C_{ijk} approaches a finite upper limit as t increases, i.e.

$$\lim_{t \to 0} C_{ijk} \neq \infty$$

12.3.2.4   Changes in the concentrations of organochlorine insecticides in tissues after exposure has ceased

It has been found that once the exposure to an insecticide is terminated, the concentration of that compound in the tissues declines, sometimes quite slowly. Declines in concentrations have been reported in the case of DDT in the adipose tissue of rhesus monkeys (Durham et al., 1963), rats (Ortega et al., 1956) and cows (McCully et al., 1966). The concentrations of HEOD in the tissues of rats (Robinson et al., 1969), and in the blood of man (Hunter et al., 1969; Jager, 1970), also declined when exposure was terminated.

Various relationships have been reported between the concentration in the tissues and the time since exposure was terminated. The change in the concentration of DDT in the adipose tissue of rhesus monkeys, rats and cows deviates from a simple exponential decline: and McCully et al. (1966) fitted their results to a power relationship of the type:

$$C = bt^a$$

where C is the concentration in the adipose tissue, t is the time since exposure was terminated, and a and b are constants. The changes of the concentrations of HEOD in the whole blood and liver of rats also deviate from a simple exponential decline, whereas that in the adipose tissue did not deviate significantly from such a relationship. However, despite these divergences in the forms of the relationships, the following general relationship is consistent with the available data:

$$C_{ijk} = f_{ijk}(t') \qquad\qquad (12.7)$$

where C_{ijk} is the concentration of the kth insecticide in the ith tissue of the jth species, and t' is the time since exposure was terminated.

The declines of the concentrations of insecticides in the tissues after termination of exposure are due, at least in part, to loss of the compound from the body of the animal, either by elimination of the unchanged compound or as degradation products. Experiments with ^{14}C-labelled compounds have indicated that the great majority of the eliminated ^{14}C-activity is in the form of metabolites (Korte, 1967; Baldwin, 1971). In the case of DDT, the metabolite DDA is found in the urine of rhesus monkeys (Durham et al., 1963) and man (Hayes et al., 1956; Durham et al., 1965; Roan and Morgan, 1971), and a metabolite of HEOD (the 9-hydroxy derivative) has been detected in human faeces, in amounts that are roughly proportional to the estimated daily exposure (Richardson and Robinson, 1971).

12.3.3   Simulation of the dynamics of pesticides in vertebrates — the compartmental model

The discussion of the experimental results has been deliberately restricted, so far, to ascertaining that there are statistically significant empirical relationships, but there are various additional inferences that could be drawn from them.

Firstly, the changes of the concentrations of HEOD, for example, in the various body tissues change in concert with those in the blood (as exemplified by the significant correlation coefficients both during exposure and subsequent to the termination of exposure), this may be explained in several ways, but the most plausible explanation is that there is a reversible exchange of HEOD (and also the other organochlorine insecticides) between the circulating blood and the other tissues. Secondly, the decline of the concentration of organochlorine insecticides in the tissues when exposure is terminated, and the isolation and identification of metabolites in body tissues or excreta, indicate that these xenobiotic molecules are susceptible to biochemical degradation. The empirical relationships between concentration (or some function of the concentration) and time (or some function of time), indicate that the metabolic process is not a zero-order reaction (a zero-order reaction also appears unlikely on *a priori* grounds, in view of the very small substrate concentrations), i.e. the rate of change of concentration is a function of the concentration at that instant:

$$dc/dt = \emptyset(c)$$

With these inferences in mind, it seems worthwhile to consider the possibility of simulating the observed phenomenon by a mathematical model. Numerous models could be proposed but, as with the discussion of the dispersion phenomena, a consideration of models developed for other classes of chemicals might be profitable. One particular type of model has been used with great success in the study of the pharmacokinetics of drugs, lipids, metallic ions, etc.: the so-called mamillary type of compartmental model. Detailed discussions of the mathematical basis of this model have been given by Atkins (1969) and Wagner (1971). The suggestion that the mamillary model may be a suitable one for the organochlorine insecticides, was made by Robinson (1963, 1964 and 1967a) and a more formal development of the model was given in subsequent papers (Robinson and Roberts, 1968; Robinson 1969a; Robinson *et al.*, 1969).

The model, as developed so far, is based upon a number of simplifying assumptions, and two of these warrant particular comment. The organism is considered to be a constant entity in all respects except that the pesticide is distributed among the various body tissues, metabolized and eliminated, i.e. effects of ageing upon body size, or upon the distribution and relative amounts of adipose tissue, are ignored. Furthermore, it is postulated that the presence of the chemical does not cause any significant changes in the organism, such as hepatic microsomal enzyme induction. These simplifications do not invalidate the model in principle; the model could be modified to allow for such variations (the tendency for the concentrations in the tissues to reach an upper limiting value during chronic exposure, followed, apparently, by a decline, may be an example of experimental results that warrant a modification of the model). The effects of environmental factors, e.g. availability of food, upon the organism (regarded as a constant entity) are also not considered in the simple model used to date.

The basic concepts of the model are as follows. The animal body is considered to consist of an infinite number of infinitesimal elements of volume. Collections of elements of volume form sets, with the defining property: any change in the chemical potential of a substrate in any element of that set, results in an instantaneous and equal change of chemical potential of the substrate in all the other elements (it should be noted that chemical equipotential does not necessarily entail equal concentrations of the substrate). Such a set is called a compartment (Bergner, 1961a and 1961b). The volumes of the compartments are assumed to be constant. The boundaries between compartments are regarded as interfaces, and some interfaces allow a reversible movement of substrate molecules between compartments. The movement of substrate molecules across interfaces continues, until the chemical potential of the substrate in all compartments approach the same value. Unchanged substrate may be eliminated from the organism through one or more compartments. Compartments may be in series or parallel or a combination of both arrangements. Biochemical conversion of substrate molecules may occur in at least one compartment, and the simplest case corresponds to a biotransformation reaction of the first-order. The model presented is very general, even though it contains a number of simplifying assumptions. A particular one, which appears suitable for the organochlorine insecticides, may be derived as follows.

It was suggested above that there is a reversible transfer of the organochlorine insecticides between the blood and the other tissues; this would correspond to a mamillary model consisting of a central compartment with one or more peripheral compartments in parallel. Direct transfer of substrate between peripheral compartments cannot be excluded, but such transfers appear to be of minor importance, and the simplest mamillary models are those in which such transfers are assumed to be zero. Biotransformation of organochlorine insecticides in lipid phases has not been demonstrated, and it is plausible to suggest that any such biochemical degradation occurs either in the central compartment, or in some peripheral compartment other than that containing the lipid phases. The minimum number of compartments is therefore two, but it may be three or more.

In the case of the elimination of HEOD from the rat, it was suggested that a two-compartmental model was in reasonably good accordance with the experimental results (Robinson et al., 1969).

The model for HEOD, based on a constant rate of entry (q_{0l}) into the

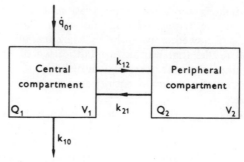

FIG. 12.1. Two compartmental model. The volumes of the central and peripheral compartments are V_1 and V_2. The quantity of pesticide entering the central compartment per unit time is $\dot{q}_{01}$; k_{12} and k_{21} are the proportionally constants for the transfer of the pesticide between the two compartments; k_{10} is the proportionality constant for the elimination of the pesticide from the central compartment.

central compartment, is shown in Fig. 12.1. The appropriate rate equations are:

$$\frac{dQ_1}{dt} = \dot{q}_{01} + k_{21}Q_2 - k_{12}Q_1 - k_{10}Q_1 \qquad (12.8)$$

$$\frac{dQ_2}{dt} = k_{12}Q_1 - k_{21}Q_2 \qquad (12.9)$$

On integration, Q_1 and Q_2 have the following values:

$$Q_1 = \frac{\dot{q}_{01}}{k_{10}} + A_e^{-m_1 t} + B_e^{-m_2 t} \qquad (12.10)$$

$$Q_2 = \frac{\dot{q}_{01} \times k_{12}}{k_{10} \times k_{21}} + C_e^{-m_1 t} + D_e^{-m_2 t} \qquad (12.11)$$

where A, B, C and D are constants for the system, and m_1 and m_2 are functions of k_{12}, k_{21} and k_{10}.

It is apparent that both Q_1 and Q_2 (and hence C_1, which equals Q_1/V, and C_{21} which equals Q_2/V_2) are asymptotic functions with upper finite limits:

$$\lim_{t \to \infty} Q_1 = \dot{q}_{01}/k_{10} \quad \text{and} \quad \lim_{t \to \infty} Q_2 = q_{01} \times k_{12}/k_{10} \times k_{21}$$

Thus, as the time of exposure increases, the concentrations in the two compartments approach a steady state, corresponding to a balance between the amount entering per unit time and the amount eliminated per unit time. Theoretically, the steady state is never achieved, but in practice, the concentrations reach values in a finite time, which cannot be differentiated experimentally from the steady state values. A *quasi* steady state is achieved in the peripheral compartment of the rat (which contains the adipose tissue) in about 50 days, for example:

When the exposure is terminated ($\dot{q}_{01} = 0$), the quantities of HEOD in the two compartments will decline; the change in concentration in the centre compartment is found in practice to correspond to the sum of two exponential functions, whereas that in the peripheral compartment corresponds to a single exponential function (the parameter C in equation (12.11) is not statistically significant).

The simplifying assumptions used in the model, make it necessary to be circumspect in making predictions based on it, nevertheless, the equations are very illuminating. The estimates of the retention ratios (η) in Table 12.4, for example, are strictly not appropriate for a model consisting of more than one compartment. Inferences about accumulation factors may be drawn from the two-compartmental model. The rate of entry by the oral route into the organism (represented by $\dot{q}_{01}$, a simplifying assumption which assumes that the rate of entry into the central compartment is independent of the ingestion process) is equal to the concentration c_0 in the diet, multiplied by the amount of food eaten in unit time ($\dot{w}_1$), and the concentration factors for the two compartments are:

$$f_1 = \frac{C_1}{c_0} = \frac{Q_1}{v_1 c_0} = \frac{\dot{w}_1}{v_1 k_{10}}$$

$$f_2 = \frac{C_2}{c_0} = \frac{Q_2}{v_2 c_0} = \frac{\dot{w}_1 k_{12}}{v_2 k_{10} k_{21}}$$

If, within each compartment, there are sub-sets in which the concentrations vary (i.e. C_1 is the mean of the concentrations in the sub-sets within the central compartment), then a number of concentration factors (concentration in tissue/concentration in diet) can be derived (Robinson, 1970a).

An important point about the compartmental model is that care must be taken in being realistic about the number of compartments that are warranted by the experimental data; it may be preferable to restrict the number of compartments to those that have some plausible physiological analogy. Fitting a large number of exponential terms may reduce the 'error' term, but the physiological significance of compartments corresponding to the additional exponential terms may be doubtful.

12.4 GENERALISED TREATMENT OF THE MECHANISMS OF TRANSPORT OF PESTICIDES IN THE BIOSPHERE

The modes of transport into, within, and out of each of the major components of the biosphere have been discussed in some detail. These major components are patently not isolated from one another, and some of the transfers between them have been mentioned above. An integrated approach is obviously necessary, although whether a practical model can be developed at present remains to be seen. Given the lack of information on many of the detailed mechanisms of transfer, and the relative paucity of quantitative data on the distribution of pesticides in the various parts of the environment, any models that are proposed are at best tentative and preliminary.

Some suggestions of a possible model which, at least in principle, involved a partially cyclical process of pesticide transport have been made (Robinson, 1967a and 1967b, 1969b). Eberhardt and his co-workers have also made some valuable suggestions on the construction of a model simulating the movement of chemicals in ecosystems (Eberhardt and Nakatami, 1967, 1968; Eberhardt, 1969; Eberhardt et al., 1970). Harrison et al. (1970), proposed a model which contains some valuable suggestions on the development of a more sophisticated model for the transport of DDT in ecosystems; unfortunately, some of their conclusions are of doubtful validity because unrealistic simplifying assumptions have been made.

The simplest and most general model is shown in Fig. 12.2. Each of the

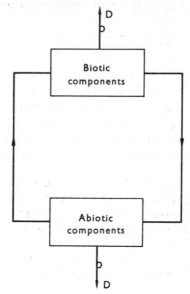

FIG. 12.2. Simple generalised model of the transport of a chemical between the major components of the biosphere.
⇢ indicates movement of the chemical between the components.
⇢ indicates degradation of the chemical.

major components can be divided into sub-systems which can, in principle, be monitored for residues of pesticides. If we know the amount of pesticide entering a sub-system, the concentration of pesticides in the elements of that sub-system and the total quantity of matter (inorganic and organic), we can calculate the retention or turn-over time for that pesticide. If, further, we can relate the movement of the pesticide into and out of that system to corresponding movements into and out of other sub-systems, then we can (theoretically at least) construct a flow diagram for the transport of that pesticide in the biosphere.

One of the major difficulties is the selection of suitable sub-systems to use in the construction of the model of the biosphere. The abiotic components can be divided into a minimum of about 5 sub-systems: two reservoirs in the atmosphere (although this ignores the slowness of inter-hemispheric transport), two reservoirs in the oceans, and the upper crust of the lithosphere. The biotic components are much more difficult to classify into sub-systems which can be used in a realistic manner to construct a model. The use of trophic levels has been suggested (Robinson, 1967a, b, c; Harrison *et al.*, 1970); models based on food-chains have been studied by Eberhardt and his colleagues, and this approach has considerable merit. The two approaches are interrelated abstractions of the same processes.

If we consider a very simple food-chain (see Fig. 12.3), we obtain some

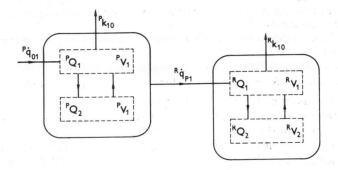

Prey(trophic level,i)   Raptor(trophic level, i+1)

FIG. 12.3.   Model of a simple food chain.

The prey species is in trophic level i, and the raptor in trophic level i+1. $P_{\dot{q}01}$ is the rate of entry of a pesticide into the prey species; $R_{\dot{q}p1}$ the rate of entry into the raptor consequent upon its eating contaminated prey.

P_{k10} and R_{k10} are the rate constants for the metabolism of the pesticide in the central compartments of prey and raptor respectively.

If the amounts of food eaten in unit time by prey and raptor are $\dot{w}_P$ and $\dot{w}_R$ respectively, then the biological-concentration factors (concentration in organism/concentration factors (concentration in organism/concentration in diet) are $\dot{w}_P/V_1 k_{10}$ and $\dot{w}_P k_{12}/V_2 k_{10} k_{21}$ for the prey organism, etc.

valuable insights into the more general trophic-level type model which is shown in Fig. 12.4. The abiotic components have been included in this model so that it represents, in a very simple and general way, a model for

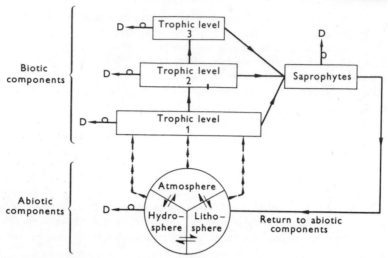

FIG. 12.4.   Schematic outside of a model of the transport of a chemical within the biosphere.

$\Rightarrow \Rightarrow$ indicates movement of the chemical by physical mechanisms from the abiotic components.

$\longrightarrow$ indicates biostransport as a consequence of the movement of the host or by predation.

$\multimap$ indicates loss of the chemical by degradation processes.

$\rightleftharpoons$ indicates transfer of the chemical between the inorganic phases.

the movement of pesticides in the biosphere.

    Harrison *et al.* (1970), in their systems study of DDT transport, derived conclusions from their model by making simplifying assumptions, and one of these in particular is of doubtful validity. Namely, that none of the organisms in a trophic level either metabolize or excrete DDT. A diagrammatic representation of their model of the transfer of a pesticide, into and out of a trophic level, is shown in Fig. 12.5. By assuming that the loss of DDT by metabolism or excretion does not occur (i.e. $\dot{p}_{met,i}$ and $c_{ex} = 0$), Harrison *et al.*, concluded that even if the input of DDT were reduced to zero at a given instant in time, it would take a period of $4T_i$ years for the concentration of DDT in each trophic level to attain an equilibrium (T_i is the average life span of the organisms in the *i*th trophic level). The assumption seems unwarranted: DDT, HEOD and other organochlorine insecticides have been shown to be metabolized by a wide variety of organisms and even if metabolism in organisms from every trophic level has not been studied, the mere demonstration of it in one or more organisms warrants any conclusions based on this assumption being regarded with suspicion. The conclusion that the time required to obtain a steady state is $4T_i$, is consequently in doubt on these grounds alone, and such evidence (unfortunately sparse) as is available from monitoring of organisms in the higher trophic levels, is inconsistent with it: Lockie *et al.* (1970) reported

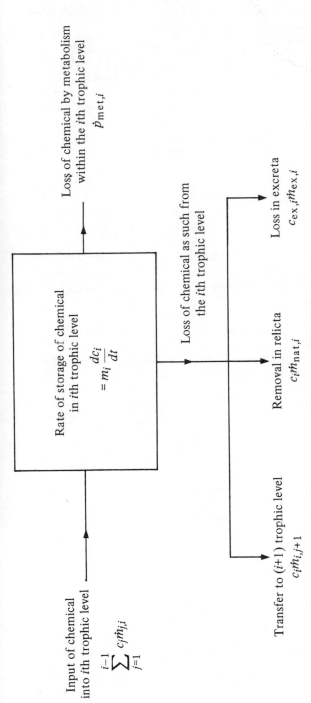

FIG. 12.5. Movement (in and out) and degradation of a chemical in a trophic level. $\dot{m}_{j,i}$ is the rate of transfer of biomass from the jth trophic level to the ith level; c_j is the mean concentration of the chemical in the biomass of the jth level; m_i and c_i are the biomass of the ith level, and the mean concentration of the chemical in that biomass respectively; $\dot{m}_{i,i+1}$ is the rate of transfer of biomass (by predation) from the ith to the $(i+1)$th trophic level; $\dot{m}_{nat,i}$ is the rate of loss of biomass from the ith level by natural causes (other than predation); $c_{ex,i}$ is the mean concentration of the chemical in excreta of organisms of the ith level; and $\dot{m}_{ex,i}$ is the rate of formation of excreta. The relationship

$$\sum_{j=1}^{i-1} c_i \dot{m}_{j,i} = m_i \frac{dc_i}{dt} + c_i \dot{m}_{i,i+1} + \dot{m}_{nat,i} + c_{ex,i}\dot{m}_{ex,i} + \dot{p}_{met,i}$$

follows from the principle of mass balance.

that the concentration of HEOD in the eggs of Scottish Golden Eagles (*Aquila chrysaetos*) has declined in recent years (their conclusions, unfortunately, appear to be unsound, but at least a continuing increase has not been demonstrated); Robinson *et al.* (1967) and Coulson *et al.* (1972) found that the concentrations of both p,p'-DDE and HEOD in the eggs of Shags (*Phalacrocorax aristotelis*) from a colony on the Farne Islands have declined significantly since about 1967. Taken in isolation, this decline does not necessarily disprove the conclusion of Harrison *et al.*, but, as there are no compelling reasons for regarding the input of HEOD into the Shag colony, as arising from an isolated pulse of dieldrin, and as there are other indications of declining concentrations of HEOD or DDT in other components of the biosphere, it is considered that their conclusions require modification, in order to allow for the effect of metabolism upon the elimination of DDT from organisms in the various trophic levels.

It is apparent from the contents of this chapter that there are considerable deficiencies in our knowledge of the fate of pesticides in the environment, particularly those which are relatively stable (namely, some of the organochlorine insecticides). As a result of the slow degradation of these compounds, by physical, chemical or biochemical mechanisms, they are potentially available for transport over large distances. However, in addition to the processes of degradation in the biosphere, there are also scavenging processes, such as washout from the atmosphere by rain. It is tentatively concluded, that the available evidence indicates a relatively localised distribution around the areas of use, rather than a global dispersion. This does not exclude the possibility of a global dispersion at a very low level (see Chapter 10), but there are formidable problems to be overcome in demonstrating the existence of such global dispersion beyond reasonable doubt.

Studies of the fate of other chemicals in the environment throw considerable light upon the possible mechanisms of transport of pesticides in the biosphere. The models developed from such studies, with appropriate adaptations, are of use in the case of pesticides. It seems probable that the continued use of even the 'persistent' pesticides will result in an approach to a steady state, in which the total annual loss by the various degradation processes will balance the annual rate of usage. However, even when the steady state is approached, it seems improbable that there will be a uniform distribution within each of the major components discussed above. In the case of a particular trophic level, for example, there will be intraspecific variations between different geographical locations (because of the tendency for localised dispersion as a result of scavenging processes), as well as interspecific variations.

The importance of reliable information on the rates of input of pesticides into particular parts of the environment, as well as into the biosphere as a whole, has been stressed. This information, together with suitable studies to monitor the changes in the concentrations of the pesticides in various parts of the environment, and the development of suitable models of the transport mechanisms in ecosystems, will enable us to continue the use of pesticides with reasonable confidence that permanent and irreversible damage to the biosphere will not result. By contrast, we should be able to detect that such potential hazards are arising, and discontinue the use of a particular chemical long before a catastrophe, local or general, is imminent.

REFERENCES

C. D. Abbott, R. B. Harrison, J. O'G. Tatton and J. Thompson, *Nature* (Lond.), **208**, 1317 (1965)

C. D. Abbott, R. B. Harrison and J. O'G. Tatton, *Nature* (Lond.), **211**, 259 (1966)

N. B. Akesson and W. E. Yates, *Annu. Rev. Ent.*, **9**, 285 (1964)

N. B. Akesson, W. E. Yates, H. H. Coutts and W. E. Burgoyne, *Agric. Aviat.*, **6**, 72 (1964)

P. Antommaria, M. Corn and L. De Maio, *Science*, **150**, 1476 (1965)

G. L. Atkins, Multicompartmental Models for Biological Systems. Methven, London (1964)

M. K. Baldwin, The Metabolism of the Chlorinated Insecticides Aldrin, Dieldrin and Endrin. Ph.D. Thesis, University of Surrey (1971)

S. Beilke and H-W. Georgii, *Tellus*, **20**, 435 (1968)

P-E. E. Bergner, *J. theor. Biol.*, **1**, 120 (1961a)

P-E. E. Bergner, *J. theor. Biol.*, **1**, 359 (1961b)

W. Bischof and B. Bolin, *Tellus*, **18**, 155 (1966)

B. Bolin and H. Stommel, *Deep Sea Res.*, **8**, 95 (1961)

C. H. Bosanquet and J. L. Pearson, *Trans. Faraday Soc.*, **32**, 1249 (1936)

H. V. Brewerton, *N.Z. J. Science*, **12**, 194 (1969)

G. T. Brooks, *Residue Reviews*, **27**, 81 (1969)

J. H. Caro and A. W. Taylor, *J. Agr. Food Chem.*, **19**, 379 (1971)

A. C. Chamberlain, *Q. Jl. R. met. Soc.*, **85**, 350 (1959)

A. C. Chamberlain and R. C. Chadwick, *Tellus*, **18**, 226 (1966)

T. T. Chow and C. C. Patterson, *Earth Planetary Sci. Letters*, **1**, 397 (1966)

J. M. Cohen and C. Pinkerton. R. F. Gould (Editor), Organic Pesticides in the Environment p. 163. Amer. Chem. Soc., Washington, D.C. (1966)

J. C. Coulson, I. R. Deane, G. R. Potts, J. Robinson and A. N. Crabtree, *Nature* (Lond.), **236**, 454 (1972)

J. Crabtree, *Q. Jl. R. met. Soc.*, **85**, 362 (1959)

H. Craig, *Tellus*, **11**, 1 (1957)

H. Graig, Earth Science and Meteoritics, p. 103. North-Holland Publishing Co. Amsterdam (1963)

W. C. Dauterman, *Bull. World Hlth. Org.*, **44**, 133 (1971)

W. B. Deichman, I. Dressler, M. Keplinger and W. E. MacDonald, *Ind. Med. Surg.*, **37**, 837 (1968)

A. C. Delaney, D. W. Parkin, J. J. Griffin, H. D. Goldberg and B. E. F. Reimann, *Geochim et Cosmoch. Acta*, **31**, 885 (1967)

W. F. Durham, J. F. Armstrong and G. E. Quinby, *Arch. environ. Hlth.*, **11**, 76 (1965)

W. F. Durham, P. Ortega and W. J. Hayes, Jr., *Arch. int. Pharmacodyn.*,
141, 111 (1963)

W. F. Durham and H. R. Wolfe, *Bull. World Hlth. Org.*, 26, 75 (1962)

L. L. Eberhardt, *J. theor. Biol.*, 24, 43 (1969)

W. C. Hanson and L. L. Eberhardt, *Hlth. Physics*, 17, 793 (1969)

L. L. Eberhardt and R. F. Nakatini, *Proc. 2nd. Nat. Symposium Radioecology*, p. 740. Ann Arbor (1967)

L. L. Eberhardt and R. F. Nakatini, *J. Fish Res. Bd Can.*, 25, 591 (1968)

L. L. Eberhardt, R. L. Meeks and T. J. Peterle, Battelle Memorial Inst.
Rep., BNWL-1297, VC-48, Richland (1970)

C. A. Edwards, Critical Reviews in Environmental Control, Chem. Rubber
Co. Cleveland (1970)

C. A. Edwards, A. R. Thompson, K. I. Beynon and M. J. Edwards, *Pestic.
Sci.*, 1, 169 (1970)

D. J. Eye, *J. Water Pollution Control Fed.*, 40, R.316 (1968)

N. Gannon, R. P. Link and G. C. Decker, *J. Agric. Food Chem.*, 7, 826 (1959)

W. D. Garrett, *Deep Sea Res.*, 14, 221 (1967)

J. L. George and D. E. H. Frear, *J. Appl. Ecology*, 3 (Suppl.), 155 (1966)

E. D. Goldberg, P. Butler, P. Meier, D. Menzel, R. W. Risebrough and L. F.
Stickel. Chlorinated Hydrocarbons in the Marine Environment,
National Acad. Sci., Washington, D.C. (1971)

V. M. Goldschmidt, Geochemistry, Chapter 4. Clarendon Press, Oxford
(1962)

J. W. Hamaker. R. F. Gould (Editor), Organic Pesticides in the Environment, p. 122. Amer. Chem. Soc., Washington, D.C. (1966)

C. I. Harris, *J. Agric. Food Chem.*, 14, 398 (1966)

H. L. Harrison, O. L. Loucks, J. W. Mitchell, D. F. Parkhurst, C. R. Tracy,
D. G. Watts and V. J. Yannacone, Jr. *Science,* 170, 503 (1970)

W. J. Hayes, Jr., DDT, The Insecticide Dichlorodiphenyltrichloroethane
and its significance, vol. 2, chapter 6. Birkpauser Verlag, Basel (1959)

W. J. Hayes, Jr., W. F. Durham and C. Cueto, *J. Am. med. Ass.*, 162, 890
(1956)

W. J. Hayes, Jr., C. I. Pirkle and W. E. Dale, *Arch. environ. Hlth.*, 22, 119
(1971)

E. Hindin, D. S. May and G. H. Dunstan. R. F. Gould (Editor), Organic
Pesticides in the Environment, p. 132. Amer. Chem. Soc., Washington,
D.C. (1966)

R. M. Hollingworth, *Bull. Wld. Hlth. Org.*, 44, 155 (1971)

C. G. Hunter and J. Robinson, *Arch. environ. Hlth.*, 15, 614 (1967)

C. G. Hunter, J. Robinson and M. Roberts, *Arch. environ. Hlth.*, 18, 12
(1969)

E. Husserl, Logische Untersuchungen, vol. 2, pp. 19ff. Max Niemeyer,
Halle (1901)

K. W. Jager, Aldrin, Dieldrin and Telodrin: An Epidemiological and
Toxicological Study of Long-term Occupational Exposure, p. 136c.
Elsevier, Amsterdam (1970)

N. L. Jarves *et al.*, *Limnol. Oceanogr.*, 12, 88 (1967)

W. T. Keane and M. R. Zavon, *Bull. environ. Contamination Toxicol.*, 4, 1
(1969)

P. C. Kearney and D. D. Kaufman, Degradation of Herbicides. Marcel

Dekker and Corpn., New York (1969)

C. D. Keeling, *Arch. environ. Hlth.*, **20**, 764 (1970)

C. D. Keeling and B. Bolin, *Tellus,* **19**, 566 (1967)

C. D. Keeling and B. Bolin, *Tellus,* **20**, 17 (1968)

F. Korte, Botyu Kagaku, **32**, 46 (1967)

D. Lal, Earth Science and Meteoritics, p. 115. North-Holland Publishing Co., Amsterdam (1963)

E. P. Laug and O. G. Fitzhugh, *J. Pharmacol.,* **87**, 18 (1946)

E. Lehman and M. Möller, *Bundesgesundheitblatt,* **17**, 261 (1967)

S. H. Lin, R. Sahai and H. Eyring, *Proc. natn. Acad. Sci. Wash.,* **68**, 777 (1971)

E. P. Lichtenstein, C. H. Mueller, G. R. Myrdal and K. R. Schulz *J. econ. Ent.,* **55**, 215 (1962)

J. D. Lockie, D. A. Ratcliffe and R. Balharry, *J. appl. Ecology,* **6**, 381 (1969)

C. Lorius, G. Baudin, C. Cittanova and R. Platzer, *Tellus,* **21**, 138

K. J. Macek, The Biological Impact of Pesticides in the Environment, p. 17. Oregon State University Press, Corvallis (1970)

K. P. Makhon'Ko, *Tellus,* **19**, 467 (1967)

S. Manabe and K. Bryan, *J. Atmos. Sci.,* **26**, 786 (1969)

K. A. McCully, D. C. Villeneuve, W. P. McKinley, W. E. J. Phillips and M. Hidirogtou, *J. Ass. off. agric. Chem.,* **49**, 966 (1966)

F. Matsumura and G. M. Boush, *Science,* **156**, 959 (1967)

F. Matsumura, G. M. Boush and A. Tai, *Nature, Lond.,* **219**, 965 (1969)

D. P. Morgan and C. C. Roan, *Arch. environ. Hlth.,* **22**, 301 (1971)

H. L. Motto, R. H. Daines, D. Chilko and C. K. Motto, *Environ. Sci. Technol.,* **4**, 231 (1970)

R. J. Moubry, J. M. Helm and G. R. Myrdal, *Pesticide Monitoring J.,* **1**, 27 (1968)

P. Ortega, J. Wayland Hayes, W. F. Durharn and A. Mattson, DDT in the diet of the rat. Monograph No. 43, U.S. Dept. Health Education and Welfare, Washington, D.C. (1956)

J. C. Pales and C. D. Keeling, *J. geophys. Res.,* **70**, 6077 (1965)

P. O. Park and C. E. McClone, *Trop. Agriculture, Trin.,* **43**, 133 (1966)

F. Pasquill, Atmospheric Diffusion. Van Nostrand, London (1962)

C. C. Patterson, *Arch. environ. Hlth.,* **11**, 344 (1965)

T. J. Peterle, *Nature, Lond.,* **224**, 620 (1969)

H. Reiquam, *Science,* **170**, 318 (1970a)

H. Reiquam, *Atmos. Environment,* **4**, 233 (1970b)

R. Revelle, Restoring the Quality of the Environment, Appendix Y4. Report of the Presidents Science Advisory Committee, Washington, D.C. (1965)

L. A. Richardson, J. R. Lane, W. S. Gardner, J. T. Peeler and J. E. Campbell, *Bull. environ. Contamination Toxicol.,* **2**, 207 (1967)

A. Richardson and J. Robinson, *Xenobiotica,* **1**, 213 (1971)

R. W. Risebrough, J. R. Huggett, J. J. Griffin and E. D. Goldberg, *Science,* **159**, 1233 (1968)

C.C. Roan, D.Morgan and E.M. Paschal, *Arch. environ. Hlth.,* **22**, 309 (1971)

J. Robinson, *Proc. 3rd International Mtg. on Forensic Immunology, Medicine, Pathology and Toxicology,* London (1963)

J. Robinson, *Symposium of the Roy. Institute Chemistry on Problems of Food Additives*, Manchester, October (1964)

J. Robinson, *Nature, Lond.*, **215**, 33 (1967a)

J. Robinson, *Proc. 4th Insecticide Fungicide Conference*, **1**, 36 (1967b)

J. Robinson, *Proc. 6th International Congress of Plant Protection*, Vienna (1967c)

J. Robinson, *Canad. med. Ass. J.*, **100**, 180 (1969a)

J. Robinson, Chemical Fallout, p. 113. C. C. Thomas, Springfield (1969b)

J. Robinson, *Bird Study*, **17**, 195 (1970a)

J. Robinson, The Biological Impact of Pesticides in the Environment, p. 54, Oregon State University Press, Corvallis (1970b)

J. Robinson, *Proc. Soc. Anal. Chem.*, **8**, 57 (1971a)

J. Robinson, *Chemistry in Britain*, **7**, 472 (1971b)

J. Robinson, A. Richardson and K. E. Elgar, *152nd Mtg. Amer. Chem. Soc.*, New York (1966)

J. Robinson, A. Richardson, A. N. Crabtree, J. C. Coulson and G. R. Potts, *Nature* (Lond.), **214**, 1307 (1967)

J. Robinson and M. Roberts, Physico-chemical and Biophysical Factors affecting the Activity of Pesticides, p. 106. Monograph No. 29, Society Chemistry and Industry, London (1968)

J. Robinson and M. Roberts, *Food Cosmetic Toxicol.*, **7**, 501 (1969)

J. Robinson, M. Robert, M. Baldwin and A. I. T. Walker, *Food Cosmetic Toxicol.*, **7**, 317 (1969)

R. L. Rudd, Pesticides and the Living Landscape. University of Wisconsin Press, Madison (1964)

R. S. Scorer, *International J. Air Pollution*, **1**, 198 (1959)

D. B. Seba and E. F. Corcoran, *Pesticides Monitoring J.*, **3**, 190 (1969)

W. L. Sladen, C. M. Menzie and W. L. Reichel, *Nature, Lond.*, **210**, 670 (1966)

B. I. Sparr, W. G. Appleby, D. M. DeVries, J. V. Osman, J. M. McBride and G. L. Foster, Organic Pesticides in the Environment, p. 146. *Amer. Chem. Soc.*, Washington, D.C. (1966)

O. G. Sutton, Micrometeorology. McGraw-Hill, New York (1953)

H. V. Sverdrup, The Earth as a Planet, Chapter 5. University of Chicago Press (1954)

E. C. Tabor, *Trans. N. Y. Acad. Sci.*, **28**, 569 (1966)

K. R. Tarrant and J. O'G. Tatton, *Nature, Lond.*, **219**, 725 (1968)

M. Tatsumoto and C. C. Patterson, *Nature, Lond.*, **199**, 350 (1963a)

M. Tatsumoto and C. C. Patterson, Earth Science and Meteoritics, p. 74. North-Holland Publishing Co., Amsterdam (1963)

J. O'G. Tatton and J. H. A. Ruzicka, *Nature, Lond.*, **215**, 346 (1967)

A. R. Thompson, C. A. Edwards, M. J. Edwards and K. I. Beynon, *Pestic. Sci.*, **1**, 174 (1970)

C. M. Tu, J. R. W. Mites and C. R. Harris, *Life Sci.*, **7**, 311 (1968)

H. C. Urey, The Planets: Their Origin and Development. Yale University Press, New Haven (1952)

J. G. Wagner, Biopharmaceutics and relevant Pharmacokinetics. Hamilton Press Inc., Hamilton (1971)

A. I. T. Walker, D. E. Stevenson, J. Robinson, E. Thorpe and M. Roberts, *Toxicol. appl. Pharmacol.*, **15**, 345 (1969a)

A. I. T. Walker, C. H. Neill, D. E. Stevenson and J. Robinson, *Toxicol.*

appl. Pharmacol., **15**, 69 (1969b)

S. R. Weibel, R. B. Weidner, J. M. Cohen and A. G. Christianson, *J. amer. Water Works Ass.*, **58**, 1075 (1966)

G. A. Wheatley and J. A. Hardman, *Nature, Lond.*, **207**, 486 (1965)

H. Windom, J. Griffin and E. D. Goldberg, *Environ. Sci. Technol.*, **1**, 923 (1967)

G. M. Woodwell, C. F. Wurster and P. A. Isaacson, *Science N.Y.*, **156**, 821 (1967)

W. E. Yates, N. B. Akesson and H. H. Coutts, *Trans. Amer. Soc. agric. Engrs.*, **9**, 389 (1966)

D. Yeo, N. B. Akesson and H. H. Coutts, *Nature* (Lond.), **183**, 131 (1959)

D. Yeo and B. W. Thompson, *Nature, Lond.*, **172**, 168 (1953)

Chapter 13

Degradation of
Pesticide Residues in the Environment

F. Matsumura

Department of Entomology,
University of Wisconsin,
Madison, Wisconsin

13.1   CONCEPT OF TERMINAL RESIDUES

The degradation of pesticide residues in the environment can be affected by various factors. Generally speaking, the single most important factor contributing to the speed of degradation is the chemical nature of the pesticide itself. For persistent pesticides, for instance DDT, it is difficult to find the physical or biological means to degrade such a compound. On the other hand, if the pesticide is labile, many physical and biological processes can be found to degrade the compound.

The emphasis of this chapter is on the stable and persistent pesticides

494

which cause environmental problems by accumulating in various eco-systems and locations, because of their chemical stabilities and often accompanying affinities to organic matter.

It is important to remember that, for these pesticides, degradation of the original compounds does not always mean an immediate elimination of the hazard. On the contrary, it is often one finds that an equally or more toxic metabolic product appears after the original compound has dis-appeared. Formation of photodieldrin from dieldrin, by the action of both sunlight (Rosen *et al.*, 1966 and Robinson *et al.*, 1966) and micro-organisms (Matsumura *et al.*, 1970), exemplifies such processes. Many of the conversion products are stable and cause just as many residual problems as the original compounds. These breakdown compounds are often called 'terminal residues'. DDD (TDE), DDE, dicofol, etc., are terminal residues of DDT, for instance, while dieldrin, photoaldrin, photo-dieldrin, etc., are terminal residues of dieldrin.

13.2 FACTORS INFLUENCING RESIDUAL CHARACTERISTICS

For a given compound, there are many factors influencing its residual characteristics. A pesticide that has fallen onto the soil surface, for instance, can be: (a) absorbed by soil constituents; (b) leached by rain water; (c) picked up by plants and animals; (d) evaporated either directly or with water vapour; or (e) carried away by wind, all of which can reduce the level of the pesticide at the site. It is important to note, however, that most of the processes mentioned here represent merely translocation phenomena. When one thinks of the environment as a whole (i.e. one enclosed system), the total amount of the pesticide is not decreased by the above-mentioned processes. The processes that actually play important roles in decreasing the total amount of the actual pesticidal residues, are the ones mediated by micro-organisms, animals, plants and sunlight. Other factors, such as pH, heat, catalytic agents in the soil and soil enzymes, also play important roles in degrading relatively unstable pesticides, e.g. simazine and atrazine by soil pH (Hamilton and Moreland, 1962 and Armstrong *et al.*, 1967), and malathion by pH and possibly soil enzymes (Tiedje and Alexander, 1967 and Getzin and Rosefield, 1968).

For the stable pesticides, only under extreme conditions (such as high temperature and high alkalinity against BHC) should such factors have any significant effect upon the metabolic fate of the chemicals (Yoshida and Castro, 1970 and Bradbury, 1963, etc.).

13.3 DEGRADATION BY MICRO-ORGANISMS AND SOILS

The major groups of soil micro-organisms are actinomycetes, fungi and bacteria (Alexander, 1965). They can degrade pesticides through β-oxida-tion, ether-cleavage, ester and amide hydrolysis, oxidation of alcohols and aldehydes, dealkylation, hydroxylation, hydrohalogenation, epoxidation, reductive dehalogenation and N-dealkylation, etc. The most noticeable characteristics of degradation systems in micro-organisms are the reductive systems. Examples can be found in the reductive dechlorination reaction of DDT to form TDE (Plimmer *et al.*, 1968), production of aldrin from dieldrin (Matsumura *et al.*, 1968), in contrast to a reverse process of epoxidation of aldrin which commonly occurs, and conversion of parathion to aminoparathion (Menzel *et al.*, 1961). The ability to enzymatically cleave the aromatic ring itself is limited to micro-organisms. Utilization of pesticidal analogues, or their basic skeletons as a sole carbon source, is also a unique property of microbial metabolism. However, micro-organisms lack an efficient conjugation system which would convert xenobiotics into excretable forms as in the case of higher animals. Furthermore, micro-organisms do not generally possess distinct microsomal mixed function oxidase systems.

13.3.1 Organochlorine insecticides
13.3.1.1 *DDT analogues*
The major microbial metabolic steps are the reductive dechlorination reaction and the oxidative system. For instance, DDT is degraded mainly by the former route (Fig. 13.1) to yield a series of dechlorinated analogues, namely TDE (Finley and Pillmore, 1963), DDMS and DDNS.* Two other enzymatic reactions influence the metabolism the oxidative system to either hydroxylate at the number 2 carbon to yield dicofol and FW-152 from DDT and TDE (Matsumura and Boush, 1968), and the dehydrochlorination system to yield DDE, DDMU and DDNU (Matsumura *et al.*, 1971).

Another important degradation pathway of organochlorine insecticides is the process of dehydrochlorination. Both BHC and DDT can be degraded by this route.

Cleavage of the aromatic rings to form aliphatic carbons, or ringopening of cyclic hydrocarbons, is not a common metabolic reaction. Focht and Alexander (1971) demonstrated that the chlorine atoms at p,p'-positions provide the stability of DDT, inasmuch as its analogues without these chlorine atoms degrade quickly to yield phenol and benzoic acid. By the same token, substitution of these chlorine atoms with various alkoxy and alkylthio groups greatly enhance the biodegradation probabilities (Kapoor *et al.*, 1970; Mendel *et al.*, 1967).

13.3.1.2 *BHC*
BHC has long been known to disappear comparatively quickly in the soil. Both microbial actions and soil alkalinity could play important roles in

---

* The common naming of the DDT metabolites are not according to the official method of nomenclature. DDMS roughly stands for 2,2-dichlorodiphenyl-1-monochlorinated saturated ethane, DDNS for 2,2-dichlorodiphenyl-1-nonechlorinated saturated ethane and DDMU for 2,2-dichlorodiphenyl-1-monochlorinated unsaturated ethylene.

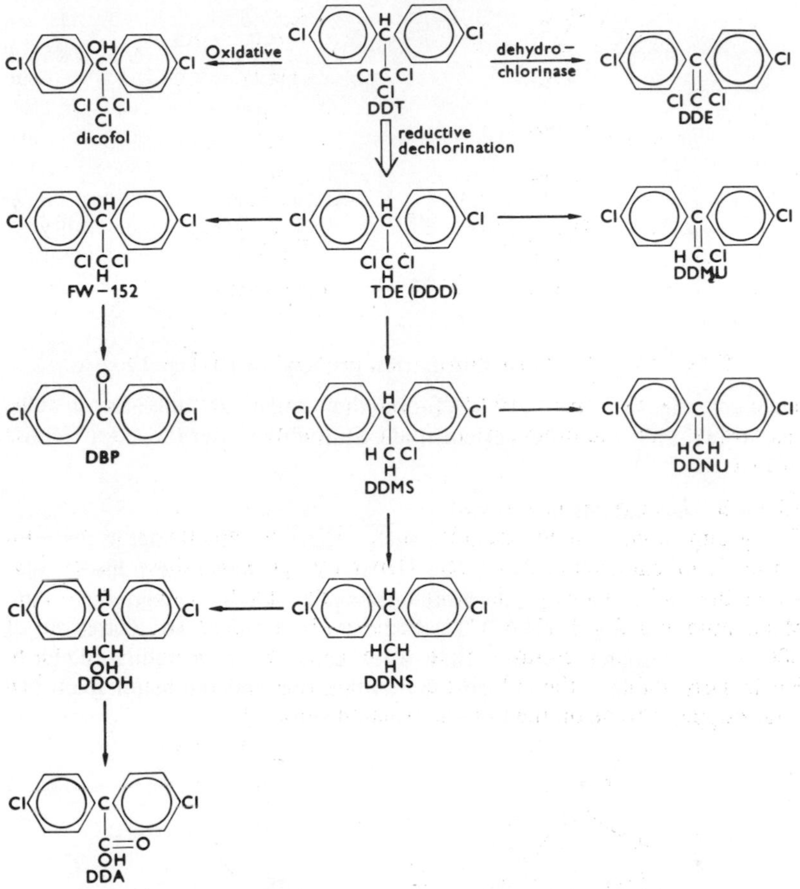

FIG. 13.1.   General degradation pattern of DDT by micro-organisms.

degrading BHC (Bradbury, 1963). The dehydrochlorination process mentioned above (Fig. 13.2) also appears to play an important role (Yule *et al.*, 1967). The major metabolic product of such a process of gamma-BHC is gamma-pentachlorocyclohex-1-ene (γ-PCCH). Other dehydrochlorination products, such as 1,2,3,5-tetrachlorobenzene (Allan, 1955), could also form as the result of further microbial attack on PCCH. It is generally acknowledged that BHC can be degraded much more easily under anaerobic than in aerobic conditions (Raghu and MacRae, 1966; Yoshida and Castro, 1970; Sethunathan, 1970). For this reason BHC disappears rather rapidly in submerged paddy fields. The latter authors, however, found that the major metabolic product produced by *Clostridium* sp. micro-organisms isolated from rice fields was not PCCH.

The most important aspect of BHC degradation in the environment is the difference in the rate of disappearance between the BHC isomers. A

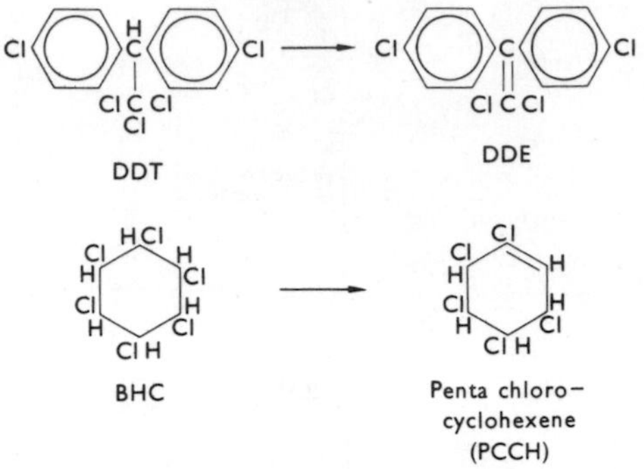

FIG. 13.2.   Dehydrochlorination process for DDT and BHC.

Japanese scientist (Goto, 1970) found that β-BHC accumulated in soils, rice straws, milk and other agricultural commodities after the extensive use of BHC.

13.3.1.3   *Cyclodiene insecticides*

This group of compounds includes such important insecticides as dieldrin, aldrin, heptachlor, chlordane, etc. Generally speaking, these insecticides are stable and not many micro-organisms are capable of degrading them. Matsumura and Boush (1967), for instance, found only 10 isolates out of 600 soil microbial cultures that were capable of degrading dieldrin. Particularly stable is the chlorine-containing ring and the major microbial attack appears to be on the non-chlorinated rings.

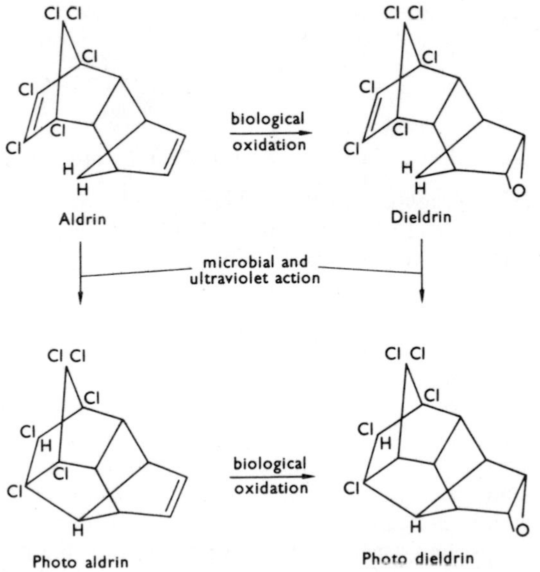

FIG. 13.3.   Environmental conversion of aldrin.

The most thoroughly studied microbial metabolic process of cyclo-
dienes is the expoxidation reaction which converts aldrin to dieldrin,
heptachlor to heptachlor epoxide and isodrin to endrin (Kiigemagi *et al.*,
1958 and Lichtenstein and Schulz, 1960). This is an exceedingly common
oxidation process in many biological systems. Micro-organisms also appear
to be capable of reducing dieldrin to form aldrin: a reaction which has not
been reported in higher animals and plants.

Other characteristic reactions associated with cyclodiene metabolism
are rearrangement processes such as the intramolecular bridge formation
(e.g. photodieldrin and photoaldrin formation (Fig. 13.3)), (Rosen *et al.*,
1966; Robinson *et al.*, 1966) and the rearrangement process of epoxy-ring
to form ketones, aldehydes and alcohols (Fig. 13.4) (Matsumura *et al.*,
1968, 1971). The significance of such rearrangement reactions is not
known. However, it is not likely that the micro-organisms are able to
derive any energy from the processes. The rearrangements could, at least,
give the micro-organisms the chance to degrade the compounds further. In

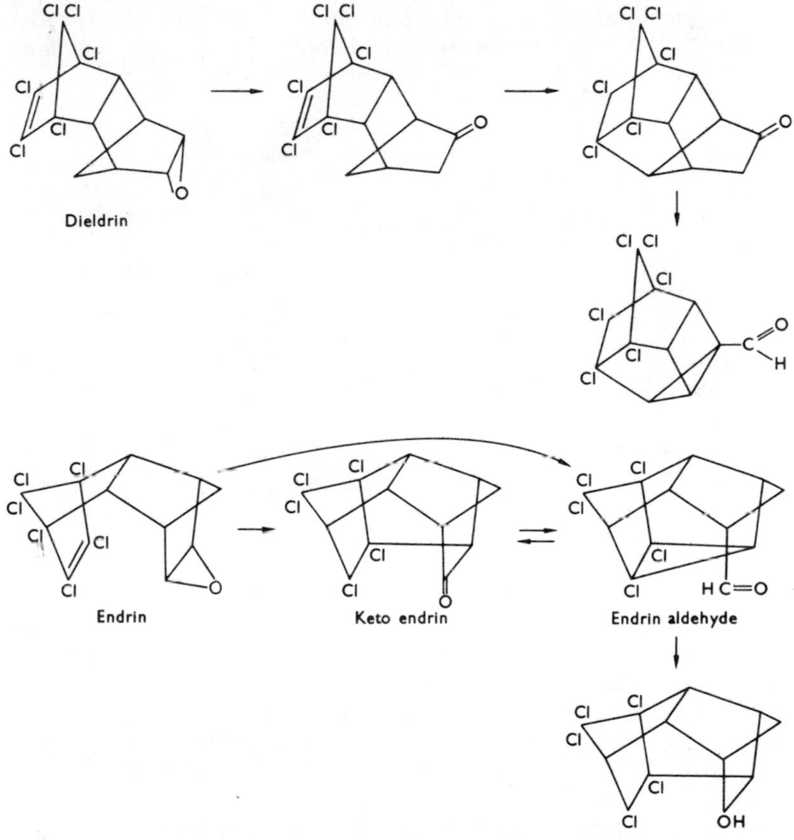

FIG. 13.4.   Microbial rearrangement processes for dieldrin and endrin.

most cases, such reactions yield more polar compounds than the parent molecules. Another important process of cyclodiene degradation is the hydrolysis reaction of the epoxy ring to yield dihydroxy analogues (Wedemyer, 1968 and Matsumura *et al.*, 1968), and also a combined effect of chemical and microbial action to produce 1-hydroxy-2,3-epoxy-chlordane from heptachlor epoxide (Miles *et al.*, 1969).

13.3.2 Organophosphate and carbamate insecticides

The most common reaction mechanism evolved by micro-organisms to degrade these insecticides is the hydrolysis process through esterases (Matsumura and Boush, 1966, 1968). The oxidative processes which are important in higher animals in degrading these compounds are less frequently observed. This is probably because of the lack of defined mixed function oxidase systems in micro-organisms.

The presence of reductive degradation systems among the micro-organisms is, however, quite apparent. For instance, the major degradation route for parathion (*O,O*-diethyl-*O*-*p*-nitrophenyl thiophosphate) is the formation of aminoparathion (*O,O*-diethyl-*O*-*p*-aminophenyl thiophosphate) (Fig. 13.5) (Cook, 1957 and Ahmed *et al.*, 1958).

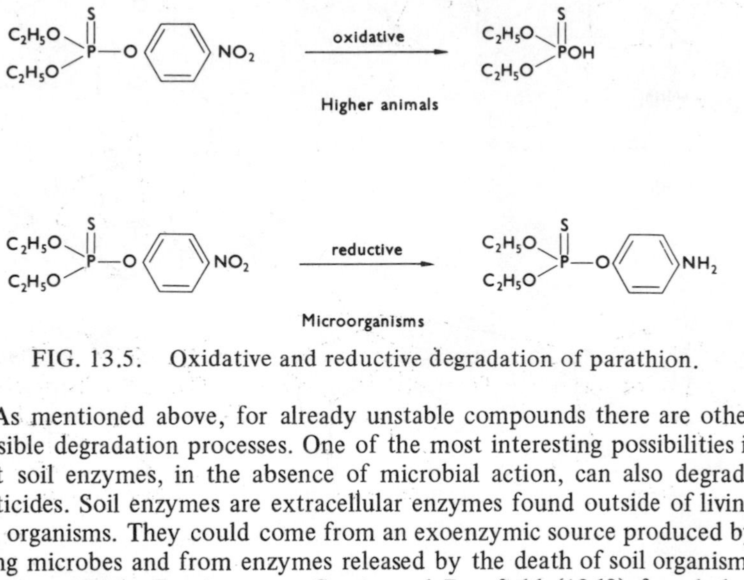

FIG. 13.5.   Oxidative and reductive degradation of parathion.

As mentioned above, for already unstable compounds there are other possible degradation processes. One of the most interesting possibilities is that soil enzymes, in the absence of microbial action, can also degrade pesticides. Soil enzymes are extracellular enzymes found outside of living soil organisms. They could come from an exoenzymic source produced by living microbes and from enzymes released by the death of soil organisms (Skujins, 1966). For instance, Getzin and Rosefield (1968) found that sterilisation of soil samples by autoclaving, destroyed 90% of the degradation activity of the soil, while a gamma radiation treatment (4 mrad, at 250,000 rads/hr) hardly affected its metabolic activity. The authors were able to extract a fraction with 0.2 N NaOH that could actively degrade malathion.

13.3.3   Degradation of herbicides by micro-organisms and soils

Much more is known about the metabolic activity of herbicides in soils than any other group of pesticides. This is partly because the residual effectiveness of these herbicidal pesticides has been of great economic concern in the past. Overall persistences of herbicides are in the order of 10 to 18 months for urea and triazine herbicides, 6 months for toluidine, 1-5 months for phenoxyalkanoic acids (2,4-D), and 3 months for aliphatic and carbamate herbicides (all expressed in the length of time required to decrease the levels to 0-25% of the original level, Kearney, 1970). This marks a sharp contrast to the similar values for organochlorine insecticides: e.g. 4 years for DDT, 5 years for chlordane, 2-3 years for BHC and most of the other cyclodiene insecticides. While insecticides are generally acknowledged to be more hazardous to animals because of their toxicities, some of the herbicidal residues, contaminants (such as tetrachlorobenzodioxin in 2,4,5-T preparations), and some of their metabolic products, could also cause serious environmental problems through their toxic actions against non-target plants, plankton and sometimes to animals.

13.3.3.1   *Halogenated aliphatic acids*
The major degradation route for this group of compounds by micro-organisms is dehalogenation (Kearney *et al.*, 1965). While the process is generally catalysed by hydrolytic enzymes, to release halide ions (Jensen, 1957), in the presence of some halide acceptor, degradation may proceed without being accompanied by the release of halide ions (Hirsch and Alexander, 1960). The α-substituted halogens (e.g. α-chlorinated propionic acid, see below) are more readily dehalogenated:

$$\underset{\text{Cl}}{\text{R CHCOO}^-} + H_2O \longrightarrow \underset{\text{OH}}{\text{R C COO}^-} + H^+ + Cl^-$$

than corresponding β-substituted analogues (Kearney *et al.*, 1964). The result of dehalogenation, as seen in the above reaction scheme, is replacement of the halogen atom with a hydroxy group. The reaction often yields alcohols and ketones, e.g.:

$$\underset{\text{Cl}}{\overset{\text{Cl}}{CH_3CCOO^-}} + OH^- \longrightarrow CH_3\overset{\text{O}}{\overset{\|}{C}} COO^- + 2Cl^- + H^+$$

2,2-dichloropropionate   pyruvate

13.3.3.2   *Phenylureas and phenylcarbamates*
Phenylureas can be readily degraded by soil micro-organisms (Sheets, 1964). The rate of degradation appears to be favoured in relatively

alkanoic acid (Fig. 13.9) is susceptible to microbial attack. This process reduces the herbicidal activity of the compounds. Micro-organisms reported to degrade 2,4-D through this route are, for instance, a *Flavobacterium* (MacRae and Alexander, 1963) and a soil *Arthrobacter* (Loos *et al.*, 1965). The β-oxidation process is an important activation process for this group of compounds. It occurs naturally in soil and in microbial cultures. It is likely that the herbicides in the odd-numbered aliphatic acids will become phenoxypropionic acid derivatives, while those with the even-numbered acids will form corresponding phenoxyacetic acids. Ring hydroxylation is also a common enzymatic process. For 2,4-D and MCPA both the 5 (*Aspergillus niger*) (Faulkner and Woodcock, 1964) and the 6 positions (*Pseudomonas*) appear to be susceptible for such an enzymatic attack.

13.4   DEGRADATION BY SUNLIGHT

While there are a number of physical factors known to influence the residual fates of pesticides in nature (e.g. light, air, surfaces, moisture, pH, etc.) the effect of sunlight, particularly the ultraviolet portion of sunlight, appears to make the most significant contribution. The sunlight reaching the surface of the earth does not show any shorter ultraviolet component than 300 mμ. This is because the atmosphere of the earth effectively eliminates such shortwave ultraviolet rays (Koller, 1965). It is possible, therefore, that artificial UV radiation, such as by using an intense mercury lamp, for instance, can create degradation products which are not produced by the action of the natural sunlight.

One of the most important factors affecting the rate of sunlight-degradation of pesticides and other organic chemicals, is the presence of photosensitizers. Photosensitizers are the chemicals which facilitate the transfer of the energy of light into the receptor chemicals. In the past, photolytic research work has been carried out in the presence and absence of photosensitizers. No significant qualitative differences have been found in the metabolic (photolytic) routes between experiments with and without a photosensitizer.

The initial discovery of a photosensitizer for a pesticidal compound was made by Bell (1956), who reported that riboflavin sensitized the photo-decomposition of 2,4-D. Rosen and Carey (1968) and Rosen *et al.* (1969, 1970) used both benzophenone and riboflavin-5'-phosphate (FMN) as sensitizers for their studies on photodecomposition of pesticides. Ivie and Casida (1970, 1971a) found rotenone and other pesticidal and non-pesticidal sensitizers for degradation of various insecticides. In addition to rotenone, some aromatic amines, anthraquinone (which showed the broadest spectrum) and benzophenone were found to act as good photo-sensitizers. Insecticidal combinations which synergistically acted as sensi-tizers are: abate-dieldrin, barban, fenitrothion, phenothiazine-DDT and rotenone-dieldrin. The same authors (1970b) also reported that substituted 4-chromanones and rotenone, sensitized dieldrin conversion to photodieldrin on bean leaves. Chlorophylls from spinach chloroplasts, plus rotenone, also acted as sensitizers for organophosphates, carbamates, pyrethroids, and dinitrophenol insecticides.

13.4.1 Organochlorine insecticides

There appear to be two types of major reaction processes catalysed by ultraviolet radiation affecting this group of compounds: an intramolecular rearrangement process to form isomers of the original compound (Fig. 13.10), and the dechlorination process. The former process is best

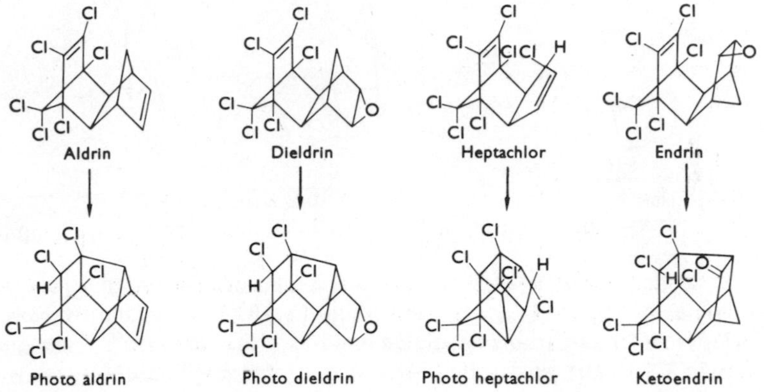

FIG. 13.10. Photochemical intramolecular rearrangement processes for cyclodiene insecticides.

exemplified by the 'cage formation' reaction of dieldrin (Rosen et al., 1966; Robinson et al., 1966). Rosen and Sutherland (1967) further confirmed that a similar photochemical reaction can take place with aldrin to form caged photoaldrin. These reaction products can be found in nature. For instance, Robinson et al. (1966) and Korte and his associates (1967) found photodieldrin on the surface of plant leaves. Lichtenstein et al. (1970) found photodieldrin in soil samples which had previously been treated with a large amount of aldrin in the field.

For endrin photolysis reactions, Rosen et al. (1966) found that ketoendrin (Fig. 13.11) forms as a result of UV irradiation. The same rearrangement product has already been reported to form from endrin as a result of thermal decomposition (Phillips et al., 1962). Zabik et al. (1971) also confirmed this photolytic pathway for endrin to show that ketoendrin, caged endrin aldehyde and alcohol are the major photolytic products of endrin.

In the presence of a photosensitizer, benzophenone, heptachlor and isodrin photochemically convert into corresponding caged isomers (Rosen et al., 1969).

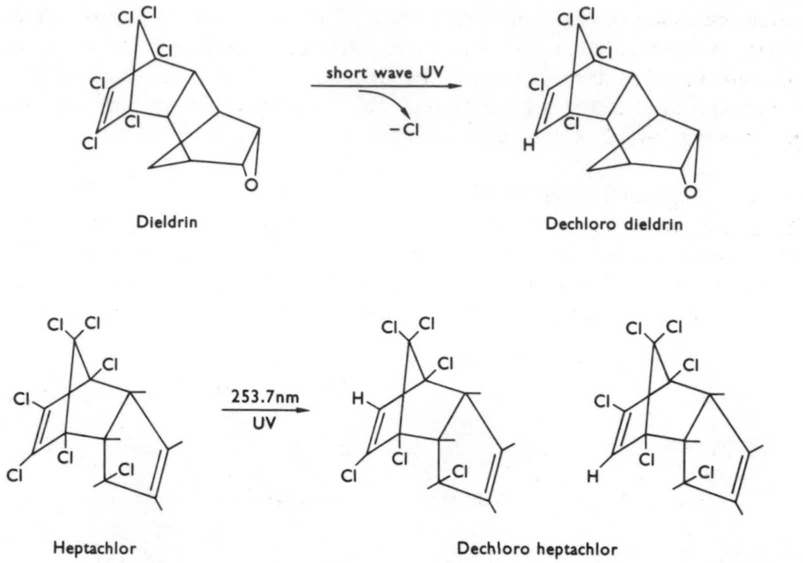

FIG. 13.11. Photodechlorination processes for dieldrin and heptachlor.

The second major photolytic reaction, a photodechlorination process, was originally reported by Henderson and Crosby (1967), and by Rosen (1967). The former authors irradiated dieldrin with shortwave UV (ranging less than 290 nm) and found that one of the olefinic chlorines was removed from the original compound, as judged by the results of NMR (nuclear magnetic resonance) and mass spectroscopic analyses. Anderson *et al.* (1968) has proposed that the photodechlorination takes place via a singlet state, while the cage formation undergoes via a triplet state. McGuire *et al.* (1970) showed that the dechlorination process could also take place on heptachlor when it was directly irradiated with a shortwave (253.7 nm) ultraviolet in hexane or cyclohexane, while irradiation of the same compound with a relatively long wave source (300 nm peak), in acetone, yielded only the caged isomer of Rosen *et al.* (1969). Benson (1971) irradiated solid and dissolved (in ethyl acetate) dieldrin, and confirmed the production of both caged (photodieldrin) and dechlorinated dieldrin isomers. Upon further irradiation of photodieldrin, two photolytic products were formed, one of which appears to be chlorohydrin of photodieldrin.

13.4.2   Aromatic pesticides
Four basic types of photochemical reactions may take place when aromatic pesticides are exposed to ultraviolet. These are: ring-substitution, hydrolysis (in aqueous solution), oxidation and polymerisation.

The most common form of ring-substitution reaction for the chlorinated aromatics is the replacement of a ring chlorine by a hydroxyl group. Examples can be seen in the photolytic degradation of 2,4-D in aqueous solution (Fig. 13.12). The final reaction product of 2,4-D is 1,2,4-benzenetriol indicating both cleavage of the ether group and replacement of ring chlorines by hydroxyl (Crosby and Tutass, 1967).

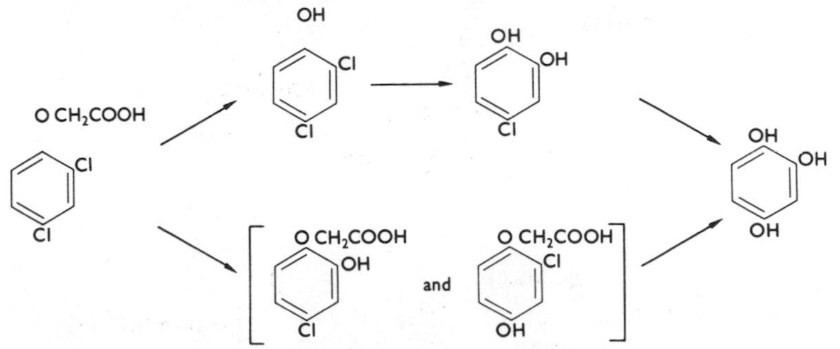

FIG. 13.12. Photochemical decomposition of 2,4-D. (From Crosby and Tutass, 1967.)

A similar reaction can take place with cyclic triazine herbicides (Fig. 13.13), where the chlorine atom on the triazine ring is known to be vulnerable (Pape and Zabik, 1970).

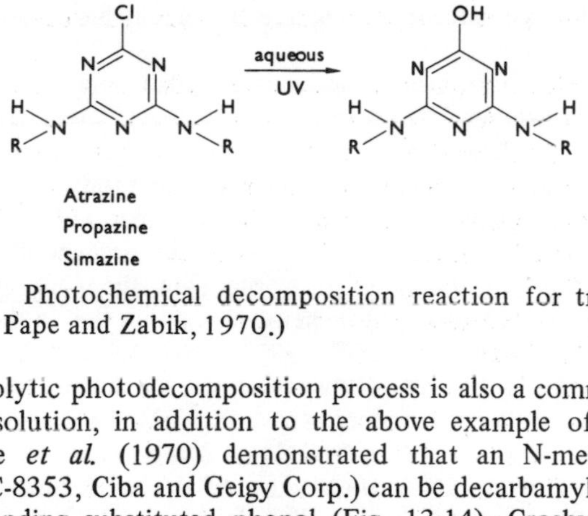

FIG. 13.13. Photochemical decomposition reaction for triazine herbicides. (From Pape and Zabik, 1970.)

The hydrolytic photodecomposition process is also a common reaction in aqueous solution, in addition to the above example of 2,4-D (Fig. 13.12). Pape et al. (1970) demonstrated that an N-methyl phenyl-carbamate (C-8353, Ciba and Geigy Corp.) can be decarbamylated to yield the corresponding substituted phenol (Fig. 13.14). Crosby and Tutass (1966) also showed that carbaryl was photochemically decomposed to methyl isocyanate and 1-naphthol. Exposure of parathion to sunlight also gave p-nitrophenol (Hasegawa, 1959), indicating a hydrolytic splitting of the P-O linkage of parathion. However, an example of non-hydrolytic cleavage of an organophosphate can be found in the photolytic degradation reaction of azinphosmethyl (O,O-dimethyl S-[4-oxo-1,2,3-benzo-triazin-3-(4H)-yl methyl] phosphorodithioate), which gives rise to

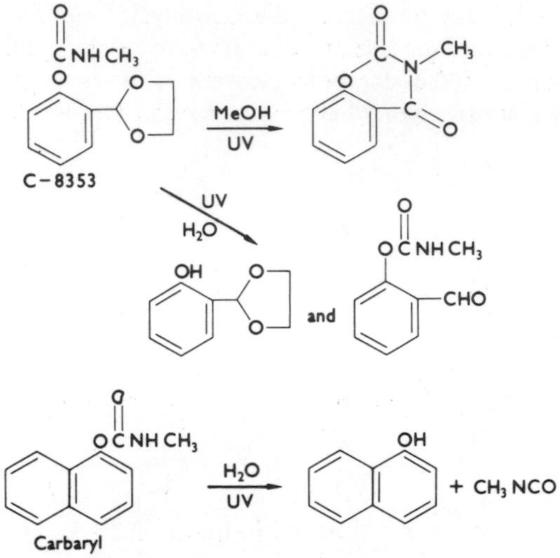

FIG. 13.14. An example of photodecomposition of N-methylphenyl-carbamates. (From Pape et al. (1970); Crosby (1969).)

O,O-dimethyl S-methyl phosphorodithioate and 4-oxo-3,4 dihydro-1,2,3-benzotriazin upon irradiation in hexane, indicating utilisation of two hydrogens from the solvent, to degrade the parent molecule (Kurihara et al., 1966).

Oxidative photochemical reactions are also important degradation processes of pesticides. Thus, chlorobenzoic acid gives rise to benzaldehyde upon UV irradiation (Crosby, 1966). Exposure of parathion to sunlight and ultraviolet radiation, results in the production of paraoxon and other oxo-analogues of parathion (Koivistoinen, 1963; Payton, 1953; Sandi, 1958 and Hasegawa, 1959). 4,4'-Dichlorobenzophenone is also formed from DDT by the aid of ultraviolet light, the reaction being partly carried out by an oxidative photochemical reaction (i.e. in the presence of oxygen) (Fleck, 1949).

Polymerization of aromatic substitutes, by the aid of ultraviolet radiation, sometimes takes place along with other initial structural changes of the pesticidal derivatives. Thus, DDT loses 2 chlorine atoms first and then forms a dimer (Fig. 13.15) in the absence of air. In the case of chlorinated aniline derivatives, it is likely that the initial reaction is the formation of diazobenzene analogue (Plimmer and Kearney, 1969; Rosen et al., 1970) which in turn, reacts with another molecule of the parent compound, through a dehydrochlorination process to form the corresponding trimer in the presence of a photosensitizer (FMN or riboflavin-5-phosphate salt) and ultraviolet light (Rosen et al., 1970). Rosen and Siewierski (1971) later synthesised 4-(3,4-dichloroanilino)-3-3',4'-trichloroazobenzene (R_1 = Cl and R_2 = H in Fig. 13.15), which is formed from 3,4-dichloro-

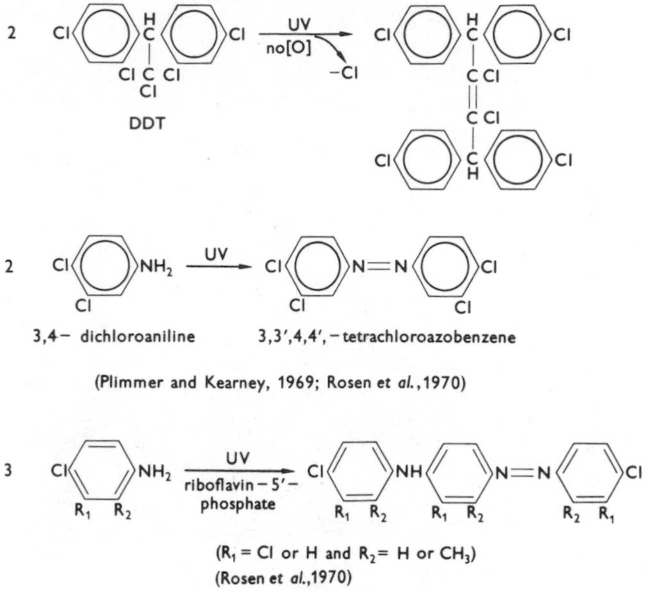

FIG. 13.15. Formation of dimers and trimers by the action of ultra-violet.

aniline, and confirmed the structure of the trimer. This trimer formation proceeded at a much slower rate than the one for the corresponding 4-chloro-4'-(4-chloroanilino) azobenzene ($R_1 = H$ and $R_2 = H$ in Fig. 13.15). The former compound appears to be stable after incubation in soil, or after irradiation by sunlight.

Combination of such complex oxidation, ring-substitution and polymerisation reactions can produce many unexpected photolytic products. An excellent example of such an interaction of different photochemical reactions, on a single pesticidal compound, can be found in the work by Munakata and Kuwahara (1969), on the degradation studies of pentachlorophenol under sunlight (Fig. 13.16). Various monomers of chlorinated quinones, hydroxyquinones and phenols are originally formed. These intermediates, and the parent compounds, react with each other to form ether linkages which yield both dimers and trimers.

13.5   DISCUSSION

Despite the complexity of degradation patterns of pesticides by environmental factors, several basic principles can be pointed out here to facilitate understanding the processes. It is important to consider first, whether such degradation products are stable enough to become 'terminal residues',

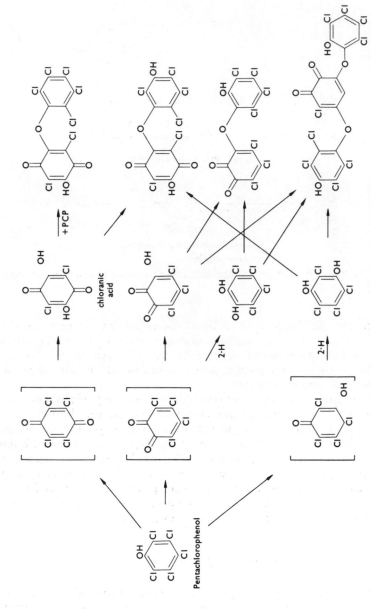

FIG. 13.16.   Photodegradation pathways for pentachlorophenol. (From Munakata and Kuwahara, 1969.)

second, whether such terminal residues have such a strong affinity to biological materials (either direct or through secondary means such as a food-chain) as to cause biological magnification phenomena, and third, whether these products are harmful to any form of biological system. That is to say, pesticides, in general, are low volume and highly biologically active environmental contaminants. Unlike fertilisers, industrial wastes, etc., pesticides do not physically alter the environments, as such, very much. Their potential hazards, therefore, must be judged in view of their biological effects.

Studies of metabolic and physical degradation activities in the environment, therefore, should be primarily concerned with the detection of potential 'terminal residues'.

It is now generally acknowledged that the toxic contaminant of 2,4,5-T is 3,4,7,8-tetrachlorodibenzo-p-dioxin. This compound exists in crude 2,4,5-T preparations at levels of 1 to several ppm. Its oral rat LD_{50} is in the order of $\mu g/kg$. Studies made by the United States Department of Agriculture (Kearney, personal communication, 1970), and by my research group indicate that it is much more stable to microbial attack than 2,4,5-T. Along with the suggestion that octachlorodibenzo-p-dioxin may form by photolysis from pentachlorophenol (Crosby and Wong, 1970), it becomes probable that chlorinated dibenzodioxins are the toxic terminal residues of both chlorinated phenoxy and phenol pesticides. Accumulation of abnormal amounts of β-BHC in the environment, particularly in biological systems, also serves as a good example of a build-up of unexpected terminal residues. Metabolic studies in the laboratory should be followed, therefore, with quantitative survey work in the field as well as the toxicological evaluation of effects of such environmental contaminants on various biological systems.

REFERENCES

M. K. Ahmed, J. E. Casida and R. F. Nichols, *J. Agr. Food Chem.*, **6**, 740 (1958)

M. Alexander, *Soil Sci. Soc. Amer. Proc.*, **29**, 1 (1965)

J. Allan, *Nature*, **175**, 1131 (1955)

C. M. Anderson, J. B. Bremmer, I. W. McCay, and R. N. Warrener, *Tetrahedron Lett.*, 1255 (1968)

D. E. Armstrong, G. Chesters and R. F. Harris, *Soil Sci. Soc. Amer. Proc.*, **31**, 61 (1967)

W. R. Benson, *J. Agr. Food Chem.*, **19**, 66 (1971)

F. R. Bradbury, *Ann. Appl. Biol.*, **52**, 361 (1963)

J. W. Cook, *J. Agr. Food Chem.*, **5**, 859 (1957)

D. G. Crosby and H. O. Tutass, *J. Agr. Food Chem.*, **14**, 596 (1966)

D. G. Crosby, Abst., *152nd Meet. Am. Chem. Soc.*, New York (1966)

D. G. Crosby, *Residue Rev.*, **25**, 1 (1969)

D. G. Crosby and A. S. Wong, *160th Meet. Am. Chem. Soc.*, Chicago (1970)

R. L. Dalton, A. W. Evans and R. C. Rhodes, *Southern Weeds Conf.*, **18**, 72 (1965)

J. K. Faulkner and D. Woodcock, *Nature* (Lond.), **205**, 865 (1964)

R. B. Finley and R. E. Pillmore, *Amer. Inst. Biol. Sci. Bull.*, **13**, 41 (1963)

E. E. Fleck, *J. Am. Chem. Soc.*, **71**, 1034 (1949)

D. D. Focht and M. Alexander, *J. Agr. Food Chem.*, **19**, 20 (1971)

H. Geissbuhler, C. Haselback, H. Aebi and L. Ebner, *Weed Res.*, **3**, 277 (1963)

L. W. Getzin and I. Rozefield, *J. Agr. Food Chem.*, **16**, 598 (1968)

M. Goto, F. Coulson and F. Korte (Editors), Environmental Safety and Quality, vol. 1. Georg Thieme Verlag, Berlin (1972)

R. H. Hamilton and D. E. Moreland, *Science N.Y.*, **135**, 373 (1962)

C. I. Harris, *Weed Res.*, **5**, 275 (1965)

G. L. Henderson and D. G. Crosby, *J. Agr. Food Chem.*, **15**, 888 (1967)

T. Hasegawa, *Fukuoka Acta Medica*, **50**, 1900 (1959)

G. D. Hill, *Weed Soc. Am. Abstr.*, 42 (1956)

P. Hirsch and M. Alexander, *Can. J. Microbiol.*, **6**, 241 (1960)

G. W. Ivie and J. E. Casida, *Science N.Y.*, **167**, 1620 (1970)

G. W. Ivie and J. E. Casida, *J. Agr. Food Chem.*, **19**, 405 (1971a)

G. W. Ivie and J. E. Casida, *J. Agr. Food Chem.*, **19**, 410 (1971b)

H. L. Jensen, *Can. J. Microbiol.*, **3**, 151 (1957)

I. P. Kapoor, R. L. Metcalf, R. F. Nystrom and G. K. Sangha, *J. Agr. Food Chem.*, **18**, 1145 (1970)

P. C. Kearney, D. D. Kaufman and M. L. Beall, Jr., *Biochem. Biophys. Res. Commun.*, **14**, 29 (1964)

P. C. Kearney, C. I. Harris, D. D. Kaufman and T. J. Sheets, *Advances in Pest Control Research*, **6**, 1 (1965)

P. C. Kearney and D. D. Kaufman, *Science N.Y.*, **147**, 740 (1965)

P. C. Kearney, *J. Agr. Food Chem.*, **13**, 561 (1965)

P. C. Kearney and T. J. Sheets, *J. Agr. Food Chem.*, **13**, 369 (1965)

P. C. Kearney, Personal Communication, Gordon Conference, Holderness, N.H. (1970)

U. Kiigemagi, H. E. Morrison, J. E. Roberts and W. B. Bollen, *J. Econ. Entomol.*, **51**, 198 (1958)

P. Koivistoinen, *Acta Agr. Scand.*, **12**, 285 (1963)

L. R. Koller, Ultraviolet Radiation, 2nd ed. J. Wiley & Sons, New York (1965)

F. Korte, Proceedings of the Commission of Terminal Residues and of the Commission on Residue Analysis, Intern. Union Pure Appl. Chem. Vienna, August (1967)

N. H. Kurihara, D. G. Crosby and H. F. Beckman, Abst. *152nd Meet. Am. Chem. Soc.*, New York (1966)

E. P. Lichtenstein and K. R. Schulz, *J. Econ. Entomol.*, **5**, 192 (1960)

E. P. Lichtenstein, K. R. Schulz, T. W. Fuhremann and T. T. Liang, *J. Agr. Food Chem.*, **18**, 100 (1970)

M. A. Loos, R. N. Roberts and M. Alexander, *Bact. Proc.*, **65**, 3 (1965)

I. C. MacRae and M. Alexander, *J. Bacteriol.*, **86**, 1231 (1963)

F. Matsumura and G. M. Boush, *Science N.Y.*, **156**, 959 (1967)

F. Matsumura and G. M. Boush, *J. Econ. Entomol.*, **61**, 610 (1968)

F. Matsumura, G. M. Boush and A. Tai, *Nature* (Lond.), **219**, 965 (1968)

F. Matsumura, K. C. Patil and G. M. Boush, *Science N. Y.*, **170**, 1206 (1970)

F. Matsumura, K. C. Patil and G. M. Boush, *Nature, 230*, 325 (1971)

F. Matsumura, V. G. Khanvilkar, K. C. Patil and G. M. Boush, *J. Agr. Food Chem., 19*, 27 (1971)

R. R. McGuire, M. J. Zabik, R. D. Schuetz and R. D. Blotard, *J. Agr. Food. Chem.,* **18**, 319 (1970)

D. B. Mendel, A. K. Klein, J. T. Chen and M. S. Walton, *J. Assoc. Offic. Agr. Chem.,* **50**, 897 (1967)

D. B. Menzel, S. M. Smit, R. Miskus and W. M. Hoskins, *J. Econ. Entomol.,* **54**, 9 (1961)

J. R. Miles, C. M. Tu and C. R. Harris, *J. Econ. Entomol.,* **62**, 1334 (1969)

K. Munakata and M. Kuwahara, *Residue Rev.*, Springer-Verlag, **25**, 13 (1969)

B. E. Pape and M. J. Zabik, *J. Agr. Food Chem.,* **18**, 202 (1970)

B. E. Pape, M. F. Para and M. J. Zabik, *J. Agr. Food Chem.,* **18**, 490 (1970)

J. Payton, *Nature,* **171**, 355 (1953)

D. D. Phillips, G. E. Pollard and S. B. Soloway, *J. Agr. Food Chem.,* **10**, 217 (1962)

J. R. Plimmer, P. C. Kearney and D. W. Vonendt, *J. Agr. Food Chem.,* **16**, 594 (1968)

J. R. Plimmer and P. C. Kearney, *158th Meet. Am. Chem. Soc.,* New York (1969)

J. Robinson, A. Richardson, B. Bush and K. E. Elgar, *Bull. Environ. Contam. Toxicol.,* **1**, 127 (1966)

J. D. Rosen, D. J. Sutherland and G. R. Lipton, *Bull. Environ. Contam. Toxicol.,* **1**, 133 (1966)

J. D. Rosen and D. J. Sutherland, *Bull. Env. Contam. Toxicol.,* **2**, 1 (1967)

J. D. Rosen, Chem. Comm., 189 (1967)

J. D. Rosen and W. F. Carey, *J. Agr. Food Chem.,* **16**, 536 (1968)

J. D. Rosen, D. J. Sutherland and M. A. Q. Khan, *J. Agr. Food Chem.,* **17**, 404 (1969)

J. D. Rosen, M. Siewierski and G. Winnett, *J. Agr. Food Chem.,* **18**, 494 (1970)

E. Sandi, *Nature* (Lond.), **181**, 499 (1958)

T. J. Sheets, *J. Agr. Food Chem.,* **12**, 30 (1964)

H. D. Skipper, C. M. Gilmour and W. R. Furtick, *Soil Sci. Soc. Amer. Proc.,* **31**, 653 (1967)

J. J. Skujins, A. D. McLaren and G. H. Peterson (Editors), Soil Biochemistry, p. 371. Marcel Dekker, Inc., New York (1966)

J. W. Smith and T. J. Sheets, *Weed Soil. Am. Abstr.,* **6**, 39 (1966)

J. M. Tiedje and M. Alexander, Abstr. *Amer. Soc. Agron. Ann. Meet.,* Washington, D.C., 94 (1967)

G. Wedemyer, *Appl. Microbiol.,* **16**, 661 (1968)

T. Yoshida and T. F. Castro, *Soil Sci. Soc. Am. Proceed.,* **34**, 440 (1970)

General References

Anon., Pesticides and their effects on soils and water. *Soil Science Society of America,* 150 (1966)

Anon., The chemical basis for action. *Rep. Amer. Chem. Soc.,* 193 (1969)

Anon., Pesticides in the soil. *Int. Symp. Mich. State Univ.,* 144 pp. (1970)

A.R.C., Third Report of the Research Committee on Toxic Chemicals (1970)

A. Bevenue and Y. Kawano, Pesticides, Pesticide Residues, Tolerances and the Law (U.S.A.), *Res. Rev.,* **35,** 103 (1971)

N. C. Brady, Agriculture and the Quality of our Environment, *American Association for the Advancement of Science,* 460 pp. (1966)

R. Carson, Silent Spring, p. 368. Houghton-Mifflin Co., Boston (1962)

C. O. Chichester, Research in Pesticides, p. 380. Academic Press, New York (1965)

C. A. Edwards, Insecticide Residues in Soils, *Res. Rev.,* **13,** 83 (1966)

C. A. Edwards, Persistent Pesticides in the Environment, p. 71. Chem. Rubber Co., Cleveland (1970)

C. A. Edwards, Soil Pollutants and Soil Animals, *Sci. Amer.,* **220** (4), 88 (1969)

C. A. Edwards and A. R. Thompson, Pesticides and the Soil Fauna, *Res. Rev.,* **45** (1973)

O.E.C.D., The Problems of Peristent Chemicals, 113 pp. Paris (1971)

B. E. Frazier, G. Chester and B. G. Lee, *Pesticides Monit. J.,* 4 (2), 67-70 (1970)

J. Frost, Earth, Air, Water, *Environment,* 11 (6), 14 (1969)

J. W. Gillett, The Biological Impact of Pesticides in the Environment, *Proc. Sym. Environ. Health Sciences Series,* 1, 210 pp. (1970)

E. D. Goldberg, P. Butler, P. Meier, D. Menzel, R. W. Risebrough and L. F. Stickel, Chlorinated Hydrocarbons in the Marine Environment, p. 42. National Academy of Sciences, Washington, D.C. (1971)

R. F. Gould, Organic Pesticides in the Environment, *American Chemical Society,* 309 pp. (1966)

F. Graham, Jr., Since Silent Spring, p. 297. Hamish Hamilton, London (1970)

F. A. Gunther, In: Scientific Aspects of Pest Control, *Nat. Acad. of Sci.–Nat. Res. Council, publ.,* **402,** 284 (1966)

F. A. Gunther, Pesticide Residues in the Total Environment, *Pure Appl. Chem.,* **21,** 355 (1970)

D. W. Johnson, *Trans. Amer. Fish. Soc.,* **97,** 398 (1968)

E. F. Knipling, *J. Econ. Entomol.,* **46,** 1 (1953)

M.A.F.F., Further Review of Certain Organochlorine Pesticides Used in Great Britain. H.M.S.O., London (1969)

E. H. Marth, *Res. Rev.,* **9,** 1 (1965)

J. R. W. Miles, *J. Agr. Food Chem.,* **16,** 620 (1968)

N. W. Miller and G. G. Berg, Chemical Fallout–Current Research on Persistent Pesticides, p. 531. C. C. Thomas, Springfield, Ill. (1969)

N. W. Moore, *Bird Study,* **12,** 222 (1965)

N. W. Moore, Pesticides in the Environment and their Effects on Wildlife, p. 311. Blackwell, Oxford (1966)

N. W. Moore, A Synopsis of the Pesticide Problem, *Adv. Ecol. Res.*, **4,** 75 (1967)

E. M. Mrak, Rep. Secretary's Commission on Pesticides and their Relationship to Environmental Health, U.S. Dept. Hlth. Educ. Wlfr. (1969)

R. C. Muirhead-Thompson, In: Pesticides and freshwater fauna, p. 248. Academic Press, London and New York (1971)

L. D. Newsom, Consequences of insecticide use on non-target organisms. *Ann. Rev. Entomol.,* **12,** 257 (1967)

D. Pimentel, Ecological effects of pesticides on non-target species, p. 220. Exec. Office of the President, Office of Science and Technology, Washington, D.C., 20506 (1971)

J. Robinson, *Canad. Med. Assoc. J.,* **100,** 180 (1969)

R. L. Rudd, Pesticides and the Living Landscape, p. 320. Faber & Faber, London (1964)

L. F. Stickel, Organochlorine Pesticides in the Environment, *U.S.D.I. Bur. sport Fish Wildlife Rept.,* **119,** 1 (1968)

U.S.D.A., *Report of Committee on persistent pesticides,* 33 (1969)

U.S.D.H.E.W., Pesticides in Soil and Water. An annotated bibliography, 90 pp. (1964)

U.S.D.H.E.W., Report of the Secretary's Commission on Pesticides and their Relationship to Environmental Health. Pts i and ii. U.S. Govt. Printing Office, Washington, D.C., 677 pp. (1969)

W.H.O., Pesticide Residues, *W.H.O. & F.A.O. Tech. Rept.,* **391,** 43 (1968)

R. White-Stevens, (Editor), Pesticides in the environment, p. 270. Marcel Dekker Inc., New York (1971)

J. L. Whitten, That We May Live, p. 251. van Nostrand, New York (1966)

General Index

A

Abdominal fat, 268

Absorption of pesticides, 6, 13, 15, 22, 24, 58, 62, 65, 66, 68, 77, 105, 123, 154, 156, 157, 158, 159, 164, 196, 300

Acaricides, 1, 12, 75, 78

Acarina, 430

Accidental contamination, 58, 270, 300, 329, 330, 353, 355

Accumulation of pesticides, 127, 128, 129, 148, 160, 175, 182, 183, 194, 201, 269, 270, 290, 296, 300, 302

Actinomycetes, 436, 495

Acute toxicity, 170, 218, 233

Additive effects of pesticides, 285

Adipose tissue, 37, 183, 194, 314, 477, 479, 483

Adsorption of pesticides, 15, 16, 17, 18, 19, 106, 149, 151, 153, 154, 156, 157, 170, 196, 380, 427

Aeolian transport of pesticides, 149, 150, 466, 467, 470

Aerial application of pesticides, 74, 352, 419, 437

Aerial contaminants, 151, 370, 371

Aerial spraying, 127, 195, 216, 410, 413, 439, 445, 465

Aerosols, 15, 373, 375, 464, 465, 471, 472

Africa, 149, 273, 403, 469

Age effects on pesticide uptake, 22, 142, 318, 322, 323

Agricultural Research Committee on Toxic Chemicals, 9

Agriolimax reticulatis, 109, 110, 112, 118, 119

Air, pesticides in, 3, 4, 5, 15, 16, 31, 42, 43, 150, 151, 313, 316, 317, 353, 365–405

Air sampling, 15

Air-sea interface, 150, 151

Air pressure, 370

Air turbulence, 155, 369, 381, 464, 465

Airborne particles, 70, 466

Airshed, 372, 375, 378

Alaska, 277, 280

Alcohol, 499

Alderfly, 140, 144, 146

Alfalfa, 61, 63, 64, 72, 73, 76, 375, 412, 415, 418, 428, 439

Alga, 136, 140, 144, 157

Alimentary canal, 89, 93, 98, 200, 205, 246, 272, 285

Allolobophora spp., 94, 97, 98, 100, 101, 102, 103, 104, 105, 106, 108, 117, 122, 127

American avocet, 264

American cockroach, 122

American kestrel, 262, 285

American Society for the Advancement of Science, 8

American widgeon, 261

Amphipod, 140, 144, 146

Analytical error, 138

Animal behaviour, effects of pesticides on, 138, 170, 202, 239, 302

Animal feed, 57, 354, 376

Animal produce, 13, 31

Animal tissues, 2, 22, 26, 45, 89, 93, 96, 100, 101, 105, 107, 108, 109, 116, 123, 125, 126, 127, 128, 329, 330, 331

Annual production of pesticides, 3, 150

Annual treatments, 418

Antarctic, 6, 403, 468, 469, 470, 474

Aphids, 62, 382

Apples, 31, 35, 38, 45, 72, 98, 99, 105, 108, 114, 116, 361

Application of pesticides, 74, 290, 300, 336, 342, 352

Chemical Index

Pesticide names taken from:
Pesticide Manual, Ed. H. Martin, third edition, 1972. British Crop Protection Council.

Trade names of pesticides have been placed in parenthesis.